CAMBRIDGE STUDIES IN ADVANCED MATHEMATICS 182

Editorial Board
B. BOLLOBÁS, W. FULTON, F. KIRWAN,
P. SARNAK, B. SIMON, B. TOTARO

TOEPLITZ MATRICES AND OPERATORS

The theory of Toeplitz matrices and operators is a vital part of modern analysis, with applications to moment problems, orthogonal polynomials, approximation theory, integral equations, bounded- and vanishing-mean oscillations, and asymptotic methods for large structured determinants, among others.

This friendly introduction to Toeplitz theory covers the classical spectral theory of Toeplitz forms and Wiener–Hopf integral operators and their manifestations throughout modern functional analysis. Numerous solved exercises illustrate the results of the main text and introduce subsidiary topics, including recent developments. Each chapter ends with a survey of the present state of the theory, making this a valuable work for the beginning graduate student and established researcher alike. With biographies of the principal creators of the theory and historical context also woven into the text, this book is a complete source on Toeplitz theory.

Nikolaï Nikolski is Professor Emeritus at the Université de Bordeaux, working primarily in analysis and operator theory. He has been co-editor of four international journals, editor of more than 15 books, and has published numerous articles and research monographs. He has also supervised 26 Ph.D. students, including three Salem Prize winners. Professor Nikolski was elected Fellow of the American Mathematical Society (AMS) in 2013 and received the Prix Ampère of the French Academy of Sciences in 2010.

CAMBRIDGE STUDIES IN ADVANCED MATHEMATICS

Editorial Board
B. Bollobás, W. Fulton, F. Kirwan, P. Sarnak, B. Simon, B. Totaro

All the titles listed below can be obtained from good booksellers or from Cambridge University Press.
For a complete series listing, visit www.cambridge.org/mathematics.

Already Published
145   M. Viana *Lectures on Lyapunov Exponents*
146   J.-H. Evertse & K. Győry *Unit Equations in Diophantine Number Theory*
147   A. Prasad *Representation Theory*
148   S. R. Garcia, J. Mashreghi & W. T. Ross *Introduction to Model Spaces and Their Operators*
149   C. Godsil & K. Meagher *Erdős–Ko–Rado Theorems: Algebraic Approaches*
150   P. Mattila *Fourier Analysis and Hausdorff Dimension*
151   M. Viana & K. Oliveira *Foundations of Ergodic Theory*
152   V. I. Paulsen & M. Raghupathi *An Introduction to the Theory of Reproducing Kernel Hilbert Spaces*
153   R. Beals & R. Wong *Special Functions and Orthogonal Polynomials*
154   V. Jurdjevic *Optimal Control and Geometry: Integrable Systems*
155   G. Pisier *Martingales in Banach Spaces*
156   C. T. C. Wall *Differential Topology*
157   J. C. Robinson, J. L. Rodrigo & W. Sadowski *The Three-Dimensional Navier–Stokes Equations*
158   D. Huybrechts *Lectures on K3 Surfaces*
159   H. Matsumoto & S. Taniguchi *Stochastic Analysis*
160   A. Borodin & G. Olshanski *Representations of the Infinite Symmetric Group*
161   P. Webb *Finite Group Representations for the Pure Mathematician*
162   C. J. Bishop & Y. Peres *Fractals in Probability and Analysis*
163   A. Bovier *Gaussian Processes on Trees*
164   P. Schneider *Galois Representations and (φ, Γ)-Modules*
165   P. Gille & T. Szamuely *Central Simple Algebras and Galois Cohomology (2nd Edition)*
166   D. Li & H. Queffelec *Introduction to Banach Spaces, I*
167   D. Li & H. Queffelec *Introduction to Banach Spaces, II*
168   J. Carlson, S. Müller-Stach & C. Peters *Period Mappings and Period Domains (2nd Edition)*
169   J. M. Landsberg *Geometry and Complexity Theory*
170   J. S. Milne *Algebraic Groups*
171   J. Gough & J. Kupsch *Quantum Fields and Processes*
172   T. Ceccherini-Silberstein, F. Scarabotti & F. Tolli *Discrete Harmonic Analysis*
173   P. Garrett *Modern Analysis of Automorphic Forms by Example, I*
174   P. Garrett *Modern Analysis of Automorphic Forms by Example, II*
175   G. Navarro *Character Theory and the McKay Conjecture*
176   P. Fleig, H. P. A. Gustafsson, A. Kleinschmidt & D. Persson *Eisenstein Series and Automorphic Representations*
177   E. Peterson *Formal Geometry and Bordism Operators*
178   A. Ogus *Lectures on Logarithmic Algebraic Geometry*
179   N. Nikolski *Hardy Spaces*
180   D.-C. Cisinski *Higher Categories and Homotopical Algebra*
181   A. Agrachev, D. Barilari & U. Boscain *A Comprehensive Introduction to Sub-Riemannian Geometry*
182   N. Nikolski *Toeplitz Matrices and Operators*
183   A. Yekutieli *Derived Categories*
184   C. Demeter *Fourier Restriction, Decoupling and Applications*

Toeplitz Matrices and Operators

NIKOLAÏ NIKOLSKI
Université de Bordeaux

Translated by
DANIÈLE GIBBONS
GREG GIBBONS

CAMBRIDGE
UNIVERSITY PRESS

Shaftesbury Road, Cambridge CB2 8EA, United Kingdom

One Liberty Plaza, 20th Floor, New York, NY 10006, USA

477 Williamstown Road, Port Melbourne, VIC 3207, Australia

314–321, 3rd Floor, Plot 3, Splendor Forum, Jasola District Centre, New Delhi – 110025, India

103 Penang Road, #05–06/07, Visioncrest Commercial, Singapore 238467

Cambridge University Press is part of Cambridge University Press & Assessment,
a department of the University of Cambridge.

We share the University's mission to contribute to society through the pursuit of
education, learning and research at the highest international levels of excellence.

www.cambridge.org
Information on this title: www.cambridge.org/9781107198500
DOI: 10.1017/9781108182577

© Nikolaï Nikolski 2020

This publication is in copyright. Subject to statutory exception and to the provisions
of relevant collective licensing agreements, no reproduction of any part may take
place without the written permission of Cambridge University Press & Assessment.

First published 2020

A catalogue record for this publication is available from the British Library

ISBN   978-1-107-19850-0   Hardback

Cambridge University Press & Assessment has no responsibility for the persistence
or accuracy of URLs for external or third-party internet websites referred to in this
publication and does not guarantee that any content on such websites is, or will
remain, accurate or appropriate.

Every effort has been made to secure necessary permissions to reproduce
copyright material in this work, though in some cases it has proved impossible
to trace copyright holders. If any omissions are brought to our notice,
we will be happy to include appropriate acknowledgements on reprinting.

Уничтожайте рукописи,
но сохраняйте то, что вы начертали сбоку,
от скуки, от неуменья, и как бы во сне.
Эти второстепенные и мимовольные создания вашей фантазии не пропадут в мире

Destroy your manuscript,
but save whatever you have inscribed in the margin
out of boredom, out of helplessness, and, as it were, in a dream.
These secondary and involuntary creations of your fantasy will not be lost in the world.

The Egyptian Stamp (1928)
Osip Mandelstam

Contents

Preface		*page* xiii
Acknowledgments for the English Edition		xvii
Acknowledgments for the French Edition		xvii
List of Biographies		xix
List of Figures		xxi

1 Why Toeplitz–Hankel? Motivations and Panorama   1
1.1 Latent Maturation: The RHP and SIOs   1
     1.1.1 Nineteenth Century: Riemann and Volterra   1
     1.1.2 Twentieth Century: David Hilbert   6
     1.1.3 George Birkhoff and Henri Poincaré   12
1.2 The Emergence of the Subject: Otto Toeplitz   16
1.3 The Classical Period   20
     1.3.1 Gábor Szegő's Revolution   20
     1.3.2 The Wiener and Hopf Integral Operators   21
     1.3.3 A New Challenge Arises: The Lenz–Ising Model   23
1.4 The Golden Age and the Drama of Ideas   23
     1.4.1 Solomon Mikhlin and the Symbolic Calculus of the SIO   23
     1.4.2 The School of Mark Krein   24
     1.4.3 Lars Onsager, and Szegő Again   24
     1.4.4 Rosenblum, Devinatz, and the Drama of Coincidence   25
1.5 The Parallel/Complementary World of Hankel, and the Post-modern Epoch of the Ha-plitz Operators   26
1.6 Notes and Remarks   30

2 Hankel and Toeplitz: Sibling Operators on the Space H^2 — 31
2.1 Three Definitions of Toeplitz Operators: The Symbol — 31
    2.1.1 The Spaces ℓ^2, L^2, and H^2 — 32
    2.1.2 Shift (or Translation) Operators — 33
    2.1.3 Matrix of an Operator — 34
    2.1.4 Toeplitz Matrices — 35
    2.1.5 Toeplitz Operators — 35
    2.1.6 Comment: Three Equivalent Definitions of Toeplitz Operators — 39
    2.1.7 Examples — 39
2.2 Hankel Operators and Their Symbols — 41
    2.2.1 Hankel Matrices — 41
    2.2.2 Hankel Operators — 41
2.3 Exercises — 50
    2.3.0 Basic Exercises: Hilbert and Hardy Spaces, and Their Operators — 50
    2.3.1 Toeplitz Operators, the Berezin Transform, and the RKT (Reproducing Kernel Thesis) — 59
    2.3.2 The Natural Projection on $\mathcal{T}_{L^\infty}$ — 62
    2.3.3 Toeplitz Operators on $\ell^p(\mathbb{Z}_+)$ and $H^p(\mathbb{T})$ — 62
    2.3.4 The Space BMO($\mathbb{T}$), the RKT, and the Garsia Norm — 64
    2.3.5 Compact Hankel Operators and the Spaces VMO($\mathbb{T}$) and QC($\mathbb{T}$) — 68
    2.3.6 Finite Rank Hankel Operators (Kronecker, 1881) — 70
    2.3.7 Hilbert–Schmidt Hankel Operators — 72
    2.3.8 The Original Proof of Sarason's Lemma 2.2.7 (Sarason, 1967) — 73
    2.3.9 Compactness of the Commutators $[P_+, M_\varphi]$ (Power, 1980) — 73
    2.3.10 The Natural Projection on Hank ($\ell^2(\mathbb{Z}_+)$) — 74
    2.3.11 Vector-Valued Toeplitz Operators — 75
    2.3.12 Some Algebraic Properties of Toeplitz/Hankel Operators — 76
2.4 Notes and Remarks — 78

3 H^2 Theory of Toeplitz Operators — 101
3.1 Fredholm Theory of the Toeplitz Algebra — 101
    3.1.1 The Role of Homotopy — 110

3.2	The Simonenko Local Principle		112
	3.2.1	Proof of Theorem 3.2.1 (Sarason, 1973)	115
	3.2.2	Examples	116
3.3	The Principal Criterion of Invertibility		117
	3.3.1	Wiener–Hopf Factorization	124
	3.3.2	Upper Bounds for $\|T_\varphi^{-1}\|$	125
	3.3.3	A Comment on Wiener–Hopf Factorization	126
	3.3.4	First Consequences of the Principal Criterion	128
3.4	Exercises		129
	3.4.0	Basic Exercises: Integral and Multiplication Operators	129
	3.4.1	Spectral Inclusions	137
	3.4.2	Holomorphic Symbols $\varphi \in H^\infty$	138
	3.4.3	Fredholm Theory for the Algebra alg $\mathcal{T}_{H^\infty + C(\mathbb{T})}$	139
	3.4.4	$H^\infty + C(\mathbb{T})$ is the Minimal Algebra Containing H^∞ (Hoffman and Singer, 1960)	141
	3.4.5	Fredholm Theory for the Algebra alg$\mathcal{T}_{PC(\mathbb{T})}$	142
	3.4.6	A Simplified Local Principle (Simonenko, 1960)	143
	3.4.7	Fred(H^2) and Local Sectoriality	143
	3.4.8	Multipliers Preserving Fred(H^2)	145
	3.4.9	The Toeplitz Algebra alg$\mathcal{T}_{L^\infty(\mathbb{T})}$: A Necessary Condition	145
	3.4.10	Hankel Operators from the Toeplitz Algebra alg$\mathcal{T}_{L^\infty(\mathbb{T})}$	145
	3.4.11	On the Equation $T_\varphi f = 1$ (Another Proof of Theorem 3.3.8)	146
	3.4.12	Is There a Regularizer of T_φ in $\mathcal{T}_{L^\infty(\mathbb{T})}$ and/or in alg$\mathcal{T}_{L^\infty(\mathbb{T})}$?	146
	3.4.13	Fredholm Theory for Almost Periodic Symbols	149
	3.4.14	Fredholm Operators T_φ with Matrix-Valued Symbols	151
	3.4.15	"Truncated" Toeplitz Operators	153
3.5	Notes and Remarks		155
4	**Applications: Riemann–Hilbert, Wiener–Hopf, Singular Integral Operators (SIO)**		**178**
4.1	The Riemann–Hilbert Problem and the SIO		178
	4.1.1	The RHP and Toeplitz Operators	179
	4.1.2	The Hilbert Transform $\mathbb{H}$ and SIOs	180
	4.1.3	Comment: Operators and Singular Integral Equations	186

4.2 Toeplitz on $H^2(\mathbb{C}_+)$ and Wiener–Hopf on $L^2(\mathbb{R}_+)$ 188
    4.2.1 On the Space $H^2(\mathbb{C}_+)$: The Paley–Wiener Theorem 189
    4.2.2 Pseudo-Measures and Wiener–Hopf Operators 191
    4.2.3 Transfer of Spectral Theory to Wiener–Hopf Operators 196
    4.2.4 Classical Wiener–Hopf Equations and Operators 197
    4.2.5 Finite Difference Operators 198
    4.2.6 Operators W_μ with Causal Measures μ 198
    4.2.7 The Hilbert SIO on $L^2(\mathbb{R}_+)$ 199
4.3 The Matrix of W_k in the Laguerre ONB 200
4.4 Wiener–Hopf Operators on a Finite Interval 203
    4.4.1 Determination of the Symbol 203
    4.4.2 W_k^a of Rank 1 204
    4.4.3 Bounding the Norm $\|W_k^a\|$ by the Best Extension 204
    4.4.4 Example: An Operator W_k^a Bounded but Without Symbol $k \in \mathcal{PM}(\mathbb{R})$ With Support in $[-a, a]$ 205
    4.4.5 Example: The Volterra Operator 207
4.5 Exercises 207
    4.5.0 Basic Exercises: From the Hilbert Singular Operator to the Riesz Transforms ("Method of Rotation") 207
    4.5.1 Sokhotsky–Plemelj Formulas 211
    4.5.2 Systems of Equations and Matrix Wiener–Hopf Operators 214
    4.5.3 Hankel Operators on $H^2(\mathbb{C}_+)$ and $L^2(\mathbb{R}_+)$ 216
    4.5.4 Laguerre Polynomials 218
    4.5.5 Compact W_k^a Operators 219
4.6 Notes and Remarks 220

5 Toeplitz Matrices: Moments, Spectra, Asymptotics 230
5.1 Positive Definite Toeplitz Matrices, Moment Problems, and Orthogonal Polynomials 230
    5.1.1 Proof of Theorem 5.1.1 (Following Stone, 1932) 235
    5.1.2 The Truncated TMP: Extension to a Positive Definite Sequence 235
    5.1.3 Truncated Toeplitz Operators 236
    5.1.4 The Operator Approach to Orthogonal Polynomials (Akhiezer and Krein, 1938) 237
    5.1.5 The Truncated TMP: The Approach of Carathéodory (1911) and Szegő (1954) 241

5.2	Norm of a Toeplitz Matrix		246
	5.2.1	Comments and Special Cases	247
	5.2.2	Proof of Theorem 5.2.1	251
	5.2.3	Proof of Lemma 5.2.2	253
5.3	Inversion of a Toeplitz Matrix		254
	5.3.1	Two Matrix Inversion Theorems	254
	5.3.2	Comments	256
5.4	Inversion of Toeplitz Operators by the Finite Section Method		258
	5.4.1	The Finite Section Method	258
	5.4.2	Theorem (IFSM for Toeplitz Operators)	261
	5.4.3	Comment: A Counter-Example of Treil (1987)	265
5.5	Theory of Circulants		265
	5.5.1	Cyclic Shift	265
	5.5.2	Definition of Circulants	267
	5.5.3	Basic Properties	267
	5.5.4	Spectrum and Diagonalization of Circulants	269
	5.5.5	An Inequality of Wirtinger (1904)	269
5.6	Toeplitz Determinants and Asymptotics of Spectra		270
	5.6.1	The First Szegő Asymptotic Formula (1915)	271
	5.6.2	Equidistribution of Sequences, after Weyl (1910)	271
	5.6.3	Asymptotic Distribution of Spectra	276
	5.6.4	Asymptotic Distribution Meets the Circulants	280
	5.6.5	The Second Term of the Szegő Asymptotics	284
	5.6.6	A Formula for Determinant and Trace	291
	5.6.7	Some Formulas for Trace$[T_\varphi, T_\psi]$ (Following Helton and Howe, and Berger and Shaw, 1973)	291
	5.6.8	Conclusion	293
5.7	Exercises		294
	5.7.0	Basic Exercises: Volumes, Distances, and Approximations	294
	5.7.1	Positive Definite Sequences and Holomorphic Functions	299
	5.7.2	Semi-Commutators of Finite Toeplitz Matrices	301
	5.7.3	Inversion of Wiener–Hopf Operators by the Finite Section Method	301
	5.7.4	When the Second Szegő Asymptotics Stabilize (Szegő, 1952)	302
	5.7.5	Cauchy Determinants (1841)	304

5.7.6	The Second Term of the Asymptotic Distribution of Spectra (Libkind (1972), Widom (1976))	306
5.7.7	The Helton and Howe Formula of Lemma 5.6.9	306
5.7.8	The Formula of Borodin and Okounkov (2000) (and Geronimo and Case (1979))	307
5.8	Notes and Remarks	308

Appendix A	**Key Notions of Banach Spaces**	329
Appendix B	**Key Notions of Hilbert Spaces**	333
Appendix C	**An Overview of Banach Algebras**	339
Appendix D	**Linear Operators**	348
Appendix E	**Fredholm Operators and the Noether Index**	359
Appendix F	**A Brief Overview of Hardy Spaces**	387

References	395
Notation	416
Index	419

Preface

*Par ma fois! . . . je dis de la prose
sans que j'en susse rien, et je vous suis
le plus obligé du monde de m'avoir appris cela.*

Good heavens! . . . I have been speaking prose
without knowing anything about it, and I am
much obliged to you for having taught me that.

Monsieur Jourdain (1670)

As in Molière's *Le Bourgeois gentilhomme* with his prose, we often speak the language of Toeplitz and Hankel matrices/operators without realising it; but it would be better if we did it knowingly and in a technically correct manner.

The introduction to Toeplitz operators and matrices proposed in this text concerns the matrix and integral transforms defined by a "kernel" (a matrix or a function of two variables) with constant diagonals. In particular, a sequence of complex numbers $(c_k)_{k \in \mathbb{Z}}$ defines a *Toeplitz matrix*

$$T = \begin{pmatrix} c_0 & c_{-1} & c_{-2} & c_{-3} & \cdots & \cdots \\ c_1 & c_0 & c_{-1} & c_{-2} & \cdots & \cdots \\ c_2 & c_1 & c_0 & c_{-1} & \cdots & \cdots \\ c_3 & c_2 & c_1 & c_0 & \cdots & \cdots \\ \cdots & \cdots & \cdots & \cdots & \cdots & \cdots \\ \cdots & \cdots & \cdots & \cdots & \cdots & \cdots \end{pmatrix}$$

and a *Hankel matrix*

$$\Gamma = \begin{pmatrix} c_{-1} & c_{-2} & c_{-3} & c_{-4} & \cdots & \cdots \\ c_{-2} & c_{-3} & c_{-4} & c_{-5} & \cdots & \cdots \\ c_{-3} & c_{-4} & c_{-5} & c_{-6} & \cdots & \cdots \\ c_{-4} & c_{-5} & c_{-6} & c_{-7} & \cdots & \cdots \\ \cdots & \cdots & \cdots & \cdots & \cdots & \cdots \\ \cdots & \cdots & \cdots & \cdots & \cdots & \cdots \end{pmatrix}.$$

An operator (mapping) that has a Toeplitz (respectively, Hankel) matrix with respect to a basis is called a *Toeplitz* (respectively, *Hankel*) operator. For deep reasons (deep in two senses: hidden deep beneath the surface of mathematical facts, and deep in terms of their power), these transforms have played an exceptional role in contemporary mathematics. The set of problems and the basis of the techniques associated with these transforms were formulated by such giants of mathematics as Bernhard Riemann, David Hilbert, Norbert Wiener, George Birkhoff, and Otto Toeplitz.

This book is an introduction to a highly dynamic domain of modern analysis based on the techniques of Hardy spaces. To make it independent we provide a thorough overview of these spaces in Appendix F, and each time we reiterate the principal facts of Hardy theory immediately before their use. A complete presentation of Hardy spaces, the closest in style to the present work, can be found in *Espaces de Hardy* (Éditions Belin, 2012) by the same author, also translated as *Hardy Spaces* (Cambridge University Press, 2019). The current text, like the first, corresponds to a course at Master's level, given several times at the Université de Bordeaux during the years 1991–2011. Numerous resolved exercises show how the techniques developed can be put into action and extend the scope of the theory.

The somewhat hidden aspect of this text is the following. It is devoted to the study of integral and matrix operators with kernels (or matrices) depending on the difference of arguments, which seems, at first glance, to be a rather specialized subject. But a second glance leads to the discovery that many classical results of analysis and its applications rely directly on the operators known as "Toeplitz" and their "siblings," the "Hankel" operators, which are so intimately associated with those of Toeplitz that the pair are often referred to as the "Ha-plitz operators/matrices." This area of analysis includes Wiener's filtering problems, the statistical physics of gases, diverse moment problems, ergodic properties of random processes, complex interpolation, etc. The goal of this text is to present the diversity of Toeplitz/Hankel techniques and to draw conclusions from the "inexplicable efficacy" of Toeplitz (and Hankel) operators.

The *prerequisites* are standard courses on functional analysis (or Hilbert/Banach spaces) along with a few elements of complex analysis and a certain familiarity with Hardy spaces. A summary/overview of all these basic notions (as well as the notation used) can be found in the Appendices at the end of the book.

More precisely, Appendices A–D provide the definitions and notation of basic analysis (at undergraduate level), whereas Appendix E provides a short but complete presentation (including proofs) of a less well-known theory of functional analysis – that of "Fredholm operators." Appendix F is a summary of the theory of Hardy spaces; the text by the same author on this subject, *Espaces de Hardy* (*Hardy Spaces*), is cited here as [Nikolski, 2019].

Within the text, there are also numerous historical references – on the subjects developed, their creators, and their diverse situations. We only hope that these "asides" will help the reader to better appreciate the mathematical methods presented and their efficacy, as well as the dramaturgy of mathematics (and mathematical life).

Each chapter contains exercises and their solutions (155 in total) at different levels. To use a musical metaphor of Israel Glazman and Yuri Lyubich [Glazman and Lyubich, 1969], they range from exercises on open strings up to virtuoso pieces using double harmonics ("double flageolet tones"). In particular, the series of exercises in each chapter (with the exception of Chapter 1) begin with Basic Exercises accessible at Master's level or for preparing students for the French *agrégations* exams (a competitive exam to attain the highest teaching diploma).

Each chapter concludes with a section entitled "Notes and Remarks," which discusses the history of the subjects treated, certain recent results, and (on occasion) some open questions; this discussion is destined primarily for the more experienced reader. For an appreciation of this type of text in a poetic form, see the maxim on page v, due to Osip Mandelstam, the greatest Russian poet of the twentieth century.

Chapter 1 plays the role of a detailed but informal introduction: a description of sources of inspiration for the (future) theory, the principal components of its current state, as well as a panorama of its applications and the history of its evolution during the twentieth century.

Chapter 2 establishes the basic contours of the theory of Hankel/Toeplitz operators on the circle $\mathbb{T} = \mathbb{R}/2\pi\mathbb{Z}$.

Chapter 3 is devoted to the spectral theory of Toeplitz operators.

Chapter 4 explains the unitary equivalence of the theory on $\mathbb{T}$ with the classical theory of Wiener–Hopf equations on $\mathbb{R}$, as well as the Riemann–Hilbert problem.

Chapter 5 treats some properties of finite Toeplitz matrices (inversion, links with trigonometric moments, the approximation of infinite Toeplitz matrices, the asymptotic distribution of spectra).

The reader will soon become aware that this handbook constitutes a rather elementary introduction to Toeplitz matrices and operators based on the theory of Hardy spaces. It can thus be regarded as a source of basic knowledge (a "First elements of ..."). Nevertheless, in principle, a student reaching the end of this book will be capable of embarking upon a project of independent research (the author can affirm this by positive experiences with numerous students). For such an endeavor, the aid of experts will be needed: this can be found in the dozens of existing monographs devoted to Toeplitz operators and matrices, to Hardy spaces, and to the "hard analysis" that was developed around them. Some of this torrent of literature is mentioned in the "Notes and Remarks" sections.

Good luck!

Acknowledgments

Acknowledgments for the English Edition

The author warmly thanks the translators Danièle and Greg Gibbons for their high-quality work, for thorough attention to all shades of meaning of the French text, and for friendly collaboration at all stages of the work.

The author is also sincerely grateful to Cambridge University Press for including the book in this prestigious series, and to the entire editorial team for highly professional preparation of the manuscript and for patience during his numerous hold-ups due to varying circumstances.

Élancourt,

February 4, 2019

Acknowledgments for the French Edition

The work on the final draft of this text was in part supported by the research project "Spaces of Analytic Functions and Singular Integrals" of the University of St. Petersburg, RSF Grant 14-41-00010.

I am truly grateful to Éric Charpentier, my friend and colleague at the Université de Bordeaux, who – as with my preceding book (*Espaces de Hardy*, Belin, 2012) – sacrificed much of his time for a thorough and effective review of the complete text. Without his invaluable and generous aid the text would never have seen the light of day.

I warmly thank Albrecht Böttcher, the great expert in the Toeplitz domain, who took the time to read the manuscript and whose profound and nourishing comments aided me in polishing the text. I am also beholden to several of my colleagues for *ad hoc* consultations on the subjects of the book, and especially to Anton Baranov of St. Petersburg for his remarks and suggestions concerning the editing.

I also address my cordial thanks to Gilles Godefroy (Institut de Mathématiques de Jussieu, Université Paris VI) who, at a delicate moment of the project, rescued it from a dead end brought on by circumstances beyond my control, by recommending the publisher Calvage et Mounet. I would of course also like to warmly thank the publisher (in particular Alain Debreil and Rachid Mneimné) for the painstaking and highly qualified attention they paid to the present manuscript, as well as for the friendly atmosphere that marked our collaboration on this occasion.

And finally, but foremost, I think of my large family – Pascale, Laure, Ivan, Jeanne, and Alekseï – who stoically supported my prolonged isolation while I struggled with my text, and who encouraged me in my moments of doubt.

Élancourt,
November 2016

Biographies

Bernhard Riemann	*page* 3	Igor Simonenko	*page* 113
Vito Volterra	5	Carl Runge	123
David Hilbert	7	George Birkhoff	170
Henri Poincaré	12	Nikolaï Luzin	184
Otto Toeplitz	17	Norbert Wiener	192
Hermann Hankel	28	Eberhard Hopf	194
Marie Hankel(-Dippe)	28	Edmond Laguerre	200
Zeev Nehari	44	Josip Plemelj	212
Mark Krein	48	Constantin Carathéodory	232
Felix Berezin	60	Gábor Szegő	244
Leopold Kronecker	71	Hermann Weyl	272
Solomon Mikhlin	78	Israel Gelfand	342
Paul Halmos	83	Erik Ivar Fredholm	360
Gaston Julia	89	Frederick Atkinson	366
Augustin Cauchy	104	Fritz Noether	378
Israel Gohberg	108	Felix Hausdorff	381

Figures

Bernhard Riemann (Mathematisches Forschungsinstitut Oberwolfach gGmbH (MFO): https://opc.mfo.de/) *page* 2
A self-caricature by Lewis Carroll (Culture Club/Hulton Archive/Getty Images) 3
Vito Volterra 5
David Hilbert (MFO) 7
Henri Poincaré 12
Otto Toeplitz (MFO) 17
Hermann Hankel 28
Figure 2.1 The function $\dfrac{t(\pi - |t|)}{|t|}$. 40
Zeev Nehari (I Have a Photographic Memory by Paul R. Halmos; ©1987 American Mathematical Society) 44
Mark Grigorievich Krein (Ukraine Mathematical Society: www.imath.kiev.ua) 48
Felix A. Berezin (Courtesy of Elena Karpel, personal archive) 60
Leopold Kronecker (MacTutor History of Mathematics Archive: www-history.mcs.st-and.ac.uk) 71
Figure 2.2 The matrix A of the Hint of §2.3.10(a) 74
Solomon Mikhlin (Vladimir Maz'ya and Tatyana Shaposhnikova, uploaded by Daniele TampieriCC BY-SA 3.0: https://en.wikipedia.org/wiki/Solomon_Mikhlin) 78
Paul Halmos (MFO) 83
Gaston Julia in later years (MacTutor History of Mathematics Archive) 89
Gaston Julia in the French army during WWI (MacTutor History of Mathematics Archive) 90
Augustin Louis Cauchy (MacTutor History of Mathematics Archive) 104

List of Figures

Israel Gohberg (MacTutor History of Mathematics Archive)	108
Igor Simonenko (Ivleva.n.s/CC BY-SA 4.0: https://en.wikipedia.org/wiki/Igor_Simonenko)	113
Figure 3.1 The disk $D(\lambda, R)$, $\lambda \to \infty$	118
Figure 3.2 The disk $\overline{D}(1, d)$ of Lemma 3.3.4	120
Figure 3.3 The domain $\Omega(r, s, \alpha)$ of Lemma 3.3.4	120
Figure 3.4 The disk $\overline{D}(\lambda, R)$ and the domain $\Omega(r, s, \alpha)$ of Lemma 3.3.4	121
Carl Runge	123
George David Birkhoff (MFO)	170
Nikolaï N. Luzin (MacTutor History of Mathematics Archive)	184
Norbert Wiener (Bettmann/Bettmann/Getty Images)	192
Eberhard Hopf (MacTutor History of Mathematics Archive)	194
Edmond N. Laguerre	200
Josip Plemelj (MacTutor History of Mathematics Archive)	212
Constantin Carathéodory (MacTutor History of Mathematics Archive)	232
Great Fire of Smyrna, 1922	233
Gàbor Szegő (MacTutor History of Mathematics Archive)	244
Figure 5.1 A banded Toeplitz matrix. The gray is a Toeplitz portion $\widehat{\varphi}(k - j)$, $0 \leq j,k \leq n$; the white is zero.	251
Hermann Weyl (MFO)	272
Israel Gelfand (Photo: Nick Romanenko, Copyright Rutgers, The State University of New Jersey)	342
Eric Ivar Fredholm (MacTutor History of Mathematics Archive)	360
Frederick V. Atkinson (MacTutor History of Mathematics Archive)	366
Fritz Noether (MFO)	378
Fritz Noether with Emmy Noether (MFO)	379
Oryol Prison, Russia	380
Felix Hausdorff	381

1
Why Toeplitz–Hankel? Motivations and Panorama

Topics

- Four cornerstones of the theory of Toeplitz operators: the Riemann–Hilbert problem (RHP), the singular integral operators (SIO), the Wiener–Hopf operators (WHO), and last (but not least) the Toeplitz matrices and operators (TMO) (strictly speaking, compressions of multiplication operators).
- The founding contributions of Bernhard Riemann, David Hilbert, George Birkhoff, Otto Toeplitz, Gábor Szegő, Norbert Wiener, and Eberhard Hopf.
- The modern and post-modern periods of the theory.

Biographies Bernhard Riemann, Vito Volterra, David Hilbert, Henri Poincaré, Otto Toeplitz, Hermann and Marie Hankel.

1.1 Latent Maturation: The RHP and SIOs

The most ancient form of a "Toeplitz problem," which was not identified as such for a hundred years (!), is the Riemann, or Riemann–Hilbert, problem.

1.1.1 Nineteenth Century: Riemann and Volterra

Bernhard Riemann submitted his thesis in 1851 (under the direction of Gauss) and presented his inaugural dissertation entitled "Grundlagen für eine allgemeine Theorie der Funktionen einer veränderlich complexen Grösse" (also available in [Riemann, 1876]). Its principal value lay in its pioneering introduction of geometrical methods to the theory of functions, and in the objects that we now know under the names of Riemann surfaces, conformal mappings, and variational techniques. Moreover, among the 22 sections of this 43-page text (in today's format) there was a short Section 19 containing what is known (following Hilbert) as the 'Riemann problem," one of the cornerstones of the future theory of Toeplitz operators. In this Section 19 Riemann used almost no formulas, and he summed up himself, somewhat abstractly:

19. Überschlag über die hinreichenden und notwendigen Bedingungen zur Bestimmung einer Funktion complexen Argumenten innerhalb eines gegebenen Grössengebiet.

19. An outline of the necessary and sufficient conditions for the determination of a function of a complex variable in the interior of a given domain.

Bernhard Riemann (1826–1866) was a German mathematician, an ingenious creator whose contributions continue to fertilize mathematics 150 years after his passing. Riemann's ideas transformed complex analysis, geometry, and number theory, and also provided a strong impetus to real harmonic analysis and mathematical physics. Three of Riemann's four most influential works appeared as "qualification texts": his doctoral thesis (Göttingen, 1851, under the direction of Gauss), containing the theory of *Riemann surfaces* and conformal mappings, as well as what is now known as the *Riemann (boundary) problem* (RP); his *Habilitation* thesis (1853) devoted to trigonometric series (with the *Riemann integral* as a tool); and the famous *Habilitationsvortrag* (inaugural *Habilitation* conference, 1854) entitled "Über die Hypothesen, welche der Geometrie zu Grunde liegen," which stimulated profound interactions between geometry and physics, leading to Einstein's general theory of relativity. These three masterpieces were published posthumously. The fourth work was "Über die Anzahl der Primzahlen unter einer gegebenen Grösse" (1859), on the distribution of prime numbers, containing – among other subjects – the famous *Riemann hypothesis* (RH) on the zeros of the ζ function in the complex plane (this publication was also "obligatory," as Riemann was obliged to present an article to the Berlin Academy as a new corresponding member). Riemann's works became – and remain – absolutely fundamental to the mathematics and physics of the nineteenth to twenty-first centuries.

1.1 Latent Maturation: The RHP and SIOs

An astronomical number of publications are dedicated to the development of Riemann's ideas and results. For a presentation aimed at the general public, see, for example, *Bernhard Riemann 1826–1866: Turning Points in the Conception of Mathematics* by Detlef Laugwitz (Birkhäuser, 2008), or *Riemann: Le géomètre de la nature* by Rossana Tazzioli (Pour la Science, no. 12, 2002), or *Riemann* by Hans Freudenthal in the *Complete Dictionary of Scientific Biography* (2008). As noted in this last source, "Riemann's evolution was slow and his life short." He only produced 15 mathematical publications, but these rare works defined an epoch. Riemann's name is associated with almost a hundred important concepts, such as *Riemannian geometry*, the *Cauchy–Riemann equations*, *Riemann surfaces*, the *Riemann integral*, the *Riemann conformal mapping theorem*, the *Riemann–Hilbert method (problem)*, the *Riemann hypothesis*, the *Riemann–Lebesgue lemma*, the *Riemann sphere*, the *Riemann–Roch theorem*, etc. In particular, as is well known, Riemannian geometry was decisive in the creation of general relativity – and also in the inspiration of mathematician Charles Dodgson (better known by his literary pseudonym Lewis Carroll: see the sketched self-portrait) for his brilliant *Alice's Adventures in Wonderland* (1865) and *Through the Looking-Glass* (1871).

Riemann's career was slow and difficult right from the start, beginning even with the choice of mathematics as the subject of his studies: on his arrival in 1846 at the University of Göttingen (quite provincial at the time) he was forced by his father to enrol in the faculty of theology and was not able to switch to mathematics until he received his father's permission in 1847. There, Riemann flourished in the Gauss/Weber seminar, where mathematics and physics were intimately interlinked. It was only in 1854 that Riemann gave his first lectures in Göttingen. In 1857, he participated in a competition for a professor's position at the École Polytechnique of Zürich, but, because of his difficulties in oral expression, he lost to his colleague and friend Richard Dedekind (1831–1916). Riemann became a tenured professor at Göttingen only in 1859 (after the death of Per Gustav

Dirichlet, 1805–1859), and always suffered from a lack of students (his famous course on Abelian functions was only attended by three students, including Dedekind). A deterioration of his health (latent tuberculosis?) often forced him to seek refuge in Italy (1862–1866), and according to one of his students (in May 1861) Riemann was often "weak and fatigued" to the point where, at times, he "could not succeed in proving even the simplest results." The level of appreciation of his innovative results by his colleagues was also very low: in the 1860s, Enrico Betti (1823–1892), who had warmly welcomed Riemann to Italy, pointed out that "the works of Riemann are practically unknown to the scientific community" because of "the concision and the obscurity of the style of this eminent geometer"; Karl Weierstrass (1815–1897) underestimated the results of Riemann's thesis; in England, he remained "almost unknown," and "in France and Italy his works were often studied but not well understood" (according to Rossana Tazzioli (2002)). Effectively, Riemann's mathematical style was barely accessible and he often laid himself open to criticism. For example, one can mention the statement of the Riemann problem (neither made explicit, nor linked to the context: see §1.1.1, §4.6 for details), or the famous article "Über die Anzahl ..." (1859), "the most ingenuous and fruitful" according to Edmund Landau (1877–1938), which was written – according to Detlef Laugwitz – "in such a manner that it is not easy to understand how he reached his solution; we easily recognize, in this manner of erasing his traces, a student of Gauss!" These reactions of his colleagues of course contributed to the deterioration of Riemann's health: again in 1857, a prolific year for Riemann, according to Dedekind, he was "hypochondriacal to the extreme, mistrustful of others and of himself," and in 1863 (again according to Dedekind) he was already in a "sad state" with depressive tendencies. Riemann died while travelling in Italy on Lake Maggiore (1866) without even reaching his 40th birthday.

Riemann married Elise Koch in 1862; they had one daughter.

In fact, this Section 19 contains the very first statement of a now famous problem; the author did not relate it to the other questions in his thesis and did not follow it up. In modern terms (especially due to Hilbert, see §1.1.3 below), the problem is as follows:

1.1 Latent Maturation: The RHP and SIOs

Given a bounded domain G in the complex plane $\mathbb{C}$ and real functions a, b, c on the boundary ∂G, find a holomorphic function $f \in \text{Hol}(G)$, $f = u + iv$ ($u = \text{Re } f$, $v = \text{Im } f$) such that its boundary values satisfy $au + bv = c$ on ∂G.

Riemann saw this as a generalization of the *Dirichlet problem*: find a function u satisfying $\Delta u = 0$ in G (where $\Delta = \partial^2/\partial x^2 + \partial^2/\partial y^2$ is the Laplacian in $\mathbb{R}^2$) and $u = c$ (a given function) on ∂G. A restatement of the Riemann problem (and the naming of it as such) was introduced by David Hilbert in the twentieth century (1905) in his famous courses (Göttingen, 1901–1910) on integral equations (see §1.1.3 below), and it is in this exact form that we will consider and resolve it in Chapter 4. At this stage, there is as yet no question of any Toeplitz matrices/operators.

We must add that before the decisive intervention of Hilbert, an article by Vito Volterra [1882] appeared in which the Riemann problem was clearly stated and discussed, but it went almost totally unnoticed. Volterra's name will appear again in our outline of the historic evolution of Toeplitz theory (see §1.3.2 below).

Vito Volterra (1860–1940) was an Italian mathematician, known for his research in partial differential equations, in real analysis, as well as in the theory of integral operators (Volterra operators) and integral-differential operators. He was among those who paved the way between the mathematics of the nineteenth and twentieth centuries. Moreover, Volterra was one of the creators (with Lotka, but independently) of the mathematical theory (the dynamics and the equilibrium) of communities of antagonistic species (predator/prey), and in particular, of the *Lotka–Volterra equation*. His theory of "functions of lines" (which Hadamard (1865–1963) named "functionals"), which dates to 1887, inaugurated modern functional analysis. Volterra authored the very first publication on the Riemann problem (see §1.1.1), as well as 235 other research articles. He is the author of half a dozen monographs, including his theory of the struggle for life [Volterra, 1931].

Volterra studied in Pisa, at the University and at the Scuola Normale Superiore, where he submitted his thesis in 1882 (under the direction of Enrico Betti), and then made a successful career as professor and chairman in Pisa, Turin and Rome. (His work in Turin was especially disturbed by a conflict with another great mathematician, Giuseppe Peano, known for his intolerance for the slightest weakness in the mathematical arguments of others, as well as for his project "Formulario Mathematico" (to code all of mathematics in symbols of logic and to teach it according to this source), which exasperated his colleagues.) Volterra is the only mathematician to have been invited four times to present a plenary conference at the ICM (1900, 1908, 1920, 1928). He became a member of several Academies (Royal Societies of London and Edinburgh, Accademia dei Lincei, and others) and received several distinctions; he founded a number of Italian research organizations (such as the Consiglio Nazionale delle Ricerche (1923) of which Volterra was the first president). Today, there are around a dozen mathematical objects bearing Volterra's name.

Volterra's career was greatly disrupted during the Fascist period in Italy. As one of the 12 Italian professors (out of a total of 1250) who refused to pledge allegiance to the Fascist government in 1931, he was fired from his position of professor and expelled from the Accademia dei Lincei; after the shameful "Manifesto della razza" (1938), Volterra (as well as his two sons who already had university positions) was expelled as a Jew from the Instituto Lombardo di Scienze e Lettere. He died of phlebitis at his home in Rome.

In 1900 Volterra married his cousin Virginia Almagià, and they had six children.

1.1.2 Twentieth Century: David Hilbert

David Hilbert played a multifaceted role in the evolution of Toeplitz operators: he made the first real advance for the Riemann problem (abbreviated RP), he launched "Problem 21" (of which the RP represents the principal part) in his celebrated 1900 list of 23 unresolved problems for the twentieth century, he indicated the principal technique for the solution of the RP (the singular integral operators – SIO), and finally (last but not least!) he suggested to Otto Toeplitz that he consider Laurent matrices/forms (which was to lead to the Toeplitz operators) as an illustration of his brand new spectral theory presented in his courses of 1901–1910 on integral operators.

1.1 Latent Maturation: The RHP and SIOs

David Hilbert (1862–1943) was a German mathematician whose works and personality exercised a decisive influence on all of the mathematics of the twentieth century. His contributions to number theory (via the theory of class fields in the famous *Zahlbericht*, "Report on numbers" (1897), written at the request of the German Mathematical Society), the axiomatization of geometry, integral equations and functional analysis, mathematical physics, the calculus of variations, and mathematical logic are fundamental. His speech to the second International Congress of Mathematicians (Paris, 1900), containing 23 unresolved problems across all domains, determined the development of mathematics for the following decades.

Several disciplines were quite simply created by the pen of David Hilbert, such as proof theory and metamathematics, or "spectral theory" (the name given by Hilbert to his theory of bounded self-adjoint operators; 25 years later, commenting on the quantum physics of Max Born, Werner Heisenberg, and Erwin Schrödinger, he remarked "I even called it 'spectral analysis' without any presentiment that it would later find applications to actual physical spectra").

His founding course on integral equations given at the University of Göttingen 1901–1908 (first published by his student Hermann Weyl in 1908, then in Hilbert's own book in 1912) contains what is known today as *Hilbert spaces*, the *Hilbert transform*, the *Hilbert inequality*, and the "general Riemann problem" (known, after Hilbert's intervention, as the *Riemann–Hilbert problem*: see §1.1.1 and Chapter 4 below). Hermann Weyl (student and then successor of Hilbert in his position, and himself also a master of analysis of the twentieth century) provided an overview and analysis of Hilbert's works by dividing them into five periods: see [Weyl, 1944].

Hilbert created the "Göttingen school," a community without precedent in the history of mathematics, with 69 students who had submitted a thesis under his direction, and which featured a plethora of key figures of twentieth-century mathematics, such as Otto Blumenthal, Felix Bernstein,

Sergei Bernstein, Richard Courant, Alfréd Haar, Erich Hecke, Ernst Hellinger, Erhard Schmidt, Hugo Steinhaus, Hermann Weyl, and Adolf Hurwitz. Many others, without having formally been his doctoral students, spent long periods at Hilbert's seminar in Göttingen, including Emmy Noether, Harald Bohr, Max Born (Nobel Prize 1954), Emanuel Lasker (world chess champion, 1894 and 1921), Alonzo Church, John von Neumann, Otto Toeplitz, Hermann Weyl, Ernst Zermelo, and dozens of others.

Several dozen mathematical objects bear Hilbert's name (a few examples were given above); according to Constance Reid ([1970], page 216), "Like some mathematical Alexander, he had left his name written large across the map of mathematics."

Almost all of Hilbert's career was spent in Göttingen, apart from a short period in Königsberg (1892–1895 as a *Privatdozent*), where he married Käthe Jarosch (a long-time family friend) and had a son, Franz. After years of intense and triumphant work, Hilbert became gravely ill in 1925 (pernicious anaemia, incurable at the time) and was given only a few weeks to live (in fact, the first signs of weakness were already apparent in 1908). Hilbert was miraculously saved thanks to the efforts of his friends and colleagues (Richard Courant, Oliver Kellogg, George Birkhoff, and others) who organized a veritable human chain to bring a new experimental medicine from Harvard (at a time of crisis and cruel daily problems!). Hilbert retired in 1930, the year when his long-standing efforts finally led to the opening of the new Mathematical Institute. He continued to give his courses (one per year), but as he was already very weak, he often lost the thread of his reasoning or lost track of its final goal. It is no doubt this weakness that explains the sad fact that we find his signature on a collective letter of support for Adolf Hitler published before the referendum of 1934 (he who was already so hurt by the ethnic cleansing that had left his Institute bled dry).

Hilbert's life has given birth to a kind of mythology. His reaction to a new German currency in 1923 (in order to control the galloping inflation) was: "Impossible to resolve a problem by changing the name of an independent variable." His first article on invariants (containing among others the celebrated *Nullstellensatz*) was rejected by Paul Gordan, an expert with the *Mathematischen Annalen*, who remarked on the proof of (pure) existence of a finite number of generators: *Das ist nicht Mathematik. Das ist Theologie* ("That is not mathematics. That is theology").

1.1 Latent Maturation: The RHP and SIOs

Hilbert's credo was "A perfect formulation of a problem is already half of its solution" ([Reid, 1970], page 101). Then there is the famous *Wir müssen wissen, wir werden wissen* ("We must know, we will know"), the final motto of his retirement speech at Königsberg (1930), on the eve (!) of Kurt Gödel's announcement of his theorem of the incompleteness of the Zermelo–Fraenkel axiomatic system. (The phrase itself seems to be an allusion to a memorable German slogan, *Wir müssen siegen, und wir werden siegen* ("We must win, and we will win"), used in particular by Kaiser Wilhelm II in August 1914, in war propaganda on thousands of postcards and medals, in war songs, etc., and which probably originated from the *Zwinger Saga* (1517–1524), a medieval text about the siege of the imperial city of Goslar. The Germans were not the only ones – think of *Venceremos*, etc.)

In his courses, first published in *Nachrichten der Königliche Gesellschaft der Wissenschaften zu Göttingen* between 1904 and 1910, starting with "Grundzüge einer allgemeinen Theorie der linearen Integralgleichungen" [Hilbert, 1904], later published as a book [Hilbert, 1912], Hilbert generalized the RP and provided a formulation in terms of singular integral operators (SIO). In particular, with the use of the Riemann conformal mapping theorem (proved in his inaugural dissertation mentioned above, up to a small vague detail on the applicability of the "Dirichlet principle"), Riemann's question stated in §1.1.1 can be reduced (if G is simply connected) to $G = \mathbb{D} = \{z \in \mathbb{C}: |z| < 1\}$. Then, since $u = (f + \overline{f})/2$, $v = (f - \overline{f})/2i$, and $\overline{f}|\partial\mathbb{D}$ is the boundary value of a function $f_-(z) = \overline{f}(1/\overline{z})$ holomorphic in $\mathbb{C} \setminus \overline{\mathbb{D}}$, the problem can be reformulated in the following manner.

Given functions A, B, C on the circle $\partial\mathbb{D} = \mathbb{T} = \{z \in \mathbb{C}: |z| = 1\}$, *find two holomorphic functions $f_+ \in \mathrm{Hol}(\mathbb{D})$ and $f_- \in \mathrm{Hol}(\mathbb{C} \setminus \overline{\mathbb{D}})$ such that their boundary values satisfy*

$$Af_+ + Bf_- = C$$

on $\mathbb{T}$ (*and* $f_-(\infty) = 0$).

Under the hypothesis that A is never zero on $\mathbb{T}$ (and under some rather vague conditions on the regularity of A, B, C: it seems to be $A, B, C \in C^2$), Hilbert reduced the question to a singular integral equation (in fact by identifying $f_\pm$ with projections $P_\pm F$ of a function on $\mathbb{T}$: see Chapter 4 for details) that belonged to a family of equations already studied (in the same course),

$$a(s)\varphi(s) + \int_{[-\pi,\pi]} K(s,t)\varphi(t)\,dt = \psi(s),$$

where K is a kernel holomorphic outside the diagonal $s = t$, on which it has a simple pole, and can be written in the form

$$K(s,t) = b(s) \operatorname{ctg} \frac{s-t}{2} + N(s,t)$$

with $N \in L^2([-\pi,\pi]^2)$ (see Chapter 4 for details). Hilbert thus inferred that for $n = \operatorname{wind}(B/A) = 0$ ("winding number," the Cauchy index of the curve $B/A(\mathbb{T})$: see Definitions 3.1.2 below) a solution exists and is unique; for $n \neq 0$, it is necessary to impose $|n|$ supplementary conditions, or else obtain n independent solutions – a conclusion which anticipated the concept of the index of an operator (introduced by Noether only in 1921: see Appendix E). Finally, this technique led to a complete solution of the RP, presented in Chapter 4 below.

In the same book, Hilbert ([1912], Chapter X) applied the results about the RP to the *problem of monodromy groups* of linear ordinary differential equations (ODEs), also proposed by Riemann. Initially, the problem concerns ODEs of order n in the complex plane $\mathbb{C}$; by a change of notation, such an equation can be reduced to a system of n equations of first degree, i.e.

$$dy(z)/dz = A(z)y(z)$$

where $y = (y_1, \ldots, y_n)^{\operatorname{col}}$ and A is an $n \times n$ matrix-valued function. The system is said to be *Fuchsian* (after Lazarus Fuchs, a German mathematician, 1833–1902) if A is holomorphic in $\mathbb{C}$ except at a finite set of points z_j where it has simple poles. The poles of A provoke eventual branches (logarithmic) of y along an analytic extension on a given closed curve $\gamma: [0,1] \to \Omega$ in $\Omega = \mathbb{C} \setminus \{z_j\}$, which leads to the equality $y \circ \gamma(1) = C(y \circ \gamma(0))$ where $C = C_\gamma$ is an invertible matrix with constant elements. Clearly, if $\gamma = \gamma_1 \cdot \gamma_2$ is a composite path, then $C_\gamma = C_{\gamma_1} C_{\gamma_2}$; the image $\gamma \mapsto C_\gamma$ of the fundamental group of Ω in the group of invertible matrices $\mathbb{C}^n \to \mathbb{C}^n$ is called the *monodromy group* of the equation. Problem 21 in Hilbert's famous list of problems (1900) questions whether an arbitrary group of $n \times n$ matrices can be the monodromy group of a certain Fuchsian system.

Hilbert [1912] showed how, for $n = 2$, the monodromy problem can be reduced to a vectorial Riemann problem $f_- = af_+$ ($f_\pm = (f_\pm^1, f_\pm^2)^{\operatorname{col}}$, a a constant 2×2 matrix, $\det(a) \neq 0$), and then resolved this RP. (In the English-speaking engineering literature, this presentation of the RP is known as the "barrier problem.") Hilbert's appreciation and treatment of the RP (with the aid of the SIOs), as well as its links with the problem of monodromy groups, attracted a strong focus to the RP and gave birth to powerful techniques with a broad spectrum of applications: see §3.5 and 4.6 for a description of some of this research. Impressed by this approach, Émile Picard [1927] gave it the nickname

1.1 Latent Maturation: The RHP and SIOs

"Hilbert problem," but in the end, it was the name "Riemann–Hilbert problem" (RHP) that survived (in what follows, this is the name that will be used).

Currently, because of other important applications of the RHP, it is considered as the *first of the four "cornerstones"* of the theory of Toeplitz operators (see the topics of this chapter), since – by the intermediary of the SIO and with the participation of the WHO (Wiener–Hopf operators: see §1.3.2 below) – it gave birth to a profound spectral theory of Toeplitz operators (presented in Chapter 3). Moreover, in Chapter 4, it will be shown that the theory of *normally solvable RHPs* (in the sense of Hausdorff: see Appendix E) is simply an *equivalent form* of the spectral theory of Toeplitz operators, a fact that was late in being entirely recognized (see the Remarks in §1.4 as well as the comments and references in §4.6).

As for Problem 21 on monodromy groups (which remained tightly linked to the RHP and especially to the Birkhoff method of factorization: see §1.1.3 below), during the twentieth century, it profited from a long and rich evolution. The tone was set by Hilbert himself, who, according to his own testimonial (in a note at the bottom of the page after Theorem 32 (Chapter X) of [Hilbert, 1912]), had already described the approach to the RP with the aid of the SIOs in his course of the winter semester 1901/1902. Later, the idea was fleshed out in different frameworks by Oliver Kellogg, Ludwig Schlesinger, and Josip Plemelj [Plemelj, 1908b] ("who provided a simplified version of my solution of a special case of the RP," said Hilbert, *loc. cit.*). This last publication contains a solution of "Problem 21 in the broad sense" (i.e. if we realize a monodromy group over any ODE whose solutions are of at most polynomial growth at the points of singularity), and for a long time it was considered to also contain a solution for "Problem 21 in the strict sense" (hence, as stated by Hilbert, when admitting Fuchsian equations uniquely). This interpretation was even raised to the level of holomorphic vector bundles (by Röhrl in 1957). Nonetheless, in the 1980s it was discovered that there were gaps in the arguments of [Plemelj, 1908b]: the proof for the Fuchsian ODE works uniquely in the case where there is at least one singular point where the monodromy matrix is diagonalizable (the work of Kohn Treibich in 1983 and of Arnold and Ilyashenko in 1985). The same remark applies to Birkhoff's result [Birkhoff, 1909, 1913] which is based on a factorization method (see §1.1.3 below). In contrast with certain cases where the problem was resolved (the case of monodromy matrices in a neighborhood of the identity, or the case $n = 2$, with an arbitrary number of singularities), in 1989 Andrei Bolibruch obtained an unexpected negative answer to Problem 21, by giving a counter-example for $n = 3$ and 4 singular points. Next, he (and co-authors) found the necessary and sufficient conditions for a group to be realizable as a monodromy group of a

Fuchsian system. The literature on the subject of Problem 21 is immense: see [Anosov and Bolibruch, 1994], [Its, 2003], as well as Bolibruch's comments in pages 591–593 of [Hilbert, 1998].

1.1.3 George Birkhoff and Henri Poincaré

George Birkhoff (1884–1944) and Henri Poincaré (1854–1912) participated in the foundations of Toeplitz theory, completely independently and in different ways: Birkhoff by introducing the most important technique, known as "Birkhoff factorization" (or "Birkhoff–Wiener–Hopf") [Birkhoff, 1909, 1913], and Poincaré by generalizing the Riemann–Hilbert problem and introducing elements of the functional calculus of SIOs [Poincaré, 1910a].

George Birkhoff (see the biography in §3.5) worked on a classification of the singularities of ODEs and on the problem of monodromy groups, but his techniques and results on the factorization of functions (scalar- and matrix-valued) $f = f_- d f_+$ (see [Birkhoff, 1909, 1913]) far exceeded the limits of their initial usage and have become indispensable tools for a large number of disciplines. First of all, this remark applies to the spectral theory of Toeplitz operators (and to the RHP and SIO): for example, the invertibility of a Toeplitz operator T_φ is assured if $\varphi = \varphi_- \varphi_+$ with $\varphi_\pm$ invertible, and then $T_\varphi^{-1} = T_{1/\varphi_+} T_{1/\varphi_-}$. As this concerns a somewhat technical question, we defer its detailed treatment to Chapters 3 and 4 and to the comments in §3.5 and §4.6.

Henri Poincaré (1854–1912), a French mathematician (as well as a physicist, philosopher, and engineer), was one of the last two universal mathematicians (along with Hilbert). His highly important works had a major influence on the landscape of mathematics at the dawn of the twentieth century. Poincaré studied the three-body problem and founded the qualitative theory of systems of differential equations and chaos theory, kickstarting the modern theory of dynamical systems. Several of Poincaré's ideas defined the development of these subjects for the following decades (such as the study of limit cycles and bifurcations). In physics, Poincaré was a major precursor of special relativity: among other contributions, in 1905 he published two articles on the Lorentz transformations (containing all the mathematical elements of the theory), one before and the other after the note of Albert Einstein giving the final physical form to the subject. His results in physics were broadly recognized, in particular by an invitation to the famous first Solvay Congress in Brussels (1911) where he encountered Hendrik Lorentz, Albert Einstein, Max Planck, Ernest Rutherford, and other leading actors in the relativity/quantum revolution. Poincaré never claimed priority for special relativity (unlike several of his partisans who fought for him, including hardliners such as Edmund Whittaker (1873–1956), who attributed everything to Lorentz and Poincaré without ever mentioning the publications of Einstein). On the other hand, he extensively argued with the obscurantists about the new ideas, for example in "La Terre tourne-t-elle?" (*Bull. Soc. Astron. France* **18** (1904), 216–217) on the question of gravitational ether.

In pure mathematics, Poincaré constructed the theory of automorphic functions (invariant under a group of transformations) to resolve linear differential equations with algebraic coefficients, founded algebraic topology ("Analysis situs" 1895) – the celebrated *Poincaré conjecture* (1904) became the theorem of Grigori Perelman a hundred years later (2003–2006) – and the theory of analytic functions of several complex variables. His research on the automorphic functions (Fuchsian, as he called them) was marked by a famous "friendly (and dramatic) rivalry" between Poincaré and a great German mathematician, Felix Klein, who, by coincidence, worked in parallel on the same subject (1881–1882). Klein initiated an intense correspondence with Poincaré in search of the statement of the "great theorem of uniformization," but it was Poincaré who succeeded in finding it first. As for Klein, his health did not support this competition under such feverish and prolonged stress, and he succumbed to serious depression (1882), which troubled his life and career. It seems that Poincaré's spirit was also marked by this intense effort, as it was his research on the automorphic functions that he selected much later as an example in his famous essays on the psychological nature of mathematical discoveries:

[after long fruitless efforts] I left Caen, where I lived at the time ... The adventures of the trip helped me forget my mathematical work; on arrival at Coutances, we boarded an omnibus for I don't remember what promenade; the moment I set foot on the running board, the idea came to me, though nothing in my anterior thoughts seemed to have prepared me, that the transformations that I had used to define the Fuchsian functions were identical to those of non-Euclidean geometry ... On my return to Caen, I verified the result with a clear head in order to acquit my conscience. [Poincaré, 1908]

We must add that much later, the theory of automorphic functions became the basis for the important "Langlands program" (1967) linking harmonic analysis and number theory. In fact, to enumerate the domains where Poincaré figured in a considerable or decisive manner, one could simply make a list of the entire nomenclature of mathematics (and a large part of physics and astronomy) from the beginning of the twentieth century (even in the foundations of mathematics: according to Dieudonné [1975], Poincaré preceded Gödel and his famous incompleteness theorem). In total, the list of Poincaré's publications contains more than 600 references, including fundamental monographs: *Les méthodes nouvelles de la méchanique céleste* (in three volumes published between 1892 and 1899) and *Leçons de méchanique céleste* (1905), as well as several essays of philosophy and popularization.

Dozens of mathematical objects bear Poincaré's name. He received several scientific prizes (beginning with the prestigious prize awarded by the Swedish king Oscar II in 1887 for the three-body problem) and other distinctions (for example, he was the member of more than 30 academies). His personality and his vision of nature and science made him the target of studies by mathematicians (Gaston Darboux defines him as an intuitive), psychologists (Édouard Toulouse (1910), who noted that most mathematicians begin working from well-established principles, whereas Poincaré always began with a revision of the basic principles), historians of science (Pierre Boutroux, Poincaré's nephew, wrote in a letter to Mittag-Leffler in 1913: "he habitually neglected the details and only regarded the peaks, but could pass from one to the other with surprising promptitude"), biographers (André Belliver, 1956), and of himself:

The scientist does not study nature because it is useful; he studies it because he delights in it, and he delights in it because ... it is beautiful. If nature were not beautiful, it would not be worth knowing, and if nature were not worth knowing, life would not be worth living.

and

How is it that there are so many minds that are incapable of understanding mathematics? If mathematics invokes only the rules of logic ... And there is more: how can error be possible in mathematics? [Poincaré, 1908]

Poincaré came from an influential family of intellectuals. His father was the dean of the Faculty of Medicine of Nancy, and his first cousin was Raymond Poincaré, future President of France (1913–1920, nicknamed *Poincaré-la-guerre* after the First World War) and President of the Conseil des ministres (1922–1924, 1926–1929). When he was five years old, Poincaré fell sick with diphtheria and remained paralysed for five months. In spite of his athletic and artistic ineptitude and a failed exercise in descriptive geometry, he was classed first in the entrance exams to the École Polytechnique (1873–1875). Subsequently he went to the École des Mines, and graduated (third out of three) in 1878, and was appointed engineer third-class in Vesoul (1879). Poincaré submitted his thesis at the Sorbonne in 1879 (under the direction of Charles Hermite) and became a lecturer at the Faculté des Sciences in Caen. From 1881 onwards, he occupied various positions at the Sorbonne and the École Polytechnique; he was elected to the Académie des Sciences in 1887 (the only case of an election to all five of the departments of the Académie at that time – geometry, mechanics, physics, geography, and navigation; he became President in 1906) and to the Académie Française in 1908, as well as to the Bureau des Longitudes (1893). During the Dreyfus affair, which shook France during the years 1894–1906 (with numerous repercussions long after), Poincaré played a certain role, by addressing a letter to the second trial in Rennes criticizing the "scientific" analysis of the *bordereau* (analysis which "accused" Dreyfus), then – at the demand of the Cour de Cassation (Court of Appeals) in 1904 – by producing a report (co-signed with Gaston Darboux and Paul Appell) on the mathematical errors of this analysis.

In 1881 Poincaré married Louise Poulain d'Andecy, great-granddaughter of Étienne Geoffroy Saint-Hilaire (1772–1844), zoologist and member of the Académie des Sciences; they had four children.

A vast literature is dedicated to Poincaré's scientific heritage, as well as to descriptions of his life, his evolution, and his entourage. See for example [Dieudonné, 1975] and [Verhulst, 2012]. The list of references for a biography of Poincaré on the website of the University of St. Andrews (www-history.mcs.st-and.ac.uk/Biographies/Poincare.html) contains 100 bio-bibliographical references (including Aleksandrov, Appell, Armand Borel, Émile Borel, Bell, Brelot, Browder, Dantzig, Darboux, Dugac, Gray, Griffiths, Hadamard, Valiron, and more).

Poincaré, in his mathematical theory of tides, tackled a problem similar to that of Riemann–Hilbert; it concerns the determination of a harmonic function u ($\Delta u = 0$ in a domain G) satisfying a condition on the boundary ∂G,

$$a(s)\frac{\partial u}{\partial n} + b(s)\frac{\partial u}{\partial s} + c(s)u = d(s),$$

where a, b, c, d are given functions and $\partial u/\partial n$, $\partial u/\partial s$ are the normal and tangential derivatives of u at the boundary. He reduced the question to a singular integral equation, and for its solution introduced a kind of functional calculus including an important formula, called the Poincaré–Bertrand formula (see §4.6 below), which played a stimulating role in the discovery of symbolic calculus (of Mikhlin and others; see §3.5 and §4.6 for a discussion and references).

1.2 The Emergence of the Subject: Otto Toeplitz

There are legends connecting the appearance of the theory of *Toeplitz operators* with an article by Otto Toeplitz in 1911 and that of *Hankel matrices* with the works of Hermann Hankel in the 1860s. In fact, neither the first nor the second of these attributions stands up to close examination. Toeplitz never studied "Toeplitz operators,"

$$T = (t_{i-j})_{i,j\geq 0}: \ell^2(\mathbb{Z}_+) \to \ell^2(\mathbb{Z}_+),$$

but uniquely those of Laurent,

$$L = (t_{i-j})_{i,j\in\mathbb{Z}}: \ell^2(\mathbb{Z}) \to \ell^2(\mathbb{Z}).$$

As for Hankel, he certainly examined (in passing) matrices of the form $(a_{i+j})_{i,j\geq 0}$ (said to be of "Hankel structure"), but never drew any consequences or significant advantages from this structure.

More precisely, Toeplitz's article "Zur Theorie der quadratischen und bilinearen Formen von unendlichvielen Veränderlichen, part I: Theorie des L-Formen"[1] [Toeplitz, 1911a] contains a spectral theory of Toeplitz *forms* (Tp, p) and Laurent operators $L = (t_{i-j})_{i,j\in\mathbb{Z}}$ on the space $\ell^2(\mathbb{Z})$ (and in fact, also of *Laurent quadratic forms* according to the language used in [Toeplitz, 1911a]); the latter can be reduced, after the intervention of the Fourier transform, to operations of multiplication

$$f \mapsto \varphi f, \quad \varphi(x) = \sum_{j\in\mathbb{Z}} t_j e^{ijx}$$

---

[1] The "part II" implicitly promised by the title never actually appeared.

1.2 The Emergence of the Subject: Otto Toeplitz

in $L^2(0, 2\pi)$. To close the question, note that the *"finite* Toeplitz matrices" $(t_{i-j})_{i,j=0}^n$ are nonetheless presented in [Toeplitz, 1911a], in a note at the bottom of page 355, which simply says that when it comes to the theory of *forms* $x \mapsto (Ax, x)$, there is no difference between the "Toeplitz case" $A = T$ and the "Laurent case" $A = L$, given the "Toeplitz structure" of matrices invariant under translation. In particular, Toeplitz [1911a] showed that the upper bounds of the bilinear forms (Tx, y) and (Lx, y) over the unit balls of the corresponding spaces are the same, and hence – in modern notation – the formula

$$\|T\| = \left\|\sum_{j \in \mathbb{Z}} t_j z^j\right\|_\infty ;$$

see Theorem 2.1.5 below and comments on §2.1.5 and Lemma 2.1.4 in Section 2.4. On the contrary, there is a (huge) difference between the *Laurent operators*,

$$Lx = \left(\sum_{j \in \mathbb{Z}} t_{i-j} x_j\right)_{i \in \mathbb{Z}}, \quad x \in l^2(\mathbb{Z}),$$

and *those of Toeplitz*,

$$Tx = \left(\sum_{j \in \mathbb{Z}_+} t_{i-j} x_j\right)_{i \in \mathbb{Z}_+}, \quad x \in l^2(\mathbb{Z}_+).$$

Otto Toeplitz (1881–1940), a German mathematician, was one of the creators of the spectral theory of linear operators. He submitted his doctoral thesis on algebraic geometry in 1905 at Breslau (Wrocław in Poland after 1945), and then collaborated with the circle of Hilbert's brilliant students in Göttingen, alongside Max Born, Richard Courant, and Ernst Hellinger (to become his closest friends). With the latter, Toeplitz realized, over several years, a vast project of fundamental articles on integral equations (published in 1928) where, among others, the concept of normal operator was introduced. At the time of the Nazis' arrival to power (1933), he was a professor at the University of Bonn; targeted by anti-Jewish legislation, he was not immediately removed from his functions because of an exception concerning positions attributed before 1914, and retained his employment for two more years. When he was fired in 1935, Toeplitz founded a private school in Bonn for Jewish children excluded from the German educational system. However, the *Kristallnacht* ("Night of Broken Glass") of November 9–10, 1938 (a nationwide pogrom ordered by the Nazis in response to the assassination of a German diplomat in Paris by a young Polish Jew of German origin, Herschel Grynszpan) convinced him to emigrate. He settled in Palestine in 1939 and participated in the modernization of the University of Jerusalem (founded in 1925); he died in 1940 of tuberculosis. One of his three children, Uri, an eminent musician (he was for a long time the first flute of the Israel Philharmonic Orchestra), wrote a book on the history of his family: *And Words are Not Enough: From Mathematics in Germany to Music in Israel; A Jewish Family History 1812–1998*.

Toeplitz was very interested in the history of science; he created a "genetic approach" to analysis and how to teach it. With Hans Rademacher (1892–1969), he authored a famous book of popularization, *Von Zahlen und Figuren* (1930), translated as *The Enjoyment of Math* (Princeton University Press, 1966), which attracted generations of youngsters to the in-depth study of mathematics. One of the key figures of functional analysis of the twentieth century, Toeplitz severely criticized Stefan Banach's *Théorie des opérations linéares* (a "cult" text on the subject) for its "much too abstract character," but then, in response, created with Gottfried Köthe an even more general theory of topological vector spaces. Today, Toeplitz's name is especially cited for Toeplitz matrices and operators (even though he never studied the latter: see the explanations in the text), as well as for his interpretation of moment problems in the language of positive definite matrices.

He is also the author of the *Toeplitz conjecture* in plane geometry (1911): "every Jordan curve contains an inscribed square" (i.e. such that all the vertices belong to the curve); although it has been verified for numerous special cases, the problem remains open in general (2014).

In fact, the Laurent operators represent – at best – an example (anecdotal) of the spectral theorem of normal operators, whereas those of Toeplitz $T = (t_{i-j})_{i,j \geq 0}$ (hence, the compressions P_+LP_+ of the L to $\ell^2(\mathbb{Z}_+)$) generate a profound and rich theory, in the framework of both pure analysis and its applications (just to cite a few keywords, we can mention the RHP (above), the problems of trigonometric moments, the Wiener filters, the Wiener–Hopf

1.2 The Emergence of the Subject: Otto Toeplitz

equations (see below), etc.). Of course, today, to strongly insist on the differences between matrices/forms $x \mapsto (Ax, x)$ (which are best adapted to the study of the metric properties of A) and the operators $A: x \to Ax$ they define (which allow the study of the dynamics A^n, $n \geq 0$, as well as functions $f(A)$, equations $Ax - \lambda x = y$, i.e. the spectrum of A, etc.), is a bit like "storming an open door," but at the time of Toeplitz [1911a] these nuances were not yet as clear, and operator theory did not yet exist. In any case, the name *Toeplitz operator* is now inseparable from the mappings of type T defined above.

The article "Zur Theorie der quadratischen und bilinearen Formen von unendlichvielen Veränderlichen" [Toeplitz, 1911a], where the subject of this book had its origin, was part of his *Habilitation* thesis, submitted in Göttingen. Toeplitz stayed in Göttingen between 1906 and 1913, arriving just at the moment when Hilbert concluded his theory of self-adjoint integral operators in a series of courses given between 1901 and 1905. Under the influence of this brand new theory (and no doubt, with a few direct hints from the master), Toeplitz looked for examples of its application. During his stay in Göttingen, he wrote five articles on spectral theory, including [Toeplitz, 1911a]. In modern language, this article shows that the spectrum $\sigma(L)$ of a Laurent operator is the range $\varphi(\mathbb{T})$ of the circle $\mathbb{T}$ where $\varphi(e^{ix}) = \sum_{j \in \mathbb{Z}} t_j e^{ijx}$ (for very smooth "symbols" φ). To calculate the spectrum based on the spectra of matrices $(t_{i-j})_{-n \leq i,j \leq n}$ or, equivalently, $(t_{i-j})_{0 \leq i,j \leq 2n}$, he used an approximation of L by *circulant matrices (or circulants)* (for which an explicit diagonalization is available: see Chapter 5 below), which anticipated Szegő's theorems of spectral distribution: see §1.3 and Chapter 5 below. As we will see in Chapter 3, the spectrum $\sigma(T)$ of the corresponding (true) Toeplitz operator is in fact much larger: it is necessary to add the interior of curves where the Cauchy index is different from zero, $\{\lambda \in \mathbb{C}: \text{wind}(\lambda - \varphi) \neq 0\}$. To finish with the opposition forms/operators, note that, precisely in the case of Toeplitz matrices, it plays a a kind of mathematical joke: if the inverses of the truncations $T_{\varphi,n}^{-1}$ (defined by the *forms* $x \mapsto (T_\varphi x, x)$, $x \in \mathcal{P}_n$) of a Toeplitz operator T_φ converge, the limit is indeed the inverse operator T_φ^{-1} (see Lemma 5.4.1), but the *spectra* $\sigma(T_{\varphi,n})$ always converge (in mean) to the spectrum of a Laurent operator $\sigma(L_\varphi) = \varphi(\mathbb{T})$ (see Theorems 5.6.3–5.6.6)!

Toeplitz made other important discoveries about *Toeplitz forms*,

$$\sum_{i,j \geq 0} t_{i-j} \bar{x}_i x_j,$$

finite or otherwise (the name itself seems to have been introduced by Ernst Fischer [1911]), in particular, about their role in the *trigonometric moment problem*: find a measure μ whose moments $\int_\mathbb{T} z^k \, d\mu = c_k$ are given for certain values of k. The language of matrices proposed by Toeplitz (see Chapter 5 for details) served later as a model for several other moment problems. It is curious to note that the reciprocal relationship between Toeplitz matrices and moment problems, have, with time, been paradoxically reversed: today, the moment problem represents a field of application of Toeplitz theory (among dozens of others, and it is very far from being the most important), whereas at the beginning of the twentieth century, the moment problem, arising in physics and mechanics, and inaugurated as separate mathematical subject in the nineteenth century by Pafnuty Chebyshev and Thomas Stieltjes (in fact the name itself was introduced by Stieltjes), was already of major importance, and immediately hoisted Toeplitz matrices to the same level of importance.

1.3 The Classical Period

1.3.1 Gábor Szegő's Revolution

Gábor Szegő (1895–1985: see the biography in §5.1) was solely responsible for the basis of spectral theory for finite Toeplitz matrices (forms), beginning with a paper he wrote at the age of 19 [Szegő, 1915]. Specifically, his famous formula

$$D_n/D_{n-1} = \mathrm{dist}^2_{L^2(\mu)}(z^n, \mathcal{P}_{n-1})$$

linked the Toeplitz matrices to the theory of weighted polynomial approximation; here μ is a positive Borel measure on the circle $\mathbb{T}$, $\mathcal{P}_{n-1} = \mathrm{Vect}(z^k : 0 \leq k < n)$ is the space of polynomials of degree $< n$, and $D_n = \det(T_{\mu,n})$ the determinant of the Toeplitz matrix $T_{\mu,n} = (\widehat{\mu}(j-k))_{0 \leq k,j \leq n}$ formed by the Fourier coefficients of μ. By taking the limit $n \to \infty$ for $\mu = w \cdot m$, with m the Lebesgue measure, Szegő found his first asymptotic formula for the distribution of spectra: see §5.6.1 below. This was only the beginning of the close connections made between Toeplitz forms/matrices and the theory and techniques of Hardy spaces, newly born at this point in time.

A two-part work, "Beiträge zur Theorie der Toeplitzsche Formen," [Szegő, 1920, 1921a] established the fundamentals of the modern theory of orthogonal polynomials and the intrinsic links of this new theory with the spectral properties of Toeplitz matrices. In particular, the elegant technique presented in §5.1.4 and §5.1.5 showing how orthogonal polynomials can be associated with a Toeplitz matrix (even though this was only later developed by Akhiezer and Krein) has its origin in these two articles by Szegő. The identification (under appropriate circumstances) of the zeros of orthogonal polynomials with the eigenvalues of the corresponding Toeplitz matrices also has roots in this founding work; in turn it later drove interactions with the techniques of potential theory, thus provoking an immense mutual enrichment of these two theories: see [Saff and Totik, 1997]. We refer to [Simon, 2005b] for explanations and clarifications of these links.

For a second series of Szegő's implications in Toeplitz theory [Szegő, 1952, 1954], also revolutionary in character, see the comments in §1.4 below.

1.3.2 The Wiener and Hopf Integral Operators

In fact, the true beginning of the history of *Toeplitz operators* is particularly associated with the names Norbert Wiener and Eberhard Hopf and their founding work, "Über eine Klasse singulären Integralgleichungen" [Wiener and Hopf, 1931]. For Wiener (see the biography in §4.2.2), the subject arose naturally during his studies of causal signals and the best quadratic predictions for random processes. For Hopf (see the biography in §4.2.2), the equations which now bear the name Wiener–Hopf were linked to the problem of radiative equilibrium, already a subject of his interest for several years.

The integral equations studied in [Wiener and Hopf, 1931] were

$$\lambda f - W_k f = g,$$

$$W_k f(x) = \int_0^\infty k(x-y) f(y) \, dy, \quad x > 0,$$

where k is an exponentially decreasing function (a "kernel") and the functions f (unknown) and g (given) are permitted to grow in such a manner that the integrals converge absolutely. The mappings W_k are clearly continuous analogs of those defined by Toeplitz matrices and can be transformed into them by a discretization,

$$\frac{1}{h} \sum_{j \geq 0} k((i-j)h) f(jh), \quad i \geq 0.$$

In contrast with Toeplitz, whose motivations were purely mathematical, Wiener and Hopf, [1931] were led to the subject via applications. More precisely, as already mentioned, Hopf had already studied the equations of stellar radiative equilibrium (see also the biography in §4.2.2), in particular, the classic equation of Milne from 1921:

$$B(x) = \frac{1}{2} \int_0^\infty B(y) E_1(|x-y|) \, dy, \quad x > 0,$$

where B is the radiation density and $E_1(x) = \int_0^1 e^{-x/t} \frac{dt}{t}$ the integral exponential function. Wiener came to it in a different manner, through his studies of generalized harmonic analysis and electronic filters (see Wiener [1966] for a survey of his approach and for references) where a causal stationary filter has the structure of the restriction of a convolution $f \mapsto (k * f)|\mathbb{R}_+$ ("causal" meaning $k(t) = 0$ for $t < 0$). Later, while formalizing his reflections of the years 1931–1941 into a brand new theory of optimal quadratic prediction of random processes, he arrived at the same equation (which, in this context, links the auto-correlation and the bi-correlation of a process) as a criterion for optimality (see e.g. [Wiener, 1966], page 27). Moreover, Wiener pointed out (see [Wiener, 1966], page 88) that it is exactly here (where the variable has the sense of time) that the principal application of the "Wiener–Hopf equations" (WHE) is encountered, and not in Milne-type problems (where the variable is spatial). This is confirmed by the role played today by the WHE and their associated techniques in the theory of random processes (see the numerous comments in §4.6 and §3.5, and the comments on Exercise 3.4.14).

It remains to add that many previous works contained *de facto* the "Wiener–Hopf equations" (without explicitly identifying them as such: see the declaration of Monsieur Jourdain at the beginning of the Preface). This was the case in pure mathematics (in particular, in Hilbert's work on the RHP) as well as in applications: in engineering (Oliver Heaviside and his operational calculus), in theoretical physics (Lord Rayleigh), and even in biology! For example, Volterra [1931] introduced a mathematical theory for the "struggle for life," where an equation of coexistence in a population with delayed effects or "post-actions") is equivalent to

$$g(t) = \lambda f(t) - \int_0^t k(t-x) f(x) \, dx, \quad t > 0.$$

The technique for resolving the WHE, invented in [Wiener and Hopf, 1931], was also revolutionary: by applying the Fourier transform (or the Laplace transform) $\Phi = \mathcal{F}(\lambda \delta_0 - k)$, under certain conditions the "symbol" Φ can be factorized into two factors $\Phi = \Phi_- \Phi_+$ respectively extendable in the half-planes $\mathbb{C}_\pm = \{z \in \mathbb{C}: \pm \operatorname{Im} z > 0\}$ to holomorphic functions without zeros (this part of the method was already partially discovered by Birkhoff [1909, 1913] in his study of the RHP). If we set $k_\pm = \mathcal{F}^{-1} \Phi_\pm$, then the $W_{k_\pm}$ are invertible and $(\lambda I - W_k)^{-1} = W_{k_+}^{-1} W_{k_-}^{-1}$. See Chapters 3 and 4 for the details of this technique of "Birkhoff–Wiener–Hopf factorization," as well as for a short description of its immense subsequent development both from the point of view of the mathematical "technique" as well as the applications.

The links between the Toeplitz operators and those of Wiener–Hopf are, in fact, much tighter than could be imagined staying on a fairly superficial level of analogies and/or discretizations, but these were only discovered 30 years later, much to everyone's surprise and confusion. For details, see §1.4.4 below.

1.3.3 A New Challenge Arises: The Lenz–Ising Model

This small subsection serves to note that at about the same time, a physical theory appeared that, 30 years later, would play an enormous role in the acceleration and stimulation of Toeplitz theory. This was a model of Wilhelm Lenz (1888–1957) for spontaneous ferro-magnetism (hence in absence of external fields) of a grid of particles on $\mathbb{Z} \times \mathbb{Z}$ where only the neighboring particles interact. The fundamental problem is to find what is known as the "thermodynamic limit" of the system, and in particular, to determine the spontaneous magnetization and the "critical temperature" (called the "Curie point," or phase transition point). In 1924, Ernst Ising (1900–1998), a thesis student of Lenz, resolved the case dim = 1 of the model (the case where the particles lie on $\mathbb{Z}$) where he showed that in fact there is no phase transition at all.

More than a quarter of a century later (1949–1952), Lars Onsager (1903–1976, winner of the Nobel Prize in Chemistry 1968) showed that in the case dim = 2 (with particles on $\mathbb{Z} \times \mathbb{Z}$), the problem can be reduced to an asymptotic question of Toeplitz determinants. For the rest of the story, see §1.4.3 below.

1.4 The Golden Age and the Drama of Ideas

1.4.1 Solomon Mikhlin and the Symbolic Calculus of the SIO

A new phase in the history of our quadruple theory (RHP–SIO–WHO–TMO) was reached in the context of the SIO with the invention of the *symbol* of a composite operator of SIO. In 1936, Solomon Mikhlin (see the biography in §2.4), while studying (for a particular example) some operators composed with the Cauchy singular integral operator $\mathbb{S}$,

$$T = \sum_i \prod_j M_{a_{ij}} \mathbb{S} M_{b_{ij}}, \quad \mathbb{S}f(z) = \text{p.v.} \int_\mathbb{T} \frac{f(\zeta)}{\zeta - z} \frac{d\zeta}{\pi i}, \quad z \in \mathbb{T},$$

where $M_a \colon f \to af$ is the operator of multiplication by a, observed that for continuous coefficients a_{ij}, b_{ij}, certain spectral properties of T depend uniquely on $\text{Sym}(T) = \sum_i \prod_j a_{ij} b_{ij}$, which he named the *symbol of* T [Mikhlin, 1936]. The Sokhotsky–Plemelj formulas $P_\pm = \frac{1}{2}(\text{id} \pm \mathbb{S})$ (see §4.5.1), show that we obtain the same class of operators T when replacing $\mathbb{S}$ with the Hilbert transform $\mathbb{H}$, or with the Riesz projections $P_\pm$. During the years 1950–1960, Mikhlin himself, with the participation of many others, arrived at a true "symbolic calculus" (a homomorphism)

$$T \mapsto \text{Sym}(T)$$

which allowed the construction of a complete "Fredholm theory" (i.e. a spectral theory modulo the compact operators: see Appendix E for details). This is presented in Chapter 3 below.

1.4.2 The School of Mark Krein

This school (see the biography of Mark Krein in §2.2), based in Odessa during the years 1930–1980, invested enormously in the theory of Toeplitz and Hankel operators (as well as in the WHO and SIO), beginning with the introduction of *techniques of Banach algebras* to the theory (these were initiated by Gohberg and Krein at the end of the 1950s, and then – on the other side of the Atlantic – by Coburn and Douglas between 1960 and 1970). For the elements of spectral theory developed with the aid of these techniques, see Chapter 3, and for numerous references and historical comments, see §3.5, as well as the biographies in Chapters 2 and 3 concerning several actors in this history. Note here a single detail: the first appearance of a "true" *Toeplitz operator* $T: \ell^2(\mathbb{Z}_+) \to \ell^2(\mathbb{Z}_+)$,

$$Tx = \left(\sum_{j \geq 0} c_{i-j} x_j\right)_{i \geq 0},$$

in order to study its spectral properties, also took place in Odessa, in 1948 [Rappoport, 1948a,b].

1.4.3 Lars Onsager, and Szegő Again

The "delayed-action mathematical bomb" mentioned in §1.3.3 above exploded between 1948 and 1952, first in the community of physicists, and then contaminating the camp of mathematicians. More precisely, during a colloquium at Cornell University in 1948, Lars Onsager stupefied his colleagues by presenting a solution to the two-dimensional Lenz–Ising model, with formulas for the magnetization limit and the critical temperature (of phase transition). The method was based on vertiginous calculations of asymptotic spectra of certain Toeplitz matrices, and his strategy was just the reverse of what would today be regarded as "normal" logic: he interpreted the equations of eigenvalues of large Toeplitz matrices as a "limit case" of a Milne integral equation, then applied the Birkhoff–Wiener–Hopf factorization method to return to matrices. Clearly this approach is hyper-sensitive to a proper calculus of symbols, which was not really accessible for this model. In search of a mathematical colleague competent for this question (and able – as he wrote – to "fill out the holes in

1.4 The Golden Age and the Drama of Ideas

the mathematics and show the epsilons and deltas and all of that"), Onsager spoke to Shizuo Kakutani (1911–2004), a famous Japanese–American applied mathematician, who knew Szegő. Szegő immediately adjusted his calculation [Szegő, 1915] and arrived at the "strong Szegő theorem," Theorem 5.6.8, but under quite restrictive hypotheses: for positive symbols $0 < a \leq \varphi \leq b < \infty$, smooth enough so that $(\log(\varphi))' \in \text{Lip}(\alpha)$, $0 < \alpha < 1$, we have

$$\lim_n \frac{D_n(\varphi)}{G(\varphi)^{n+1}} = \exp\left(\sum_{k\geq 1} k \widehat{\log(\varphi)}(k) \cdot \widehat{\log(\varphi)}(-k)\right),$$

where

$$D_n(\varphi) = \det(\widehat{\varphi}(i-j))_{0\leq i,j\leq n}, \quad G(\varphi) = \exp\left(\int_{\mathbb{T}} \log(\varphi)\,dm\right),$$

and the $\widehat{f}(k)$ are the Fourier coefficients of a function f.

This was exactly the formula that Onsager was missing, so that, on seeing it, he wrote "the mathematicians got there first" (a comment in [Deift, Its, and Krasovsky, 2013] on this occasion: "In the long history of mathematics and physics, it is most unusual for a physicist to be scooped out of a formula by a mathematician!"). But the victory was not complete: for the calculation of other characteristics of the Lenz–Ising model (especially, for the correlations, extremely interesting for physics), the positive symbols were not sufficient; a version of the second Szegő theorem truly applicable to the model was found by Widom [1976] (Theorem 5.6.8 below, already mentioned) and also later by Johansson [1988] (without the condition $\log(\varphi) \in L^\infty$).

The importance of the applications in physics and the unusually close and spontaneous interactions between these disciplines gave rise to a veritable explosion of research on the asymptotic properties of Toeplitz matrices and their diverse applications. A survey article by Deift, Its, and Krasovsky [2013] (already cited) provides a good overview of the subject (we have used this source for this book); for numerous other references see §5.8 below.

1.4.4 Rosenblum, Devinatz, and the Drama of Coincidence

In this subsection, we wish to highlight an aspect of the evolution of the "Toeplitz field" during its century of existence – an important, dramatic, and especially inexplicable circumstance. It concerns the breadth of the contents of "Toeplitz theory," including, as we have seen, (i) the Riemann–Hilbert problem and its equivalent form as a singular integral operator, then (ii) the Wiener–Hopf operators, and (iii) the Toeplitz matrices and operators. From the start and up to the end of the 1960s, the subjects (i)–(iii) were certainly considered to be similar or parallel, but quite independent. Several discoveries were made within one or the other of these theories (often demanding significant effort) and then transposed to the others, also requiring non-negligible technical efforts. The community of mathematicians that worked in this area of analysis consisted of hundreds of renowned experts, including in their ranks such geniuses as Hilbert, Wiener, and Krein. But then, suddenly (and without much in the way of consequences), it was shown that "RHP = SIO" [Rappoport, 1948a], and later that "WHO = TMO" [Rosenblum, 1965; Devinatz, 1967]. This latter identification is particularly "blatant": pure and simple, the matrix of a Wiener–Hopf operator in $L^2(\mathbb{R}_+)$ with respect to the orthonormal basis of Laguerre polynomials $(\mathcal{L}_k)_{k\in\mathbb{Z}_+}$ is a Toeplitz matrix! Hence, there is no difference, they are the same operators, and there you go! It seems quite inexplicable that this equivalence was spotted so late (1965): the link between the trigonometric basis $(e^{ikt})_{k\in\mathbb{Z}_+}$ and the Laguerre basis involves nothing more than nineteenth-century mathematics (a change of variables, a conformal mapping of the half-plane $\mathbb{C}_+$ onto the disk $\mathbb{D}$, and a Fourier transform). Only a certain collective blindness can be suspected (costing significant efforts now recognized as useless – a drama for the community!). For more details, see §3.5 and §4.6.

1.5 The Parallel/Complementary World of Hankel, and the Post-modern Epoch of the Ha-plitz Operators

At the end of the 1940s and the beginning of the 1950s, the "Toeplitz field" underwent rapid and important progress (Rappoport, Krein, Wintner, Hartman). However, as a separate theory of major importance, the Toeplitz operators (i.e. the integral or matrix transforms representing restrictions to a half-line, or a "half-group" $\mathbb{Z}_+$, $\mathbb{Z}_+^n$, of convolution operators – operators with kernels depending on the difference of variables) were only classified in the survey articles by Krein (1958) and then by Krein (1958). From then on, the spectral theory of Toeplitz operators took on its (quasi-) definitive form as we know it today and as is presented here (at times somewhat abbreviated). In this theory, a parallel and/or complementary role to those of Toeplitz is played by the *Hankel operators/matrices*, blood brothers (or sisters) of the Toeplitz transforms.

While speaking of Hankel, we begin by mentioning that a historical aberration similar to that concerning the Toeplitz operators (see §1.2 above) arose in the nomenclature "Hankel matrices/operators" – in his doctoral thesis "Über

1.5 The Parallel/Complementary World of Hankel

eine besondere Klasse der symmetrischen Determinanten" (Göttingen, 1862), Hankel certainly considered matrices of the form

$$H = (a_{i+j})_{i,j\geq 0}$$

(said to be of "Hankel structure") and their determinants, but he did not extract any real consequence or advantage. Later on, Hankel matrices continued to occasionally pop up here and there for decades, without being truly exploited (for example, in the moment problems – beginning with the founding works of Chebyshev and Stieltjes, or in the theory of stationary processes). The progressive and uninterrupted evolution of the subject began with the result of Zeev Nehari [1957] (Theorem 2.2.4 below), which described the symbols of Hankel operators bounded on $\ell^2(\mathbb{Z}_+)$. This was followed by a veritable explosion of interest (and results and applications) beginning with the works of Adamyan, Arov, and Krein [1968], partially presented in Chapters 2 and 3 below. Today, it is easier to enumerate the domains *where there are no* Hankel operators, rather than those where their role is decisive or dominant. The Hankel matrices and operators are recognized as the principal and indispensable tool for optimization problems of a variety of types – from Nevanlinna–Pick complex interpolation to the optimal control of dynamical systems, passing through weighted Fourier series and stationary processes. Moreover, in a somewhat surprising way, it was discovered that "almost every" bounded operator is a Hankel operator with respect to an appropriate basis.

The Hankel and Toeplitz operators are inseparable "siblings" given their complementary positions in the matrix of the same multiplication operator:

$$M_\varphi f = \varphi f, \quad f = \sum_{k\geq 0} a_k z^k, \quad \varphi = \sum_{j\in\mathbb{Z}} c_j z^j,$$

$$M_\varphi = \begin{pmatrix} \cdots & \cdots & \cdots & \cdots \\ c_2 & c_1 & c_0 & \cdots \\ c_1 & c_0 & c_{-1} & \cdots \\ \hline c_0 & c_{-1} & c_{-2} & \cdots \\ c_{-1} & c_{-2} & c_{-3} & \cdots \\ c_{-2} & c_{-3} & c_{-4} & \cdots \\ c_{-3} & c_{-4} & c_{-5} & \cdots \\ \cdots & \cdots & \cdots & \cdots \end{pmatrix} = \begin{pmatrix} \text{Toeplitz} \\ \text{Hankel} \end{pmatrix} \qquad (1.1)$$

(the top matrix is Toeplitz if we reverse it properly from bottom to top); this leads to several matrix and operator identities important for the theory: see for example Lemma 2.2.9 and Exercise 2.3.12 below.

Hermann Hankel (1839–1873), a German mathematician, studied with Möbius, and then with Riemann, Weierstrass, and Kronecker. After he submitted his thesis "Über eine besondere Classe der symmetrischen Determinanten" in Leipzig in 1862 (and his HDR in 1863), he was named professor at the University of Tübingen in 1869. He worked on Grassmann's multilinear algebra, on integration theory (preparing the way for measure theory), and on transforms of functions of complex variables. We know him today for the *Hankel transform* (on functions in a Euclidean space depending uniquely on $\|x\|$), the *Hankel function* (= Bessel function of the third kind), and – especially – the *Hankel matrices (and operators)* that appeared in his thesis. The latter are defined as the matrices $A = \{a(i,j) : 1 \leq i, j \leq n\}$ whose elements are constant on the diagonals perpendicular to the principal diagonal, i.e. $a(i,j) = \varphi(i+j)$ where φ is a function of a single variable. Today, these matrices (and their continuous analogs) are ubiquitous in several analytic disciplines and their applications (harmonic analysis, the theory of holomorphic interpolation, optimal control, random processes, signal processing, etc.).

Hankel invested himself heavily in the study of the history of mathematics (in particular, by illuminating the role of Bernard Bolzano in the theory of infinite series) and left influential texts on this subject.

Stricken with meningitis, Hankel died at 34 from a cerebral haemorrhage while on vacation with his wife **Marie (Hankel-Dippe)**. Much later, Marie Hankel became an emblematic figure in the feminist movement and in linguistic circles. Having learned the artificially constructed language Esperanto at the age of 61, she organized the World Esperanto Congress in Dresden (1908). She was recognized as the first poetess in the history of Esperanto, and won First Prize at the Barcelona Congress (1909) for her poem *La simbolo de l'amo* ("The symbol of love") – an extract:

> Rozujo sovaĝa, simbolo de l'amo,
> En vintro vi staras sen ia ornamo!
> En kampo vi nuda ĉagrenas dezerte,
> Rozujo, malvarmo mortigos vin certe.

1.5 The Parallel/Complementary World of Hankel

The theories of Hankel/Toeplitz are intertwined to the point that we often speak of "Ha-plitz operators." Nonetheless, we will see, among other things, that there are contrasting differences between Toeplitz and Hankel (a compact Toeplitz operator is null, while the theory of compact Hankel operators is very rich, etc.), which is a bit paradoxical, given that an $n \times n$ Toeplitz matrix T can be transformed into a Hankel matrix H by a simple multiplication by a unitary matrix J (known as symplectic):

$$H = JT, \quad J = \begin{pmatrix} 0 & 0 & \ldots & 0 & 1 \\ 0 & 0 & \ldots & 1 & 0 \\ \ldots & \ldots & \ldots & \ldots & \ldots \\ 0 & 1 & \ldots & 0 & 0 \\ 1 & 0 & \ldots & 0 & 0 \end{pmatrix}. \qquad (1.2)$$

During the years 1980–2010 the techniques of Toeplitz and Hankel matrices/operators underwent a profound and rapid development, so that the subject has become extraordinarily vast and it is difficult to imagine treating it fully in a reasonably sized text. We refer the reader to the dozens of remarkable monographs devoted to this theory and cited below in the Notes and Remarks at the end of each chapter of this book. In our brief surveys on the actual state of the theory (in the Notes and Remarks in Chapters 2–4) we will see that after the reunification of PRS–SIO–WHO–TO, the development passed through three remarkable transformations known as "algebraization," "essentialization," and "localization," which completely changed the landscape of the theory (all three are presented in Chapters 3 and 4). This was followed by a far-reaching extension of the concept of "Toeplitz operator" to cover all transformations of the type $PM_\varphi P$ where M_φ is a multiplication operator and P an orthogonal projection onto a certain subspace (these generalizations were already introduced by Szegő in the 1920s). New and broad fields of applications were added to the traditional fields of filtering and random processes by the discovery of links with integrable systems and the (new wave) orthogonal polynomials, but we do not touch on these advanced subjects; see [Hilbert, 1998] (comments), [Deift, 2000], and [Its, 2003].

One particularity of the Ha-plitz theory not to be overlooked is that it is so profoundly implanted in the harmonic analysis of classes of functions (such as L^p (Lebesgue), H^p (Hardy), $B_{p,q}^\alpha$ (Besov/Dirichlet), etc.) that it can be regarded as an operatorial interpretation of harmonic analysis. Many of these links are presented in this handbook; for the others we provide an abundance of commented references, especially in the Notes and Remarks of Chapters 2–5.

This handbook is a short and elementary introduction to the "Toeplitz and Ha-plitz world," and we do not attempt to describe this vast subject exhaustively in the few pages of this historical overview. The book is, finally, in front of the reader, and it is for him or her to now discover the extraordinary universe of Toeplitz/Hankel. The author entreats the reader to not judge him too severely if his brush was not capable of faithfully transcribing all the nuances of the magnificent artwork of Toeplitz/Hankel theory.

1.6 Notes and Remarks

The author learned at the last minute that the title of this chapter almost repeats the first phrase of the article "Une introduction aux opérateurs de Toeplitz" by Colin de Verdière [1995] (incidentally an excellent overview of Toeplitz operators (in a broad sense) over manifolds).

The spectacular impact of the Lenz–Ising model in statistical physics to the history of Toeplitz theory, briefly described in §1.3.3 and §1.3.3 (see also §5.6.5 and §5.8), exercised (and continues to do so) a major influence on the research concerning the Toeplitz matrices: see §5.8 for a discussion and references. Note that the activity around this model during the last few years is such that about 800 articles per year are published on the topic! In the physics literature the habit is to speak of the "Ising model" (without mentioning Wilhelm Lenz); however, Ernst Ising published only a single article stemming from his thesis on the problem (posed by Lenz), before disappearing from the horizon of physics.

We conclude with a remark concerning the final paragraph of Hilbert's biography. The author is grateful to Gilles Godefroy for pointing out that Hilbert's allusion to *Wir müssen wissen, wir werden wissen* is in fact a saying dating back to the First World War, 1914–1918. In fact, as it happens (and as is mentioned in the text), it goes back to the Middle Ages.

2
Hankel and Toeplitz: Sibling Operators on the Space H^2

Topics

- Toeplitz matrices and operators, Toeplitz operators on $l^2(\mathbb{Z}_+)$.
- The existence of the symbol, the discrete Hilbert transform.
- The Hankel matrices and operators: the existence of the symbol (Nehari), compact and finite rank Hankel operators.
- The Berezin transform, the Garsia norm, the spaces BMO, VMO, $H^\infty + C$ and QC, the compact commutators.

Biographies Zeev Nehari, Mark Krein, Felix Berezin, Leopold Kronecker, Solomon Mikhlin, Paul Halmos, Gaston Julia.

In this chapter we work principally in the Hardy space $H^2(\mathbb{T})$ and the Lebesgue spaces $L^2(\mathbb{T})$, $\ell^2(\mathbb{Z})$, $\ell^2(\mathbb{Z}_+)$; see Appendix F for the main properties of $H^2(\mathbb{T})$.

2.1 Three Definitions of Toeplitz Operators: The Symbol

See Appendices C–E for an overview and the standard terminology for bounded linear operators. Here we summarize only the bare minimum, along with a few special notions. By default, unless otherwise noted, everything takes place on a Hilbert space (see Appendix B).

In particular, for Hilbert spaces H and K, we let $L(H, K)$ denote the space of bounded operators from H to K equipped with the operator norm (see Appendix D), with the abbreviated notation $L(H, H) = L(H)$.

2.1.1 The Spaces ℓ^2, L^2, and H^2

The privileged surroundings for the development of the theory of Toeplitz and Hankel operators are the Lebesgue space $\ell^2(\mathbb{Z}_+)$ and the Hardy space $H^2 = H^2(\mathbb{T})$:

$$\ell^2(\mathbb{Z}_+) =: \left\{ x = (x_j)_{j\geq 0} : x_j \in \mathbb{C}, \sum_j |x_j|^2 =: \|x\|^2 < \infty \right\},$$

$$H^2(\mathbb{T}) =: \{ f \in L^2(\mathbb{T}) : \widehat{f}(k) = 0, k < 0 \}$$
$$= \operatorname{span}_{L^2(\mathbb{T})}(z^k : k \in \mathbb{Z}_+) = \operatorname{clos}_{L^2(\mathbb{T})} \mathcal{P}_a,$$

where $L^2(\mathbb{T})$ is the Lebesgue space on the circle $\mathbb{T} = \{z \in \mathbb{C} : |z| = 1\}$ with respect to the normalized Lebesgue measure m (defined by $m(\{e^{ix} : a \leq x \leq b\}) = (b-a)/2\pi$ for $0 \leq b - a \leq 2\pi$), $\widehat{f}(k)$ denotes a Fourier coefficient of an integrable function f,

$$\widehat{f}(k) = \int_{\mathbb{T}} f(z) z^{-k} \, dm(z), \quad k \in \mathbb{Z},$$

where $\mathcal{P}_a$ is the set of polynomials in z,

$$\mathcal{P}_a = \operatorname{Vect}(z^k : k \geq 0),$$

the notation $A =: B$ means "equality by definition" (and it is always clear from the context who defines whom), and – finally – Vect, clos, and span are, respectively, the linear hull, the closure, and the closed linear hull. For all of these notations and definitions, see also Appendices A, B, and E.

The spaces $L^2(\mathbb{T})$, H^2 and $\ell^2(\mathbb{Z}_+)$ are Hilbert spaces (Appendix B) equipped with the respective scalar products:

$$(f, g) = \int_{\mathbb{T}} f \overline{g} \, dm, \quad (x, y) = \sum_{j \geq 0} x_j \overline{y}_j.$$

The orthonormal bases most commonly used are as follows:

(1) $(z^k)_{k \in \mathbb{Z}}$ (in $L^2(\mathbb{T})$),
(2) $(z^k)_{k \in \mathbb{Z}_+}$ (in $H^2(\mathbb{T})$),
(3) $\left(\frac{1}{\sqrt{\pi}(z+i)} \left(\frac{z-i}{z+i} \right)^k \right)_{k \in \mathbb{Z}_+}$ in $H^2(\mathbb{C}_+)$, $\mathbb{C}_+ = \{z \in \mathbb{C} : \operatorname{Im}(z) > 0\}$,
(4) $(e_k)_{k \in \mathbb{Z}_+}$ (in $\ell^2(\mathbb{Z}_+)$), with $e_k = (\delta_{jk})_{j \geq 0}$ where δ_{jk} is the Kronecker δ function,
(5) $(l_k)_{k \in \mathbb{Z}_+}$, $l_k(x) = \frac{1}{k!} e^{x/2} (x^k e^{-x})^{(k)}$ in $L^2(\mathbb{R}_+)$ (*Laguerre functions*).

2.1 Three Definitions of Toeplitz Operators: The Symbol

(For examples (3) and (5), see Lemma 4.3.1 below.) In particular, the set $\mathcal{P}$ of trigonometric polynomials is dense in $L^2(\mathbb{T})$,

$$\mathcal{P} = \text{Vect}(z^k : k \in \mathbb{Z}), \quad \text{clos}_{L^2(\mathbb{T})} \mathcal{P} = L^2(\mathbb{T}).$$

The orthogonal projection of the space $L^2(\mathbb{T})$ onto the subspace $H^2(\mathbb{T})$ (*Riesz projection*) is denoted by P_+:

$$P_+\left(\sum_{k \in \mathbb{Z}} \widehat{f}(k) z^k\right) = \sum_{k \in \mathbb{Z}_+} \widehat{f}(k) z^k$$

(see also Appendix F).

The (discrete) Fourier transform $\mathcal{F}$,

$$\mathcal{F}f = (\widehat{f}(k))_{k \in \mathbb{Z}},$$

is a unitary operator (see Appendix C) from $L^2(\mathbb{T})$ to $\ell^2(\mathbb{Z})$, and from H^2 to $\ell^2(\mathbb{Z}_+)$ (by convention $\ell^2(\mathbb{Z}_+)$ is the subspace of $\ell^2(\mathbb{Z})$ formed by the sequences (x_j) having $x_j = 0$ for $j < 0$).

Another space important for our subject is $H^\infty(\mathbb{T})$,

$$H^\infty(\mathbb{T}) =: L^\infty(\mathbb{T}) \cap H^2(\mathbb{T}),$$

which is in fact a closed sub-algebra of the algebra $L^\infty(\mathbb{T})$ (also called the *Hardy algebra*); see Appendix F for details.

2.1.2 Shift (or Translation) Operators

For a sequence $x = (x_k)_{k \in \mathbb{Z}}$, let

$$Sx = (y_k)_{k \in \mathbb{Z}}, \quad y_k = x_{k-1} \quad (k \in \mathbb{Z}),$$

and for $x = (x_k)_{k \in \mathbb{Z}_+}$

$$Sx = (y_k)_{k \in \mathbb{Z}_+}, \quad y_0 = 0, \quad y_k = x_{k-1} \quad (k \geq 1).$$

Here S is called the *shift (or translation)* operator.

Lemma 2.1.1

(1) $Se_k = e_{k+1}$ where $e_k = (\delta_{jk})_j$.
(2) S *is a unitary operator on* $\ell^2(\mathbb{Z})$, $S^* = S^{-1}$ *hence* $S^* e_k = e_{k-1}$ *for every k*.
(3) *The restriction* $S|\ell^2(\mathbb{Z}_+)$ *is an isometry of* $\ell^2(\mathbb{Z}_+)$ *to itself and* $(S|\ell^2(\mathbb{Z}_+))^* e_0 = 0$, $(S|\ell^2(\mathbb{Z}_+))^* e_k = e_{k-1}$ *for* $k \geq 1$.
(4) $\mathcal{F}^{-1} S \mathcal{F}$ *is multiplication by the "independent variable"* z *on* $\mathbb{T}$ (= *the identity mapping* $z(\zeta) = \zeta$, *for all* ζ), *hence a unitary operator on* $L^2(\mathbb{T})$,

$$\mathcal{F}^{-1} S \mathcal{F} f = zf, \quad (\mathcal{F}^{-1} S \mathcal{F})^* f = \bar{z} f \quad (\forall f \in L^2(\mathbb{T})).$$

(5) The operator $\mathcal{F}^{-1}S\mathcal{F}|H^2$ is an isometry of H^2,

$$\mathcal{F}^{-1}S\mathcal{F}f = zf, \quad (\mathcal{F}^{-1}S\mathcal{F}|H^2)^*f = P_+(\bar{z}f) \quad (\forall f \in H^2(\mathbb{T})),$$

or, with the representation $H^2 = H^2(\mathbb{D})$ (where $\mathbb{D} = \{z \in \mathbb{C}: |z| < 1\}$ is the unit disk; see Appendix E for details concerning the Hardy space H^2),

$$(\mathcal{F}^{-1}S\mathcal{F}|H^2)^*f = \frac{f - f(0)}{z} \quad (\forall f \in H^2(\mathbb{D})).$$

Proof These are simple calculations following from the definitions, left to the reader. ∎

Notation By convention, any operator among

$$S, \quad S|\ell^2(\mathbb{Z}_+), \quad \mathcal{F}^{-1}S\mathcal{F}, \quad \text{or} \quad \mathcal{F}^{-1}S\mathcal{F}|H^2$$

is called the *shift (or translation) operator* and is denoted by S, but indicating each time the space on which it operates. The operators

$$S^*, \quad (S|\ell^2(\mathbb{Z}_+))^*, \quad (\mathcal{F}^{-1}S\mathcal{F})^*, \quad (\mathcal{F}^{-1}S\mathcal{F}|H^2)^*$$

are called the *backward (or inverse) shifts* and are denoted by S^*.

2.1.3 Matrix of an Operator

Let $\mathcal{E} = (e_j)_{j \in J}$ and $\mathcal{E}' = (e'_i)_{i \in I}$ be orthonormal bases in the Hilbert spaces H and K, respectively (the sets of indices are not necessarily the same), and let

$$T: \text{Vect}(\mathcal{E}) \to K$$

be a linear mapping. The *matrix of T with respect to* $\mathcal{E}, \mathcal{E}'$ is

$$(t_{ij}) = ((Te_j, e'_i)),$$

so the jth column represents the coefficients of the decomposition $Te_j = \sum_i (Te_j, e'_i)e'_i$ over the basis $\mathcal{E}'$. When $H = K$ and $\mathcal{E} = \mathcal{E}'$, we speak of the *matrix of T with respect to* $\mathcal{E}$.

Conversely, a matrix (t_{ij}) on $I \times J$ such that, for every $j \in J$,

$$\sum_i |t_{ij}|^2 < \infty$$

generates a linear mapping of the linear hull $\text{Vect}(\mathcal{E})$ in K:

$$T: \text{Vect}(\mathcal{E}) \to K, \quad T\left(\sum a_j e_j\right) = \sum_i \left(\sum_j t_{ij} a_j\right) e'_i.$$

Clearly the continuity and the norm of such an operator T,

$$\|T\| = \sup\{\|Tx\|: x \in \text{Vect}(\mathcal{E}), \|x\| \le 1\},$$

depend only on the matrix (t_{ij}), and not on the choice of the orthonormal bases $\mathcal{E}, \mathcal{E}'$.

Moreover, if $U : \text{Vect}(\mathcal{E}) \to K$ is another linear mapping, then $T = U$ if and only if their matrices coincide: $(t_{ij} = u_{ij}$ for every i, j).

2.1.4 Toeplitz Matrices

A *matrix* (t_{ij}) on a set of indices $I = J = \mathbb{Z}_+$, or on $I = J = \mathbb{Z}$, is said to be *Toeplitz* (or, respectively, Laurent) if t_{ij} depends only on the difference of the indices:

$$t_{ij} = t_{i+k,j+k}$$

for every admissible i, j, k.

t_0	t_{-1}	t_{-2}	t_{-3}	t_{-4}	$\cdots$
t_1	t_0	t_{-1}	t_{-2}	t_{-3}	$\ddots$
t_2	t_1	t_0	t_{-1}	t_{-2}	$\ddots$
t_3	t_2	t_1	t_0	t_{-1}	$\ddots$
t_4	t_3	t_2	t_1	t_0	$\ddots$
$\vdots$	$\ddots$	$\ddots$	$\ddots$	$\ddots$	$\ddots$

2.1.5 Toeplitz Operators

A linear mapping T in a Hilbert space H equipped with an orthonormal basis $\mathcal{E} = (e_j)_{j\geq 0}$ and defined (at least) on $\text{Vect}(\mathcal{E})$ is said to be a *Toeplitz operator* if its matrix with respect to $\mathcal{E}$ is Toeplitz. Particularly important are the following cases:

(1) $H = \ell^2(\mathbb{Z}_+)$, $\mathcal{E} = (e_k)_{k\in\mathbb{Z}_+}$,
(2) $e_k = (\delta_{jk})_{j\geq 0}$,
(3) $H = H^2(\mathbb{T})$, $\mathcal{E} = (z^k)_{k\in\mathbb{Z}_+}$,
(4) $H = L^2(\mathbb{R}_+)$, $\mathcal{E} = (l_k)_{k\in\mathbb{Z}_+}$, $l_k(x) = \dfrac{1}{k!} e^{x/2}(x^k e^{-x})^{(k)}$ (*Laguerre functions*),
(5) $H = H^2(\mathbb{C}_+)$,

$$\mathcal{E} = \left(\frac{1}{\sqrt{\pi}(z+i)}\left(\frac{z-i}{z+i}\right)^k\right)_{k\in\mathbb{Z}_+}.$$

When speaking of *Toeplitz operators on one of these spaces* we always implicitly assume the choice of the bases $\mathcal{E}$ above. The Toeplitz operators on $L^2(\mathbb{R}_+)$ are also known as *Wiener–Hopf operators*: see Chapter 4 below.

The Laurent operators are defined similarly when using bases indexed on $\mathbb{Z}$.

Lemma 2.1.2 (the functional equation)

I. *Let* $H = \ell^2(\mathbb{Z}_+)$, *or* $H = H^2(\mathbb{T})$, *and* $T : \text{Vect}(\mathcal{E}) \to H$ *a linear mapping (where* $\mathcal{E} = (e_k)_{k \in \mathbb{Z}_+}$ *as above). The following assertions are equivalent.*

(1) *T is a Toeplitz operator.*
(2) $T = S^*TS$, *where S is the translation (shift) operator on H, and* S^* *is its adjoint.*

II. (Version $H = \ell^2(\mathbb{Z})$) *Let* $H = \ell^2(\mathbb{Z})$, *or* $H = L^2(\mathbb{T})$, *and* $T : \text{Vect}(\mathcal{E}) \to H$ *a linear mapping (where* $\mathcal{E} = (e_k)_{k \in \mathbb{Z}}$; $e_k = z^k$ ($k \in \mathbb{Z}$ *in the case of* $L^2(\mathbb{T})$)). *The following assertions are equivalent.*

(1) *T is a Laurent operator.*
(2) $T = S^*TS$, *where S is the translation (shift) operator on H, and* S^* *is its adjoint (left translation).*
(3) $Tf = f \cdot T1$ *for every* $f \in \text{Vect}(\mathcal{E}) = \mathcal{P}$ *(the set of trigonometric polynomials).*

Proof I. Let $(t_{ij}) = ((Te_j, e_i))$ be the matrix of T. Then

$$(S^*TSe_j, e_i) = (TSe_j, Se_i) = (Te_{j+1}, e_{i+1}) = t_{i+1, j+1}$$

defines a matrix of S^*TS, and the result follows.

II. The same calculation holds for (1) $\Leftrightarrow$ (2). The equivalence (2) $\Leftrightarrow$ (3) is evident since $z^k T1 = S^k T1 = TS^k 1 = Tz^k$, $k \in \mathbb{Z}$. ∎

Example 2.1.3 Let $\varphi \in L^\infty(\mathbb{T})$ and

$$T_\varphi f = P_+(\varphi f), \quad f \in H^2(\mathbb{T}),$$

see §2.1.1 for P_+. Then T_φ is a bounded Toeplitz operator $H^2(\mathbb{T}) \to H^2(\mathbb{T})$ and $\|T_\varphi\| \leq \|\varphi\|_\infty$.

Indeed, for every $k, j \geq 0$,

$$(T_\varphi z^j, z^k) = (\varphi z^j, z^k) = \widehat{\varphi}(k - j),$$

and hence the matrix of T_φ is Toeplitz. The inequality

$$\|T_\varphi f\|_2 = \|P_+(\varphi f)\|_2 \leq \|\varphi f\|_2 \leq \|\varphi\|_\infty \|f\|_2$$

is evident.

The principal result of this section (Theorem 2.1.5 below) affirms that every Toeplitz operator is of the form T_φ.

Lemma 2.1.4 *Let* $\varphi \in L^2(\mathbb{T})$ *and let* $M_\varphi \colon \mathcal{P} \to L^2(\mathbb{T})$ *be a multiplication operator,* $M_\varphi f = \varphi f$, $f \in \mathcal{P}$. *The following assertions are equivalent.*

(1) $\varphi \in L^\infty(\mathbb{T})$.
(2) $M_\varphi \colon \mathcal{P} \to L^2(\mathbb{T})$ *is bounded (where* $\mathcal{P}$ *is equipped with the norm* $\|\cdot\|_2$).

Moreover, $\|M_\varphi\| = \|\varphi\|_\infty$.

Proof (1) $\Rightarrow$ (2) and $\|M_\varphi\| \leq \|\varphi\|_\infty$ were already seen in Example 2.1.3.

For the implication (2) $\Rightarrow$ (1), given $f \in L^2(\mathbb{T})$, we apply Fatou's lemma to a polynomial approximation $\lim_n \|f - f_n\|_2 = 0$ such that $f(z) = \lim_n f_n(z)$ a.e. on $\mathbb{T}$ (for example, by taking for f_n a suitable subsequence of the partial sums of the Fourier series of f),

$$\int_\mathbb{T} |\varphi|^2 |f|^2 \, dm = \int_\mathbb{T} |\varphi|^2 \lim_n |f_n|^2 \, dm \leq \varliminf_n \int_\mathbb{T} |\varphi|^2 |f_n|^2 \, dm$$
$$\leq \|M_\varphi\|^2 \varliminf_n \int_\mathbb{T} |f_n|^2 \, dm = \|M_\varphi\|^2 \|f\|^2.$$

By taking $f = \chi_E$, where $E = \{\zeta \in \mathbb{T} : |\varphi(\zeta)| \geq \|\varphi\|_\infty - \epsilon\}$, $\epsilon > 0$ (we assume $\varphi \neq 0$), we obtain

$$0 < (\|\varphi\|_\infty - \epsilon)^2 m(E) \leq \|M_\varphi\|^2 m(E),$$

hence (letting $\epsilon \to 0$) $\|\varphi\|_\infty \leq \|M_\varphi\|$. ∎

Theorem 2.1.5 (symbol of a Toeplitz operator) *Let* $T \colon H^2(\mathbb{T}) \to H^2(\mathbb{T})$ *be a bounded Toeplitz operator. There exists a unique function* $\varphi \in L^\infty(\mathbb{T})$ *(the symbol of* T*) such that* $T = T_\varphi$.

Moreover,

$$\|T_\varphi\| = \|\varphi\|_\infty, \quad (T_\varphi)^* = T_{\overline{\varphi}},$$

and

$$\lim_n \|\varphi f - \overline{z}^n T P_+ z^n f\|_2 = 0, \quad \forall f \in L^2(\mathbb{T}).$$

Proof We set $a_k = (T1, z^k)$ for $k \geq 0$ and $a_k = (z^{-k}, T^*1)$ for $k < 0$, and show $(a_k)_{k \in \mathbb{Z}} \in \ell^2$ and that

$$\varphi = \sum_{k \in \mathbb{Z}} a_k z^k$$

is the desired symbol. For this, we calculate the Fourier coefficients of the function $z^{-n}Tz^n$ for $n \geq 0$ by using Lemma 2.1.2: for $k \geq 0$,

$$(z^{-n}Tz^n, z^k) = (P_+ z^{-n}Tz^n, z^k) = (S^{*n}TS^n 1, z^k) = (T1, z^k) = a_k,$$

for $-n \leq k < 0$,

$$(z^{-n}Tz^n, z^k) = (z^{-n-k}Tz^{n+k}z^{-k}, 1) = (S^{*(n+k)}TS^{n+k}z^{-k}, 1) = (Tz^{-k}, 1) = a_k,$$

and (evidently) $(z^{-n}Tz^n, z^k) = 0$ for $k < -n$. Consequently,

$$z^{-n}Tz^n = \sum_{k \geq -n} a_k z^k \quad \text{and} \quad \|z^{-n}Tz^n\|^2 = \sum_{k \geq -n} |a_k|^2 \leq \|T\|^2 + \|T^*1\|^2.$$

Passing to the limit for $n \to \infty$, we obtain $(a_k)_{k \in \mathbb{Z}} \in \ell^2$ and hence $\varphi \in L^2(\mathbb{T})$, $\varphi = \sum_{k \in \mathbb{Z}} a_k z^k$. Moreover, for every $k \in \mathbb{Z}$, we have

$$\lim_n \|\varphi z^k - z^{-n}Tz^n z^k\|^2 = \lim_n \|\varphi - z^{-n-k}Tz^{n+k}\|^2 = \lim_n \sum_{j < -(n+k)} |a_j|^2 = 0.$$

By linearity, this extends to every trigonometric polynomial $p \in \mathcal{P}$:

$$\lim_n \|\varphi p - z^{-n}Tz^n p\| = 0,$$

and thus

$$\|\varphi p\|_2 = \lim_n \|z^{-n}Tz^n p\|_2 \leq \|T\| \cdot \|p\|_2 \quad (\forall p \in \mathcal{P}).$$

Lemma 2.1.4 gives $\varphi \in L^\infty(\mathbb{T})$ and $\|\varphi\|_\infty \leq \|T\|$. By Lemma 2.1.2, for every $p \in \mathcal{P}_a$, we have $Tp = S^{*n}TS^n p$, and hence

$$\|T_\varphi p - Tp\|_2 = \lim_n \|P_+(\varphi - z^{-n}Tz^n)p\|_2 \leq \lim_n \|\varphi p - z^{-n}Tz^n p\|_2 = 0,$$

which means that $T = T_\varphi$.

The uniqueness of φ is clear since $(T_\varphi z^j, z^k) = \widehat{\varphi}(k - j)$, $(k, j \in \mathbb{Z}_+)$. ∎

In §3.1 we extend the "symbolic mapping" $T_\varphi \to \varphi$ to a homomorphism of the Toeplitz algebra $\text{alg}\mathcal{T}_{L^\infty(\mathbb{T})}$.

Corollary 2.1.6 *For every $\varphi \in L^\infty(\mathbb{T})$ and $f \in L^2(\mathbb{T})$,*

$$T_\varphi P_+(z^n f) = z^n \varphi f + o(1) \quad \text{when } n \to \infty,$$

2.1 Three Definitions of Toeplitz Operators: The Symbol

and hence T_φ is asymptotically a multiplication M_φ:

$$T_\varphi(z^n f) = \varphi \cdot (z^n f) + o(1) \quad \text{when } n \to \infty, \, \forall f \in H^2(\mathbb{T}).$$

2.1.6 Comment: Three Equivalent Definitions of Toeplitz Operators

We can conclude from Theorem 2.1.5 that for a bounded mapping $T : H^2(\mathbb{T}) \to H^2(\mathbb{T})$, the three following properties are equivalent, and define a Toeplitz operator.

(1) The matrix of T is Toeplitz.
(2) $T = S^*TS$.
(3) $T = P_+ M_\varphi | H^2(\mathbb{T})$ where $M_\varphi f = \varphi f$, $f \in L^2(\mathbb{T})$ and $\varphi \in L^\infty(\mathbb{T})$.

2.1.7 Examples

(1) The shift operator S and its adjoint S^* are Toeplitz operators, with matrices of "Jordan infinite block" type: $S = (t_{ij})$, $t_{ij} = \delta_{i,j+1}$; $S^* = (t_{ij})$, $t_{ij} = \delta_{i+1,j}$.

(2) "Classical" Toeplitz operators (Toeplitz, 1911; Wiener, c. 1930). These are integral operators on $\ell^2 = \ell^2(\mathbb{Z}_+)$ or $\ell^1 = \ell^1(\mathbb{Z}_+)$ with a kernel k depending only on the difference of its arguments:

$$C_k : x = (x_j)_{j \geq 0} \mapsto \left(\sum_{j \geq 0} k(i-j) x_j \right)_{i \geq 0},$$

where $k \in \ell^1(\mathbb{Z})$, $\|C_k\| \leq \|k\|_{\ell^1} = \sum_{n \in \mathbb{Z}} |k(n)| < \infty$.

(3) The **"analytic" Toeplitz operators** $T_\varphi : H^2 \to H^2$ are the multiplication operators $T_\varphi f = \varphi f$ ($f \in H^2$) where $\varphi \in H^\infty$.

For the spectrum $\sigma(T_\varphi)$, we clearly have

$$\sigma(T_\varphi) = \{\lambda \in \mathbb{C} : (\lambda - \varphi) \notin (H^\infty)^{-1}\} = \text{clos}(\varphi(\mathbb{D})).$$

For a (not very explicit) description of its essential spectrum $\sigma_{\text{ess}}(T_\varphi)$ we refer to Appendix E.9(7), and for the special case of a continuous symbol $\varphi \in C_a(\mathbb{D}) =: H^\infty \cap C(\mathbb{T})$ we refer to to Corollary 3.1.7.

(4) The **general Toeplitz operators on** ℓ^2 are C_k where $k(n) = \widehat{\varphi}(n)$ ($n \in \mathbb{Z}$) and $\varphi \in L^\infty(\mathbb{T})$; C_k is unitarily equivalent to T_φ. Formally, the spectrum $\sigma(T_\varphi)$ is determined in Theorem 3.3.6, but the description is not terribly transparent and requires an interpretation in each special context in order to be useful.

(5) Hilbert transform, the discrete version. This is the transform $\mathbb{H}^d$ on the spaces $\ell^2(\mathbb{Z}_+)$ or $\ell^2(\mathbb{Z})$ with the *Cauchy matrix* $(1/(l - j))_{j \neq l}$, i.e. defined as the convolution operation C_k (see examples (4) and (2)) with $k(n) = 1/n$ ($n \neq 0$), $k(0) = 0$, hence

$$\mathbb{H}^d x = \left(\sum_{j \geq 0, j \neq l} \frac{x_j}{l - j} \right)_{l \geq 0}, \quad x = (x_j)_{j \geq 0} \in \ell^2(\mathbb{Z}_+),$$

or

$$\mathbb{H}^d x = \left(\sum_{j \in \mathbb{Z}, j \neq l} \frac{x_j}{l - j} \right)_{l \in \mathbb{Z}}, \quad x = (x_j)_{j \in \mathbb{Z}} \in \ell^2(\mathbb{Z}).$$

Since $1/n = \widehat{\varphi}(n)$ ($n \neq 0$), where

$$\varphi(e^{it}) = i \frac{t}{|t|} (\pi - |t|) \ (-\pi < t < \pi),$$

we obtain that $\mathbb{H}^d$ is bounded on $\ell^2(\mathbb{Z}_+)$ (or $\ell^2(\mathbb{Z})$) and (in both cases)

$$\|\mathbb{H}^d\| = \|\varphi\|_\infty = \pi.$$

See Figure 2.1.

(6) The **Wiener–Hopf integral operators** on $L^2(\mathbb{R}_+)$,

$$W_k f(x) = \int_{\mathbb{R}_+} k(x - y) f(y) \, dy, \quad x \in \mathbb{R}_+,$$

where $k \in L^1(\mathbb{R})$, are Toeplitz with respect to the orthonormal basis of Laguerre functions $\sqrt{2} l_n(2x)$, where $l_n(x) = e^{-x/2} L_n(x)$ (L_n is the nth Laguerre polynomial), $n = 0, 1, \ldots$; see §4.3 below for explanations and the calculation.

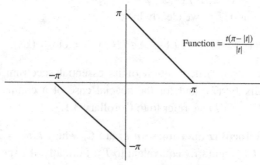

Figure 2.1 The function $\dfrac{t(\pi - |t|)}{|t|}$.

2.2 Hankel Operators and Their Symbols

The definition and the preliminaries of the theory of Hankel operators follow the same lines as those of the Toeplitz operators, but with different conclusions: the similarities and differences can be explained by the fact that these types of operators can be seen as complementary halves of the same multiplication operator (see the Introduction).

2.2.1 Hankel Matrices

A *matrix* (t_{ij}) over the set of indices $I = J = \mathbb{Z}_+$ is said to be *Hankel* if t_{ij} depends only on the sum of the indices:

$$t_{ij} = t_{i+k, j-k}$$

for every admissible i, j, k.

t_0	t_1	t_2	t_3	t_4	$\cdots$
t_1	t_2	t_3	t_4	t_5	$\cdot\cdot\cdot$
t_2	t_3	t_4	t_5	t_6	$\cdot\cdot\cdot$
t_3	t_4	t_5	t_6	t_7	$\cdot\cdot\cdot$
t_4	t_5	t_6	t_7	t_8	$\cdot\cdot\cdot$
$\vdots$	$\cdot\cdot\cdot$	$\cdot\cdot\cdot$	$\cdot\cdot\cdot$	$\cdot\cdot\cdot$	$\cdot\cdot\cdot$

2.2.2 Hankel Operators

A linear mapping Γ on a Hilbert space H equipped with an orthonormal basis $\mathcal{E} = (e_j)_{j \geq 0}$, and defined (at least) on $\text{Vect}(\mathcal{E})$ is said to be a *Hankel operator* if its matrix with respect to $\mathcal{E}$ is Hankel. Particularly important are the cases

$$H = \ell^2(\mathbb{Z}_+), \quad \mathcal{E} = (e_k)_{k \in \mathbb{Z}_+}, \quad e_k = (\delta_{jk})_{j \geq 0},$$
$$H = H^2(\mathbb{T}), \quad \mathcal{E} = (z^k)_{k \in \mathbb{Z}_+},$$
$$H = H^2(\mathbb{C}_+), \quad \mathcal{E} = \left(\frac{1}{\sqrt{\pi}(z+i)} \left(\frac{z-i}{z+i} \right)^k \right)_{k \in \mathbb{Z}_+}.$$

When speaking of *Hankel operators on the space* $\ell^2(\mathbb{Z}_+)$ we always implicitly assume the choice of the basis $\mathcal{E}$ above.

For the space H^2 we at times modify the definition, and call a *Hankel operator* a linear mapping

$$\mathcal{H} : \text{Vect}((z^k)_{k \in \mathbb{Z}_+}) \to H^2_- = L^2(\mathbb{T}) \ominus H^2,$$

whose matrix with respect to the bases $(z^k)_{k\in\mathbb{Z}_+}$ and $(z^{-k-1})_{k\in\mathbb{Z}_+}$ is Hankel. We will immediately see the extraordinary utility of such a modification.

Lemma 2.2.1 (the functional equation) *Let $H = H^2(\mathbb{T})$, $\mathcal{E} = (z^k)_{k\in\mathbb{Z}_+}$, and let $\Gamma \colon \operatorname{Vect}(\mathcal{E}) \to H^2$ be a linear mapping. The following assertions are equivalent.*

(1) *Γ is a Hankel operator.*
(2) *$S^*\Gamma = \Gamma S$, where S is the translation (shift) operator on H^2, and S^* is its adjoint.*
(3) *The mapping $\mathcal{H} \colon \operatorname{Vect}(\mathcal{E}) \to H^2_-$ where*

$$\mathcal{H} = J\Gamma \quad \text{and} \quad Jf(z) = \bar{z}f(\bar{z}),$$

is Hankel. The equation of (2) can be written for $\mathcal{H}$ as

$$P_-S\mathcal{H} = \mathcal{H}S.$$

The same holds for Hankel operators $\ell^2(\mathbb{Z}_+)$, $\mathcal{E} = (e_k)_{k\in\mathbb{Z}_+} \to \ell^2(\mathbb{Z}_+)$, $\mathcal{E} = (e_k)_{k\in\mathbb{Z}_+}$.

Proof Let $(t_{ij}) = ((\Gamma e_j, e_i))$ be the matrix of Γ. Then

$$(S^*\Gamma e_j, e_i) = (\Gamma e_j, S e_i) = (\Gamma e_j, e_{i+1}) = t_{i+1,j}$$

and

$$(\Gamma S e_j, e_i) = (\Gamma e_{j+1}, e_i) = t_{i,j+1},$$

hence the equivalence of (1) and (2). The same computation works for (1) and (3). ∎

Example 2.2.2 Let $\varphi \in L^\infty(\mathbb{T})$ and

$$H_\varphi f = P_-(\varphi f), \quad f \in H^2(\mathbb{T}),$$

where $P_- = I - P_+$ is the orthogonal projection onto H^2_- in $L^2(\mathbb{T})$ (see §2.1.1 for P_+) and then

$$\Gamma_\varphi =: JH_\varphi, \quad \text{where } Jf(z) =: \bar{z}f(\bar{z}) \quad (\forall f).$$

Then, H_φ (respectively, Γ_φ) is a bounded Hankel operator $H^2 \to H^2_-$ (respectively, $H^2 \to H^2$) and

$$\|H_\varphi\| = \|\Gamma_\varphi\| \le \|\varphi\|_\infty.$$

Indeed, for every $k, j \ge 0$, we have

$$(H_\varphi z^j, z^{-k-1}) = (\varphi z^j, z^{-k-1}) = \widehat{\varphi}(-k-j-1),$$

2.2 Hankel Operators and Their Symbols

and hence the matrix of H_φ is Hankel. The inequality for the norm is evident. The same is true for Γ_φ (since J is a unitary mapping transforming an orthonormal basis $(z^{-k})_{k\geq 1}$ into $(z^{k-1})_{k\geq 1}$).

The main result of this section (Theorem 2.2.4 below) affirms that every Hankel operator is of the form H_φ. The proof is based on certain properties of factorization of functions in the Hardy space $H^1(\mathbb{T})$ (see Appendix F), on the Hahn–Banach theorem, as well as on the following elementary lemma.

Lemma 2.2.3

(1) *The mapping $\varphi \mapsto H_\varphi$ of $L^\infty(\mathbb{T})$ into $L(H^2, H^2_-)$ defined in Example 2.2.2 is linear and contracting ($\|H_\varphi\| \leq \|\varphi\|_\infty$).*
(2) $H_\varphi = 0 \Leftrightarrow \varphi \in H^\infty$; $H_\varphi = H_\psi \Leftrightarrow \varphi - \psi \in H^\infty$.
(3) $\|H_\varphi\| \leq \mathrm{dist}_{L^\infty}(\varphi, H^\infty)$.

H_φ can be replaced everywhere with $\Gamma_\varphi \colon H^2 \to H^2, \Gamma_\varphi = JH_\varphi$.

Proof (1) is clear by Example 2.2.2.

(2) follows from the fact that $H_\varphi = 0 \Leftrightarrow (H_\varphi z^j, z^{-k-1}) = \widehat{\varphi}(-k - j - 1) = 0$, $\forall k, j \geq 0$.

(3) is a consequence of (1) and (2): for every $\psi \in H^\infty$, $\|H_\varphi\| = \|H_{\varphi+\psi}\| \leq \|\varphi + \psi\|_\infty$, hence $\|H_\varphi\| \leq \inf\{\|\varphi + \psi\|_\infty : \psi \in H^\infty\}$. ∎

Theorem 2.2.4 (Nehari, 1957) *Let $\mathcal{H}$ be a Hankel operator, $\mathcal{H} \in L(H^2, H^2_-)$. Then there exists $\varphi \in L^\infty(\mathbb{T})$ such that $\mathcal{H} = H_\varphi$ and*

$$\|\mathcal{H}\| = \mathrm{dist}_{L^\infty}(\varphi, H^\infty) = \|\varphi\|_\infty.$$

Proof By Lemma 2.2.1, $P_- S\mathcal{H} = \mathcal{H}S$ and then by induction $P_- S^n \mathcal{H} = \mathcal{H}S^n$ for every $n \in \mathbb{Z}_+$, which implies that for every polynomial $p \in \mathcal{P}_a$

$$\mathcal{H}p = P_-(p\mathcal{H}1).$$

Since $\mathcal{H}1 \in L^2(\mathbb{T})$, the sides of this equation are continuous in p as mappings $(\mathcal{P}_a, \|\cdot\|_\infty) \to H^2_-$, and thus can be extended to a mapping $C_a(\mathbb{D}) \to H^2_-$ (see Appendix F).

We consider the bilinear form associated with this mapping:

$$(\mathcal{H}f, \overline{g})_{L^2} = (P_-(f\mathcal{H}1), \overline{g})_{L^2},$$

where $f, g \in C_a(\mathbb{D})$ and $g(0) = 0$ (which means $\overline{g} \in H^1_- \cap C(\mathbb{D})$: see Appendix F). Let $F \in C_a(\mathbb{D})$, $F(0) = 0$. There exists a factorization $F = F_1 F_2$ such that

$F_1, F_2 \in C_a(\mathbb{D})$, $F_2(0) = 0$ and $\|F\|_{H^1} = \|F_1\|_{H^2}\|F_2\|_{H^2}$ (see Appendix F.2). Hence

$$\left|\int_{\mathbb{T}} F \cdot \mathcal{H}1 \, dm\right| = |(F_1 \mathcal{H}1, \overline{F}_2)_{L^2}| = |(\mathcal{H}F_1, \overline{F}_2)_{L^2}|$$

$$\leq \|\mathcal{H}\| \cdot \|F_1\|_{H^2}\|F_2\|_{H^2} = \|\mathcal{H}\| \cdot \|F\|_{H^1},$$

which means that the linear functional $L: F \mapsto \int_{\mathbb{T}} F \cdot \mathcal{H}1 \, dm$ defined on a dense subspace of $H_0^1 = \{F \in H^1 : F(0) = 0\}$ is continuous. Moreover, $L(F_1 F_2) = (\mathcal{H}F_1, \overline{F}_2)_{L^2}$ and hence, by extending L by continuity, we obtain for every $f \in H^2, g \in H_0^2$,

$$L(fg) = (\mathcal{H}f, \overline{g})_{L^2}.$$

By the Hahn–Banach theorem and the duality $(L^1)^* = L^\infty$ (see [Rudin, 1986]), there exists a function $\varphi \in L^\infty(\mathbb{T})$ such that

$$\int_{\mathbb{T}} F \cdot \mathcal{H}1 \, dm = \int_{\mathbb{T}} F \varphi \, dm \quad (\forall F \in C_a(\mathbb{D})), \quad \|\varphi\|_\infty = \|L\| \leq \|\mathcal{H}\|.$$

Thus,

$$(\mathcal{H}f, \overline{g})_{L^2} = \int_{\mathbb{T}} fg\varphi \, dm = (P_-(f\varphi), \overline{g})_{L^2} \quad (\forall f \in H^2, \forall g \in H_0^2),$$

so that $\mathcal{H} = H_\varphi$ and $\|\varphi\|_\infty \leq \|\mathcal{H}\| = \|H_\varphi\|$. Using Lemma 2.2.3(3), we obtain $\|H_\varphi\| = \text{dist}_{L^\infty}(\varphi, H^\infty) = \|\varphi\|_\infty$. ∎

Zeev Nehari (1915–1978), a Palestinian/Israeli and American mathematician, is known for his work on univalent functions and the associated extremal problems, on Riemann surfaces, and on the oscillation properties of differential equations. He wrote an influential book called *Conformal Mapping* (1952, 1975), and his name was given to the *Nehari manifolds*. However, his most famous result (Theorem 2.2.4, the founding theorem of the theory of Hankel operators) was a subject of occasional interest to him, and remained isolated in the set of Nehari's interests.

Nehari was born in Berlin under the name of Willi Weisbach, emigrated to Palestine in 1933, hebraized his name, and had to change jobs several times (he worked in a kibbutz, then as a tourist guide, etc.). He entered the Hebrew University of Jerusalem (1935), where he submitted his thesis (in 1941, under the direction of Michael Fekete; at the time, Hebrew was the only language accepted at the university, but Nehari, being particularly gifted at languages, was already fluent in six or seven). In 1947, he moved to the United States, where he worked at Harvard University (with Stefan Bergman), then at Washington University in St. Louis, and finally at Carnegie Mellon University until his premature death.

Examples 2.2.5

(a) The **general Hankel matrices** define integral operators on $\ell^2 = \ell^2(\mathbb{Z}_+)$ with kernels $k(i, j) = k(i + j)$ depending on the sum of the arguments

$$\mathcal{H}_k : x = (x_j)_{j \geq 0} \mapsto \left(\sum_{j \geq 0} k(i+j) x_j \right)_{i \geq 0}.$$

By Theorem 2.2.4, $\mathcal{H}_k$ is bounded if and only if there exists $\varphi \in L^\infty(\mathbb{T})$ such that $k(n) = \widehat{\varphi}(-n-1)$, $n \in \mathbb{Z}_+$, and then $\|\mathcal{H}_k\| = \text{dist}_{L^\infty}(\varphi, H^\infty)$.

(b) **The Hilbert inequality.** For every $x = (x_j)_{j \geq 0} \in \ell^2(\mathbb{Z}_+)$,

$$\sum_{j,l \geq 0} \frac{x_j \overline{x}_l}{j + l + 1} \leq \pi \|x\|_2^2,$$

and the constant π is sharp. Indeed, with the notation of Example 2.1.7(5) we have $1/n = \widehat{\varphi}(n)$ ($n \neq 0$), where

$$\varphi(e^{it}) = i \frac{t}{|t|} (\pi - |t|) \; (-\pi < t < \pi)$$

(see Example 2.1.7(5)), and hence, by Lemma 2.2.3,

$$(\Gamma_\varphi f, f)_{H^2} = -\sum_{j,l \geq 0} \frac{x_j \overline{x}_l}{j + l + 1}, \quad \text{where } f = \sum_{j \geq 0} x_j z^j,$$

and $\|\Gamma_\varphi\| = \text{dist}_{L^\infty}(\varphi, H^\infty) \leq \|\varphi\|_\infty = \pi$. In fact, we can verify that $\|\Gamma_\varphi\| \geq \pi$, and hence $\|\Gamma_\varphi\| = \pi$ is the best constant in the Hilbert inequality (for example, by choosing $x^\epsilon = (1/((j+1)^{\epsilon+1/2}))_{j \geq 0}$ and letting $\epsilon \to 0$).

To conclude our preliminary study of Hankel operators, we will need a lemma, which we precede with a general remark.

Remark 2.2.6 *For every $f \in C(\mathbb{T})$,*

$$\text{dist}_{L^\infty}(f, H^\infty) = \text{dist}_{L^\infty}(f, C_a(\mathbb{T})),$$

where $C_a(\mathbb{T}) = C(\mathbb{T}) \cap H^\infty$.

Indeed, the inequality $\text{dist}_{L^\infty}(f, H^\infty) \leq \text{dist}_{L^\infty}(f, C_a(\mathbb{T}))$ is clear, and for the opposite we use a well-known property of the Poisson means $f_r = \sum_{n \in \mathbb{Z}} \widehat{f}(n) r^{|n|} z^n$ of a function $f \in C(\mathbb{T})$: $\lim_{r \to 1} \|f - f_r\|_\infty = 0$ (see [Rudin, 1986]). This implies, for every $h \in H^\infty$,

$$\|f - h\|_\infty \geq \lim_{r \to 1} \|(f - h)_r\|_\infty \geq \lim_{r \to 1} (\|f - h_r\|_\infty + \|f - f_r\|_\infty)$$
$$= \lim_{r \to 1} \|f - h_r\|_\infty \geq \text{dist}_{L^\infty}(f, C_a(\mathbb{T})),$$

so that $\text{dist}_{L^\infty}(f, H^\infty) \geq \text{dist}_{L^\infty}(f, C_a(\mathbb{T}))$. ∎

Lemma 2.2.7 (Sarason, 1967) *The vector space $H^\infty + C(\mathbb{T})$,*

$$H^\infty + C(\mathbb{T}) = \{f + g : f \in H^\infty, g \in C(\mathbb{T})\},$$

is a closed sub-algebra of $L^\infty(\mathbb{T})$.

Proof (Garnett, 1981) Suppose $\varphi \in \text{clos}_{L^\infty}(H^\infty + C(\mathbb{T}))$, and let $h_n \in H^\infty$, $f_n \in C(\mathbb{T})$ be such that

$$\|\varphi - h_n - f_n\|_\infty < 2^{-(n+1)}, n = 1, 2, \ldots.$$

Then, $\text{dist}_{L^\infty}(f_{n+1} - f_n, H^\infty) \leq \|f_{n+1} - f_n + h_{n+1} - h_n\|_\infty < 2^{-n}$, and by Remark 2.2.6 there exists

$$g_n \in C_a(\mathbb{T}) = H^\infty \cap C(\mathbb{T})$$

such that

$$\|f_{n+1} - f_n - g_n\|_\infty < 2^{-n}, \quad n = 1, 2, \ldots.$$

Since $\sum_{n \geq 0} \|f_{n+1} - f_n - g_n\|_\infty < \infty$ (we set formally $f_0 = g_0 = 0$), the partial sums

$$\sum_{n=0}^{k} (f_{n+1} - f_n - g_n) = f_{k+1} - \sum_{n=0}^{k} g_n$$

converge uniformly on $\mathbb{T}$ to a function $f \in C(\mathbb{T})$. Since $\lim_k \|\varphi - h_{k+1} - f_{k+1}\|_\infty = 0$, the sequence $h_{k+1} - \sum_{n=0}^{k} g_n$ also converges, to an $h \in H^\infty$. Clearly, $\varphi = h + f \in H^\infty + C(\mathbb{T})$, and hence $H^\infty + C(\mathbb{T})$ is closed.

As the set $\bigcup_{n \geq 0} \overline{z}^n H^\infty = \mathcal{P} + H^\infty$ is a dense sub-algebra in $H^\infty + C(\mathbb{T})$, the latter is a closed sub-algebra. ∎

2.2 Hankel Operators and Their Symbols 47

Theorem 2.2.8 (compact Hankel operators) *Let $H_\varphi\colon H^2 \to H^2_-$ be a Hankel operator, with $\varphi \in L^\infty(\mathbb{T})$. Then:*

(1) *(Adamyan, Arov, and Krein, 1968)*

$$\|H_\varphi\|_{\text{ess}} = \operatorname{dist}_{L^\infty}(\varphi, H^\infty + C(\mathbb{T})),$$

(2) *(Hartman, 1958)*

$$H_\varphi \in \mathfrak{S}_\infty(H^2, H^2_-) \Leftrightarrow \varphi \in H^\infty + C(\mathbb{T}),$$

where $\|\cdot\|_{\text{ess}}$ is the essential norm of H_φ (see Appendix D), $\mathfrak{S}_\infty$ is the set (an ideal) of all compact operators, and

$$H^\infty + C(\mathbb{T}) = \{f + g\colon f \in H^\infty,\ g \in C(\mathbb{T})\}.$$

Proof (1) First observe that for any trigonometric polynomial $p \in \mathcal{P}$, the operator H_p is bounded and of finite rank, more precisely $\operatorname{rank}(H_p) \leq \deg(p)$ (indeed, if $n = \deg(p)$, then $H_p|z^n H^2 = 0$). By approximation, $\|H_f - H_p\| \leq \|f - p\|_\infty$, we obtain $H_f \in \mathfrak{S}_\infty(H^2, H^2_-)$ for every $f \in C(\mathbb{T})$, and hence

$$H_\varphi \in \mathfrak{S}_\infty(H^2, H^2_-) \quad \text{for every } \varphi \in H^\infty + C(\mathbb{T}).$$

Consequently,

$$\|H_\varphi\|_{\text{ess}} \leq \inf\{\|H_\varphi - H_f\|\colon f \in H^\infty + C(\mathbb{T})\} \leq \operatorname{dist}_{L^\infty}(\varphi, H^\infty + C(\mathbb{T})).$$

Moreover, if $K \in \mathfrak{S}_\infty(H^2, H^2_-)$, then for every $n \in \mathbb{Z}_+$, $\|H_\varphi + K\| \geq \|H_\varphi S^n + KS^n\|$ (S is the shift operator on H^2) and also

$$\lim_n \|KS^n\| = \lim_n \|S^{*n}K^*\| = 0$$

(K^* is compact and $\lim_n \|S^{*n}f\|_2 = 0$ for every $f \in H^2$). Since $H_\varphi S^n = H_{\varphi z^n}$ (verify!), by Theorem 2.2.4 there exists $h_n \in H^\infty$ such that $\|H_\varphi S^n\| = \|H_{\varphi z^n}\| = \|\varphi z^n - h_n\|_\infty = \|\varphi - \bar{z}^n h_n\|_\infty$, where $\bar{z}^n h_n \in H^\infty + C(\mathbb{T})$. It follows that

$$\|H_\varphi + K\| \geq \overline{\lim_n} \|H_\varphi S^n + KS^n\| = \overline{\lim_n} \|\varphi - \bar{z}^n h_n\|_\infty \geq \operatorname{dist}_{L^\infty}(\varphi, H^\infty + C(\mathbb{T})),$$

and then

$$\|H_\varphi\|_{\text{ess}} \geq \operatorname{dist}_{L^\infty}(\varphi, H^\infty + C(\mathbb{T})).$$

(2) The implication $\Leftarrow$ has already been verified. For the converse, it suffices to remark that $H_\varphi \in \mathfrak{S}_\infty(H^2, H^2_-)$ implies

$$0 = \|H_\varphi\|_{\text{ess}} = \operatorname{dist}_{L^\infty}(\varphi, H^\infty + C(\mathbb{T})),$$

and hence $\varphi \in H^\infty + C(\mathbb{T})$ thanks to Lemma 2.2.7. ∎

Mark Grigorievich Krein (in Russian Марк Григорьевич Крейн, 1907–1989), a Russian (Soviet) mathematician, was one of the key figures in twentieth-century analysis, renowned for his research on extremal problems, the spectral theory of vibrating strings, and operator theory (closely linked with problems of mathematical physics, such as scattering theory). Several of his results have become classics and appear in the textbooks under names such as the *Krein–Milman theorem* (on extremal points), the *Krein–Rutman theorem* (on operators preserving a cone), the *Markov–Krein theorem*, the *Adamyan–Arov–Krein theorem* (on Hankel operators), as well as *Krein spaces* (of indefinite inner product), the *Tannaka–Krein duality*, and the *Krein condition* (for problems of indeterminate moments). He published roughly 300 articles (many of which had a profound influence over the very long term) and a dozen monographs of irreproachable quality. He created the "Odessa School," a mathematical school with a very large spectrum of research; it suffices to mention a few of his very best students such as David Milman, Mark Naimark, Israel Glazman, Mikhail Livshits, Vladimir Potapov, Yurii Berezansky, Israel Gohberg, Vladimir Matsaev, Vadym Adamyan, Damir Arov, and Ilya Spitkovsky – more than 60 in total. Krein was a corresponding member of the Ukrainian Academy of Sciences (as early as 1939: see the comment below), an honorary member of the American Academy of Arts and Sciences (1968), a member of the US National Academy of Sciences (1979), Plenary Speaker at the International Congress of Mathematicians in 1966, and recipient of the Wolf Prize (shared with Hassler Whitney, 1982) and the N. Krylov Prize (of the Ukrainian Academy, 1978).

Krein was a "Soviet Jewish" mathematician, and his career and life were difficult, fraught with obstacles, discrimination, and humiliations. On his return to Odessa at the end of the German occupation in 1944, he encountered a considerable rise in state anti-Semitism, and even though he was a renowned researcher and corresponding member of the Academy, he was sacked from the university. After two years, following protests by the students against the poor level of teaching, he was reinstated in his duties, only to be definitively fired in 1948 under the

2.2 Hankel Operators and Their Symbols

> ridiculous pretext of "often being absent from the university for personal reasons." In 1952 Krein was ousted from his (part-time) position at the Kiev Institute of Mathematics. During the official campaign against "rootless cosmopolitans" (a circumlocution to designate the Jews) it was difficult for him to find a job, but Krein was saved by the courage of the Rector of the Odessa Civil Engineering Institute (P. L. Eremenok) who offered him a position as professor (1954). The oppression continued: Krein was never given permission to travel abroad (not even to receive his Wolf Prize), his students had to face constraints right from the start of their careers (beginning with the submission of their theses), etc. Years and decades of public hostility ended up ruining Krein's health: his final years were again darkened by severe depression. A bitter joke sums up the atmosphere of this dark and shameful period: "Why is the Ukrainian Academy the very best in the world? Because even Mark Krein only merits being a corresponding member." The opinion of the international community was different, as witnessed by the quotation in the preface of the book *Scattering Theory for Automorphic Functions* by Peter Lax and Ralph Phillips (Princeton University Press, 1976): "We dedicate this monograph to Mark G. Krein, one of the mathematical giants of our age, as a tribute to his extraordinarily broad and profound contributions. Like all analysts, we owe him a great deal."

The intuitive feeling that the Hankel and Toeplitz operators are in a way "complementary" (as we have already suggested in the Introduction) is confirmed and made precise by the following proposition, to be used further on.

Lemma 2.2.9 (Toeplitz/Hankel complementarity) *Let $\varphi, \psi \in L^\infty(\mathbb{T})$.*

(1) $[T_\varphi, T_\psi] =: T_\varphi T_\psi - T_{\varphi\psi} = -H_{\bar\varphi}^* H_\psi = -\Gamma_{\bar\varphi}^* \Gamma_\psi$.

(2) *For every $f \in H^2$, $\|T_\varphi f\|_2^2 + \|H_\varphi f\|_2^2 = \|\varphi f\|_2^2$.*

(3) *If φ is unimodular ($|\varphi| = 1$ a.e. on $\mathbb{T}$) and T_φ is invertible, then*

$$\|T_\varphi^{-1}\|^{-2} + \|H_\varphi\|^2 = 1;$$

in particular, $\|H_\varphi\| = \|H_{\bar\varphi}\| \ (= \mathrm{dist}_{L^\infty}(\varphi, H^\infty) = \mathrm{dist}_{L^\infty}(\bar\varphi, H^\infty))$.

Proof (1) For every $f \in H^2$,

$$(T_\varphi T_\psi - T_{\varphi\psi})f = P_+\varphi P_+\psi f - P_+\varphi(P_+ + P_-)\psi f$$
$$= -P_+\varphi P_-\psi f = -H_{\bar\varphi}^* H_\psi f.$$

(2) This is the Pythagorean theorem.

(3) $\|T_\varphi^{-1}\|^{-2} = \inf_{\|f\|=1} \|T_\varphi f\|_2^2 = 1 - \sup_{\|f\|=1} \|H_\varphi f\|_2^2 = 1 - \|H_\varphi\|^2$. ∎

2.3 Exercises

2.3.0 Basic Exercises: Hilbert and Hardy Spaces, and Their Operators

We begin with a few supplementary properties of Hilbert spaces and their operators; certain of these properties are mentioned in Appendices A–D, to which we refer for the basic definitions and notation. In particular, if H is a Hilbert space and $E \subset H$ a closed subspace, we let P_E denote the orthogonal projection onto E.

(a) Orthogonal projection as an operator. *Let $P\colon H \to H$ be a bounded linear mapping on a Hilbert space H. Show that the following properties are equivalent.*

(1) *There exists a closed subspace $E \subset H$, such that $P = P_E$.*
(2) $P^2 = P$ *and* $P = P^*$ *(P is self-adjoint)*.
(3) $P^2 = P$ *and* $\|P\| \le 1$.

SOLUTION: (1) ⇒ (2) Clearly P_E is linear and satisfies $P_E^2 = P_E$. It is self-adjoint since, for every $x, y \in H$, we have $P_E x \perp (I - P_E)y$ and $(I - P_E)x \perp P_E y$, and hence $(P_E x, y) = (P_E x, P_E y + (I - P_E)y) = (P_E x, P_E y) = (P_E x + (I - P_E)x, P_E y) = (x, P_E y)$.

(2) ⇒ (3) Indeed, $\|Px\|^2 = (Px, Px) = (Px, P^*x) = (Px, x) \le \|Px\| \cdot \|x\|$. Hence, for every $x \in H$, $\|Px\| \le \|x\|$, thus $\|P\| \le 1$.

(3) ⇒ (1) Set $E = PH = \text{Ker}(I - P)$. Since for every $x \in H$ we have $x = Px + (I - P)x$, it suffices to show that $x \perp y$ for every $x \in PH$, $y \in (I - P)H$. For every $\lambda \in \mathbb{C}$, we have $\|x\|^2 = \|P(x + \lambda y)\|^2 \le \|x + \lambda y\|^2 = \|x\|^2 + |\lambda|^2 \|y\|^2 + 2\,\text{Re}(x, \lambda y)$, hence $0 \le |\lambda|^2 \|y\|^2 + 2\,\text{Re}(x, \lambda y)$. By setting $\lambda = t(x, y)$, $t \in (-\infty, 0)$ and letting $t \to 0$, we obtain $(x, y) = 0$.

(b) Functional calculus $C_\mathbb{R}(\sigma(A))$ for the self-adjoint operators. *Here we let $C_\mathbb{R}(\sigma(A))$ denote the algebra of continuous real functions on the spectrum $\sigma(A)$.*

(1) *The norm and the spectral radius of self-adjoint operators. Show that for a self-adjoint operator $A\colon H \to H$ ($A^* = A$), $\|A\| = r(A)$.*

Hint Recall that by definition $r(A) = \max_{\lambda \in \sigma(A)} |\lambda|$ ($\sigma(A)$ the spectrum of A), and by Gelfand's formula, $r(A) = \lim_n \|A^n\|^{1/n}$: see Appendix C.

2.3 Exercises

SOLUTION: For every operator $A \in L(H)$, we have $\|A^*A\| \leq \|A^*\| \cdot \|A\| = \|A\|^2$, and also $\|A\|^2 = \sup_{\|x\|\leq 1} \|Ax\|^2 = \sup_{\|x\|\leq 1}(Ax, Ax) = \sup_{\|x\|\leq 1}(A^*Ax, x) \leq \|A^*A\|$, so that $\|A^*A\| = \|A\|^2$. If $A^* = A$, then $\|A^2\| = \|A\|^2$, and by reiterating, $\|A^{2^n}\| = \|A\|^{2^n}$. By Gelfand's formula, $r(A) = \|A\|$.

(2) *The spectrum of a self-adjoint operator. Show that the spectrum of a self-adjoint operator $A \in L(H)$, $A^* = A$ is real, $\sigma(A) \subset \mathbb{R}$, and if, in addition, A is positive ($A \geq 0$: i.e. $(Ax, x) \geq 0$ for every $x \in H$), then $\sigma(A) \subset \mathbb{R}_+ = [0, \infty)$.*

SOLUTION: If $\lambda \in \mathbb{C}$, $\lambda = x + iy$, $y = \text{Im}(\lambda) \neq 0$ and $h \in H$, then $\|(A - \lambda I)h\|^2 = \|(A - xI)h\|^2 + |y|^2\|h\|^2 - 2\text{Re}((A - xI)h, iyh) = \|(A - xI)h\|^2 + |y|^2\|h\| \geq |y|^2\|h\|$, and similarly for $\|(A - \lambda I)^*h\|^2$, which implies that $\lambda \notin \sigma(A)$, hence $\sigma(A) \subset \mathbb{R}$. If $A \geq 0$ and $x < 0$, then $\|(A - xI)h\|^2 = \|Ah\|^2 + x^2\|h\|^2 - 2\text{Re}(Ah, -xh) \geq x^2\|h\|^2$ (for every h), again giving $x \notin \sigma(A)$, and thus $\sigma(A) \subset \mathbb{R}_+$.

(3) *Functional calculus for the self-adjoint operators. Show that for every self-adjoint operator $A \in L(H)$ and any real polynomial f, we have $\|f(A)\| = \max_{\lambda \in \sigma(A)} |f(\lambda)|$, and then that the mapping $f \mapsto f(A)$ can be extended to an isometric homomorphism of $C_\mathbb{R}(\sigma(A))$ to $L(H)$. Every $f(A)$ is self-adjoint, and if $f \geq 0$ on the spectrum $\sigma(A)$, then $f(A) \geq 0$.*

SOLUTION: If f is a real polynomial, then clearly $f(A)$ is self-adjoint, and hence by (1) above $\|f(A)\| = \max_{\lambda \in \sigma(f(A))} |\lambda|$. However, by the spectral mapping theorem (Appendix C.1), $\sigma(f(A)) = f(\sigma(A))$, giving the formula. Since $\sigma(A)$ is a compact subset of $\mathbb{R}$, the polynomials are dense in $C_\mathbb{R}(\sigma(A))$, and thus $f \mapsto f(A)$ can be extended by continuity to a homomorphism. Every $f(A)$, $f \in C_\mathbb{R}(\sigma(A))$, is self-adjoint, as it is a limit of self-adjoint operators, and if $f \geq 0$ on $\sigma(A)$, then $f^{1/2} \in C_\mathbb{R}(\sigma(A))$ and hence $f(A) = B^2 \geq 0$ where $B = f^{1/2}(A)$ ($(B^2x, x) = (Bx, Bx) \geq 0$ for any $x \in H$).

(4) *Square root of a positive operator. If $A \in L(H)$ is a positive operator $A \geq 0$, then there exists a unique operator $B \geq 0$ such that $B^2 = A$.*

SOLUTION: Let $f(x) = x$, $x \in \sigma(A)$. Since $A \geq 0$, we have $f \geq 0$ on $\sigma(A)$, and hence $f^{1/2} \in C_\mathbb{R}(\sigma(A))$ and $A = (f^{1/2}(A))^2 \geq 0$.

For the uniqueness, let $B_1, B_2 \geq 0$ be such that $B_1^2 = B_2^2 = A$ and let (f_n) be a sequence of polynomials converging uniformly to $\sqrt{x}$ on $[0, c]$, $c = \max(\|B_1\|^2, \|B_2\|^2)$. Then $(f_n(x^2))$ converges to $f(x) = x$, and hence $\lim_n f_n(B_j^2) = B_j$. As $f_n(B_1^2) = f_n(B_2^2)$, we have $B_1 = B_2$.

NOTE There exist operators without square roots, for example a Jordan block $J: \mathbb{C}^2 \to \mathbb{C}^2$. If $B^2 = J$ then $\sigma(B) = \{0\}$ (by the spectral mapping theorem), hence either B is a Jordan block and $B^2 = 0$, or $B = 0$, and both alternatives contradict $B^2 = J$. However, if $A: X \to X$ is invertible and $\dim X < \infty$, then A possesses roots of every order. This is not the case in infinite dimensions: consider $Af = zf$ on the space

$$H^2(1/2 < |z| < 1) =: \left\{ f = \sum_{k \in \mathbb{Z}} c_k z^k : \sup_{1/2 < r < 1} \int_\mathbb{T} |f(r\zeta)|^2 dm(\zeta) < \infty \right\}$$

(m is the Lebesgue measure on $\mathbb{T}$). However, as in (4), every normal operator A ($A^*A = AA^*$) on a Hilbert space possesses roots of every order.

(c) Diverse facts on the translation (shift) operators. *The operators of translation (shift) $S: l^2(\mathbb{Z}_+) \to l^2(\mathbb{Z}_+)$ and inverse translation (backward shift) $(S | l^2(\mathbb{Z}_+))^*$ are defined in §2.1.2 and §2.1.3. Their unitary images in the Hardy space H^2 are given by $Sf = zf$ and $S^*f = (f - f(0))/z$, $f \in H^2$. In fact, the translations play a principal role in the theory of Hardy spaces (see Appendix F) and of Toeplitz/Hankel operators. Here we add a few supplementary features to a long list of properties of these operators used in this and subsequent chapters.*

(1) *The spectrum.* Show that a number $\lambda \in \mathbb{C}$ is an eigenvalue of S^* if and only if $|\lambda| < 1$. Moreover, if $|\lambda| < 1$ and $n \geq 1$, then

$$\text{Ker}(S^* - \bar{\lambda}I)^n = \text{Vect}\left(\frac{z^j}{(1 - \bar{\lambda}z)^{j+1}} : 0 \leq j < n\right).$$

The spectrum of S^* (and of S) is the closed disk $\sigma(S^*) = \sigma(S) = \bar{\mathbb{D}}$.

SOLUTION: We have $\|S^*\| = \|S\| = 1$ (S is an isometry of H^2), and hence $\sigma(S^*) = \sigma(S) \subset \bar{\mathbb{D}}$. Moreover, if $S^*f = \bar{\lambda}f$, then $f - f(0) = \bar{\lambda}zf$, thus

$$f = f(0)k_\lambda \quad \text{where } k_\lambda = (1 - \bar{\lambda}z)^{-1} = f(0) \sum_{k \geq 0} \bar{\lambda}^k z^k.$$

If $|\lambda| < 1$, then $k_\lambda \in H^2$ (the *reproducing kernel of H^2*: see Appendix F.2); if not, then $k_\lambda \notin H^2$. It follows that $\sigma_p(S^*)$ (the point spectrum = the set of eigenvalues) is $\mathbb{D}$ and, for every $\lambda \in \mathbb{D}$, $\text{Ker}(S^* - \bar{\lambda}I) = k_\lambda \cdot \mathbb{C}$. Then, observing that for $j \geq 1$ we have

$$(S^* - \bar{\lambda}I)(z^j k_\lambda^{j+1}) = z^{j-1} k_\lambda^j,$$

and thus $z^j k_\lambda^{j+1} \in \text{Ker}(S^* - \bar{\lambda}I)^n$ for $0 \leq j < n$, we obtain

$$\text{Ker}(S^* - \bar{\lambda}I)^n \supset \text{Vect}\left(\frac{z^j}{(1 - \bar{\lambda}z)^{j+1}} : 0 \leq j < n\right).$$

The equality follows from a dimensional argument since $\dim \text{Ker}(S^* - \bar{\lambda}I)^n \leq n \cdot \dim \text{Ker}(S^* - \bar{\lambda}I) = n$.

2.3 Exercises

(2) *Beurling's theorem on the invariant subspaces (1949).* Let $E \subset H^2$ be a closed subspace of H^2, $E \neq \{0\}$, invariant with respect to S: $SE \subset E$. Show that there exists a function $\varphi \in H^2$ (unique up to a multiplicative constant) such that $|\varphi| = 1$ a.e. on $\mathbb{T}$ (an inner function: see Appendix F.2) and $E = \varphi H^2$. Conversely, a subspace of the form φH^2 is S-invariant.

SOLUTION: [Helson, 1964] When S is an isometry, SE is a closed subspace and $SE \neq E$ (otherwise, $E = \bigcap_{n\geq 1} S^n E \subset \bigcap_{n\geq 1} S^n H^2 = \{0\}$). Let $\varphi \in E \ominus SE$ (orthogonal complement) and $\|\varphi\|_2 = 1$. Then, $\varphi \in E$ and $\varphi \perp S^n \varphi$, $n = 1, 2, \ldots$, giving $0 = (z^n \varphi, \varphi) = \int_{\mathbb{T}} |\varphi|^2 z^n dm$ ($n \geq 1$; m is the Lebesgue measure on $\mathbb{T}$), and hence all the Fourier coefficients of the integrable function $|\varphi|^2$ are zero, except that of index 0. This means that $|\varphi|^2 = $ const., and since $\|\varphi\|_2 = 1$, we must have $|\varphi| = 1$ a.e. on $\mathbb{T}$. Clearly, for every $n \geq 1$, $z^n \varphi \in E$ and hence $\varphi H^2 \subset E$ (φ is unimodular, thus the mapping $f \mapsto \varphi f$ is isometric on $L^2(\mathbb{T})$). In fact, we have the equality $E = \varphi H^2$, since if $f \in E$, $f \perp \varphi H^2$, then $f \perp z^n \varphi$ ($\forall n \geq 0$) and (by the definition of φ) $\varphi \perp z^n f$ ($\forall n \geq 1$). Hence, all the Fourier coefficients of $\varphi \overline{f}$ are zero, i.e. $\varphi \overline{f} = 0$, and then $f = 0$ (since $|\varphi| = 1$ a.e.).

The uniqueness of φ follows from the fact that if, for inner functions φ_1, φ_2, we have $\varphi_1 H^2 = \varphi_2 H^2$, then $\varphi_1/\varphi_2 \in H^2$ and $\varphi_2/\varphi_1 = \overline{\varphi_1}/\overline{\varphi_2} \in H^2$, and by the definition of H^2 as a subspace of $L^2(\mathbb{T})$, we obtain that all the Fourier coefficients of φ_1/φ_2 are zero except for the one of index 0, hence $\varphi_1/\varphi_2 = $ const.

(3) *Finite-dimensional S^*-invariant subspaces.* Let $E \subset H^2$ be a finite-dimensional subspace such that $S^*E \subset E$. Show that there exists a (unique) function $k: \mathbb{D} \to \mathbb{Z}_+$ such that $\sum_{\lambda \in \mathbb{D}} k(\lambda) = \dim E \ (< \infty)$ and

$$E = \text{Vect}\left(\frac{z^j}{(1 - \overline{\lambda}z)^{j+1}} : 0 \leq j < k(\lambda), \lambda \in \mathbb{D}\right).$$

SOLUTION: The restriction $S^*|E$ defines a linear mapping on a finite-dimensional space, $S^*|E: E \to E$, and by Jordan's theorem, to every eigenvalue $\overline{\lambda} \in \sigma(S^*|E) \subset \mathbb{D}$ corresponds an integer $k(\lambda) \geq 1$ such that

$$\text{Ker}((S^*|E) - \overline{\lambda}I)^{k(\lambda)} = \text{Ker}((S^*|E) - \overline{\lambda}I)^{k(\lambda)+1}$$

(we take the least such $k(\lambda)$) and

$$E = \text{Vect}(\text{Ker}((S^*|E) - \overline{\lambda}I)^{k(\lambda)} : \overline{\lambda} \in \sigma(S^*|E)).$$

However, $\text{Ker}((S^*|E) - \overline{\lambda}I)^{k(\lambda)} \subset \text{Ker}(S^* - \overline{\lambda}I)^{k(\lambda)}$ and, by (1) above, $k(\lambda) = \dim \text{Ker}(S^* - \overline{\lambda}I)^{k(\lambda)} \geq \dim \text{Ker}((S^*|E) - \overline{\lambda}I)^{k(\lambda)} \geq k(\lambda)$, hence

$$\text{Ker}((S^*|E) - \overline{\lambda}I)^{k(\lambda)} = \text{Ker}(S^* - \overline{\lambda}I)^{k(\lambda)} = \text{Vect}\left(\frac{z^j}{(1 - \overline{\lambda}z)^{j+1}} : 0 \leq j < k(\lambda)\right).$$

It remains only to define $k(\lambda) = 0$ for $\overline{\lambda} \in \mathbb{D} \setminus \sigma(S^*|E)$.

NOTE Since for a bounded operator on a Hilbert space $T: H \longrightarrow H$ and a closed subspace $E \subset H$ we always have $(TE \subset E) \Leftrightarrow (T^*E^\perp \subset E^\perp)$ (where $E^\perp$ denotes the orthogonal complement of E: see Appendix B), we can compare theorems (2) and (3) above. To this end, note that for every $f \in H^2$ and every $\lambda \in \mathbb{D}$, we have

$$f(\lambda) = \left(f, \frac{1}{1-\bar{\lambda}z}\right)_{H^2},$$

and, differentiating with respect to λ, $f^{(j)}(\lambda) = (f, z^j/((1-\bar{\lambda}z)^{j+1}))_{H^2}$. Hence, for a subspace E represented in the form of theorem (3), we have

$$E^\perp = \{f \in H^2 : f^{(j)}(\lambda) = 0,\ 0 \le j < k(\lambda),\ \lambda \in \mathbb{D}\}.$$

In other words, $E^\perp$ is made up of functions of H^2 having a zero of multiplicity at least $k(\lambda)$ at any point λ such that $\bar{\lambda} \in \sigma(S^*|E)$. It is easy to see that such a subspace is of the form $E^\perp = BH^2$, where

$$B = \prod_{\lambda \in \mathbb{D}} \left(\frac{\lambda - z}{1 - \bar{\lambda}z}\right)^{k(\lambda)},$$

an inner function called a finite *Blaschke product*.

(4) *The commutants of S and S^*.* Let $A: H^2 \to H^2$ be a bounded operator. Show that:

- $AS = SA$ if and only if there exists $\varphi \in H^2 \cap L^\infty = H^\infty$ such that $A = M_\varphi|H^2$, hence $Af = \varphi f$, $\forall f \in H^2$;
- $AS^* = S^*A$ if and only if $Af = P_+(\bar{\varphi}f)$, $\forall f \in H^2$, where $\varphi \in H^\infty$.

SOLUTION: Since $AS = SA \Leftrightarrow S^*A^* = A^*S^*$, it suffices to describe the commutant of S. If $AS = SA$ and k_λ is the reproducing kernel of (1) above, then $S^*A^*k_\lambda = A^*S^*k_\lambda = \bar{\lambda}A^*k_\lambda$, hence $A^*k_\lambda \in \text{Ker}(S^* - \bar{\lambda}I)$ and by (1) there exists a complex number, say $\overline{\varphi(\lambda)}$, such that $A^*k_\lambda = \overline{\varphi(\lambda)}k_\lambda$. Clearly for every $\lambda \in \mathbb{D}$, $|\overline{\varphi(\lambda)}| \le \|A^*\| = \|A\|$, and then $A1(\lambda) = (A1, k_\lambda) = (1, A^*k_\lambda) = (1, \overline{\varphi(\lambda)}k_\lambda) = \varphi(\lambda)$, hence $\varphi = A1 \in H^2$ (and consequently, $\varphi \in H^2 \cap L^\infty = H^\infty$). It only remains to note that, as before, for every $f \in H^2$ and every $\lambda \in \mathbb{D}$, $(Af)(\lambda) = (Af, k_\lambda) = (f, A^*k_\lambda) = (f, \overline{\varphi(\lambda)}k_\lambda) = \varphi(\lambda)f(\lambda)$, thus $Af = \varphi f$.

The converse (if $\varphi \in H^\infty$, then $f \mapsto \varphi f$ is a bounded operator that commutes with S) is evident by the identification $H^2 = H^2(\mathbb{D})$: see Appendix F.1.

NOTE We could obtain an alternative proof of (4) by adapting the reasoning of Lemmas 2.1.2–2.1.4.

(d) The von Neumann inequality (1951). *Let $A: H \to H$ be a linear contraction of the Hilbert space H ($\|A\| \le 1$), and let $p \in \mathcal{P}_a$ be a complex polynomial. Show that*

$$\|p(A)\| \le \max_{|z| \le 1} |p(z)| = \|p\|_\infty.$$

Hint By regarding the difference $\|(\lambda I - A)x\|^2 - \|(I - \bar{\lambda}A)x\|^2$, show first that $\|B(A)\| \le 1$ for every finite Blaschke product B (see Note of (3) for the definition); then apply Schur's theorem ((e) below).

SOLUTION: For $\lambda \in \mathbb{D}$ and $x \in H$, we have

$$\|(\lambda I - A)x\|^2 - \|(I - \bar{\lambda}A)x\|^2$$
$$= |\lambda|^2\|x\|^2 + \|Ax\|^2 - 2\operatorname{Re}\lambda(x, Ax) - \|x\|^2 - |\lambda|^2\|Ax\|^2 + 2\operatorname{Re}\lambda(x, Ax)$$
$$= (1 - |\lambda|^2)(\|Ax\|^2 - \|x\|^2) \le 0.$$

Denoting $y = (I - \bar{\lambda}A)x$ and

$$b_\lambda = \frac{\lambda - z}{1 - \bar{\lambda}z},$$

and using the holomorphic functional calculus (Appendix C.1), we obtain

$$\|b_\lambda(A)y\| = \|(\lambda I - A)x\| \le \|(I - \bar{\lambda}A)x\| = \|y\|,$$

and hence $\|b_\lambda(A)\| \le 1$. Consequently, $\|B(A)\| \le 1$ for every finite Blaschke product $B = \prod_{j=1}^n b_{\lambda_j}$ (see the comment in (c(3)) above). In what follows, we apply this fact to rA, $0 < r < 1$, in place of A.

Clearly it suffices to show that $\|p(A)\| \le 1$ for every polynomial p satisfying $\|p\|_\infty < 1$. If p is such a polynomial, then for $0 < r < 1$ sufficiently close to 1, we have $\|p_{1/r}\|_\infty \le 1$ where $p_{1/r}(z) = p(z/r)$. By applying Schur's theorem (e) below, we can find a sequence (B_n) of finite Blaschke products such that $\lim_n B_n(z) = p_{1/r}(z)$ uniformly on the compact subsets of $\mathbb{D}$. When the operator rA has its spectrum in the compact set $|z| \le r < 1$, by continuity of the holomorphic functional calculus for the uniform convergence in a neighborhood of the spectrum (Appendix C.1), we obtain $\lim_n \|B_n(rA) - p_{1/r}(rA)\| = 0$, hence $\|p(A)\| = \|p_{1/r}(rA)\| \le 1$.

(e) Approximation by finite Blaschke products (Schur, 1917). *Let $f \in \mathcal{P}_a$ be a polynomial, $\|f\|_\infty \le 1$. Show that there exists a sequence (B_n) of finite Blaschke products such that $\lim_n B_n(z) = f(z)$ uniformly on the compact subsets of $\mathbb{D}$.*

Hint (Rudin, 1969) Consider the rational function

$$B = B_{n,f} = \frac{f + z^n}{1 + z^n f_*}$$

where $f_*(z) = \overline{f}(1/\bar{z}) = \sum_{k=0}^{N} \bar{a}_k z^{-k}$ if $f = \sum_{k=0}^{N} a_k z^k$.

SOLUTION: (following Rudin, 1969) If $n \geq N$, $z^n f_*$ is a polynomial, and for $|z| < 1$ we have $|z^n f_*(z)| < \max_{|z|=1} |z^n f_*| \leq 1$, then $B_{n,f}$ is a rational function holomorphic in $\mathbb{D}$. Moreover, for $|z| = 1$, we have

$$|B_{n,f}(z)| = \frac{|f(z) + z^n|}{|\bar{z}^n + \overline{f(z)}|} = 1,$$

which implies that $B_{n,f}$ is an inner function. The canonical factorization of inner functions (into three factors: see Appendix F.2) shows that a rational inner function is a finite Blaschke product. When $\lim_n z^n = 0$ (uniformly on the compact subsets of $\mathbb{D}$), we obtain $\lim_n B_{n,f}(z) = f(z)$.

NOTE In fact, Schur's theorem holds for every $f \in H^\infty$, $\|f\|_\infty \leq 1$, which is an immediate consequence of (e): it suffices to consider the Fejér polynomials f_k of f ($\|f_k\|_\infty \leq 1$, $\lim_k f_k(z) = f(z)$) and the finite Blaschke products $B_k = B_{n_k, f_k}$, with $n_k \uparrow \infty$ quickly enough.

(f) The von Neumann inequality is sharp for the finite rank operators.
The inequality (d) can be reinforced for the finite rank contracting operators $A: H \to H$ (in particular, for the case $\dim H < \infty$), and more generally, for the algebraic operators (as always, on a Hilbert space), i.e. for $A: H \to H$ possessing an annihilating polynomial $p \neq 0$, $p(A) = 0$. If A is algebraic (and contracting), denote p_A its minimal polynomial, $p_A = \prod_{\lambda \in \sigma(A)} (z - \lambda)^{k(\lambda)}$. If necessary making an approximation $rA \to A$, $0 < r < 1$, $r \to 1$, we can suppose that

$$\sigma(A) \subset \mathbb{D}.$$

(1) *The quotient algebra $H^\infty / p_A H^\infty$. Given a function $f \in H^\infty$, its quotient norm (more precisely, the norm of its quotient class $f + p_A H^\infty$) is*

$$\|f\|_{H^\infty / p_A H^\infty} = \min \left\{ \|g\|_\infty : \frac{g - f}{p_A} \in \mathrm{Hol}(\mathbb{D}) \right\}.$$

Note that $\inf \|g\|_\infty$ taken over all the functions g, having at the points $\lambda \in \sigma(A)$ the same Taylor expansion of degree $k(\lambda) - 1$ as f, is truly attained: if (g_n) is a sequence realizing this $\inf$ ($\lim_n \|g_n\|_\infty = \inf \|g\|_\infty$), we can find a subsequence (g_{n_j}) uniformly convergent on the compact subsets of $\mathbb{D}$ (Montel's theorem), and the limit function g satisfies $\|g\|_\infty = \inf \|g\|_\infty$. If the eigenvalues of A are all simple ($k(\lambda) = 1$, $\forall \lambda \in \sigma(A)$), then $\|f\|_{H^\infty/p_A H^\infty}$ is the norm of the best interpolant g, $g|\sigma(A) = f|\sigma(A)$.

(2) *A stronger form of the von Neumann inequality.* If $A: H \to H$ is an algebraic contraction (having $\sigma(A) \subset \mathbb{D}$), show that

$$\|f(A)\| \leq \|f\|_{H^\infty/p_A H^\infty} \quad (\forall f \in H^\infty).$$

SOLUTION: This is evident by (d) since, for every $h \in H^\infty$, $\|f(A)\| = \|(f + p_A h)(A)\| \leq \|f + p_A h\|_\infty$.

NOTE This form of the von Neumann inequality is sharp among the contracting operators T annihilated by p_A ($p_A(T) = 0$). More precisely, if $E \subset H^2$ is a subspace of ((c)(3)) above defined by the "divisor" $k(\cdot)$ of the polynomial p_A and $T = (S^*|E)^* = P_E S|E$ then $\|T\| \leq 1$, $p_A(T) = 0$ and, for every polynomial f, we have

$$\|f(T)\| = \|f\|_{H^\infty/p_A H^\infty}$$

(a consequence of the commutant lifting theorem (Sarason, 1967); see [Nikolski, 2002b] C.2.4.3).

(g) The von Neumann inequality in Banach spaces. *The situation described in (d)–(f) above changes radically if we consider the linear contractions $A: X \to X$ on a general Banach space. As a principal tool we use the analytic Wiener space*

$$X = W_a(\mathbb{T}) = \mathcal{F}l^1(\mathbb{Z}_+) = \left\{ f = \sum_{k \geq 0} \widehat{f}(k) z^k : \|f\|_W = \sum_{k \geq 0} |\widehat{f}(k)| < \infty \right\}$$

(see Example 2.1.7(2) and Exercise 2.3.3(b,c) above).

(1) *Let X be a Banach space and $T: X \to X$ a linear contraction, $\|T\| \leq 1$. Show that for every $\varphi \in W_a(\mathbb{T})$, $\varphi(T)$ is well-defined as*

$$\varphi(T) = \sum_{k \geq 0} \widehat{\varphi}(k) T^k,$$

the mapping $\varphi \mapsto \varphi(T)$ is an algebra homomorphism $W_a(\mathbb{T}) \to L(X)$ coinciding with the polynomial calculus on $\mathcal{P}_A$, and $\|\varphi(T)\| \leq \|\varphi\|_W$.

In the case where $T = S$, $S: W_a(\mathbb{T}) \to W_a(\mathbb{T})$ the translation operator $Sf = zf$ on the space $W_a(\mathbb{T})$, we have $\|\varphi(S)\| = \|\varphi\|_W$ for every $\varphi \in W_a(\mathbb{T})$.

SOLUTION: Since $\|T^k\| \leq 1$, clearly the series $\sum_{k \geq 0} \widehat{\varphi}(k) T^k$ converges absolutely, and

$$\left\| \sum_{k \geq 0} \widehat{\varphi}(k) T^k \right\| \leq \sum_{k \geq 0} |\widehat{\varphi}(k)| \cdot \|T^k\| \leq \|\varphi\|_W.$$

For the case of a polynomial φ, the sum $\sum_{k\geq 0}\widehat{\varphi}(k)T^k$ coincides with $\varphi(T)$, which shows that the "calculus" $\varphi \mapsto \varphi(T)$ is well-defined on W_a, contracting and multiplicative $((\varphi_1\varphi_2)(T) = \varphi_1(T)\varphi_2(T))$.

In the case of $S : W_a(\mathbb{T}) \to W_a(\mathbb{T})$, in addition, $\|\varphi(S)\| \geq \|\varphi(S)1\|_W = \|\varphi\|_W$.

NOTE It is well known that the norm $\|\cdot\|_\infty$ is much weaker over the polynomials $\mathcal{P}_a$ than the norm $\|\cdot\|_W$: there exists a sequence of polynomials (p_n) such that $\sup_n \|p_n\|_\infty < \infty$, $\lim_n \|p_n\|_W = \infty$. For example,

$$p_n = \sum_{k=}^{n} \frac{1}{k}(z^{n-k} - z^{n+k}),$$

where $\|p_n\|_\infty \leq 5$, $\|p_n\|_W \geq \log(n+1)$: see [Nikolski, 2019], §5.6.2(c). Consequently, the result of (1) shows that for a contraction T on a general Banach space, no upper bound of the type $\|\varphi(T)\| \leq C \cdot \sup_{|z|\leq 1}|\varphi(z)|$ ($\forall \varphi \in \mathcal{P}_a$) is possible. We will look for an upper bound on a larger disk, $\|\varphi(T)\| \leq \sup_{|z|\leq R}|\varphi(z)| = \|\varphi\|_{H^\infty(R\mathbb{D})}$ ($\forall \varphi \in \mathcal{P}_a$), where $R > 1$. It is clear by (1) that the question is equivalent to an estimate $\|\varphi\|_W \leq \|\varphi\|_{H^\infty(R\mathbb{D})}$ ($\forall \varphi \in \mathcal{P}_a$).

(2) (Bohr, 1914) *Let $R \geq 1$. Show that $\|\varphi\|_W \leq \|\varphi\|_{H^\infty(R\mathbb{D})}$ ($\forall \varphi \in \mathcal{P}_a$) if and only if $R \geq 3$. For every $R > 1$ and every polynomial $\varphi \in \mathcal{P}_a$, we have*

$$\|\varphi\|_W \leq \left(\frac{R^2}{R^2-1}\right)^{1/2} \|\varphi\|_{H^\infty(R\mathbb{D})}.$$

Hint Use the method of (d)–(e) above, showing first that $\|b_\lambda(z/R)\|_W \leq 1$ for every $\lambda \in \mathbb{D}$ and $R \geq 3$.

SOLUTION: (Katsnelson and Matsaev, 1967 (see Notes and Remarks)) Note that the desired estimate $\|\varphi\|_W \leq \|\varphi\|_{H^\infty(R\mathbb{D})}$ ($\forall \varphi \in \mathcal{P}_a$) is equivalent to $\|\varphi_{1/R}\|_W \leq \|\varphi\|_{H^\infty(\mathbb{D})}$ ($\forall \varphi \in \mathcal{P}_a$) where $\varphi_{1/R}(z) = \varphi(z/R)$. By following the Hint, let $\varphi = b_\lambda(z)$,

$$b_\lambda = (\lambda - z)(1 - \bar{\lambda}z)^{-1} = \frac{1}{\lambda}\left(\frac{|\lambda|^2 - \bar{\lambda}z}{1 - \bar{\lambda}z}\right) = \frac{1}{\lambda}(1 - (1-|\lambda|^2)\sum_{k\geq 0}\bar{\lambda}^k z^k),$$

$\lambda \in \mathbb{D}$, which gives

$$N(t) =: \|(b_\lambda)_{1/R}\|_W = \frac{Rt + 1 - 2t^2}{R - t},$$

where $t = |\lambda| \in (0,1)$. The function N is well-defined on $[0,R)$, $N(1) = 1$ and is monotonically increasing on $[0,1]$ if and only if $N'(t) \geq 0$, $\forall t \in (0,1)$. However,

$$N'(t) = 2 - \frac{R^2 - 1}{(R-t)^2} \geq 0 \ (\forall t \in (0,1)) \Leftrightarrow N'(1) \geq 0 \Leftrightarrow R \geq 3.$$

Consequently, if $R \geq 3$, then $\|(b_\lambda)_{1/R}\|_W \leq 1$ for every $\lambda \in \mathbb{D}$, and if $R < 3$, then N' changes sign on $(0,1)$ and $\max_{0\leq t \leq 1} N(t) > 1$. Hence, the condition $R \geq 3$ is necessary for the upper bound $\|\varphi\|_W \leq \|\varphi\|_{H^\infty(R\mathbb{D})}$ ($\forall \varphi \in \mathcal{P}_a$).

Its sufficiency follows from Schur's theorem (e) above: first, for every finite Blaschke product $\varphi = B = \prod_{j=1}^n b_{\lambda_j}$,

$$\|\varphi_{1/R}\|_W = \left\| \prod_j (b_{\lambda_j})_{1/R} \right\|_W \leq \prod_j \|(b_{\lambda_j})_{1/R}\|_W \leq 1,$$

and then, using (e) as in the solution of (d), we obtain $\|\varphi_{1/R}\|_W \leq 1$ for any polynomial $\varphi \in \mathcal{P}_a$ having $\|\varphi\|_\infty \leq 1$.

An upper bound of $\|\varphi_{1/R}\|_W$ with a certain constant is evident:

$$\|\varphi_{1/R}\|_W = \sum_{k \geq 0} R^{-k} |\widehat{\varphi}(k)| \leq \left(\sum_{k \geq 0} |\widehat{\varphi}(k)|^2 \right)^{1/2} (1 - R^{-2})^{-1/2} \leq (1 - R^{-2})^{-1/2} \|\varphi\|_\infty.$$

(3) *Deduce that the following assertions are equivalent.*

- *For any contraction of a Banach space* $T: X \to X$, $\|\varphi(T)\| \leq \|\varphi\|_{H^\infty(R\mathbb{D})}$ $(\forall \varphi \in \mathcal{P}_a)$.

- $R \geq 3$.

- *For every contraction* $T: X \to X$ *and any* $R > 1$,

$$\|\varphi(T)\| \leq \left(\frac{R^2}{R^2 - 1} \right)^{1/2} \|\varphi\|_{H^\infty(R\mathbb{D})} \quad (\forall \varphi \in \mathcal{P}_a).$$

SOLUTION: Evident by (1) and (2).

2.3.1 Toeplitz Operators, the Berezin Transform, and the RKT (Reproducing Kernel Thesis)

Let $\varphi \in H^2$ and $T_\varphi f = P_+(\varphi f)$ for $f \in H^\infty$. Show that the following assertions are equivalent.

(1) $\varphi \in L^\infty(\mathbb{T})$.
(2) T_φ is bounded as a mapping $(H^\infty, \|\cdot\|_2) \to H^2$.
(3) $\|T_\varphi k_z\|_2 \leq c \|k_z\|_2$ ($\forall z \in \mathbb{D}$) where $c > 0$ and where

$$k_z(\zeta) = \frac{1}{1 - \bar{z}\zeta} \quad (\zeta \in \mathbb{T})$$

is the reproducing kernel of the space H^2.
(4) The Berezin transform of T_φ,

$$b_{T_\varphi}(z) =: \left(T_\varphi \frac{k_z}{\|k_z\|_2}, \frac{k_z}{\|k_z\|_2} \right)_2, \quad z \in \mathbb{D},$$

is bounded on $\mathbb{D}$.

NOTE The family of inequalities (3) for an operator T on H^2 (in this exercise for $T = T_\varphi$) is known as the *reproducing kernel thesis* (RKT) (see [Nikolski, 2002a]). The RKT is evidently necessary for T to be bounded, but for several operators linked to the structure of the space H^2 it is also sufficient (it is important to understand for exactly which ones). For comments on the Berezin transform and the RKT see §2.4.

SOLUTION: (1) $\Rightarrow$ (2) $\Rightarrow$ (3) $\Rightarrow$ (4) are evident. For (4) $\Rightarrow$ (1), verify that

$$b_{T_\varphi}(z) = \int_\mathbb{T} \frac{1 - |z|^2}{|1 - \bar{z}\zeta|^2} \varphi(\zeta)\, dm(\zeta),$$

and then apply Fatou's theorem (Appendix F.1).

Felix A. Berezin (in Russian Феликс Александрович Березин, 1931–1980) was a Russian (Soviet) mathematician, founder of a branch of mathematics called "supermathematics" (algebra and analysis over anticommutative variables; the term itself is spontaneous and imperfect), and one of the pioneers of a "boom" of new interactions between mathematics and theoretical physics (1965–1975), in a period which changed the face of mathematical physics. Berezin exercised a profound influence on a large group of Russian (Soviet) mathematicians including Yakov Sinai, Ludvig Faddeev, Ilya Piatetskii-Shapiro, Viktor Maslov, Alexandre Kirillov, Viktor Palamodov, Dimitry Leites, Mikhail Shubin, and many others. The mathematical tools that he invented continue to play a central role in supermathematics: the *Berezin transform* (see §2.4), the *Berezinian* (the analog of the Jacobian for anti-commutative variables), and the *Berezin integral*. The latter (introduced in the 1960s) allowed the formulation of quantum field theory with fermions by path integrals and has become a principal instrument for the theories of supersymmetry and superstrings, central themes in modern high-energy and solid-state physics. Berezin published a large number of influential texts, including *The Method of Second Quantization*

(1965) and *Introduction to Superanalysis* (1987, posthumous). Berezin's revolutionary ideas were not properly valued by his contemporaries.

Berezin came from a family (on his mother's side) traumatized by Stalin's Great Terror during the years 1930–1940, and he kept his childhood fears for life. Nevertheless, at the age of 16, when registering for his first identity card, he declared himself as Jewish (via his mother), despite the official's recommendation that he should declare himself Russian (via his father) to avoid the anti-Jewish restrictions already apparent in the USSR. It was not long before these took effect: at his admittance to the University of Moscow (1948), Berezin was rejected by the Faculty of Physics (which was already subject to severe restrictions) but was accepted by Mathematics. Studying under the tutorship of Eugene Dynkin, and then in the famous research seminar of Israel Gelfand, Berezin finished in 1953 as the most brilliant student of his class, but he was not accepted into Graduate School. He prepared his thesis (under the direction of Gelfand) while teaching for three years at a high school. He was then accepted into the Faculty of Mathematics and Mechanics of Moscow (1956), thanks to the politics of Khrushchev's "Thaw" and to the efforts of Israel Gelfand and Ivan Petrovsky (who was also a renowned mathematician, at the time rector of the university). He was never given tenure as a professor. However, it was difficult to deprive him of the atmosphere of enthusiasm and effervescence of the "Golden Age" of mathematics in Russia, when, in particular, mathematicians and physicists began to understand each other. In his unfinished work, Berezin was always pursuing higher goals, seeking to understand the nature of quantization, and he firmly believed that the long-sought unified theory of fields could be expressed in terms of superanalysis.

Berezin died following an accident on a tempestuous river in the north of Russia during a voyage of "wild tourism." There is a rich literature describing his career and his mathematics: see Mikhail Shifman, *Felix Berezin: The Life and Death of the Mastermind of Supermathematics* (World Scientific, 2007) and Robert Minlos, "Brève biographie scientific de F. A. Berezin," http://smf4.emath.fr/Publications/Gazette/2006/110/smf_gazette_110_30-44.pdf.

2.3.2 The Natural Projection on $\mathcal{T}_{L^\infty}$

Recall that $\mathcal{T}_{L^\infty}$ is the set of bounded Toeplitz operators (matrices) on the space H^2 (or $\ell^2(\mathbb{Z}_+)$) which is clearly a closed subspace of $L(H^2)$.

(a) *By using the seminorm p on the space $\ell^\infty(\mathbb{Z}_+)$,*

$$p(x) =: \overline{\lim_n} \frac{1}{n+1} \left| \sum_{k=0}^{n} x_k \right|, \quad x = \{x_k\}_{k \geq 0} \in \ell^\infty(\mathbb{Z}_+),$$

and the Hahn–Banach theorem (see [Rudin, 1986]), show that there exists a linear functional $G \in (\ell^\infty)^*$ (called a generalized Banach limit*) such that:*

(1) *if* $\lim_k x_k$ *exists then* $G(x) = \lim_k x_k$,
(2) $|G(x)| \leq \overline{\lim}_k |x_k|, \forall x \in \ell^\infty$,
(3) $G(S^*x) = G(x), G(Sx) = G(x) \; (\forall x \in \ell^\infty(\mathbb{Z}_+))$,
(4) $\|G\| = 1; x_k \geq 0(\forall k) \Rightarrow G(x) \geq 0$.

(b) *Given an operator $A \in L(H^2)$, define T by*

$$(Tx, y) = G(\{(S^{*k}AS^k x, y)\}_{k \geq 0}) \quad (x, y \in H^2),$$

where G is a generalized limit of (a).

(1) *Show that* $T \in \mathcal{T}_{L^\infty}(H^2)$.

(2) *Show that* $P: A \mapsto T =: P(A)$ *is a projection ($P^2 = P$) of $L(H^2)$ onto the subspace $\mathcal{T}_{L^\infty}$, $\|P\| = 1$, $P|\mathfrak{S}_\infty(H^2) = 0$, and that $A \geq 0$ (A positive definite) $\Rightarrow$ $P(A) \geq 0$.*

Hint Consider the matrix elements (Tz^j, z^i) and use properties (1) and (3) of (a).

2.3.3 Toeplitz Operators on $\ell^p(\mathbb{Z}_+)$ and $H^p(\mathbb{T})$

(a) *Let $T \in L(H^p)$, $1 < p < \infty$. Show that the following assertions are equivalent.*

(1) *There exists $\varphi \in L^\infty(\mathbb{T})$ such that $T = T_\varphi$ where $T_\varphi f = P_+(\varphi f), f \in H^p$.*
(2) $S^*TS = T$.

Hint Note that the Riesz projection P_+ is continuous on $L^p(\mathbb{T})$, $1 < p < \infty$ (see [Nikolski, 2019], Exercise 2.8.4). Then, for (1) $\Rightarrow$ (2), verify that the calculations of Lemma 2.1.2 and Example 2.1.3 remain valid for H^p, and for (2) $\Rightarrow$ (1) repeat the arguments of Lemma 2.1.4.

Remark For $p \neq 2$, we cannot guarantee the equality $\|T_\varphi\| = \|\varphi\|_\infty$, but only the bounds $\|\varphi\|_\infty \leq \|T_\varphi\| \leq \|P_+\| \cdot \|\varphi\|_\infty$ where $\|P_+\| = \|P_+ : L^p \to L^p\|$: see [Nikolski, 2019] for the numerical value of the last constant.

(b) Multipliers (convolvers) of $\ell^p(\mathbb{Z})$, $1 \leq p \leq \infty$. Let k be a function on $\mathbb{Z}$; k is called a convolver (multiplier) of $\ell^p(\mathbb{Z})$ if the mapping

$$C_k : x \mapsto x * k =: \left(\sum_{j \in \mathbb{Z}} k(i-j) x_j\right)_{i \in \mathbb{Z}}$$

is well-defined for the sequences with compact support $x = (x_j)_{j \in \mathbb{Z}}$ and satisfies

$$\|x * k\|_{\ell^p(\mathbb{Z})} \leq c \|x\|_{\ell^p(\mathbb{Z})},$$

where $c > 0$ is a constant. Then C_k can be extended to a bounded mapping $\ell^p(\mathbb{Z}) \to \ell^p(\mathbb{Z})$ for which we keep the same notation; the norm $\|C_k\|$ is called the multiplier norm of k,

$$\|C_k\| = \|k\|_{\text{Mult}(\ell^p(\mathbb{Z}))};$$

$\text{Mult}(\ell^p(\mathbb{Z})) = M_p$ is the space of multipliers equipped with the norm $\|\cdot\|_{\text{Mult}(\ell^p(\mathbb{Z}))}$. Show that:

(1) $M_p \subset \ell^p(\mathbb{Z})$, M_p is a commutative Banach algebra with unity $e_0 = (\delta_{0j})_{j \in \mathbb{Z}}$ (with convolution $k_1 * k_2$ as operation),
(2) $M_p = M_{p'}$ where $1/p + 1/p' = 1$ (with equality of norms),
(3) if $1 \leq p \leq q \leq r \leq \infty$ and $k \in M_p \cap M_r$ then $k \in M_q$ and $\|k\|_{M_q} \leq \|k\|_{M_p}^t \|k\|_{M_r}^{1-t}$, where $1/q = t/p + (1-t)/r$.

Hint Apply the *Riesz–Thorin convexity theorem*: if $T : L^a \to L^a$ is a bounded linear operator for $a = p, r$, then it is also for $a = q$, and we have $\|T\|_{L^q \to L^q} \leq \|T\|_{L^p \to L^p}^t \|T\|_{L^r \to L^r}^{1-t}$ (see [Dunford and Schwartz, 1958], §VI.10.11).

(4) For every p, $1 < p < \infty$, show that $M_p \subset \mathcal{F}L^\infty(\mathbb{T})$ (where as before $\mathcal{F}f = (\widehat{f}(k))_{k \in \mathbb{Z}}$ for a function $f \in L^1(\mathbb{T})$) and $\|\mathcal{F}^{-1}k\|_\infty \leq \|k\|_{M_p}$, $\forall k \in M_p$.

Hint Recall that $\mathcal{F}^{-1}k = \sum_{j \in \mathbb{Z}} k(j) z^j$, $|z| = 1$, where the Fourier series converges, for example, for the norm of $L^2(\mathbb{T})$. For the solution of the problem, apply (3) to p and $r = p'$, $1/p + 1/p' = 1$.

(5) Show that $M_1 = M_\infty = \ell^1(\mathbb{Z})$, $M_2 = \mathcal{F}L^\infty(\mathbb{T})$.

> SOLUTION: With the Hints, everything is evident: the properties of convolution are familiar from a course in integration (see [Rudin, 1986]). For (2), with the bilinear duality between $L^p - L^{p'}$ we have $C_k^* = C_k$; (3) and (4) are resolved in the Hints

(taking into account (5)); for (5), $M_1 = M_\infty = \ell^1(\mathbb{Z})$ since $\ell^1(\mathbb{Z})$ is a convolution Banach algebra, and $M_2 = \mathcal{F}L^\infty(\mathbb{T})$ is proved in Lemma 2.1.4.

(c) Toeplitz (Wiener–Hopf) operators on $\ell^p = \ell^p(\mathbb{Z}_+)$**.** Let $T \in L(\ell^p)$, $1 \le p \le \infty$. Show that the following assertions are equivalent.

(1) There exists $k \in M_p$ such that $T = T_\varphi$, $\varphi = \mathcal{F}^{-1}k$ where $T_\varphi x = P_+(k * x)$, $x \in \ell^p$.

(2) $S^*TS = T$. Moreover, $\|T_\varphi\| = \|k\|_{M_p}$.

Hint For (1) $\Rightarrow$ (2), verify that the calculations of Lemma 2.1.2 and Example 2.1.3 remain valid for ℓ^p, and for (2) $\Rightarrow$ (1) modify the arguments of Lemma 2.1.4, taking into account (in particular) the fact that $\mathcal{F}^{-1}(k * x) = \mathcal{F}^{-1}k \cdot \mathcal{F}^{-1}x$ (for every x with compact support, hence for every polynomial $f = \mathcal{F}^{-1}x \in \mathcal{P}$) and $\|\mathcal{F}(f\mathcal{F}^{-1}k)\|_{\ell^p(\mathbb{Z})} \le c\|\mathcal{F}f\|_{\ell^p(\mathbb{Z})}$ ($\forall f \in \mathcal{P}$) signifies that k is a multiplier and $\|k\|_{M_p} \le c$.

2.3.4 The Space BMO($\mathbb{T}$), the RKT, and the Garsia Norm

(a) The space BMO($\mathbb{T}$), *by definition, is*

$$\mathrm{BMO}(\mathbb{T}) = L^\infty(\mathbb{T}) + \mathbb{H}L^\infty(\mathbb{T}) = \{f + \mathbb{H}g : f, g \in L^\infty(\mathbb{T})\},$$

where $\mathbb{H}f = i\tilde{f}$ is the *operation of harmonic conjugation* (up to the factor i) studied in [Nikolski, 2019]. Recall that $\mathbb{H}z^k = z^k$ for $k > 0$, $\mathbb{H}z^k = -z^k$ for $k < 0$ and $\mathbb{H}1 = 0$, and that $\mathbb{H}$ can be extended to a bounded linear mapping on $L^p(\mathbb{T})$, $1 < p < \infty$. However, for $p = 1, \infty$ (or for $C(\mathbb{T})$) this is not the case. The norm on BMO($\mathbb{T}$) is defined by

$$\|F\|_{\mathrm{BMO}} = \inf(\max(\|f\|_\infty, \|g\|_\infty) : F = f + \mathbb{H}g),$$

and its subspaces of analytic and co-analytic functions BMOA and BMO_ by

$$\mathrm{BMOA} = \mathrm{BMO}(\mathbb{T}) \cap H^1(\mathbb{T}), \quad \mathrm{BMO}_- = \mathrm{BMO}(\mathbb{T}) \cap H^1_-,$$

where $H^1_- = \{f \in L^1(\mathbb{T}): \widehat{f}(k) = 0 \forall k \in \mathbb{Z}_+\}$.

(1) *Show that* $\mathrm{BMO}(\mathbb{T}) = P_-L^\infty(\mathbb{T}) + P_+L^\infty(\mathbb{T})$, $\mathrm{BMOA} = P_+L^\infty(\mathbb{T})$ *and* $\mathrm{BMO}_- = P_-L^\infty(\mathbb{T})$, *with an equivalent norm*

$$\|F\|_* = \inf\{\max(\|f\|_\infty, \|g\|_\infty) : F = P_-f + P_+g\}.$$

Hint For the functions f with $\widehat{f}(0) = 0$, we have $\mathbb{H} = 2P_+ - I$, $P_+ = (\mathbb{H}+I)/2$.

(2) *Show that*

$$\text{BMOA} = \{\varphi \in H^2 : H_{\bar{z}\bar{\varphi}} \text{ is bounded}\},$$
$$\text{BMO} = \{\varphi \in L^2 : H_{\bar{z}\bar{\varphi}} \text{ and } H_\varphi \text{ are bounded}\}$$

with an equivalent norm

$$\|\varphi\|_{*\text{BMO}} = \max(\|H_{\bar{z}\bar{\varphi}}\|_\infty, \|H_\varphi\|_\infty).$$

Hint Use (1) and Theorem 2.2.4.

(3) *Let* $\varphi \in L^2(\mathbb{T})$. *Then* $H_\varphi \in L(H^2, H_-^2)$ *if and only if* $P_-\varphi \in \text{BMO}_-$.

Hint Use (1) and the fact that $H_\varphi f = H_\psi f$ for every $f \in \mathcal{P}_a$ if and only if $P_-\varphi = P_-\psi$.

(4) *Show that* $(H^1)^* = \text{BMO}_-$ *with respect to the bilinear duality*

$$l_F(h) = \langle F, h \rangle = \int_\mathbb{T} F h \, dm$$

(with equality of the norms $\|l_F\| = \|F\|_*$*)*.

SOLUTION: For $F \in \text{BMO}_- = P_- L^\infty(\mathbb{T})$, $F = P_-\varphi (\varphi \in L^\infty(\mathbb{T}))$ we have $|l_F(h)| = |\int_\mathbb{T} \varphi h \, dm| \leq \|\varphi\|_\infty \|h\|_1$. By minimizing over φ, we obtain $\|l_F\| \leq \|F\|_*$. Conversely, a continuous linear functional $l \in (H^1)^*$ admits a representation $l = l_\varphi$ with $\varphi \in L^\infty$ and $\|l\| = \text{dist}_{L^\infty}(\varphi, H^\infty)$ (see the proof of Theorem 2.2.4). Clearly $l = l_F$, where $F = P_-\varphi$ and $\|F\|_* = \text{dist}_{L^\infty}(\varphi, H^\infty) = \|l\|$.

(b) The Garsia norm and the RKT for Hankel operators. *Let* $\lambda \in \mathbb{D}$, *let* $k_\lambda = (1 - \bar{\lambda}z)^{-1}$ *be the reproducing kernel of the space* H^2 *(see [Nikolski, 2019]), and let* $\varphi \in H_0^2 = \{\varphi \in H^2 : \varphi(0) = 0\}$. *Then, given a function* $f \in L^1(\mathbb{T})$, *let* $f(z)$ $(z \in \mathbb{D})$ *denote its (harmonic) Poisson extension to the disk (see [Nikolski, 2019]),*

$$f(z) =: f * P_z =: f * P(z) = \int \frac{1 - |z|^2}{|z - \zeta|^2} f(\zeta) \, dm(\zeta),$$

where

$$P(z) = \frac{1 - |z|^2}{|1 - z|^2} \quad \text{(Poisson kernel)}.$$

Show that

$$\|H_{\bar{\varphi}} k_\lambda\|_2^2 / \|k_\lambda\|_2^2 = |\varphi|^2(\lambda) - |\varphi(\lambda)|^2,$$

and $G(\varphi) \leq \|\varphi\|_*$ *where*

$$G(\varphi)^2 =: \sup_{\lambda \in \mathbb{D}} (|\varphi|^2(\lambda) - |\varphi(\lambda)|^2).$$

Show that $G(\cdot)$ *is a norm (the Garsia norm) on*

$$\mathrm{BMOA}_0 = \{\varphi \in \mathrm{BMOA}: \varphi(0) = 0\}.$$

SOLUTION: We have $\|k_\lambda\|_2^2 = (1 - |\lambda|^2)^{-1}$ and

$$\|H_{\overline{\varphi}}k_\lambda\|_2^2 = \|P_-(\overline{\varphi}k_\lambda)\|_2^2 = \|P_-(\overline{\varphi}k_\lambda - \overline{\varphi(\lambda)}k_\lambda)\|_2^2$$

$$= \left\|P_-\frac{\overline{\varphi} - \overline{\varphi(\lambda)}}{\overline{z} - \overline{\lambda}}\overline{z}\right\|_2^2 = \int_{\mathbb{T}}\left|\frac{\varphi - \varphi(\lambda)}{z - \lambda}\right|^2 dm(z), \quad \text{hence}$$

$$\|H_{\overline{\varphi}}k_\lambda\|_2^2/\|k_\lambda\|_2^2 = |\varphi - \varphi(\lambda)|^2(\lambda) = |\varphi|^2(\lambda) + |\varphi(\lambda)|^2 - 2\operatorname{Re}(\overline{\varphi}\varphi(\lambda))(\lambda)$$

$$= |\varphi|^2(\lambda) - |\varphi(\lambda)|^2.$$

The properties of the norm are evident because $\varphi \mapsto \|H_{\overline{\varphi}}k_\lambda\|_2$ is a seminorm and the family $(k_\lambda)_{\lambda \in \mathbb{D}}$ generates the space H^2, $H^2 = \operatorname{span}_{H^2}(k_\lambda: \lambda \in \mathbb{D})$: see [Nikolski, 2019].

Remark In fact, the norm G is equivalent to the original norm on BMOA_0 (Garsia, 1971); see for example [Koosis, 1980] and point (c) below. The complete proof depends on the John–Nirenberg inequality and on other less elementary techniques. Nonetheless, it is included in the following important theorem, which provides a local description of BMO, whose principal portion ((1) ⇔ (2)) is due to Fefferman [1971] and whose proof is beyond our scope.

(c)* A local description and duality (Fefferman, 1971). *Let* $F \in L^1(\mathbb{T})$. *The following assertions are equivalent.*

(1) $F \in \mathrm{BMO}$.
(2) $N(F) =: \sup_{I \subset \mathbb{T}} |F - F_I|_I < \infty$, *where*

$$F_I = \frac{1}{m(I)}\int_I F\, dm$$

is the mean of F *over an arc (interval)* $I \subset \mathbb{T}$.
(3) $G(F)^2 = \sup_{\lambda \in \mathbb{D}}(|F|^2(\lambda) - |F(\lambda)|^2) < \infty$.
(4) $F \in (H^1_{\mathrm{Re}})^*$, *i.e. the linear functional* $l_F: g \mapsto \int_{\mathbb{T}} Fg\, dm$ $(g \in \mathcal{P})$ *is continuous for the norm* $\|\cdot\|_{H^1_{\mathrm{Re}}}$ *of the space* H^1_{Re} *(called "real H^1") defined as follows:*

$$H^1_{\mathrm{Re}} =: L^1(\mathbb{T}) \cap \mathbb{H}(L^1(\mathbb{T})) = \{g \in L^1(\mathbb{T}): \mathbb{H}g \in L^1(\mathbb{T})\},$$

equipped with the norm $\|g\|_{H^1_{\mathrm{Re}}} = \|g\|_1 + \|\mathbb{H}g\|_1$, *or equivalently*

$$H^1_{\mathrm{Re}} = \{g \in L^1(\mathbb{T}): \mathbb{H}g \in L^1(\mathbb{T})\} = H^1_- + H^1 = \{g \in L^1(\mathbb{T}): P_+g, P_-g \in L^1(\mathbb{T})\}.$$

(5) $H_F \in L(H^2, H^2_-), H_{\overline{F}} \in L(H^2, H^2_-)$.

SOLUTION: (other than the equivalence (1) $\Leftrightarrow$ (2) for which we refer to [Koosis, 1980] or [Garnett, 1981]). (1) $\Leftrightarrow$ (5) is (2) of (a). (5) $\Rightarrow$ (3) is (b).

(1) $\Rightarrow$ (4) For $F = f + \mathbb{H}g$ with $f, g \in L^\infty(\mathbb{T})$, we have ($\forall p \in \mathcal{P}$)

$$\left| \int_\mathbb{T} Fp\, dm \right| = \left| \int_\mathbb{T} (f + \mathbb{H}g)p\, dm \right| \leq \left| \int_\mathbb{T} fp\, dm \right| + \left| \int_\mathbb{T} \mathbb{H}g \cdot p\, dm \right|$$

$$= \left| \int_\mathbb{T} fp\, dm \right| + \left| \int_\mathbb{T} g \cdot \mathbb{H}p\, dm \right| \leq \|f\|_\infty \times \|p\|_1 + \|g\|_\infty \times \|\mathbb{H}p\|_1,$$

and by minimizing the maximum $\max(\|f\|_\infty, \|g\|_\infty)$ we obtain

$$\left| \int_\mathbb{T} Fp\, dm \right| \leq \|F\|_{\mathrm{BMO}} \|p\|_{H^1_{\mathrm{Re}}},$$

and the result follows.

(4) $\Rightarrow$ (1) We leave to the reader the exercise of verifying the equivalence of the different definitions of H^1_{Re} (use again $\mathbb{H} = 2P_+ - I$, $P_+ = (\mathbb{H}+I)/2$) (for the functions with $\widehat{f}(0) = 0$), we obtain that the restrictions $l_F|H^1$, $l_F|H^1_-$ are continuous, and thus by point (1) of (a), we have $P_-F \in P_-L^\infty = \mathrm{BMO}_-$ and $P_+F \in P_+L^\infty = \mathrm{BMOA}$, so that $F = P_-F + P_+F \in P_-L^\infty + P_+L^\infty = \mathrm{BMO}$.

It only remains to verify that (3) $\Rightarrow$ (2). Note first that since $\|\cdot\|_{L^1(\mathbb{T})} \leq \|\cdot\|_{L^2(\mathbb{T})}$, we have

$$G(F) \geq (|F|^2(\lambda) - |F(\lambda)|^2)^{1/2} = (|F - F(\lambda)|^2(\lambda))^{1/2} \geq |F - F(\lambda)|(\lambda).$$

Then, let $I \subset \mathbb{T}$ be an arc and let $\lambda \in \mathbb{D}$ be such that $\lambda/|\lambda|$ is the center of I and $(1 - |\lambda|) = 2\pi m(I)$ (if $2\pi m(I) > 1$, then $\lambda = 0$). Then for $\zeta \in I$ we have $|\zeta - \lambda/|\lambda|| \leq 2\pi m(I)$, and hence $|\zeta - \lambda| \leq (1 - |\lambda|) + 2\pi m(I) = 2(1 - |\lambda|)$ and then

$$P_\lambda(\zeta) = \frac{1 - |\lambda|^2}{|\zeta - \lambda|^2} \geq \frac{1}{4(1 - |\lambda|)} = \frac{1}{8\pi m(I)},$$

which implies

$$G(F) \geq |F - F(\lambda)|(\lambda) = \int_\mathbb{T} \frac{1 - |\lambda|^2}{|\zeta - \lambda|^2} |F(\zeta) - F(\lambda)|\, dm(\zeta)$$

$$\geq \int_I \frac{1}{8\pi m(I)} |F - F(\lambda)|\, dm = \frac{1}{8\pi} |F - F(\lambda)|_I,$$

and then

$$|F_I - F(\lambda)| = \left| \frac{1}{m(I)} \int_I (F - F(\lambda))\, dm \right| \leq |F - F(\lambda)|_I \leq 8\pi G(F),$$

$$|F_I - F|_I = \frac{1}{m(I)} \int_I |F_I - F|\, dm \leq \frac{1}{m(I)} \int_I |F(\lambda) - F|\, dm + |F_I - F(\lambda)|$$

$$\leq |F(\lambda) - F|_I + 8\pi G(F) \leq 16\pi G(F).$$

Hence $N(F) \leq 16\pi G(F) < \infty$.

Remark The statement (2) of (c) explains the notation BMO, an acronym for *bounded mean oscillation*, the name initially given to this space by John and Nirenberg [1961]; the definition of the space BMO in the form (2) of (c) was proposed by John [1961] shortly before. The calculation of the last portion ((3) ⇒ (2)) is a typical example (but for beginners) of the "approach by real analysis" of the study of function classes.

2.3.5 Compact Hankel Operators and the Spaces VMO($\mathbb{T}$) and QC($\mathbb{T}$)

(a) $H_\varphi \in \mathfrak{S}_\infty$ **and the space** VMO($\mathbb{T}$). By definition,

$$\text{VMO}(\mathbb{T}) = C(\mathbb{T}) + \mathbb{H}C(\mathbb{T}) = \{f + \mathbb{H}g : f, g \in C(\mathbb{T})\},$$

where $\mathbb{H}f = i\tilde{f}$, as before, is the operation of harmonic conjugation studied in [Nikolski, 2019].

VMO is a subspace of BMO equipped with the induced norm. We define

$$\text{VMOA} = \text{VMO}(\mathbb{T}) \cap H^1(\mathbb{T}), \quad \text{VMO}_- = \text{VMO}(\mathbb{T}) \cap H^1_-.$$

(1) *Show that* VMO($\mathbb{T}$) = $P_-C(\mathbb{T}) + P_+C(\mathbb{T})$, VMOA = $P_+C(\mathbb{T})$ *and* VMO$_-$ = $P_-C(\mathbb{T}) = C(\mathbb{T})/C_a(\mathbb{T})$ *(with the identification $P_-f \mapsto f + C_a$).*

Hint For the functions f with $\widehat{f}(0) = 0$, we have $\mathbb{H} = 2P_+ - I$, $P_+ = (\mathbb{H} + I)/2$, and for VMO$_-$ use $\|f + C_a\|_{C/C_a} = \|P_- f\|_{P_-C}$.

(2) *Show that* VMOA = $\{\varphi \in H^2 : H_{\overline{z\varphi}} \in \mathfrak{S}_\infty\}$, VMO = $\{\varphi \in L^2 : H_{\overline{z\varphi}} \in \mathfrak{S}_\infty, H_\varphi \in \mathfrak{S}_\infty\}$.

Hint Use (1) and Theorem 2.2.8(2).

(3) *Let* $\varphi \in L^2(\mathbb{T})$. *Then* $H_\varphi \in \mathfrak{S}_\infty$ *if and only if* $P_-\varphi \in $ VMO$_-$.

Hint Use (1) and the fact that $H_\varphi f = H_\psi f$ for every $f \in \mathcal{P}_a$ if and only if $P_-\varphi = P_-\psi$.

(4) *Show that* $H^1 = ($VMO$_-)^*$ *(with respect to the bilinear duality).*

Hint Since VMO$_-$ = $P_-C = C/C_a$ (see (1)) it suffices to regard the solution of Exercise 2.3.8 below.

(b)* *A local description of* VMO *(Sarason, 1975).* Let $F \in L^1(\mathbb{T})$. The following assertions are equivalent.

(1) $F \in$ VMO.

(2) $\lim_{\delta \to 0}(\sup_{I \subset \mathbb{T}, m(I) < \delta} |F - F_I|_I) = 0$.

(3) $\lim_{\delta \to 0}(\sup_{\lambda \in \mathbb{D}, |\lambda| > 1-\delta}(|F|^2(\lambda) - |F(\lambda)|^2) = 0$.
(4) $H_F \in \mathfrak{S}_\infty$, $H_{\overline{F}} \in \mathfrak{S}_\infty$.

SOLUTION: (other than the equivalence (1) $\Leftrightarrow$ (2) for which we refer to [Koosis, 1980] or [Garnett, 1981]).

(1) $\Leftrightarrow$ (4) is (2) of (a).

(4) $\Rightarrow$ (3) since by Exercise 2.3.4(b),

$$\|H_{P_-F} k_\lambda\|_2^2 / \|k_\lambda\|_2^2 = |P_-F|^2(\lambda) - |(P_-F)(\lambda)|^2$$

(with a similar formula for $P_-\overline{F}$), and as $k_\lambda / \|k_\lambda\|_2 \to 0$ weakly in H^2 for $|\lambda| \to 1$ (easy to check; or see [Nikolski, 2019]) and the operators H_F and $H_{\overline{F}}$ are compact, the result follows (see Appendix D).

(3) $\Rightarrow$ (2) follows the proof of the implication (3) $\Rightarrow$ (2) of Exercise 2.3.4(c): it suffices to remember that in this proof $(1 - |\lambda|) \asymp 2\pi m(I)$.

Remark In particular, point (2) explains the notation VMO (for *vanishing mean oscillation* [Sarason, 1975]).

(c) The algebra $QC(\mathbb{T}) =: (H^\infty + C(\mathbb{T})) \cap (H_-^\infty + C(\mathbb{T}))$ **(Sarason algebra).** *Show that*

(1) (Douglas, 1972) QC *is a closed sub-algebra of* $L^\infty(\mathbb{T})$, $QC \neq C(\mathbb{T})$ (*and even* $QC \cap H^\infty \neq C_a(\mathbb{T})$), *and*
(2) (Sarason, 1975)
$$QC = VMO \cap L^\infty.$$

SOLUTION: QC is a closed algebra by Sarason's Lemma 2.2.7. By Theorem 2.2.8(2) and (2) of (a) above, $QC \subset VMO \cap L^\infty$. By the same Theorem 2.2.8(2) and the implication (1) $\Rightarrow$ (4) of (b), we have

$$VMO \cap L^\infty \subset (H^\infty + C) \cap (H_-^\infty + C) = QC,$$

hence $QC = VMO \cap L^\infty$.

For an example of $F \in (QC \setminus C(\mathbb{T}))$, recall (see Exercise 2.3.4(a)) that there exists a real function $f \in C(\mathbb{T})$ such that $\mathbb{H}f = i\tilde{f} \notin C(\mathbb{T})$. Observing that the function $G = e^{f+i\tilde{f}}$ is in H^∞, we obtain

$$F = G e^{-f} \in H^\infty \cdot C(\mathbb{T}) \subset H^\infty + C(\mathbb{T})$$

and similarly $\overline{F} = e^{-(f+i\tilde{f})} e^f \in H^\infty + C(\mathbb{T})$, hence $F = e^{i\tilde{f}} \in QC$. If we suppose that $F \in C(\mathbb{T})$, then there exists $n \in \mathbb{Z}$ and (real) $\varphi \in C(\mathbb{T})$ such that $e^{int} F(e^{it}) = e^{i\varphi(e^{it})}$, and hence

$$\tilde{f}(e^{it}) + nt = \varphi(e^{it}) + 2\pi k(e^{it}) \quad (t \in (-\pi, \pi))$$

where $k(e^{it}) \in \mathbb{Z}$ for every t. Applying Lemma (d) below we obtain $k = $ const., $n = 0$, and thus $\tilde{f} \in C(\mathbb{T})$, which is a contradiction; hence $F \in QC \setminus C(\mathbb{T})$. Clearly $G \in (QC \cap H^\infty) \setminus C(\mathbb{T})$.

Remark The notation QC is for *quasi-continuous*.

(d) Lemma (on the choice of the argument). *Let $n \in \mathbb{Z}$ and let $f, \varphi \in C(\mathbb{T})$ be real functions such that*

$$\tilde{f}(e^{it}) + nt = \varphi(e^{it}) + 2\pi k(e^{it}) \quad (t \in (-\pi, \pi))$$

where $k(e^{it}) \in \mathbb{Z}$ for every t. Then, $n = 0$ and $k = $ const.

SOLUTION: Without loss of generality, we can assume that k takes on the value 0. Let $a = e^{i\pi k(\cdot)}$, $b = \exp(i(\tilde{f} - \varphi)) = e^{-int}$ and let T_a, T_b be the corresponding Toeplitz operators. Then, T_b is invertible by Theorem 3.3.6 (or its Corollary 3.3.9; everything is independent, there is no vicious circle!), and since $T_b = S^{*n}$ if $n \geq 0$ and $T_b = S^{|n|}$ if $n \leq 0$, we obtain $n = 0$. Now, if k takes on an odd value, we obtain by Theorem 3.3.8 that $\sigma(T_a) = [-1, 1]$, which is not possible: T_a is invertible (for the same reason as before, since $\pi k = (\tilde{f} - \varphi)/2$ and $f, \varphi \in C(\mathbb{T})$). The same reasoning applied to $k/2$, then to $k/4$, etc., shows that $k = 0$.

A description of finite rank Hankel operators is due to Leopold Kronecker and has been known since the nineteenth century.

2.3.6 Finite Rank Hankel Operators (Kronecker, 1881)

Let H_f be a Hankel operator. Then

$$\text{rank}(H_f) =: \dim(\text{Ran}(H_f)) < \infty \Leftrightarrow P_- f \text{ is a rational function,}$$

and moreover $\text{rank}(H_f) = \deg(P_- f)$ (= *the number of poles in the disk* $\mathbb{D}$).

SOLUTION: If $\text{rank}(H_f) < \infty$, there exists a linear combination $\sum_{k=0}^{N} a_k H_f z^k = 0$ with $a_N \neq 0$ and $N = \text{rank}(H_f)$. Hence $P_-(fq) = 0$, $q = \sum_{k=0}^{N} a_k z^k$, and thus $0 = P_-(P_-f + P_=f)q = P_-((P_-f)q)$, meaning that $(P_-f)q = p \in H^2$. However, p is a polynomial of degree $\deg(p) \leq \deg(q) - 1$ since $\deg(P_+((P_-f)z^k)) \leq k - 1$ for every $k = 1, 2, \ldots$. We obtain $P_-f = p/q$ with

$$\deg(p/q) =: \max(\deg(p), \deg(q)) = \text{rank}(H_f)$$

(with all the poles in $\mathbb{D}$, as this is the case for $P_-f = \sum_{k<0} \widehat{f}(k) z^k$).

Conversely, the above reasoning can be reversed, thus $\deg(P_-f) =: N < \infty$ implies $\sum_{k=0}^{N} a_k H_f z^k = 0$, $a_N \neq 0$, hence $H_f z^N \in \text{Vect}(H_f z^k : 0 \leq k < N)$. However, by Lemma 2.2.1, $H_f S = P_- S H_f$ and $H_f S^k = P_- S^k H_f$ for every $k = 0, 1, \ldots$, which gives

$$\sum_{k=0}^{N} a_k H_f z^{k+1} = P_- S \sum_{k=0}^{N} a_k H_f z^k = 0,$$

hence $H_f z^{N+1} \in \text{Vect}(H_f z^k : 0 \le k < N)$, and then, by induction,

$$H_f z^n \in \text{Vect}(H_f z^k : 0 \le k < N)$$

for any $n \ge N$, and thus $\text{rank}(H_f) \le N$.

Leopold Kronecker (1823–1891) was a German mathematician, Jewish, and a student of Ernst Kummer and Per Gustav Dirichlet, and colleague and antagonist of Karl Weierstrass and Georg Cantor. He is known for his work in algebraic number theory and in the linear and multi-linear algebra initiated by Cayley and Grassmann. He constructed the theory of determinants and proposed a theory of divisors similar to that of ideals introduced at the same time by Richard Dedekind. While studying elliptic functions, Kronecker formulated a conjecture (he spoke of it as his *liebster Jugendtraum* or "most cherished dream of youth") that was later transformed by Hilbert into the twelfth problem of his famous list of problems for the twentieth century. Several mathematical objects bear Kronecker's name, in particular, the *Kronecker delta* and the *Kronecker (direct) matrix product*.

Kronecker's academic career was far from standard: after his thesis, submitted in 1845 under the direction of Dirichlet, Kronecker did not apply for a position but rushed off to manage a family property (for 10 years!). In about 1855 he returned to mathematics, but as he had become quite well-to-do, he was not obliged to seek a position, and could work independently. He was elected to the Berliner Akademie der Wissenschaften (1861). From 1870 onwards, Kronecker developed a "finitist (constructivist) vision" of mathematics that recognized as "real" only the objects obtained by a finite number of operations beginning with integers. It was Kronecker who stated "God made the integers, all else is the work of man" (*Die ganzen Zahlen hat der liebe Gott gemacht, alles andere ist Menschenwerk*, according to Heinrich Weber, *Jahresber. Deutschen Math.-Vereinigung* **2** (1893), page 19). He rejected any (pure) existence proof and any transfinite construction, in particular, irrational numbers, the continuum, etc., and hence was violently opposed to Weierstrass, Dedekind, and Cantor. His academic influence never ceased to grow (in particular, from 1880, Kronecker was the chief editor of *Crelle's Journal*, the principal mathematical journal of the time), which led him to attempt to delay the publication of the works of his opponents (Heine, Cantor). Hilbert qualified him as a *Verbotsdiktator* ("prohibiting dictator"). Some even hold him partly responsible for Cantor's depression and an attempt to send Weierstrass into "mathematical exile" in Switzerland (1885). Kronecker's constructivist point of view led to the intuitionism of Poincaré, Weyl, and Brouwer.

In addition to the Berlin Academy, Kronecker was elected to the Académie des Sciences de Paris, the Royal Society (London) and the St. Petersburg Academy of Science.

2.3.7 Hilbert–Schmidt Hankel Operators

Recall (see Appendix D) that the set $\mathfrak{S}_2 = \mathfrak{S}_2(\ell^2(\mathbb{Z}_+))$ *of Hilbert–Schmidt operators (matrices) on* $\ell^2(\mathbb{Z}_+)$ *is, by definition, the set of matrices* $A = (a_{ij})_{i,j \geq 0}$ *such that*

$$\|A\|_{\mathfrak{S}_2}^2 =: \sum_{i,j} |a_{ij}|^2 < \infty,$$

where $\|\cdot\|_{\mathfrak{S}_2}$ *is a (Hilbert) norm on* $\mathfrak{S}_2$ *and* $\mathfrak{S}_2(\ell^2(\mathbb{Z}_+))$ *is an ideal of* $L(\ell^2(\mathbb{Z}_+))$. *Show that* $\Gamma_\varphi \in \mathfrak{S}_2(\ell^2(\mathbb{Z}_+))$ *if and only if*

$$\|\Gamma_\varphi\|_{\mathfrak{S}_2}^2 = \frac{1}{\pi} \int \int_{\mathbb{D}} |\Phi'(x+iy)|^2 \, dx \, dy = \sum_{k \geq 0} (k+1)|\widehat{\varphi}(-k-1)|^2 < \infty,$$

where $\Phi = zJP_-\varphi = \sum_{k \geq 0} \widehat{\varphi}(-k-1) z^{k+1}$.

Hint Direct calculation.

SOLUTION:

$$\|\Gamma_\varphi\|_{\mathfrak{S}_2}^2 = \sum_{k \geq 0} \|\Gamma_\varphi e_k\|^2 = \sum_{k \geq 0} \sum_{j \geq 0} |\widehat{\varphi}(-k-j-1)|^2 = \sum_{n \geq 0} (n+1)|\widehat{\varphi}(-n-1)|^2$$

$$= \sum_{n \geq 0} (n+1)|\widehat{\Phi}(n+1)|^2 = \frac{1}{\pi} \int_0^1 \int_0^{2\pi} |\Phi'(re^{it})|^2 r \, dt \, dr.$$

2.3.8 The Original Proof of Sarason's Lemma 2.2.7 (Sarason, 1967)

Hint To show Lemma 2.2.7, use the following consequence of the Hahn–Banach theorem (see [Rudin, 1986]): for every closed subspace $E \subset X$ of a Banach space, the dual E^* and $(X/E)^*$ can be canonically identified as $E^* = X^*/E^\perp$ and $(X/E)^* = E^\perp$, where $E^\perp = \{f \in X^* : f|E = 0\}$.

SOLUTION: By the Hint, $(C/C_a)^* = C_a^\perp = H_0^1$, where $C = C(\mathbb{T})$, $C_a = C_a(\mathbb{T}) = C \cap H^\infty$, $H_0^1 = zH^1$, and $(H_0^1)^* = L^\infty(\mathbb{T})/(H_0^1)^\perp = L^\infty(\mathbb{T})/H^\infty$ (see [Nikolski, 2019] for the identification of $C_a^\perp$ with H_0^1 (Riesz brothers' theorem)). Clearly the canonical isometric embedding

$$j : C/C_A \to (C/C_a)^{**} = L^\infty/H^\infty$$

is $j(f + C_a) = f + H^\infty$, which implies that the inverse image $\Pi^{-1} j(C/C_a)$ with respect to the canonical projection $\Pi : L^\infty(\mathbb{T}) \to L^\infty(\mathbb{T})/H^\infty$ is a closed subspace of $L^\infty(\mathbb{T})$, and that

$$\Pi^{-1} j(C/C_a) = \{F \in L^\infty(\mathbb{T}) : F + H^\infty \in j(C/C_a)\}$$
$$= \{F \in L^\infty(\mathbb{T}) : F + H^\infty = f + H^\infty, \text{ for some } f \in C(\mathbb{T})\}$$
$$= C(\mathbb{T}) + H^\infty.$$

2.3.9 Compactness of the Commutators $[P_+, M_\varphi]$ (Power, 1980)

Let $\varphi \in L^2(\mathbb{T})$ and let $[P_+, M_\varphi] = P_+ M_\varphi - M_\varphi P_+$ be the commutator of the Riesz projection P_+ and the multiplication operator M_φ (on the space $L^2(\mathbb{T})$). Show that

$$[P_+, M_\varphi] \in \mathfrak{S}_\infty(L^2(\mathbb{T})) \Leftrightarrow \varphi \in \text{VMO}.$$

SOLUTION: We write

$$[P_+, M_\varphi] = P_+ M_\varphi - M_\varphi P_+ = P_+ M_\varphi P_+ + P_+ M_\varphi P_- - P_+ M_\varphi P_+ - P_- M_\varphi P$$
$$= P_+ M_\varphi P_- - P_- M_\varphi P_+.$$

Given that the subspaces $H^2 = P_+ L^2(\mathbb{T})$ and $H_-^2 = P_- L^2(\mathbb{T})$ are orthogonal, we have

$$[P_+, M_\varphi] \in \mathfrak{S}_\infty(L^2(\mathbb{T})) \Leftrightarrow (P_+ M_\varphi P_-, P_- M_\varphi P_+ \in \mathfrak{S}_\infty),$$

and by (3) of Exercise 2.3.5(a), $P_+ \varphi \in \text{VMOA}$ and $P_- \varphi \in \text{VMO}_-$. Since VMO = $\text{VMO}_- + \text{VMOA}$, a direct sum, the result follows.

2.3.10 The Natural Projection on Hank ($\ell^2(\mathbb{Z}_+)$)

On the set of matrices $A = (a_{ij})_{i,j\geq 0}$, a *natural projection onto the Hankel matrices* (c_{i+j}) is defined by

$$PA = (c_{i+j}), \quad c_n = \frac{1}{n+1} \sum_{0 \leq j \leq n} a_{j,n-j} \quad (n \in \mathbb{Z}_+).$$

(a) *Show with an example the existence of* $A \in L(\ell^2(\mathbb{Z}_+))$ *such that* $PA \notin L(\ell^2(\mathbb{Z}_+))$ *(and hence P is not bounded on $L(\ell^2(\mathbb{Z}_+))$).*

Hint Consider a matrix $A = (a_{ij})_{i,j\geq 0}$ with $a_{00} = 1$ and $a_{ij} = 1$ for i, j satisfying $i + j = 3 \cdot 2^k$ and $2^k \leq j < 2^{k+1}$, $k = 0, 1, \ldots$, and all other elements zero; see Figure 2.2.

SOLUTION: Indeed, the matrix of the Hint defines a unitary mapping of $\ell^2(\mathbb{Z}_+)$, since for every j, $Ae_j = e_{\omega(j)}$, where ω is a permutation of $\mathbb{Z}_+$: $\omega(0) = 0$, $\omega(j) = 3 \cdot 2^k - j$ for $2^k \leq j < 2^{k+1}$, $k \geq 0$. Moreover, $c_0 = 1$ and $c_n = 2^k/(3 \cdot 2^k + 1)$ for $n = 3 \cdot 2^k$ ($k \geq 0$), and $c_n = 0$ for all other n; consequently, $\|P(A)e_0\|^2 = \sum_{n\geq 0} c_n^2 = \infty$, and the result follows.

(b) *Show that* $\|P: \mathfrak{S}_2 \to \mathfrak{S}_2\| = 1$, *and thus P is an orthogonal projection of $\mathfrak{S}_2$ onto its Hankel subspace* $\mathfrak{S}_2 \cap \text{Hank}(\ell^2(\mathbb{Z}_+))$ *(see Exercise 2.3.7 for a reminder for $\mathfrak{S}_2(\ell^2(\mathbb{Z}_+))$).*

Hint Cauchy's inequality.

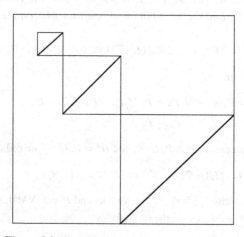

Figure 2.2 The matrix A of the Hint of §2.3.10(a).

SOLUTION: Let $A \in \mathfrak{S}_2$, $A(a_{ij})$. By Exercise 2.3.7,

$$\|PA\|_{\mathfrak{S}_2}^2 = \sum_{n\geq 0}(n+1)|c_n|^2 = \sum_{n\geq 0}(n+1)\left|\frac{1}{n+1}\sum_{0\leq j\leq n} a_{j,n-j}\right|^2$$
$$\leq \sum_{n\geq 0}\frac{1}{n+1}(n+1)\sum_{0\leq j\leq n}|a_{j,n-j}|^2 = \sum_{i,j}|a_{ij}|^2 = \|A\|_{\mathfrak{S}_2}^2,$$

and hence $\|P\colon \mathfrak{S}_2 \to \mathfrak{S}_2\| \leq 1$. We know (see Appendix D) that a contracting projection on a Hilbert space (such as $\mathfrak{S}_2$) is orthogonal.

2.3.11 Vector-Valued Toeplitz Operators

We refer the reader to Appendix F.6 for the definitions and notation for vector- and matrix-valued Hardy spaces, in particular, the spaces $H^2(\mathbb{T}, \mathbb{C}^N) \subset L^2(\mathbb{T}, \mathbb{C}^N)$ and $H^\infty(\mathbb{T}, L(\mathbb{C}^N)) \subset L^\infty(\mathbb{T}, L(\mathbb{C}^N))$.

(a) *Let $\varphi \in L^2(\mathbb{T}, L(\mathbb{C}^N))$ and let M_φ be the multiplication operator defined by*

$$M_\varphi f(\zeta) = \varphi(\zeta)f(\zeta), \quad \zeta \in \mathbb{T}$$

for the trigonometric polynomials $f \in L^2(\mathbb{T}, \mathbb{C}^N)$ ($\widehat{f}(n) \neq 0$ for a finite set of $n \in \mathbb{Z}$). Show that M_φ can be extended to a bounded operator $L^2(\mathbb{T}, \mathbb{C}^N) \to L^2(\mathbb{T}, \mathbb{C}^N)$ if and only if $\varphi \in L^\infty(\mathbb{T}, L(\mathbb{C}^N))$, and in this case $\|M_\varphi\| = \|\varphi\|_\infty$ (see Appendix F.6 for the definition).

Hint Adapt the proof of Lemma 2.1.4.

(b) *Let $T\colon H^2(\mathbb{T}, \mathbb{C}^N) \to H^2(\mathbb{T}, \mathbb{C}^N)$ be a bounded operator. Show that the following assertions are equivalent.*

(1) $S^*TS = T$, where $S = M_z$ ($Sf = zf$) and $S^*f = (f - f(0))/z$ ($f \in H^2(\mathbb{T}, \mathbb{C}^N)$).
(2) There exists $\varphi \in L^\infty(\mathbb{T}, L(\mathbb{C}^N))$ such that

$$T = T_\varphi =: P_+ M_\varphi | H^2(\mathbb{T}, \mathbb{C}^N).$$

The function φ is uniquely defined by T, and we have $\|T_\varphi\| = \|\varphi\|_\infty$.

Hint Repeat the proof of Theorem 2.1.5.

2.3.12 Some Algebraic Properties of Toeplitz/Hankel Operators

Let $\varphi, \psi \in L^\infty(\mathbb{T})$. Establish the following properties of the operators T_φ, T_ψ, and $\Gamma_\varphi = JH_\varphi$, Γ_ψ. Recall that

$$Jf(z) = \bar{z}f(\bar{z}) \quad (z \in \mathbb{T})$$

(see Example 2.2.2), $J^2 = \mathrm{id}$, $J(fg) = z(Jf)(Jg)$, $JP_+J = P_-$, and denote $\widetilde{\varphi} = \varphi(\bar{z}) = zJ\varphi$.

(a) (Widom, 1976) $\Gamma_\varphi T_\psi + T_{\widetilde{\varphi}} \Gamma_\psi = \Gamma_{\varphi\psi}$.

SOLUTION: By definition, for every $f \in H^2$, we have

$$\Gamma_\varphi T_\psi f + T_{\widetilde{\varphi}} \Gamma_\psi f = JP_-\varphi P_+ \psi f + P_+(zJ\varphi)JP_-\psi f,$$

$$P_+(zJ\varphi)JP_-\psi f = JJP_+ JJ(zJ\varphi)JP_-\psi f = JP_-\varphi JJP_-\psi f$$

$$= JP_-\varphi P_-\psi f,$$

hence $\Gamma_\varphi T_\psi f + T_{\widetilde{\varphi}} \Gamma_\psi f = JP_-\varphi P_+ \psi f + JP_-\varphi P_-\psi f = JP_-\varphi\psi f = \Gamma_{\varphi\psi}$.

(b) (Brown and Halmos, 1964)

(1) *Product of two Toeplitz operators: $T_\varphi T_\psi$ is a Toeplitz operator if and only if $\bar{\varphi} \in H^\infty$ or (non-exclusively) $\psi \in H^\infty$; if this is the case, then $T_\varphi T_\psi = T_{\varphi\psi}$.*

SOLUTION: Denote $a_k = \widehat{\varphi}(k)$, $b_j = \widehat{\psi}(j)$ ($k, j \in \mathbb{Z}$). For the matrix elements of the product $(T_\varphi T_\psi)_{ij} = c_{ij} = \sum_{k \geq 0} a_{i-k} b_{k-j}$, we obtain

$$c_{i+1,j+1} = \sum_{k \geq 0} a_{i+1-k} b_{k-j-1} = a_{i+1} b_{-j-1} + \sum_{k \geq 1} a_{i+1-k} b_{k-j-1} = a_{i+1} b_{-j-1} + c_{ij}.$$

If $T_\varphi T_\psi$ is Toeplitz, then $a_{i+1} b_{-j-1} = 0$ for every $i, j \geq 0$, and hence either $a_{i+1} = 0$ for every $i \geq 0$ or $b_{-j-1} = 0$ for every $j \geq 0$, and the result follows. The converse (and the formula $T_\varphi T_\psi = T_{\varphi\psi}$) is in Lemma 2.2.9(1).

(2) T_φ^{-1} *is again a Toeplitz operator if and only if $\bar{\varphi} \in H^\infty$ or $\varphi \in H^\infty$ (and then $1/\varphi \in L^\infty(\mathbb{T})$).*

SOLUTION: By (1), if $T_\varphi T_\psi = \mathrm{id} = T_1$, we have $\psi = 1/\varphi \in L^\infty$, and the rest is evident.

(3) $T_\varphi T_\psi = 0$ *if and only if $\varphi = 0$ or $\psi = 0$.*

SOLUTION: By (1), $\varphi\psi = 0$ and we have $\bar{\varphi} \in H^\infty$ or $\psi \in H^\infty$. If $\varphi \neq 0$, ψ vanishes on a set of positive m measure, hence $\psi = 0$ (see Appendix F).

(4) $\Gamma_\varphi \Gamma_\psi = 0$ *if and only if $\Gamma_\varphi = 0$ (hence $\varphi \in H^\infty$) or (non-exclusively) $\Gamma_\psi = 0$ (hence $\psi \in H^\infty$).*

SOLUTION: By Lemma 2.2.9(1), $\Gamma_\varphi \Gamma_\psi = T_{\widetilde{\varphi}\psi} - T_{\widetilde{\varphi}} T_\psi$ where $\widetilde{\varphi} = \varphi(\bar{z})$, and then by (1), either $\varphi \in H^\infty$, or $\psi \in H^\infty$, and the result follows.

(5) *Commuting Toeplitz operators:* $[T_\varphi, T_\psi] =: T_\varphi T_\psi - T_\psi T_\varphi = 0$ *if and only if* (α) $\varphi, \psi \in H^\infty$, *or* ($\beta$) $\overline{\varphi}, \overline{\psi} \in H^\infty$, *or* ($\gamma$) φ, ψ *are linear functions of each other* ($\exists A \neq 0, B \neq 0: A\varphi + B\psi + C = 0$).

SOLUTION: With the notation $(T_\varphi T_\psi)_{ij} = c_{ij}$ of the preceding solution, a simple computation gives $c_{i+1,j+1} = a_{i+1}b_{-j-1} + c_{ij}$, and hence if $T_\varphi T_\psi = T_\psi T_\varphi$, then

$$a_{i+1}b_{-j-1} = b_{i+1}a_{-j-1} \quad \text{for every } i, j \geq 0.$$

Case (α) is equivalent to $a_{-j-1} = b_{-j-1} = 0$, $\forall j \geq 0$, and case (β) is equivalent to $a_{i+1} = b_{i+1} = 0$, $\forall i \geq 0$. If neither (α) nor (β) are satisfied, then there exist $J \geq 0$ and $I \geq 0$ such that $a_{I+1} \neq 0$, $b_{-J-1} \neq 0$ and $0 \neq b_{-J-1}/a_{-J-1} = b_{I+1}/a_{I+1} =: \lambda$, which implies (with the aid of the same equation) $a_{i+1}\lambda = b_{i+1}$ for every i, and then $b_{-j-1} = \lambda a_{-j-1}$ for every $j \geq 0$. Consequently, $\psi - b_0 = \lambda(\varphi - a_0)$, hence ($\gamma$) and the result.

The sufficiency of (α)–(γ) is evident.

(6) *Normal Toeplitz operators:* T_φ *is normal if and only if it is a linear function of a self-adjoint operator (hence, if and only if* $\varphi = Ah + B$ *where h is a real-valued function and* $A, B \in \mathbb{C}$).

SOLUTION: Let $\varphi = a + ib$ where a, b are real. Then, $T_a = (T_\varphi + T_\varphi^*)/2$, $T_b = (T_\varphi - T_\varphi^*)/2i$ and T_φ is normal if and only if T_a commutes with T_b. As a real function is in H^∞ if and only if it is constant, we obtain the result from case (γ) of (5).

(c)* *Other results.*

(1) (Power, 1984) ($\Gamma_\varphi^n = 0$ *for some* $n \geq 1$) $\Leftrightarrow \Gamma_\varphi = 0 \Leftrightarrow \varphi \in H^\infty$.

(2) (Aleman and Vukotic, 2009) *If* $T_{\varphi_1} T_{\varphi_2} \ldots T_{\varphi_n} = 0$, *then there exists j such that* $\varphi_j = 0$.

(3) rank$(\Gamma_\varphi \Gamma_\psi) < \infty$ *if and only if* rank$(\Gamma_\varphi) < \infty$ *or (non-exclusively)* rank$(\Gamma_\psi) < \infty$ *(Axler, Chang, and Sarason, 1978).*

Moreover [Nikolski, 1986],

$$\text{rank}(\Gamma_\varphi \Gamma_\psi) = \min\{\text{rank}(\Gamma_\varphi), \text{rank}(\Gamma_\psi)\}.$$

See §2.4 for precise references.

(d) (Janson, 1985 (see Notes and Remarks)) $H_\varphi^* H_\varphi = H_\psi^* H_\psi \Leftrightarrow P_-\varphi = \epsilon P_-\psi$ *where* $|\epsilon| = 1$ ($\Leftrightarrow H_\varphi = \epsilon H_\psi$).

SOLUTION: The property is clear if $P_-\varphi = 0$. Let $m \geq 1$, $l \geq 1$ such that $P_-\varphi = \sum_{k \geq m} a_k \bar{z}^k$, $a_m \neq 0$ and $P_-\psi = \sum_{k \geq l} b_k \bar{z}^k$, $b_l \neq 0$. For every $n, k \geq 0$, we have $(H_\varphi z^n, H_\varphi z^k) = (H_\psi z^n, H_\psi z^k)$, and by Lemma 2.2.1 and Example 2.2.2 $(H_\varphi 1, \bar{z}^n P_- z^k H_\varphi 1) = (H_\psi 1, \bar{z}^n P_- z^k H_\psi 1)$, hence

$(H_\varphi 1, \bar{z}^n(I - \bar{z}^k P_- z^k)H_\varphi 1) = (H_\psi 1, \bar{z}^n(I - \bar{z}^k P_- z^k)H_\psi 1)$ $(\forall n, k \geq 0)$.

For $0 \leq k < m$ we have $(I - \bar{z}^k P_- z^k)H_\varphi 1 = 0$, and $(I - \bar{z}^m P_- z^m)H_\varphi 1 = a_m \bar{z}^m$. Thus, after a similar computation for $H_\psi 1$, it follows that $l = m$ and $|a_m|^2 = |b_m|^2$. Applying the same equation for $k = m$, $n = 0, 1, \ldots$, we obtain $\bar{a}_m(H_\varphi 1, \bar{z}^{n+m}) = \bar{b}_m(H_\psi 1, \bar{z}^{n+m})$ ($\forall n \geq 0$), and the result follows.

2.4 Notes and Remarks

The central result of this chapter is the discovery of the symbols of Toeplitz (c_{i-j}) and Hankel (c_{i+j}) matrices (operators), i.e. their representation in the form $T_\varphi = P_+ M_\varphi P_+$ and $\Gamma_\varphi = J(P_- M_\varphi P_+)$, respectively. The Toeplitz symbol is uniquely defined, and isometric: $\|T_\varphi\| = \|\varphi\|_\infty$. However, for a Hankel symbol, it is unique up to addition by a function of H^∞, with a "quotient isometry" $\|\Gamma_\varphi\| = \inf_{\psi \in H^\infty} \|\varphi - \psi\|_\infty$. From the last formula, we can suspect that there are extensive applications of Hankel operators to extremal problems: interpolation, H^∞ control theory, the prediction of random processes, etc.; see [Nikolski, 2002a,b], [Peller, 2003].

The concept of *symbol of an operator* (and the word itself) was introduced by Solomon Mikhlin (or Mihlin: see his biography) in the 1930s in his study of singular integral operators $T = M_a \mathbb{H} M_b$ on the real axis $\mathbb{R}$ (where $\mathbb{H}$ is the *Hilbert transform* (continuous variable version) which is a convolution with the Cauchy kernel $1/x$).

Solomon Mikhlin (in Russian Соломон Григорьевич Михлин, 1908–1990) was a Russian mathematician (in the period of the USSR), Professor at the University of Leningrad (St. Petersburg), and one of the pioneers of the mathematical theory of linear elasticity, of singular integrals, and of numerical analysis. In the latter domain, Mikhlin tackled such a large spectrum of problems that today he is often considered one of the founders of modern numerical analysis, along with Boris Galerkin, Alexander Ostrowski, John von Neumann, Walter Ritz, and Mauro Picone. In pure mathematics, Mikhlin's most well-known innovation is, of course, his conception of

2.4 Notes and Remarks

the "symbol of a singular integral operator" (1936), a notion that was later extensively applied to much more general objects (in particular, to Toeplitz operators: see Theorem 2.1.5) and finally resulted in the creation of the theory of pseudo-differential operators. The "Mikhlin theorem of Fourier multipliers" takes its place on the "Great Podium" of the three best-known multiplier theorems – those of Marcinkiewicz, Mikhlin, and Hörmander.

Mikhlin was one of the very few to have escaped the system of anti-Jewish obstruction of the Stalin (and post-Stalin) Soviet regime, and succeeded in a brilliant classical career: he received his Master's degree at the University of Leningrad in 1929 (under the direction of Vladimir Smirnov, one of the key figures of complex analysis between the two World Wars), obtained the grade of *docteur ès sciences* (1935) and a post as professor (1937) at the Institute of Seismology, and then at the University of Leningrad (St. Petersburg) (1944), where he also founded and directed the Laboratory of Numerical Analysis. Like most intellectuals, Mikhlin was morally opposed to the regime, and always kept his distance, claiming (according to Vladimir Maz'ya) "They just have power, but we have theorems. Here is our strength!" Mikhlin wrote half a dozen monographs (now classics) and was elected member of the Academie Leopoldina (Germany, 1970) and of the Accademia Nazionale dei Lincei (Italy, 1981). More complete and diverse information can be found on the websites http://en.wikipedia.org/wiki/Solomon_Mikhlin and http://dm47.com/Sbornik_iimm_mihlin.html (very interesting historical and autobiographical anecdotes of Mikhlin himself, in Russian).

His survey article "Singular integral equations" [Mikhlin, 1948] (a landmark for the subject for the next several decades) already contains a well-developed "theory of symbols," as well as references to previous works, including those from the beginning of the twentieth century, by Erik Ivar Fredholm, Henri Poincaré, Fritz Noether, and Torsten Carleman. For some comments on this classical part of the evolution of the theory (linked especially to the notion of index of an operator, which itself represents a chapter in today's spectral theory – non-existent at the time), see the Notes and Remarks of Appendix E. We recall here only a single notable predecessor to Mikhlin's calculus, namely the *Poincaré–Bertrand formula* on the superpositions of singular integrals (see §4.6, the comment to Corollary 4.1.4) which *de facto* allowed working with the commutators, semi-commutators, and symbols (without defining them) of SIO and which continues to play an important role in many applications.

Toeplitz [1911a] constructed the spectral theory of Laurent matrices $L = (c_{i-j})_{i,j\in\mathbb{Z}}$ on the space $\ell^2(\mathbb{Z})$ in terms of the symbol $\varphi = \sum_{j\in\mathbb{Z}} c_j e^{ijx}$ (without using this word, of course). The principal result states that, if φ is an analytic function of $x \in \mathbb{R}$ (i.e. $|c_j| \le c\, e^{-\epsilon|j|}$ with appropriate $\epsilon > 0$ and $c > 0$), then the spectrum is a curve:

$$\sigma(L) = \varphi([0, 2\pi]).$$

It should be noted (yet again) that in [Toeplitz, 1911a] there are no Toeplitz operators T_φ in the sense given today, but only those of Laurent L_φ. A big difference between these two types of operators is that the symbolic calculus (especially the "quantization") of Laurent $\varphi \mapsto L_\varphi$ is a homomorphism (L_φ and L_ψ commute for all φ, ψ, and are thus *normal matrices*), whereas that of Toeplitz $\varphi \mapsto T_\varphi$ is not. Incidentally, it seems that Toeplitz was strongly attached to this commutativity, to the point that even when he examined finite truncations $(c_{i-j})_{k\le i, j \le l}$ of Laurent or Toeplitz matrices (these are the same for both!), where in general the commutativity is lost, he limited himself to *circulants* (see §5.5 below), which commute among themselves.

For a contemporary and complete description of the abstract theory of symbols see [Krupnik, 1984]. The symbolic calculus has become particularly effective and important for differential and pseudo-differential operators, especially for problems arising in mathematical physics where the symbol plays the role of a classic observable and the operator itself of a quantum observable (and the passage $\varphi \mapsto A_\varphi$ provides the "quantization"): see [Berezin, 1989], [Shubin, 1987], [Boutet de Monvel and Guillemin, 1981], and [Hörmander, 1985].

See also the Notes and Remarks of Chapter 3.

A second aspect of the theory of Ha-plitz operators, which the reader may already have noticed, is that the three equivalent definitions of Toeplitz and Hankel are all different in nature and, in principle, lead to different generalizations. Here are a few observations.

(1) *The analytic form of the definition*, $T = T_\varphi$ for Toeplitz and $\Gamma = JH_\varphi$ for Hankel, can be generalized in other functional Hilbert spaces to the study of operators of the form PMP (Toeplitz) and PJMP ("small Hankel") where J is an involution similar to those on the unit circle $f \to f(\bar z)$ and $P = P_+$ is the orthogonal projection onto the "analytic part" of the space in question, while M is a multiplication operator. Another frequent form of generalization is $(I - P_+)MP_+$ ("big Hankel"). In most cases, no matrices of Toeplitz (or of Hankel) appear. Typical examples are the *Bergman spaces*,

$$L^2_a(\mathbb{D}) = \mathrm{Hol}(\mathbb{D}) \cap L^2(\mathbb{D}, dx\,dy),$$

and *Dirichlet spaces*,

$$\mathcal{D} = \{f : f' \in L^2_a(\mathbb{D})\},$$

as well as their weighted versions (frequently with the weight $(1 - |z|)^\alpha$). Clearly the sense of the words "Toeplitz operator" varies from one space to another, and similarly for spaces of functions of several variables or on differential manifolds; see the sources already cited ([Shubin, 1987]; [Boutet de Monvel and Guillemin, 1981]; [Hörmander, 1985]) as well as [Zhu, 2007] and [Hedenmalm, Korenblum, and Zhu, 2000]. Outside of realizations in the spaces of analytic functions the results are few, but curious [Devinatz and Shinbrot, 1969]. For the Hankel operators the situation is even more confused, as the harmonic analysis of the past decade systematically uses bilinear forms, called "Hankel forms,"

$$\{f, g\} \to (fg, \varphi)_H,$$

depending on the product of arguments f and g ($f, g \in H$, H is not an algebra!). However – take care! – if the space H is not an L^2 space (perhaps weighted), this form is very different from the bilinear form of a small Hankel operator $P_+ J M_{\bar\varphi} P_+$, $\{f, g\} \to (f\bar\varphi, Jg)_H$. See also the comments on Theorem 2.2.4 below.

(2) *The operator identities* $S^*TS = T$ *for Toeplitz, and* $S^*\Gamma = \Gamma S$ *for Hankel* have also given rise to several generalizations; see for example [Douglas, 1969a] or [Barria and Halmos, 1982], who developed a theory of "asymptotic Toeplitz operators," i.e. such that, for the strong operator topology, the limit $\lim_n S^{*n}TS^n$ exists (which is a standard Toeplitz operator). The set of such T is a vast operator algebra, where the symbol can be correctly defined, and contains the algebra generated by all the Toeplitz and Hankel operators (see Definitions 3.1.2 for alg $\mathcal{T}_{L^\infty(\mathbb{T})}$). In general, the operator algebras (C^* or von Neumann) are privileged locations for the study of operator equations.

(3) *The Toeplitz matrices on spaces different from* $\ell^2(\mathbb{Z}_+)$ *and* H^2 are important for harmonic analysis: for example, the same $T_\varphi = P_+ M_\varphi$ (P_+ is the same Riesz projection as before) on weighted spaces $L^2(\mathbb{T}, w \cdot m)$, where $w \in L^1(\mathbb{T})$ and where the exponentials e^{ikx} do not necessarily form an orthogonal basis; see [Zhu, 2007], [Nikolski, 2002a], [Nikolski, 2002b], and [Hedenmalm, Korenblum, and Zhu, 2000] for a few results in this direction. The operators having a *Hankel matrix* with respect to some *different orthonormal bases* are in the same line of ideas, and almost every operator on Hilbert spaces can be found in this way – in any case, every self-adjoint operator with simple spectrum (without multiplicity) and trivial kernel (see [Nikolski, 2002a] for the exact statement), along with many others.

In short, the Ha-plitz operators and related techniques are inseparable from the major problems of modern harmonic analysis, and are strongly involved in the theory of singular integrals, the theory of the spaces BMO, H^p, H^1_{Re}, VMO, and others, without mentioning numerous applied theories.

Formal credits, subject by subject.

§2.1.3. Matrices versus operators. The theory of operators arose from the theory of matrices and of bilinear and quadratic forms. Up to the beginning of the 1950s, specialists preferred to speak of the "spectral theory of infinite matrices"; the founding article on Toeplitz matrices [Toeplitz, 1911a] was no exception. The understanding that the essence of an operator is the mapping, or action, itself, and not a particular analytic representation (matrix, integral, composition, or any other), came much later.

§2.1.5. Toeplitz operators. We have already observed that Toeplitz (in [Toeplitz, 1911a] as well as in other publications) did not study "Toeplitz operators" but "Laurent operators." Today this designation is in any case ubiquitous, and we follow it. But the history and evolution of this aberration in terminology are curious, and the cause is hidden in the seminal work of Toeplitz [1911a]: Toeplitz worked with *bilinear forms*, and at the level of forms there is no difference between "Laurent" and "Toeplitz." However, the mappings (operators) $M_\varphi = L_\varphi$ and $T_\varphi = P_+ M_\varphi P_+$ are very different! Thus the two forms

$$\sum_{i,j\in\mathbb{Z}} c_{i-j} x_i \bar{x}_j \quad \text{and} \quad \sum_{i,j\geq 0} c_{i-j} x_i \bar{x}_j$$

were called "Toeplitz forms." The same then happened with the determinants $\det(c_{i-j})_{-n\leq i,j\leq n}$ and $\det(c_{i-j})_{0\leq i,j\leq n}$ (one of the first, very influential, publications of this type was that of Szegő [1915] with the very first asymptotic formula for Toeplitz determinants). When the time came to study the mappings defined by the matrices $(c_{i-j})_{i,j\geq 0}$, these were also associated with Toeplitz's name. One of the first articles where they are denoted "Toeplitz operators" is [Hartman and Wintner, 1950a] (nonetheless with traces of the alternative "matrices/operators").

Lemma 2.1.2. The functional equation. The equation $T = S^* T S$ characterizes the Toeplitz operators (if it concerns the set of indices $\mathbb{Z}_+$) but also those of Laurent (on the set of indices $\mathbb{Z}$; S^* is the backward shift on $\mathbb{Z}$). The necessity of the equation was already recognized in [Toeplitz, 1911a] (see note at the bottom of page 355), but it seems that the sufficiency, as well as the comprehension of the importance of this equation in general, had to wait for 50 years, until Brown and Halmos [1963].

Paul Halmos (1916–2006), an American mathematician, was born in Hungary and only moved to the United States at the age of 13. According to witnesses, he always spoke with a strong Hungarian accent, but this did not stop him becoming famous for a long series of mathematical books and survey articles, irreproachable for their linguistic and mathematical style. These include the well-known *How to Write Mathematics* (1970) and *How to Talk Mathematics* (1974). He even won the 1983 Steele Prize for Exposition. His professional career was scattered across a dozen American universities, starting with the famous Princeton Institute for Advanced Study, where he showed up in 1938 without having a position, and spontaneously proposed an extra-curricular course entitled "Elementary Theory of Matrices." After its success he was offered a teaching position, and was even named mathematical secretary for John von Neumann, member of the IAS. As researcher, Halmos left his mark in logic, measure theory, operators, and ergodic theory. Among other topics, in operator theory he introduced a number of useful concepts (or even research themes) such as "unitary dilation," "subnormal operator," and "quasi-triangular operator." He had a particularly luminous and fluid style of mathematical expression, capturing the reader with his enthusiasm and fascination for the problem, and communicating the irresistible desire to attempt to solve it. His paradigm of mathematics was especially based on aesthetic criteria, which led, at its best, to texts that long served as beacons for the subject, such as the famous "Ten problems in Hilbert space" (*Bull. Amer. Math. Soc.* **76** (1970)). Less happily, the apology of aesthetics led him to provocative concepts such as "Applied mathematics is bad mathematics" (in *Mathematics Tomorrow*, Springer, 1981). Finally, Halmos wrote about himself in *I Want to be a Mathematician: An Automathography* (Springer, 1985): "I was, in I think decreasing order of quality, a writer, an editor, a teacher, and a research mathematician" (don't forget his very high standards of quality).

Lemma 2.1.4. The equalities $\|M_\varphi\| = \|T_\varphi\| = \|\varphi\|_\infty$ appear already in [Toeplitz, 1911a].

Theorem 2.1.5. The existence of the symbol dates back to Mikhlin [1948] and is discussed at the beginning of these Notes and Remarks. The proof in the text and the idea of the asymptotic equation $T_\varphi P_+(z^n f) = z^n \varphi f + o(1)$ (Corollary 2.1.6) is taken from [Brown and Halmos, 1963]. The proof can easily be adapted to the case of vector-valued Toeplitz operators, either abstractly, using the equation $V^*TV = T$ in a Hilbert space where V is a pure isometry of arbitrary (including infinite) multiplicity dim ker(V^*) (see Appendix D), or directly for an operator $T: H^2(E) \to H^2(E)$, where E is an auxiliary Hilbert space and $H^2(E)$ the Hardy space of functions with values in E, and $S_E^* T S_E = T$ where $S_E f = zf$. We thus obtain $T = T_\varphi = P_+ M_\varphi$ where $\varphi \in L^\infty(\mathbb{T}, L(H))$, the space L^∞ with values in $L(H)$. For details, with precise definitions, see [Rosenblum and Rovnyak, 1985] or [Nikolski, 2002a].

Example 2.1.7(2). Wiener, and subsequently Wiener and Hopf, worked mainly on the half-line $\mathbb{R}_+$ with integral operators: see the comments in §1.3.2 and in Chapter 4. The *Hilbert transform* appeared in his famous course of 1907–1908 on integral equations [Hilbert, 1912], but in the continuous version

$$\text{p.v.} \int \frac{f(y)}{x-y}\, dy$$

and with an estimate of the norm $\leq 2\pi$ (later improved to $\leq \pi$ by Hardy, Littlewood, and Pólya [1934]).

§2.2.1 and Lemma 2.2.1. Hankel operators. As already seen in Chapter 1, the destiny of Hankel operators is even stranger than that of their Toeplitz analogs. In the form of matrices, they appeared in Hankel's thesis (1861), and then resurfaced in several extremal problems and moment problems (Chebyshev, Stieltjes, Szegő) without playing a predominant role. From a modern point of view, there is nothing surprising about such "negligence": as long as the interest in Hankel matrices remained limited to the case of finite matrices and was reduced chiefly to the calculation of determinants (which is always the case for the theory of moments or for orthogonal polynomials: see for example [Szegő, 1959]), there is no need for a particular theory of Hankel matrices, since the link between Toeplitz T_n and Hankel H_n is performed by a unitary symplectic matrix $J_n = (\delta_{i,n-j})$: $T_n = J_n H_n$. However, with the appearance of new problems, arising especially in control theory and the prediction of random processes, we needed to know much more.

The new life of "Hankel matrices/operators" began with Nehari's Theorem 2.2.4. Note also (as the reader has perhaps already noticed in Chapter 1) that the Hankel operators almost never appear in some kind of equation (in contrast to the Toeplitz operators), and hence are not very interesting for their "dynamics," their spectrum, etc. However, the "metrical properties" of a Hankel operator H, i.e. the spectral properties of its *Hermitian square* H^*H, are critical for a number of problems in analysis and their application (see [Nikolski, 2002a,b], [Peller, 2003]).

Theorem 2.2.4. Nehari's theorem: the existence of the symbol. The theorem was proved by Nehari [1957]. An explanation for the passage by the space $C_a(\mathbb{D})$ in the proof: we cannot shorten the reasoning by limiting ourselves to polynomials $f, g \in \mathcal{P}_a$ because we do not know how to factorize a polynomial $F \in \mathcal{P}_a$ into

$$F = F_1 F_2, \quad \text{where } F_1, F_2 \in \mathcal{P}_a \quad \text{and} \quad \|F\|_1 = \|F_1\|_2 \|F_2\|_2.$$

The absence of such a factorization in several variables (hence, in the polydisk $\mathbb{D}^n$ and the unit ball $\mathbb{B}^n$ of $\mathbb{C}^n$) is one of the principal difficulties in finding a good version of Nehari's theorem in $\mathbb{C}^n$. The role of the factorization is played by a "weak factorization," where every function $F \in H^1$ is the sum of a series $F = \sum_{k \geq 1} F_k G_k$ satisfying $\sum_{k \geq 1} \|F_k\|_{H^2} \|G_k\|_{H^2} < \infty$, and $A(F) =:$ inf of these last sums over all such representations of F is an equivalent norm on the space H^1: $\|F\|_1 \leq A(F) \leq a_n \|F\|_1$. For the case of the ball $H^2 = H^2(\mathbb{B}^n)$, the weak factorization is shown in [Coifman, Rochberg, and Weiss, 1976], and for the polydisk $H^2 = H^2(\mathbb{D}^n)$ in [Ferguson and Lacey, 2002] ($n = 2$) and [Lacey and Terwilleger, 2009] ($n > 2$). In the same publications, these remarkable results were used to obtain a complete analog of Theorem 2.2.4: a symbol $\varphi \in H^2$ generates a *bounded Hankel form* (which in this case is also the form of a "small" Hankel operator)

$$\{f, g\} \to (\overline{\varphi} f, Jg)_{L^2} =: (H_{\overline{\varphi}} f, g)_{H^2}$$

on the space H^2 (where $Jg(\zeta) = g(\overline{\zeta})$) if and only if there exists $\psi \in L^\infty$ such that $P_+ \psi = \varphi$ (P_+ is the orthogonal projection of L^2 onto H^2); and then

$$\|H_{\overline{\varphi}}\| \leq \inf \|\psi\|_\infty \leq b_n \|H_{\overline{\varphi}}\|.$$

Nehari showed that for $n = 1$, the constants are $a_n = b_n = 1$, but in general, on the polydisk $\mathbb{D}^n$, we have $a_n \geq b_n \geq (\pi^2/8)^{n/4}$ (proved in [Ortega-Cerdà and Seip, 2012], based on a posthumous publication of Helson [2010]).

This weak factorization is linked, and even equivalent, to the description of bounded "Hankel forms," in the sense defined previously, on the *Dirichlet space* $\mathcal{D}$ above. First of all, the "true" Hankel operators were described.

More precisely, Rochberg and Wu [1993] showed that a small Hankel operator $P_+JM_{\bar\varphi}P_+$ with co-analytic symbol, $\varphi \in X_\alpha$, is continuous on $\mathcal{D}$ if and only if $\mu_\varphi = |\varphi'(z)|^2\,dz\,d\bar z$ is a *Carleson measure for $\mathcal{D}$*, i.e. we have the embedding $\mathcal{D} \subset L^2(\mathbb{D}, \mu_\varphi)$. These $\mathcal{D}$-Carleson measures are characterized by a condition of type $\mu_\varphi(\Delta) \leq c \cdot \mathrm{Cap}(\Delta \cap \mathbb{T})$ where Cap is the Riesz–Bessel capacity of order $1/2$, associated with the kernel $|z - \zeta|^{-1/2}$ on $\mathbb{T}$ and Δ runs over the finite unions of "Carleson squares (or windows)" (Cap(E) is comparable to $\log(1/\gamma(E))$ where $\gamma(E)$ is the logarithmic capacity, or the "transfinite diameter" of E: see for example [Landkof, 1972], [Adams and Hedberg, 1996]). See also [Wu, 1998] for a survey article. Twenty years later, and with much more advanced techniques, an Italian–American team [Arcozzi, Rochberg, Sawyer, and Wick, 2010] proved the same theorem for the "Hankel forms" in $\mathcal{D}$: the upper bound of a form $(fg, \varphi)_\mathcal{D}$ (for $\|f\|_\mathcal{D} \leq 1$, $\|g\|_\mathcal{D} \leq 1$) is comparable to the embedding norm of $\mathcal{D} \subset L^2(\mathbb{D}, \mu_\varphi)$. See also the survey article by Arcozzi, Rochberg, Sawyer, and Wick [2011]. It should be noted that in 2002 a similar result was obtained, but with different techniques, by Maz'ya and Verbitsky [2002] in the context of the spectral theory of the Schrödinger operator: if

$$\mathcal{D} = L^{2,1}(\mathbb{R}^n) = \{f\colon \nabla f \in L^2(\mathbb{R}^n)\} \quad \text{and} \quad (f,g)_\mathcal{D} = \int \nabla f \overline{\nabla g}\,dx,$$

then for every φ the $\sup\{|(fg, \varphi)_\mathcal{D}|\colon \|f\|_\mathcal{D} \leq 1, \|g\|_\mathcal{D} \leq 1\}$ is comparable to the embedding norm

$$L^{2,1}(\mathbb{R}^n) \subset L^2(\mathbb{R}^n, |(-\Delta)^{1/2}\varphi|^2\,dx).$$

We conclude this comment on the Dirichlet space $\mathcal{D}$ with a comparison to Nehari's Theorem 2.2.4. On one hand, the Hardy space H^2 can be written as a weighted Dirichlet space

$$H^2 = \{f\colon f' \in L^2_a(\mathbb{D}, (1-|z|^2)\,dz\,d\bar z)\},$$

and on the other, according to Fefferman and Stein [1972] (see also, for example, [Koosis, 1980], [Garnett, 1981]), a holomorphic function φ is in BMOA if and only if

$$\mu_{\varphi,1} =: |\varphi'(z)|^2(1 - |z|^2)\,dz\,d\bar z$$

is a *Carleson measure* (i.e. $H^2 \subset L^2_a(\mathbb{D}, \mu_{\varphi,1})$). Hence, we can read Nehari's theorem as "$H_{\bar\varphi}$ is bounded in H^2 if and only if $\mu_{\varphi,1}$ is a Carleson measure" (the latter are characterized by the condition $\mu_{\varphi,1}(\Delta) \leq c \cdot m(\Delta \cap \mathbb{T})$ for every

"Carleson square Δ"). The same factorization $H^1 = H^2 \cdot H^2$ (and again the duality $(H_0^1(\mathfrak{S}_1))^* = L^\infty(\mathbb{T}, L(H))/H^\infty(L(H))$) is a new obstacle (surmountable, but painfully) for a generalization to the case of matrix and operator values (such a generalization is not whimsical curiosity, it is exactly what is necessary for applications to control theory: see [Nikolski, 2002a,b] and [Peller, 2003]).

There exists another proof of Theorem 2.2.4, exempt from the difficulties indicated, based on a "step-by-step extension algorithm" of Adamyan, Arov, and Krein [1968, 1971] (sometimes abbreviated AAK). The algorithm relies on the following "four-block lemma."

Let H_i, K_i ($i = 1, 2$) be four Hilbert spaces and let $T_{ij}: H_j \to K_i$, $(i, j) \neq (1, 1)$ be three bounded operators. Define $M_X: H_1 \oplus H_2 \to K_1 \oplus K_2$ as a matrix with four blocks,

$$M_X = (T_{ij})_{1 \leq i, j \leq 2} = \begin{pmatrix} X & T_{12} \\ T_{21} & T_{22} \end{pmatrix},$$

where $T_{11} = X: H_1 \to K_1$. Then

$$\min_X \|M_X\| = \max\{\|(T_{12}, T_{22})^{\text{col}}\|, \|(T_{21}, T_{22})\|\}.$$

For the proof see Lemma 5.2.3 as well as [Adamyan, Arov, and Krein, 1968, 1971], [Parrott, 1978], and [Nikolski, 2002a], and also the comments after the AAK algorithm below.

The Adamyan–Arov–Krein algorithm. Let $\Gamma = (c(i + j))_{i,j \in \mathbb{Z}_+}$ be a bounded Hankel operator (matrix) and let $\ell_n^2(\mathbb{Z})$, $n = 0, 1, 2, \ldots$, be a nested chain of subspaces of $\ell^2(\mathbb{Z})$,

$$\ell_n^2(\mathbb{Z}) = \{x \in \ell^2(\mathbb{Z}): x_k = 0, k < -n\}, \quad n = 0, 1, 2, \ldots$$

The algorithm consists of extending the operator Γ by induction to operators $\Gamma_n = (c(i + j))_{i, j \geq -n}$ on $\ell_n^2(\mathbb{Z})$ in such a way that

$$P_n \Gamma_{n+1} P_n = \Gamma_n, \quad \|\Gamma_{n+1}\| = \|\Gamma\| \quad \text{and} \quad S^* \Gamma_{n+1} = \Gamma_{n+1} S \quad \text{(in a natural sense)},$$

where P_n is the orthogonal projection onto $\ell_n^2(\mathbb{Z})$.

(1) *The extension.* Step 1. If Γ_n is already defined, we first enlarge its *target space*, i.e. we look for a mapping

$$\Gamma'_{n+1}: \ell_n^2(\mathbb{Z}) \to \ell_{n+1}^2(\mathbb{Z})$$

verifying the equations $P_n \Gamma'_{n+1} = \Gamma_n$ and $S^* \Gamma'_{n+1} = \Gamma'_{n+1} S$. These equations leave only one single matrix element of Γ'_{n+1} to be defined freely, $\gamma_{-n,-n-1} =: x$.

$X = ?$	$c(-2n)$	$c(-2n+1)$	$c(-2n+2)$	$c(-2n+3)$	$\cdots$
$c(-2n)$	$c(-2n+1)$	$c(-2n+2)$	$c(-2n+3)$	$\cdot\cdot\cdot$	
$c(-2n+1)$	$c(-2n+2)$	$c(-2n+3)$	$\cdot\cdot\cdot$		
$c(-2n+2)$	$c(-2n+3)$	$\cdot\cdot\cdot$			
$c(-2n+3)$	$\cdot\cdot\cdot$				
$\vdots$					

Hence Γ'_{n+1} is a matrix M_X with four blocks with a block $X = x$ of dimension 1 (multiplication by x in $\mathbb{C}$) and $(T_{12}, T_{22})^{\mathrm{col}} = \Gamma'_{n+1} P_{n-1}$, $(T_{21}, T_{22}) = P_n \Gamma'_{n+1}$. However, the matrices $\Gamma'_{n+1} P_{n-1}$ and $P_n \Gamma'_{n+1}$ coincide with Γ_n, and hence by the four-block lemma there exists $x =: c(-2n-1) \in \mathbb{C}$ such that $\|\Gamma'_{n+1}\| = \|\Gamma'_{n+1} P_{n-1}\| = \|P_n \Gamma'_{n+1}\| = \|\Gamma_n\| = \|\Gamma\|$.

Step 2. Given Γ'_{n+1}, we enlarge the *origin space*, i.e. we seek $\Gamma_{n+1} : \ell^2_{n+1}(\mathbb{Z}) \to \ell^2_{n+1}(\mathbb{Z})$ verifying the equations $\Gamma_{n+1} P_n = \Gamma'_{n+1}$ and $S^* \Gamma_{n+1} = \Gamma_{n+1} S$. The same reasoning as in step 1 provides the desired Γ_{n+1}. ∎

(2) The passage to the limit and the proof of Theorem 2.2.4 do not present any problems, since for sequences x, y of finite support, the bilinear form of the limit $(\Gamma_\infty x, y)$ coincides with $(\Gamma_n x, y)$ for an n large enough. The limit Γ_∞ satisfies $\|\Gamma_\infty\| = \|\Gamma\|$ and $S^* \Gamma_\infty = \Gamma_\infty S$, where S^*, S are the bilateral shifts. Hence, $\Gamma_\infty = (c(i+j))_{i,j \in \mathbb{Z}}$ and $(\Gamma_\infty x, y) = (\Gamma x, y)$ for every $x, y \in \ell^2(\mathbb{Z}_+) = \ell^2_0(\mathbb{Z})$. By Lemma 2.1.2 and Lemma 2.1.4 there exists $\varphi \in L^\infty$ such that $\widehat{\varphi}(k) = c(k)$ ($\forall k \in \mathbb{Z}$) and $\|\varphi\|_\infty = \|\Gamma\|$. This completes the proof of Nehari's theorem. ∎

A great advantage of this proof is that it hardly depends at all on the multiplicity and works equally well for Hankel operators on vector-valued spaces, in particular for the non-square "rectangular" case $\mathcal{H}\colon \ell^2(\mathbb{Z}_+, E_1) \to \ell^2(\mathbb{Z}_+, E_2)$, where E_1, E_2 are two Hilbert spaces with perhaps different dimensions. The links between the matrix- and operator-valued Hankel operators and the space BMO are not as simple as in the scalar case: see the comment to Exercise 2.3.4 below. To finish with Theorem 2.2.4 and the approach of AAK, note that the idea of the "four-block lemma" dates back to Krein [1947a,b] (but formally appeared only in [Parrott, 1978]) and is based on a lemma of Julia [1944] stating that for every contraction between two Hilbert spaces $A\colon H \to K$ ($\|A\| \le 1$), the "Julia matrix"

$$U_A = \begin{pmatrix} -A & (I - AA^*)^{1/2} \\ (I - A^*A)^{1/2} & A^* \end{pmatrix}$$

defines a unitary mapping from $H \oplus K$ to $K \oplus H$. We present a proof of this last fact, as well as of the four-block lemma, in Lemmas 5.2.3 and 5.2.4 below (following [Nikolski, 2002a]).

Gaston Julia (1893–1978), a French mathematician, was a co-founder of complex dynamics (with Pierre Fatou), and an important figure in twentieth-century analysis, especially in complex analysis. His novel results on the iteration of rational mappings of the Riemann sphere (a theme proposed in 1915 for the Grand Prix de l'Académie des Sciences awarded in 1918; only three entries were received, those of Julia, Lattès, and Pincherle, and it was Julia who won the contest) were welcomed with great enthusiasm, but the mathematical weathervane had changed direction and his results on iterations were forgotten for half a century.

They became famous with the emergence of Benoit Mandelbrot's theory of fractals in *Les objets fractals: Forme, hasard et dimension* (1974) where the Julia (or Fatou–Julia) sets, similar to those of Mandelbrot, and their properties were the pillars of the new chaos theory. It is curious that the wave of media frenzy about fractals raised Julia sets to the pinnacle of universal popularity, to such a point that on February 3, 2004 the search engine Google published a pretty vignette, *Gaston Julia's 111th Birthday Google Doodle*!

A similar fate befell a long series of more than 40 short notes by Julia in the decade 1940–1950 on operators on Hilbert spaces, where he proposed various original techniques that were not properly understood at the time, but which later served as the basis of the Szőkefalvi-Nagy–Foias theory of dilatations, as well as the theory of optimal completions of matrices – a vast field encompassing the operators and their numerous applications, touching on, among other subjects, *Julia operators*, *sparse Julia matrices*, *Julia colligations*, etc. The first element of this theory is Julia's lemma (see also the comments concerning the AAK approach to Nehari's theorem, and also on Lemmas 5.2.3 and 5.2.4 and §5.8). Nonetheless, even at the time, some of the results of this series were already a bit obsolete and, according to Francis Murray, a close collaborator of John von Neumann, in his review for *Math. Reviews* (MR0018343) "It seems to the reviewer, also, that a reader might be led into believing that results obtained by other mathematicians are due to the author."

As was the destiny of his mathematical legacy, so was Julia's career spectacular and dramatic. He was admitted in first place to the École Normale Supérieure in 1911 (he was also first at the École Polytechnique) and in 1914 obtained his Bachelor's degree and passed the *agrégation* competition. During his studies in Paris, he also devoted much time to music.

The day after the declaration of war, Julia was called up by the army, and after a brief period of training, was sent to defend the plateau of the Chemin des Dames as a second lieutenant of the infantry. The first time he came under fire, on January 25, 1915, he was severely wounded in the face, but remained at his post until the attack was repulsed. Incapable of speaking, he gave orders by writing. The offensive was said to be particularly ferocious in that it was planned as a "birthday present" (January 27) for Kaiser Wilhelm II.

2.4 Notes and Remarks

For three long years Julia suffered several unsuccessful but painful operations (while occupying the "recesses" with his research activities), but finally, he was forced to wear a leather mask for the rest of his life. (As an aside, note that the "egalitarian" politics of recruitment at the time cost the French intellectual elite heavily: for example, 41% of the students recruited from the École Normale Supérieure lost their lives, compared with 16.8% of soldiers mobilized – a shocking proportion in its own right! – cited in Michèle Audin: *Fatou, Julia, Montel*, Springer, 2009, page 22.) In 1917 Julia submitted his thesis, greatly inspired by Charles Hermite (who died in 1901) and exceptionally long at 293 pages, before a jury composed of Émile Picard, Henri Lebesgue, and Pierre Humbert. Afterwards, he held different positions at the École Normale Supérieure, the École Polytechnique, and the Sorbonne (professor in 1925). Among the honors and recognitions he received were two nominations to courses at the Fondation Peccot of the Collège de France (1918 and 1920), his election to the Académie des Sciences (1934), titles of the Légion d'honneur (from *Chevalier* in 1915 to *Grand Officier* in 1950). But the students who took his class at the École Polytechnique in 1950s were merciless, and dubbed him "Mr. No Nose."

Gaston Julia married Marianne Chausson (daughter of the romantic composer Ernest Chausson, who died in 1899) and they had six children.

A few dissonant notes are nonetheless present in this extraordinary portrait, beginning with the famous competition of 1915–1918 where Julia had a strong competitor, Pierre Fatou, who had obtained the same results as those of Julia, at practically the same time (with differences in submissions of only a few days or weeks before or after), and in a form even more definitive. However, for some unknown reason Fatou refrained from participating in the competition. As we know, at the time, the Académie did not arrive at a Judgment of Solomon, which has left traces of incomprehension even to today. Here is the recent opinion of John Milnor (*Dynamics in One Complex Variable*, Princeton, 2006, page 40; also cited by Audin, *loc. cit.*, page 210): "The most fundamental and incisive contributions were those of Fatou himself. However, Julia was a determined competitor and tended to get more credit because of his status as a wounded war hero." Worse still, much later, during the 1940–1944 occupation, Julia allowed himself to establish relations with the German mathematicians: in 1942, he even attended a seminar in Berlin (where, according to witnesses, "he compromised himself by declaring himself to be a 'friend of National Socialism' ''': see Chevassus-au-Louis, *La Recherche* **372** (2004), page 43). By coincidence, it was this speech, adapted into an article and published in German in *Abh. Preuss. Akad. Wiss. Math.-Nat. Kl.* 1942 (1943) (a journal directed by Ludwig Bieberbach) that provoked the remark by Murray mentioned above. However, one could always remember an "(anti)symmetric" event on the other side of the front, 25 years earlier: Hilbert's speech in May 1917 (at a time of "patriotism" and unrestrained propaganda on both sides!) on the influence of the ideas of Gaston Darboux, which was later adapted into an article and published in French. But perhaps, as in the Russian proverb: "The wolfhound is right, but not the cannibal."

The reader can find more information about Gaston Julia (both mathematical and personal) in the book by Michèle Audin cited above.

To finish with Hankel operators and their symbols, note that there are descriptions of bounded $H_\varphi \colon H^p(\mathbb{D}) \to H^q_-(\mathbb{D})$ for pairs $0 < p,q \le \infty$ (Peller, Tolokonnikov, and others: see [Nikolski, 2002a], A.1.7.9 for an overview of the subject). For bounded Hankel symbols ("small" and "big") between the weighted spaces $\ell^2_a(\mathbb{Z}_+, w_n) - \ell^2_a(\mathbb{Z}_+, v_n)$ and $\ell^2_a(\mathbb{Z}_+, w_n) - L^2_a(\mathbb{D}, \mu)$ (including for the Bergman space $L^2_a(\mathbb{D}, dz\,d\bar{z})$) – under certain conditions – see [Treil and Volberg, 1994] (the conditions are always expressed in terms of corresponding embedding theorems); see also [Nikolski, 2002a], A.1.7 for an overview.

Example 2.2.5(b). The *Hilbert inequality*, initially published in [Hilbert, 1912], was extensively discussed in [Nikolski, 2019]. One can add (thanks to an observation by Hervé Queffélec) that in Hilbert's article, it was established with the constant 2π in place of π (perhaps a slip of the pen?), which was then reproduced in numerous publications, before being corrected, without comment, in [Hardy, Littlewood, and Pólya, 1934].

Lemma 2.2.7. Sarason's lemma is taken from [Sarason, 1967], where it appears with the proof used in Exercise 2.3.8. Another independent proof is due to Adamyan, Arov, and Krein [1968]. The elegant proof of Lemma 2.2.7 was published in [Sarason, 1973a] and realizes an idea of Lawrence Zalcman (personal communication). A version of the same reasoning can also be found in [Garnett, 1981]. A generalization for operatorial values can be found in [Page, 1970] (with a vector analog of Theorem 2.2.8): $H^\infty(L(H,K)) + C(\mathbb{T}, \mathfrak{S}_\infty(H,K))$ is closed in $L^\infty(\mathbb{T}, L(H,K))$ where H, K are Hilbert spaces. Later, Rudin proved a very general axiomatic version providing a criterion of closure for the sum

$E + F$ of subspaces E, F of a Banach function space X: see [Koosis, 1980], §VII.A.3 for the proof.

The fact that $H^\infty + C$ is not only a subspace but *a sub-algebra* of L^∞ revealed its importance for applications to the Fredholm theory of Toeplitz operators (see §3.2,§3.4), which led to intensive studies of the closed algebras A between H^∞ and L^∞,

$$H^\infty \subset A \subset L^\infty(\mathbb{T});$$

such an A is called a *Douglas algebra*. The highlights are as follows.

(1) $A = H^\infty + C$ is the smallest Douglas algebra: if $A \neq H^\infty$ then $A \supset H^\infty + C$ (Hoffman and Singer: see [Garnett, 1981], chapter 9 §1).

(2) For every Douglas algebra A there exists a multiplicative semigroup of inner functions F such that

$$A = \mathrm{span}_{L^\infty}(\overline{F} \cdot H^\infty) = \mathrm{span}_{L^\infty}(\overline{f}g : f \in F, g \in H^\infty)$$

[Chang, 1976; Marshall, 1976]. This last, profound, theorem confirmed a conjecture of Douglas that circulated in the early 1970s. The theorem contains two previously known special cases, $A = H^\infty + C$ (with $F = (z^n : n \in \mathbb{Z}_+)$: this is Lemma 2.2.7) and $A = L^\infty$ (where F is the family of inner functions: this is the Douglas–Rudin theorem [Douglas and Rudin, 1969]).

See [Garnett, 1981] and [Nikolski, 1986] for more details and proofs.

Theorem 2.2.8. The Adamyan–Arov–Krein theorem and the Hartman theorem are proved, respectively, in [Adamyan, Arov, and Krein, 1968] and [Hartman, 1958]; a vector-valued analog of the latter is proved in [Page, 1970].

Lemma 2.2.9. The key formula for the (future) "algebraization" of the theory of Toeplitz operators (see §3.5 for an explanation), that of Lemma 2.2.9(1), appeared in [Sarason, 1973b]. Its counterpart in the language of SIO (singular integral operators), the *Poincaré–Bertrand formula* (see §4.6 for comments), dates back to [Hardy, 1908]; it was, and continues to be, largely used in Mikhlin's "symbolic calculus" of SIO (for more details, see §4.6 and the comments on Corollary 4.1.4).

Exercise 2.3.0. Basic exercises. Parts (a) and (b) are standard; for more details see also Appendix D and the references mentioned there. For additional information, the original references and the later developments on the translation operators and the inequalities of von Neumann (Exercise 2.3.0(d)) and Bohr

(Exercise 2.3.0(g)), as well as the other subjects of parts (c)–(g), we refer to [Nikolski, 2019, 2002a,b].

Exercise 2.3.1. Berezin transform and the RKT. The Berezin transform

$$b_T(z) = (TK_z, K_z)$$

of an operator T on a reproducing kernel Hilbert space (RKHS) with the kernel k_z, $K_z = k_z/\|k_z\|$, was introduced by Berezin in his construction of the "second quantization" in quantum theory [Berezin, 1972, 1989]. Today, it takes its place in a wide theory joining theoretical physics and mathematics, known as "supermathematics," where the role of functions is played by elements of Grassmann algebras. In the framework of classical analysis, the transform shows itself particularly efficient in the spaces of Hardy, Bergman, and Bargmann–Fock (where it was originally defined by Berezin), especially for operators "linked to the structure of the space" (singular integral operators, Ha-plitz operators, embedding operators, pseudo-differential operators, "Fourier-type" operators, etc.): see [Hedenmalm, Korenblum, and Zhu, 2000] and [Zhu, 2007]. For the relationship of the Berezin transform with Toeplitz and Hankel operators, see [Peetre, 1989].

The *reproducing kernel thesis* (RKT) is based on the square Hermitian Berezin transform $b_{T^*T}(z) = \|TK_z\|^2$. The idea of the RKT was introduced in [Havin and Nikolski, 2000] to formalize a frequent observation in a reproducing kernel space: "for an operator T to be bounded it suffices that the function $b_{T^*T}(z)$ be bounded." Besides the information that it carries in itself for each operator for which it is verified, the RKT raises a question: given a reproducing kernel space, describe the operators for which the RKT is truly valid. See also [Nikolski, 2002a] for supplementary details, as well as the comments on Exercise 2.3.4 below.

Exercise 2.3.2. Natural projections on $\mathcal{T}_{L^\infty}$. This is essentially von Neumann's construction of "conditional expectation" developed in the framework of operator algebras.

Exercise 2.3.3. The H^p spaces, Fourier multipliers and (Wiener–Hopf) Toeplitz operators on ℓ^p. The results presented in this exercise are classical. No other algebras of multipliers M_p outside of the cases of $p = 1, 2, \infty$ (see Exercise 2.3.3(5)) possess a transparent and/or easy description, but there exist a good number of very useful theorems giving sufficient conditions to have $k \in M_p$: the theorems of Mikhlin (1956), Marcinkiewicz (1939), Hörmander (1960), Stechkin (1950), Calderón–Zygmund (1952), and others. Most of these results apply to multipliers of the spaces $L^p(\mathbb{R}^n)$ and $L^p(\mathbb{T}^n)$ (with the same definition $\mathcal{F}M_\varphi\mathcal{F}^{-1}L^p \subset L^p$, where $\mathcal{F}$ is the Fourier transform on the group in question) and are based on the Littlewood–Paley decomposition technique (1931). See [Duoandikoetxea, 2001], [Edwards and Gaudry, 1977], and [Grafakos, 2008], to mention only a small part of the existing literature. A "transfer principle" links the multipliers on the different groups: see [Coifman and Weiss, 1977]. In particular, a continuous function φ on $\mathbb{T}$ (or, according to the terminology of Exercise 2.3.3, its sequence of Fourier coefficients $\mathcal{F}^{-1}\varphi$) is a multiplier of $\ell^p(\mathbb{Z})$ if and only if the traces of φ on the cyclic subgroups $(\varphi(\zeta^k))_{k\in\mathbb{Z}}$ are multipliers of $L^p(\mathbb{T})$ ($\forall \zeta \in \mathbb{T}$).

For a development of the theme of Exercise 2.3.3(c) see [Böttcher and Silbermann, 1990]. We can mention that the (Wiener–Hopf) Toeplitz operators between the different spaces $\ell^p(\mathbb{Z}_+)$, $\ell^q(\mathbb{Z}_+)$ can always be reduced to multipliers: $C_k\colon \ell^p(\mathbb{Z}_+) \to \ell^q(\mathbb{Z}_+)$ is bounded $\Leftrightarrow k \in \mathrm{Mult}(\ell^p(\mathbb{Z}), \ell^q(\mathbb{Z}))$, and in particular $k = 0$ if $p > q$ (this is an observation by Hörmander [1960]). The corresponding fact for $T_\varphi\colon H^p \to H^q$ is practically evident: if $\varphi \neq 0$, then $p \geq q$, and by inverse Hölder inequality, $\varphi \in L^s$ where $1/s = 1/q - 1/p$.

Exercise 2.3.4 and Exercise 2.3.5. Spaces BMO, VMO, *and* QC. The definition of the space BMO by John [1961] and its treatment in [John and Nirenberg, 1961] represent a grand opening of the theory of Hardy spaces towards new techniques and new classes of functions, especially in several variables. Ten years later, the incisive character of this invention was confirmed by the duality $(H^1_{\mathrm{Re}})^* = $ BMO of Fefferman [1971] (the discovery earning him the Salem Prize (1971) and the Fields Medal (1978)) and by the role of the techniques developed in the construction of Hardy spaces of several variables [Fefferman and Stein, 1972]. For the corresponding general theory see [Stein, 1993] and [Duoandikoetxea, 2001], and in the context of a single complex variable see [Garnett, 1981] and [Koosis, 1980]. Today, the spaces of BMO type play a predominant role in a variety of contexts: there are versions for martingales [Petersen, 1977], on homogeneous spaces, or even in metric spaces [Hytönen, 2009].

The Garsia norm $G(\cdot)$ was probably defined in [Garsia, 1971], and then extensively used. Note also that there exist diverse versions of BMO with matrix/operator values; one of the most useful is the "strong BMO" space $\mathrm{BMO}_s(\mathbb{T}, \mathbb{C}^n \to \mathbb{C}^n)$ ("strong" in the same sense as the strong operator topology), leading to a dimensional upper bound

$$\|H_\varphi\colon H^2(\mathbb{T}, \mathbb{C}^n) \to H^2(\mathbb{T}, \mathbb{C}^n)\| \leq c \cdot \log(n)$$

for matrix symbols φ satisfying $\|\varphi\|_{\mathrm{BMO}_s} \leq 1$, $\|\varphi^*\|_{\mathrm{BMO}_s} \leq 1$ ([Petermichl, 2000]; the sharpness of this bound up to a constant was shown in [Nazarov,

Pisier, Treil, and Volberg, 2002]). For an overview of the matrix/operator-valued BMO spaces, see [Blasco, 2004] and [Mei, 2007].

The space VMO, as well as most of its properties, was introduced in [Sarason, 1975] (with the property given in Exercise 2.3.5b(2) as definition), whereas the algebra QC = $\{f: f, \bar{f} \in H^\infty + C(\mathbb{T})\}$ was defined by Douglas [1973], but its principal properties were established in [Sarason, 1975] (including Exercise 2.3.5(b) and Exercise 2.3.5(c(1))). For more information on these two spaces, we refer to [Garnett, 1981] and [Nikolski, 2002a].

The tight links between the BMO and VMO spaces with the theory of Toeplitz and Hankel operators (of which an initial portion is given in Exercises 2.3.4 and 2.3.5) were established in the 1970s, principally by Douglas and Sarason, but also by Bonsall [1984] (for the RKT for Hankel operators ((3) ⇔ (5) of Exercise 2.3.4(c) and (3) ⇔ (4) of Exercise 2.3.5(b)), and then by many others; see [Nikolski, 1986, 2002a] and [Peller, 2003] for plenty of information and references. For the *RKT of Hankel operators*, the best numerical relations known at the moment (2013) are as follows.

(i) The inequality $G(\varphi) \leq \|H_\varphi\|$ is evident by the computation of Exercise 2.3.4(b).
(ii) The converse is not at all evident, and the best known constant was given by Treil [2012]:

$$\|H_\varphi\| \leq 2\sqrt{e}G(\varphi).$$

Moreover, this estimate was established for vector-valued Hankel operators $H_\varphi: H^2 \to H^2_-(\mathbb{T}, E)$, where E is a Hilbert space and φ a *co-analytic function*, $\varphi \in L^\infty(\mathbb{T}, E)$.
(iii) For the matrix-valued symbols $\varphi \in L^\infty(\mathbb{T}, \mathbb{C}^n \to E)$, always co-analytic, the sharp estimate in n is $\|H_\varphi\| \leq \text{const.} \cdot \log(n) \cdot G(\varphi)$ (with an appropriate definition of $G(\varphi)$): see [Nazarov, Pisier, Treil, and Volberg, 2002].

Foreseeing the "algebraization" of the theory of Toeplitz operators, presented in Chapter 3, note the important role played by the "semi-commutators" $[T_f, T_g]$,

$$[T_f, T_g] = T_f T_g - T_{fg},$$

linked to the Hankel operators by the simple formula

$$[T_f, T_g] = -P_+ M_f P_- M_g = -H^*_{\bar{f}} H_g.$$

Sarason [1994] launched a conjecture that became the touchstone for a large part of harmonic analysis for the following two decades: for functions

2.4 Notes and Remarks

$f, g \in H^2$ the product $T_f T_{\bar{g}}$, defined initially for the polynomials $\mathcal{P}_a$, is bounded $H^2 \to H^2$ (and hence so is $[T_f, T_g]$) if and only if

$$\sup_{z \in \mathbb{D}} |f|^2(z) |g|^2(z) < \infty,$$

where, as before, $|f|^2(z) = P_z * |f|^2$ is the Poisson extension of the function $|f|^2$ (in particular, $fg \in L^\infty(\mathbb{T})$). Besides other applications, the problem of continuity of the products $T_f T_{\bar{g}}$ is tightly linked to the *method of "Wiener–Hopf factorization"*: according to a theorem of Widom [1960a], a Toeplitz operator T_φ is invertible if and only if there exists a factorization $\varphi = \varphi_1 \bar{\varphi}_2$, $\varphi_i^{\pm 1} \in H^2$ such that the mapping

$$A: f \to \frac{1}{\varphi_1} P_+ \frac{1}{\bar{\varphi}_2} f = T_{1/\varphi_1} T_{1/\bar{\varphi}_2} f \quad (f \in \mathcal{P}_a)$$

can be extended to a bounded operator $H^2 \to H^2$; in this case, $A = T_\varphi^{-1}$ (see Theorem 3.3.6 below as well as the comments in §3.3.3 and §3.5).

An argument of Treil, included in [Sarason, 1994], shows that the Sarason condition is necessary. In fact, it is easy to see that the problem is equivalent to the "two-weight boundedness" of the Riesz projection P_+, namely to the continuity of

$$P_+: L^2\left(\mathbb{T}, \frac{1}{|g|^2}\right) \to L^2(\mathbb{T}, |f|^2).$$

This explains the special case $g = 1/f$, where the conjecture was known beforehand: P_+ is bounded on $L^2(\mathbb{T}, |f|^2)$ if and only if the *Muckenhoupt condition* $|f|^2 \in A_2$ is satisfied (see Appendix F.3 and [Nikolski, 2019, Chapter 4]), which coincides with

$$\sup_{z \in \mathbb{D}} |f|^2(z) \frac{1}{|f|^2}(z) < \infty$$

(and is also equivalent to the invertibility of the operator $T_{\bar{f}/f}$; in the case of the invertibility, we have $(T_{\bar{f}/f})^{-1} = T_f T_{1/\bar{f}}$: see Chapter 3 below). Nonetheless, Nazarov [1997] refuted the Sarason conjecture. After a long, intense, and profound study, the problem took on the form of the "Nazarov–Treil–Volberg conjecture" [Nazarov, Treil, and Volberg, 2004] for even more general operators T (including the Hilbert transform $\mathbb{H}$ and certain singular integral operators) where Sarason's "Muckenhoupt condition with two weights" is completed by the David and Journée conditions $T(1) \in$ BMO, $T^*(1) \in$ BMO, and also by a "test of finite energy." In this form the problem was finally resolved by Lacey [2013b]; more than 10 years earlier, in 2000, the same criteria but without the "test of finite energy" had been confirmed by Nazarov and Volberg (see [Volberg, 2003]) for $P_+, \mathbb{H}: L^2(\mathbb{T}, \mu) \to L^2(\mathbb{T}, \nu)$ with measures

μ, ν satisfying the "doubling condition" $\mu(2I) \leq c \cdot \mu(I)$, $\nu(2I) \leq c \cdot \nu(I)$ (for every interval (arc) $I \subset \mathbb{T}$).

Note that if it is not the goal to find a criterion of type RKT (i.e. to *test* an operator against a family of "simple" functions such as the reproducing kernels, or the indicator functions of intervals), the problem was resolved long ago by Cotlar and Sadosky [1979]: the Hilbert transform $\mathbb{H} = 2P_+ - I$ is bounded,

$$L^2(\mathbb{T}, \mu) \to L^2(\mathbb{T}, \nu)$$

(μ and ν two positive measures), and $\|\mathbb{H}\| \leq M$ if and only if there exists $h \in H^1(\mathbb{T})$ such that, for every Borel set $\sigma \subset \mathbb{T}$, we have

$$|(M^2\mu + \nu - h\,dm)(\sigma)| \leq (M^2\mu - \nu)(\sigma);$$

in particular, ν is necessarily absolutely continuous with respect to the Lebesgue measure m.

To conclude, note that in [Sarason, 1994] a similar conjecture was also formulated for the Bergman space $L_a^2(\mathbb{D})$; here, the question remains open as of May 2013, but for its generalized form $P_B: L_a^2(\mathbb{D}, \mu) \to L_a^2(\mathbb{D}, \nu)$ a counterexample was constructed in [Aleman, Pott, and Reguera, 2013] (P_B is the Bergman projection, the orthogonal projection of $L^2(\mathbb{D})$ onto $L_a^2(\mathbb{D})$). For more information and references on the weighted estimations, see [Volberg, 2003], [Lacey, 2013a,b], and [Nikolski, 2002a], §A.5.8.

The lemma of Exercise 2.3.5(d) is taken from [Nikolski, 1986], Appendix 4, Lemma 85.

Exercise 2.3.6. Kronecker's theorem has been known since [Kronecker, 1881]. Our elementary proof relying solely on linear algebra is not the only one known; see [Peller, 2003] and [Nikolski, 1986, 2002a] for other variants. In particular, the proof in [Nikolski, 1986] uses the invariant subspaces in the following manner:

(1) as $S^*\Gamma = \Gamma S$, the range $\Gamma(H^2)$ is an S^*-invariant subspace, and hence of the form $(\Theta H^2)^\perp = H^2 \ominus \Theta H^2$, where Θ is an inner function (see Appendix F);
(2) $\dim \Gamma(H^2) < \infty$ if and only if Θ is a finite Blaschke product, $\dim \Gamma(H^2) = \deg(\Theta)$;
(3) $\Gamma 1 = JP_-\varphi \in H^2 \ominus \Theta H^2$ and $H^2 \ominus \Theta H^2$ is made up of rational functions.

For an analog of Exercise 2.3.6 for the RKHS (*reproducing kernel Hilbert spaces*), see [Nikolski, 2002a], B.2.5.4.

2.4 Notes and Remarks

Exercise 2.3.7. On the Hankel operators in other Schatten–von Neumann classes $\mathfrak{S}_p$, see [Nikolski, 2002a,b] and [Peller, 2003].

Exercise 2.3.8. This proof is published in [Sarason, 1967].

Exercise 2.3.9. The *commutators* of singular integral operators (such as P_+ or $\mathbb{H}$) and the multiplication operators M_φ appeared in the works of Mikhlin [1948], and were then revealed as a crucial subject from the functional calculus (see [Gohberg and Krupnik, 1973] and also §3.1 and §3.4 below) up to the "local spectral theories" of Simonenko and Douglas and Sarason. Another line of development, based on the works of Calderón (see [Stein, 1993]) interpreted the commutators as a simplified approximation to the weighted estimations of the Hilbert transform $\mathbb{H}$ or Cauchy transform P_+. An explanation of this latter idea comes, for example, from the formula $e^{-\epsilon\varphi}\mathbb{H} e^{\epsilon\varphi} = \mathbb{H} + \epsilon(\mathbb{H}\varphi - \varphi\mathbb{H}) + o(\epsilon)$ as $\epsilon \to 0$ (see [Stein, 1993]). The theorem of Exercise 2.3.9 is from [Power, 1980], whereas the case of $\varphi \in C(\mathbb{T})$ was already known by Mikhlin [1948].

Exercise 2.3.10. The *projection onto* Hank. The fact that the natural projection P onto Hank$(\ell^2(\mathbb{Z}_+))$ is not bounded is due to Peller [1980], using a different proof. See also the discussion in [Nikolski, 2002a], B.1.8, B.4.7.6.

Exercise 2.3.11. *Vector-valued Toeplitz operators.* Most applications of Toeplitz operators (and of SIO or WHO, i.e. *Wiener–Hopf operators*) require a spectral theory for matrix/operator-valued symbols, since ordinarily the applications present systems of equations of the form $\sum_j T_{\varphi_{ij}} f_j = g_i$ ($1 \le i, j \le N$). The origins of this theory date back to the flagship article of Birkhoff [1909], where the factorization method $\varphi = \varphi_- D \varphi_+$ of a matrix-valued symbol φ is introduced (in the context of a system of SIO). We defer the discussion of factorizations to §3.5 (comments on Exercise 3.4.14) and §4.6; note only that today the vector-valued Toeplitz operators (or SIO, or WHO) form the privileged context of research in this domain: see [Mikhlin, 1962], [Clancey and Gohberg, 1981], [Gohberg, Goldberg, and Kaashoek, 1990], and [Böttcher, Karlovich, and Spitkovsky, 2002].

Properties (a) and (b) of Exercise 2.3.11 are standard; we will see the rest in Exercise 3.4.14, §3.5, Exercise 4.5.2, and §4.6.

Exercise 2.3.12. *Algebraic properties of Toeplitz/Hankel operators.* Formula (a) comes from [Widom, 1976], and the collection of properties (b) is borrowed from [Brown and Halmos, 1963]. Note that parts (ii) and (iv) (as well as all the others) are elementary, whereas their generalizations (c(i)) and especially (c(ii)) are far from being so.

The result of (c(i)) was proved in [Power, 1984] (see also [Peller, 2003]), that of (c(ii)) in [Aleman and Vukotic, 2009] (the fact remains correct for Toeplitz operators with matrix-valued symbols, by replacing $\varphi_j = 0$ by $\det(\varphi_j) = 0$), and (c(iii)) in [Axler, Chang, and Sarason, 1978] and [Nikolski, 1986] (the rank formula).

The result of Janson, Exercise 2.3.12(d), was published in [Nikolski, 1985].

3
H^2 Theory of Toeplitz Operators

Topics

- The Toeplitz algebra, semi-commutators and symbolic calculus, index and Cauchy topological index, homotopy of symbols.
- dist(f, H^∞) and dist$(f, H^\infty + C)$, local principle, local distances.
- The spectrum and essential spectrum of Toeplitz operators, sectorial and locally sectorial symbols, spectral inclusions.
- Wiener–Hopf factorization.
- Self-adjoint Toeplitz operators, Hankel operators in the Toeplitz algebra.

Biographies Augustin Cauchy, Israel Gohberg, Igor Simonenko, Carl Runge, George Birkhoff.

3.1 Fredholm Theory of the Toeplitz Algebra

"Fredholm theory" is a spectral theory "up to a compact operator"; it is presented in Appendix E. It happens that for Toeplitz operators, it is easier to first develop the Fredholm theory, and then to "climb back" to the initial space. As a collateral advantage, we obtain for free a similar theory for the "composite (aggregate)" operators of the Toeplitz algebra (see below).

The standard details about compact operators can be found in Appendices D and E. Recall only that for every Banach space X, the set of compact operators $\mathfrak{S}_\infty(X)$ is a bilateral closed ideal of $L(X)$, the quotient algebra

$$\mathcal{K} = L(X)/\mathfrak{S}_\infty(X)$$

is called the *Calkin algebra*, and the canonical projection $L(X) \to L(X)/\mathfrak{S}_\infty(X)$ is denoted Π,

$$\Pi(T) = T^\bullet = T + \mathfrak{S}_\infty(X), \quad T \in L(X);$$

see Appendix D. Moreover,

$$\|T\|_{\mathrm{ess}} = \|\Pi(T)\|_{\mathcal{K}} = \mathrm{dist}_{L(X)}(T, \mathfrak{S}_\infty(X)) \quad \text{and} \quad \sigma_{\mathrm{ess}}(T) = \sigma_{\mathcal{K}}(\Pi(T)).$$

The set of *Fredholm operators* on a space H is denoted by

$$\mathrm{Fred}(H);$$

see Appendix E for the definition and properties of Fred(H).

Lemma 3.1.1 (aggregate operators) *Let* $(\varphi_{ij}) \subset L^\infty(\mathbb{T})$ *be a finite family,* $T = \sum_i \prod_j T_{\varphi_{ij}}$ *an "aggregate operator" (a finite sum of finite products), and* $f \in H^2$. *Then*

$$\left(\sum_i \prod_j \varphi_{ij}\right) f = \bar{z}^n T P_+ z^n f + o(1) \quad (n \to \infty),$$

$$\left\|\sum_i \prod_j \varphi_{ij}\right\|_\infty \le \|T + K\|, \quad \forall K \in \mathfrak{S}_\infty(H^2).$$

Proof For every $\varphi, \psi \in L^\infty(\mathbb{T})$ and $f \in L^2(\mathbb{T})$, Corollary 2.1.6 implies

$$T_\varphi T_\psi P_+(z^n f) = T_\varphi P_+(z^n \psi f) + o(1) = z^n \varphi \psi f + o(1) \quad (n \to \infty),$$

and by induction

$$\left(\sum_i \prod_j \varphi_{ij}\right) f = \bar{z}^n T P_+(z^n f) + o(1) \quad (n \to \infty, \ f \in L^2(\mathbb{T})).$$

Since $\lim_n P_+(z^n f) = 0$ weakly in H^2, for $K \in \mathfrak{S}_\infty(H^2)$ we obtain

$$\lim_n \|K P_+(z^n f)\|_2 = 0 \quad (\forall f \in L^2(\mathbb{T})),$$

and hence

$$\left\|\left(\sum_i \prod_j \varphi_{ij}\right) f\right\|_2 = \lim_n \|\bar{z}^n T P_+(z^n f)\|_2$$

$$= \lim_n \|\bar{z}^n (T + K) P_+ z^n f\|_2 \le \|T + K\| \cdot \|f\|_2,$$

and the result follows. ∎

Definitions 3.1.2 The *Toeplitz algebra* $\mathrm{alg}\mathcal{T}_{L^\infty(\mathbb{T})}$ is the sub-algebra of $L(H^2)$ generated by the family of operators $\mathcal{T}_{L^\infty(\mathbb{T})} = \{T_\varphi : \varphi \in L^\infty(\mathbb{T})\}$, i.e.

3.1 Fredholm Theory of the Toeplitz Algebra

$$\mathrm{alg}\mathcal{T}_{L^\infty(\mathbb{T})} = \mathrm{clos}_{L(H^2)}\left(\sum_i \prod_j T_{\varphi_{ij}} : (\varphi_{ij}) \subset L^\infty(\mathbb{T}) \text{ a finite family}\right).$$

Similarly, given a closed unital sub-algebra $X \subset L^\infty(\mathbb{T})$, the *Toeplitz X-algebra* is the sub-algebra of $L(H^2)$ generated by the family of operators $\mathcal{T}_X = \{T_\varphi : \varphi \in X\}$, i.e.

$$\mathrm{alg}\mathcal{T}_X = \mathrm{clos}_{L(H^2)}\left(\sum_i \prod_j T_{\varphi_{ij}} : (\varphi_{ij}) \subset X \text{ a finite family}\right).$$

For the study of X-algebras $\mathrm{alg}\mathcal{T}_X$, the following operators (called "commutators" and "semi-commutators" of Toeplitz operators) play a principal role:

$$[T_\varphi, T_\psi] =: T_\varphi T_\psi - T_\psi T_\varphi \text{ (commutator)},$$

$$[T_\varphi, T_\psi) =: T_\varphi T_\psi - T_{\varphi\psi} \text{ (semi-commutator)}.$$

We can already say that the Toeplitz algebra $\mathrm{alg}\mathcal{T}_{L^\infty(\mathbb{T})}$ appears to be too large to be the subject of a developed theory (without being, nonetheless, equal to $L(H^2)$: see Exercise 3.4.9 below), but the algebras $\mathrm{alg}\mathcal{T}_X$ for $X = C(\mathbb{T})$ and $X = H^\infty + C(\mathbb{T})$ are much more tractable, and they will be studied in this chapter.

We also define the *essential range* (or the *range mod* 0) of a measurable function $\varphi : \mathbb{T} \to \mathbb{C}$ as

$$\mathrm{Ran}_\mathrm{ess}(\varphi) = \{\lambda \in \mathbb{C} : \mathrm{ess\,inf}|\varphi - \lambda| = 0\};$$

clearly $\mathrm{Ran}_\mathrm{ess}(\varphi) = \mathrm{Ran}(\varphi) = \varphi(\mathbb{T})$ if $\varphi \in C(\mathbb{T})$.

Recall also the notion of the Cauchy, or topological, index of a curve $\gamma \subset \mathbb{C}$ with respect to the origin: if $\varphi^{\pm 1} \in C(\mathbb{T})$ then there exists a unique integer $n \in \mathbb{Z}$ such that $\varphi = z^n e^{ia}$, where $a \in C(\mathbb{T})$ is a real-valued function. This integer is called the *Cauchy index of φ* (or of $\gamma = \varphi(\mathbb{T})$), or the *number of loops around* 0, and is denoted

$$\mathrm{wind}(\varphi) = \mathrm{wind}(\varphi(\mathbb{T})) = n$$

and called the *winding number*. It is well known that

$$\mathrm{wind}(\varphi) = \frac{1}{2\pi i} \int_{\varphi(\mathbb{T})} \frac{dz}{z};$$

see for example [Rudin, 1986], §10.10.

Augustin Louis (Baron) Cauchy (1789–1857), a French mathematician, was one of the dominant figures of nineteenth-century mathematics, and the founder of mathematical analysis, with almost 800 research articles to his name (the most prolific in the history of mathematics, just behind Euler). His name is associated with more than 30 mathematical objects, spread over the full set of domains of the epoch. A classic *par excellence*.

Born to a profoundly Catholic and royalist family (his father was senior aide to the very last Lieutenant General of the Paris police, who was guillotined in 1794), for all his life Cauchy held to these two convictions, the third evidently being mathematics. A graduate of the École Polytechnique and the École des Ponts et Chaussées, his first employment as an engineer (1810) was with the construction of the (defensive) military port of Cherbourg. At the time, almost everything was controlled by political events, and Cauchy's first three candidatures for an academic position were refused. During the Second Restoration he obtained a position as professor of mechanics at the École Polytechnique (1815). Then, when the Académie des Sciences ejected the Republicans Carnot and Monge, Cauchy took over one of the vacant positions (1816).

In 1830, as a "légitimiste" royalist (partisan of the Bourbons), he refused to take an oath of allegiance to the new king Louis-Philippe, and thus lost his position at the École Polytechnique. and voluntarily exiled himself to Switzerland, then Italy, and finally Bohemia. After a few more quarrels, Cauchy was tenured to the chair of mathematical astronomy at the *Faculté des Sciences de Paris* (1849). In 1852, he again refused to take an oath (this time, to Napoléon III), but kept his position. His principles also led him to found *l'Institut Catholique* (1842), *l'Œuvre d'Irlande* (1846) (to fight the famine in Ireland), *l'Œuvre d'observation du dimanche* (1854) (to force shops to close on Sunday), and *l'Œuvre des Écoles d'Orient* (1855) (for emancipation by education). He died of a cold in the night of May 23, 1857.

As is evident from this rapid description of Cauchy's career, he had a controversial character and left behind a long series of disputes – with

Niels Abel (after his visit to Cauchy in 1826), with Joseph Liouville (over a position at the Collège de France), with Jean-Marie Duhamel and Jean-Victor Poncelet (on priorities in the theory of elasticity and in projective geometry), with Bernard Bolzano (on the definition of continuity), with Adrien Marie Legendre (concerning a report by Cauchy on the revolutionary posthumous works of Abel, three years late (!) and in any case considered by historians as "superficial and unworthy of the calibre of the opponents"), etc.

Despite all these troublesome events, Cauchy produced an enormous quantity of innovative mathematics, let's say, for the period 1839–1849, at the rate of an article per week (!). Without entering into a retrospective analysis of Cauchy's creations, we give an extract of what he left us: among the basic notions, convergence and the Cauchy criterion, power series with the formula of the radius of convergence $1/\overline{\lim}\sqrt[n]{|a_n|}$ (and the concept itself), continuity and intermediate values, the product of two power series, the Cauchy–Schwarz inequality, etc., and among the more advanced concepts, holomorphy (the term itself was introduced by two of Cauchy's disciples, Charles Briot and Jean-Claude Bouquet) and the Cauchy–Riemann equations (with the Cauchy integral formula, using the Cauchy index (see Definitions 3.1.2 in the text)), contour integrals (and the Cauchy principal value) with Cauchy's inequality, the residue calculus, the Cauchy problem (in PDEs), the Cauchy–Kovalevskaya theorem, etc., as well as several notions and results in geometry, algebra, mechanics, and optics. The reader can find numerous sources of information about Cauchy on the websites www-history.mcs.st-and.ac.uk/Biographies/Cauchy.html and https://en.wikipedia.org/wiki/Augustin-Louis_Cauchy.

Theorem 3.1.3 (symbol on alg$\mathcal{T}_{L^\infty(\mathbb{T})}$) *Let* $X \subset L^\infty(\mathbb{T})$ *be a closed unital subalgebra, and for every finite family* $(\varphi_{ij}) \subset X$ *define*

$$\mathrm{Sym}\left(\sum_i \prod_j T_{\varphi_{ij}}\right) =: \sum_i \prod_j \varphi_{ij}.$$

(1) *The mapping* Sym *is well-defined and can be extended by continuity to an algebra homomorphism*

$$\mathrm{Sym}\colon \mathrm{alg}\mathcal{T}_X \to X$$

factorable through the Calkin algebra $T \mapsto \Pi(T) \mapsto \mathrm{Sym}T$, *and hence satisfying*

$$\|\mathrm{Sym}T\|_\infty \leq \|T\|_\mathrm{ess} \leq \|T\|, \quad \mathrm{Sym}(RT) = \mathrm{Sym}(R)\mathrm{Sym}(T) \quad (R, T \in X).$$

(2) *For every* $\varphi \in L^\infty(\mathbb{T})$ *and* $n \in \mathbb{Z}_+$, $\|T_\varphi^n\|_\mathrm{ess} = \|T_\varphi^n\| = \|\varphi\|_\infty^n$.

(3) *If* $T \in \mathrm{alg}\mathcal{T}_X \cap \mathrm{Fred}(H^2)$ *then* $\mathrm{Sym}(T) \in L^\infty(\mathbb{T})^{-1}$, *and*

$$\|(\mathrm{Sym}T)^{-1}\|_\infty \leq \|(\Pi(T))^{-1}\|, \quad \mathrm{Ran}_\mathrm{ess}(\mathrm{Sym}(T)) \subset \sigma_\mathrm{ess}(T).$$

Proof (1) Let $A = \{\sum_i \prod_j T_{\varphi_{ij}} : (\varphi_{ij}) \subset X \text{ a finite family}\}$ be the sub-algebra generated by the Toeplitz operators with symbols in X. The inequality

$$\left\|\sum_i \prod_j \varphi_{ij}\right\|_\infty \leq \left\|\sum_i \prod_j T_{\varphi_{ij}}\right\|$$

of Lemma 3.1.1 shows that the mapping $T \mapsto \mathrm{Sym}(T)$ ($T \in A$) can be extended to a homomorphism $A \to X$ (hence, well-defined, linear, bounded, and multiplicative). Moreover, $\|\mathrm{Sym}T\|_\infty \leq \|T\|_\mathrm{ess} = \inf_{K \in \mathfrak{S}_\infty} \|T + K\|$, hence (1).

(2) The assertion follows from the same Lemma 3.1.1:

$$\|\varphi\|_\infty^n = \|\varphi^n\|_\infty \leq \|T_\varphi^n\|_\mathrm{ess} \leq \|T_\varphi^n\| \leq \|\varphi\|_\infty^n.$$

(3) Let R be a regularizer of T (see Theorem E.7.2), $RT = I + K$, $K \in \mathfrak{S}_\infty(H^2)$. Then, for every $f \in L^2(\mathbb{T})$, when $n \to \infty$, we have

$$\|f\|_2 = \lim_n \|P_+ z^n f\|_2 = \lim_n \|(RT - K)P_+ z^n f\|_2$$
$$\leq \lim_n \|RT P_+ z^n f\|_2 \leq \overline{\lim_n} \|R\| \cdot \|\bar{z}^n TP_+ z^n f\|_2.$$

Also, the first formula of Lemma 3.1.1 extends to the operators $T \in \mathrm{alg}\mathcal{T}_X$ (by the Banach–Steinhaus theorem): for every $f \in L^2(\mathbb{T})$, $TP_+ z^n f = (\mathrm{Sym}T)z^n f + o(1)$ as $n \to \infty$. Thus

$$\|f\|_2 \leq \|R\| \cdot \|(\mathrm{Sym}T)f\|_2 \quad (\forall f \in L^2(\mathbb{T})),$$

and hence $\|(\mathrm{Sym}T)^{-1}\|_\infty \leq \|R\|$. By Theorem E.7.5, $\inf \|R\| = \|\Pi(T)^{-1}\|$ where the infimum is taken over all the regularizers of T.

The last inclusion follows from what has already been shown, applied to $T - \lambda I$, $\lambda \in \mathbb{C}$. ∎

To deduce from Theorem 3.1.3 the Fredholm theory for the "classical" Toeplitz operators (with continuous symbols), we need the following simple but decisive observations by Lewis Coburn (Lemma 3.1.4).

3.1 Fredholm Theory of the Toeplitz Algebra

Lemma 3.1.4 (Coburn, 1967)

(1) *For every* $\psi \in H^\infty, \varphi \in L^\infty(\mathbb{T})$, $T_\varphi T_\psi - T_{\varphi\psi} = 0$ *and* $T_{\bar\psi}T_\varphi - T_{\varphi\bar\psi} = 0$.
(2) *For* $n \in \mathbb{Z}$ *and* $\varphi \in L^\infty(\mathbb{T})$, $\mathrm{rank}(T_\varphi T_{z^n} - T_{\varphi z^n}) \leq |n|$, $\mathrm{rank}(T_{z^n}T_\varphi - T_{\varphi z^n}) \leq |n|$.
(3) $I - T_z T_{\bar z} = (\cdot, 1)1$ *and hence* $T_\psi(I - T_z T_{\bar z})T_{\bar\varphi} = (\cdot, \varphi)\psi$ *for every* $\varphi, \psi \in H^\infty$.
(4) $\mathfrak{S}_\infty(H^2) \subset \mathrm{alg}\mathcal{T}_{C(\mathbb{T})}$.
(5) *For every* $\varphi \in L^\infty(\mathbb{T})$, $\varphi \neq 0$, *either* $\ker T_\varphi = \{0\}$ *or* $\ker T_\varphi^* = \{0\}$. *Consequently, every operator* $T_\varphi \in \mathrm{Fred}$ *such that* $\mathrm{ind}(T_\varphi) = 0$ *is invertible*.

Proof (1) For every $f \in H^2$, $T_\psi f = P_+(\psi f) = \psi f$, and the result follows.

(2) If $n \geq 0$, then $T_\varphi T_{z^n} - T_{\varphi z^n} = 0$ (by (1)), and if $n < 0$ and $f = z^{-n}g$, where $g \in H^2$, then $T_\varphi T_{z^n} f - T_{\varphi z^n} f = T_\varphi g - T_\varphi g = 0$, and thus the range

$$(T_\varphi T_{z^n} - T_{\varphi z^n})H^2 = (T_\varphi T_{z^n} - T_{\varphi z^n})(H^2 \ominus z^{-n}H^2)$$

is a subspace of dimension at most $|n|$.

(3) We have $T_z = S$ and $T_{\bar z} = S^*$, and by Lemma 2.1.1, for every $f \in H^2$, $f - T_z T_{\bar z} f = f(0)$.

(4) For all analytic polynomials $p, q \in \mathcal{P}_a$, by (1) and (3) we have

$$T_q(I - T_z T_{\bar z})T_{\bar p} = (\cdot, p)q,$$

hence $(\cdot, p)q \in \mathrm{alg}\mathcal{T}_{C(\mathbb{T})}$, and then $(\cdot, f)g \in \mathrm{alg}\mathcal{T}_{C(\mathbb{T})}$ ($\forall f, g \in H^2$), thus $\mathfrak{F}(H^2) \subset \mathrm{alg}\mathcal{T}_{C(\mathbb{T})}$ (where $\mathfrak{F}(H^2)$ is the set of finite rank operators on H^2). As $\mathfrak{F}(H^2)$ is dense in $\mathfrak{S}_\infty(H^2)$ (see Appendix D.5), we obtain $\mathfrak{S}_\infty(H^2) \subset \mathrm{alg}\mathcal{T}_{C(\mathbb{T})}$.

(5) Suppose $f, g \in H^2$ are such that $P_+(\varphi f) = P_+(\bar\varphi g) = 0$ and $f \neq 0$. We show that $g = 0$. Indeed, the equations for f and g show that $\varphi f = \bar a$, $\bar\varphi g = \bar b$ where $a, b \in H_0^2 = zH^2$. Multiplying these equations by, respectively, $\bar g$ and $\bar f$, we obtain $\overline{fb} = ga \in H_0^1 \cap \overline{H}_0^1$ (for the inclusion $H^2 \cdot H_0^2 \subset H_0^1$ see Appendix F), and thus $fb = 0$, $ga = 0$. Since $f \neq 0$, we have $b = 0$ and hence $\varphi g = 0$; and since $\varphi \neq 0$ (in L^∞), $g = 0$ (see Appendix F). ∎

Theorem 3.1.5 (Gohberg, 1964; Coburn, 1967)

$$[T_\varphi, T_\psi), [T_\psi, T_\varphi] \in \mathfrak{S}_\infty(H^2) \quad (\forall \varphi \in L^\infty(\mathbb{T}), \forall \psi \in C(\mathbb{T})),$$

and

$$T - T_{\mathrm{sym}(T)} \in \mathfrak{S}_\infty(H^2) \quad \text{for every } T \in \mathrm{alg}\mathcal{T}_{C(\mathbb{T})}.$$

Moreover,

$$\mathrm{alg}\mathcal{T}_{C(\mathbb{T})} = \mathcal{T}_{C(\mathbb{T})} + \mathfrak{S}_\infty(H^2) \quad (\textit{direct sum}),$$

where $\mathrm{Sym}_C\colon T \mapsto \mathrm{Sym}(T)$ is a contractive surjective homomorphism of Banach algebras whose kernel is $\mathfrak{S}_\infty(H^2)$; its quotient mapping

$$\widehat{\mathrm{Sym}}_C\colon \mathrm{alg}\mathcal{T}_{C(\mathbb{T})}/\mathfrak{S}_\infty(H^2) \to C(\mathbb{T})$$

is an isometric algebra isomorphism.

Proof By Lemma 3.1.4(2), for every $\varphi \in L^\infty(\mathbb{T})$ and $\psi \in \mathcal{P}$, we have $[T_\varphi, T_\psi]$, $[T_\psi, T_\varphi] \in \mathfrak{F}(H^2)$, and thus by a uniform approximation $(\mathrm{clos}_{L^\infty}(\mathcal{P}) = C(\mathbb{T}))$ we obtain

$$[T_\varphi, T_\psi], [T_\psi, T_\varphi] \in \mathfrak{S}_\infty(H^2) \quad (\forall \varphi \in L^\infty(\mathbb{T}), \forall \psi \in C(\mathbb{T})).$$

By induction, for every finite product with $\varphi_j \in C(\mathbb{T})$, we have $(\prod_j T_{\varphi_j} - T_{\prod \varphi_j}) \in \mathfrak{S}_\infty(H^2)$, hence $T - T_{\mathrm{sym}(T)} \in \mathfrak{S}_\infty(H^2)$ for every composite operator $T = \sum_i \prod_j T_{\varphi_{ij}}$ such that $\varphi_{ij} \in C(\mathbb{T})$. By a passage to the limit, $T - T_{\mathrm{sym}(T)} \in \mathfrak{S}_\infty(H^2)$ for every $T \in \mathrm{alg}\mathcal{T}_{C(\mathbb{T})}$. The rest follows from Theorem 3.1.3 applied to $X = C(\mathbb{T})$. ∎

Israel Gohberg (in Russian Израиль Цудикович Гохберг, 1928–2009) was a Soviet (Republic of Moldavia), and later Israeli mathematician, one of the key figures in the twentieth-century spectral theory of singular integral and Wiener–Hopf operators, and creator of an international school (between Kishinëv (USSR), Tel Aviv and Amsterdam) on the subject and its applications. An extremely prolific author, Gohberg published around 500 research articles and two dozen monographs and texts, some of which have become classic reference books. He is known for the *Gohberg–Sementsul / Baxter–Hirschman formula* (for the inverse of a Toeplitz matrix: see Theorems 5.3.1 and 5.3.2 below and the comments in § 5.8), the *Gohberg–Krein theorems* (Wiener–Hopf factorization for matrix-valued symbols), and the *Gohberg–Kaashoek numbers* (in control theory). He was founder of *Integral Equations and Operator Theory* (a first-rate international journal), as well as a series of monographs, *Operator Theory: Advances and Applications* (roughly 250 volumes have been issued (2013)), and a series of conferences, IWOTA

(*International Workshop on Operator Theory and Applications*, with the 30th edition in 2019). Gohberg was elected foreign member of the Royal Netherlands Academy (1985) and received a *Doctor Honoris Causa* degree from a number of European universities (Amsterdam, Vienna, Darmstadt, Timisoara, etc.); he is a laureate of the Humboldt Prize (1992) and of several other scientific distinctions.

During the "Soviet period" of his career (up to 1974), despite his brilliant scientific and pedagogical success (around 30 theses submitted under his direction in the USSR, and more than 40 in all), Gohberg suffered a long series of restrictions and injustices linked to the internal anti-Jewish politics in the USSR, which forced him to emigrate to Israel (1974). The charismatic figure of Israel Gohberg inspired numerous biographical texts, lively and rich in content, including an article on Russian Wikipedia (very different to the English and French versions) and a posthumous book [Bart, Hempfling, and Kaashoek, 2008] edited with affection by his friends and disciples.

Theorem 3.1.6 (Gohberg, 1964; Douglas, 1968) *Let* $T \in \text{alg}\mathcal{T}_{C(\mathbb{T})}$.

(1) $T \in \text{Fred}(H^2) \Leftrightarrow \text{Sym}(T) \in C(\mathbb{T})^{-1}$.
(2) *If* $T \in \text{Fred}(H^2)$, *then* $T_{1/\text{Sym}(T)}$ *is a regularizer of* T.
(3) *If* $T \in \text{Fred}(H^2)$, *then* $\text{ind}(T) = -\text{wind}(\text{Sym}(T))$.

Proof (1) The implication $\Rightarrow$ is contained in Theorem 3.1.3. The converse follows from (2), which, in turn, is a consequence of Theorem 3.1.5: $T_{1/\text{Sym}(T)}T - I \in \mathfrak{S}_\infty(H^2)$, $TT_{1/\text{Sym}(T)} - I \in \mathfrak{S}_\infty(H^2)$.

For (3), we see that by Theorem 3.1.5 (or Lemma 3.1.4),

$$T = T_{z^n}T_{e^{ia}} + K$$

where $\varphi =: \text{Sym}(T) = z^n e^{ia}$, $n = \text{wind}(\varphi)$, $a \in C(\mathbb{T})$. By Atkinson's theorem, Theorem E.7.3,

$$\text{ind}(T) = \text{ind}(T_{z^n}) + \text{ind}(T_{e^{ia}}).$$

For T_{z^n}, we know that $T_{z^n} = S^n$ if $n \geq 0$ and $T_{z^n} = S^{*|n|}$ if $n < 0$, hence $\text{ind}(T_{z^n}) = -n$.

For $T_{e^{ia}}$, there is a homotopy with the identity I: $T(s) = T_{e^{isa}}$, $0 \leq s \leq 1$, and $T(s) \in \text{Fred}(H^2)$ for any s. By Corollary E.7.4(4), we have $\text{ind}(T(1)) = \text{ind}(T(0)) = 0$. Thus $\text{ind}(T) = -n$. ∎

Corollary 3.1.7 *For every $T \in \mathrm{alg}\mathcal{T}_{C(\mathbb{T})}$,*

$$\sigma_{\mathrm{ess}}(T) = \mathrm{Ran}(\mathrm{Sym}(T)) = \mathrm{Sym}(T)(\mathbb{T}),$$

$$\sigma(T) = \sigma_{\mathrm{ess}}(T) \cup \left(\bigcup \Omega_{\mathrm{ind}\neq 0}\right),$$

where $\Omega_{\mathrm{ind}\neq 0}$ are the connected components of the complement $\mathbb{C} \setminus \mathrm{Sym}(T)(\mathbb{T})$ where $\mathrm{wind}(\lambda - \mathrm{Sym}(T)) \neq 0$.

Indeed, the formulas follow from Theorem 3.1.6 and Lemma 3.1.4(5). ∎

3.1.1 The Role of Homotopy

Homotopy has already played an important role in the proof of Theorem 3.1.6, but we would again like to highlight it in the case of pure Toeplitz operators, for example, as in the following corollary.

Corollary 3.1.8 (continuous homotopic symbols)

(1) *If $\varphi_0, \varphi_1 \in C(\mathbb{T})^{-1}$ (the group of invertible elements of $C(\mathbb{T})$) and φ_0, φ_1 are homotopic in $C(\mathbb{T})^{-1}$, then T_{φ_0} and T_{φ_1} are simultaneously invertible, and conversely.*

(2) *Let $\varphi \in C(\mathbb{T})$. Then,*

$$T_\varphi \text{ is invertible} \Leftrightarrow \varphi = e^a \text{ where } a \in C(\mathbb{T})$$

(thus, if and only if φ is homotopic to 1 in $C(\mathbb{T})^{-1}$).

Indeed, for (1), clearly T_{φ_0} and T_{φ_1} are homotopic in $\mathrm{Fred}(H^2)$, hence, if T_{φ_0} is invertible, then $\mathrm{ind}(T_{\varphi_1}) = \mathrm{ind}(T_{\varphi_0}) = 0$, which implies the invertibility of T_{φ_1} (by Lemma 3.1.4(5)).

For (2), the inclusion $\varphi \in C(\mathbb{T})^{-1}$ is necessary for the invertibility, and under this last condition $T_\varphi \in \mathrm{Fred}(H^2)$ and the invertibility of T_φ is equivalent to $\mathrm{ind}(T_\varphi) = -\mathrm{wind}(\varphi) = 0$. By definition of the Cauchy index, $\mathrm{wind}(\varphi) = 0 \Leftrightarrow \varphi = e^a$ where $a \in C(\mathbb{T})$. ∎

Example 3.1.9 Let $\varphi \in C(\mathbb{T})$. Then $\sigma_{\mathrm{ess}}(T_\varphi) = \varphi(\mathbb{T})$. If $\varphi = p/q$ is rational (p, q are relatively prime polynomials in z), then $\lambda \in \mathbb{C} \setminus \sigma_{\mathrm{ess}}(T_\varphi)$ if and only if $\lambda - p/q \neq 0$ on $\mathbb{T}$, and in this case,

$$\mathrm{ind}(T_\varphi) = P_\lambda - N_\lambda,$$

where P_λ is the number of poles of $\lambda - p/q$ in $\mathbb{D}$ and N_λ the number of zeros in $\mathbb{D}$ (poles and zeros counted with their multiplicity).

Hence, $\lambda \in \mathbb{C} \setminus \sigma(T_\varphi)$ if and only if $\lambda - p/q \neq 0$ on $\mathbb{T}$ and $P_\lambda = N_\lambda$.

3.1 Fredholm Theory of the Toeplitz Algebra

Indeed, this is a consequence of the "argument principle" of complex analysis for the computation of wind$(\lambda - p/q) = N_\lambda - P_\lambda$ (see [Rudin, 1991]); in particular, for $|\lambda|$ large enough, $P_\lambda = N_\lambda$ by the classical Rouché theorem (meaning that the spectrum is, as always, compact). ∎

Besides the preceding consequences, Theorem 3.1.3 allows the question of essential spectrum $\sigma_{\mathrm{ess}}(T_\varphi)$ to be reduced to the case of a unimodular symbol ($|\varphi| = 1$ a.e. on $\mathbb{T}$) and then to give *a certain description of $\sigma_{\mathrm{ess}}(T_\varphi)$* (somewhat implicit). More precisely, the following theorem holds, where point (1) is simply a reiteration of Theorem 3.1.3(3).

Theorem 3.1.10 (criterion for $T_\varphi \in$ Fred [Douglas and Sarason, 1970]) *Let* $\varphi \in L^\infty(\mathbb{T})$.

(1) $T_\varphi \in \mathrm{Fred}(H^2) \Rightarrow \varphi \in L^\infty(\mathbb{T})^{-1}$ *and* $\|1/\varphi\|_\infty \leq \|(\Pi(T_\varphi))^{-1}\|$.
(2) *If* $\varphi \in L^\infty(\mathbb{T})^{-1}$, *then* $T_\varphi \in \mathrm{Fred}(H^2) \Leftrightarrow T_{\varphi/|\varphi|} \in \mathrm{Fred}(H^2) \Leftrightarrow T_{\varphi/[\varphi]} \in \mathrm{Fred}(H^2)$, *where $[\varphi]$ is the outer function with the same modulus as φ (see Appendix F.2).*
(3) *Suppose that $|\varphi| = 1$ a.e. on $\mathbb{T}$. Then,*

$$T_\varphi \in \mathrm{Fred}(H^2) \Leftrightarrow \|H_\varphi\|_{\mathrm{ess}} = \mathrm{dist}_{L^\infty}(\varphi, H^\infty + C(\mathbb{T})) < 1,$$

$$\|H_{\overline{\varphi}}\|_{\mathrm{ess}} = \mathrm{dist}_{L^\infty}(\overline{\varphi}, H^\infty + C(\mathbb{T})) < 1.$$

In the case where T_φ is Fredholm, then

$$\mathrm{dist}_{L^\infty}(\varphi, H^\infty + C(\mathbb{T})) = \mathrm{dist}_{L^\infty}(\overline{\varphi}, H^\infty + C(\mathbb{T})) =: d$$

and

$$\|(\Pi(T_\varphi))^{-1}\|^2 = \frac{1}{1 - d^2}.$$

Proof (1) is part of Theorem 3.1.3.

(2) $\varphi = \overline{h}(\varphi/|\varphi|)h$ where $h = [\varphi]^{1/2}$ (the outer function with modulus $|\varphi|^{1/2}$, see Appendix F.2) and hence $T_\varphi = T_{\overline{h}}T_{\varphi/|\varphi|}T_h$ where T_h and $T_{\overline{h}} = T_h^*$ are invertible (since $h^{\pm 1} \in H^\infty$). Similarly, $T_\varphi = T_{\varphi/[\varphi]}T_{[\varphi]}$, and the result follows.

(3) By Lemma 2.2.9, $I - T_\varphi^* T_\varphi = H_\varphi^* H_\varphi = \Gamma_\varphi^* \Gamma_\varphi$ (where $\Gamma_\varphi = JH_\varphi \colon H^2 \to H^2$) which can be projected into the Calkin algebra $\mathcal{K} = L(H^2)/\mathfrak{S}_\infty(H^2)$,

$$I - \Pi(T_\varphi)^* \Pi(T_\varphi) = \Pi(\Gamma_\varphi)^* \Pi(\Gamma_\varphi).$$

We now use a few elementary properties of the algebra $\mathcal{K}$, in particular that $\mathcal{K}$ *is a C^*-algebra*: see for example [Dixmier, 1996]. Then, a positive element

$\Pi(T_\varphi)^*\Pi(T_\varphi) = I - \Pi(\Gamma_\varphi)^*\Pi(\Gamma_\varphi)$ is invertible if and only if it is strictly positive, $\Pi(T_\varphi)^*\Pi(T_\varphi) \geq \alpha^2 I$, $\alpha > 0$, and with the best possible lower bound we obtain

$$\|(\Pi(T_\varphi)^*\Pi(T_\varphi))^{-1}\| = 1/\alpha^2.$$

On the other hand, the lower bound of $I - \Pi(\Gamma_\varphi)^*\Pi(\Gamma_\varphi)$ is equal to $1 - \beta^2$ where β^2 is the upper bound of $\Pi(\Gamma_\varphi)^*\Pi(\Gamma_\varphi)$, $\beta^2 = \|\Pi(\Gamma_\varphi)^*\Pi(\Gamma_\varphi)\| = \|\Pi(\Gamma_\varphi)\|^2$. Thus, the invertibility of $\Pi(T_\varphi)^*\Pi(T_\varphi)$ is equivalent to $\|\Pi(\Gamma_\varphi)\|^2 < 1$, or again to

$$\|\Pi(\Gamma_\varphi)\|^2 = \|H_\varphi\|^2_{\text{ess}} = \text{dist}_{L^\infty}(\varphi, H^\infty + C(\mathbb{T})) < 1$$

(see the AAK Theorem 2.2.8(1)). Similarly, we obtain that the invertibility of $\Pi(T_{\bar\varphi})^*\Pi(T_{\bar\varphi}) = \Pi(T_\varphi)\Pi(T_\varphi)^*$ is equivalent to $\|\Pi(\Gamma_{\bar\varphi})\|^2 < 1$. However, for any operator A (or an element of a C^*-algebra), the Hermitian square A^*A is invertible if and only if A is *left-invertible* (with left inverse $(A^*A)^{-1}A^*$).

It only remains to recall that A is invertible if and only if A and A^* are left-invertible.

If $A = \Pi(T_\varphi)$ is invertible, then

$$\|\Pi(T_\varphi)^{-1}\|^2 = \|(\Pi(T_\varphi)^*\Pi(T_\varphi))^{-1}\| = \frac{1}{1 - \|\Pi(\Gamma_\varphi)\|^2}.$$

The equality $\text{dist}_{L^\infty}(\varphi, H^\infty + C(\mathbb{T})) = \text{dist}_{L^\infty}(\bar\varphi, H^\infty + C(\mathbb{T}))$ follows from the fact that $\|\Pi(T_\varphi)^{-1}\| = \|(\Pi(T_\varphi)^{-1})^*\| = \|\Pi(T_{\bar\varphi})^{-1}\|$. ∎

Another means to establish the Fredholm property of a Toeplitz operator – an important "local principle" – is presented in the next section.

Remark 3.1.11 Is there a Toeplitz regularizer for a $T_\varphi \in$ Fred? If $T_\varphi \in$ Fred and T_ψ is a regularizer of T_φ, then clearly $T_\psi T_\varphi - I \in \mathfrak{S}_\infty$, and thus necessarily $\psi = 1/\varphi$. Then Theorem 3.1.6 shows that for T_φ with $\varphi \in C(\mathbb{T})$ (and even for the operators of the algebra $\text{alg}\mathcal{T}_{C(\mathbb{T})}$), the answer is "yes," and in Exercise 3.4.3 we will see that this is also the case for $T \in \text{alg}\mathcal{T}_{C(\mathbb{T})+H^\infty(\mathbb{T})}$. For more information, and, in particular, for some examples where the answer is "no," see Exercise 3.4.12 and §3.5.

3.2 The Simonenko Local Principle

The "local principle" of the Fredholm theory for Toeplitz operators is a family of propositions stating that the property $T_\varphi \in \text{Fred}(H^2)$ depends only on the "local properties" of the symbol φ (properties of restrictions of φ to small intervals). As will be seen in §3.3, this is a significant difference from the property of invertibility of T_φ. Today, several versions of the "local principle"

3.2 The Simonenko Local Principle

are known (see §3.5 for a few references), but here we limit ourselves to a basic theorem, the very first local principle invented by Igor Simonenko [1964], following the version of Douglas and Sarason [1970].

Theorem 3.2.1 (Simonenko, 1964) *Let $\varphi \in L^\infty(\mathbb{T})$ and suppose that for every $\lambda \in \mathbb{T}$ there exists $\varphi_\lambda \in L^\infty(\mathbb{T})$ such that*

(1) $T_{\varphi_\lambda} \in \text{Fred}(H^2)$,
(2) $\overline{\lim}_{\zeta \to \lambda} |\varphi(\zeta) - \varphi_\lambda(\zeta)| =: \text{dist}_\lambda(\varphi, \varphi_\lambda) = 0$.

Then, $T_\varphi \in \text{Fred}(H^2)$.

The proof depends on the following lemma and of the local distance dist_λ defined for two functions $f, g \in L^\infty(\mathbb{T})$ by

$$\text{dist}_\lambda(f, g) = \overline{\lim_{\zeta \to \lambda}} |f(\zeta) - g(\zeta)| = \lim_{\epsilon \to 0} \|f - g\|_{L^\infty(D(\lambda, \epsilon))},$$

and for an $f \in L^\infty(\mathbb{T})$ and a set $M \subset L^\infty(\mathbb{T})$ by

$$\text{dist}_\lambda(f, M) = \inf_{g \in M} \text{dist}_\lambda(f, g).$$

We leave as an exercise the following properties of dist_λ:

(1) $\lambda \mapsto \text{dist}_\lambda(f, M)$ is upper semi-continuous (and hence attains its maximum on $\mathbb{T}$);
(2) for every $f, g \in L^\infty(\mathbb{T})$, $\|f - g\|_\infty = \max_{\lambda \in \mathbb{T}} \text{dist}_\lambda(f, g)$;
(3) if the constants belong to M ($M \subset L^\infty(\mathbb{T})$), then for every $f \in L^\infty(\mathbb{T})$, $\text{dist}_\lambda(f, M) = \text{dist}_\lambda(f, M + C(\mathbb{T}))$.

Igor Simonenko (in Russian Игорь Борисович Симоненко, 1935–2008) was a Russian/Soviet mathematician, a great expert in singular integral operators (SIOs), as well as in the Riemann–Hilbert problem (RHP) and in several branches of physics and applied mechanics. His principal invention is the "local principle" (1967) for the Wiener–Hopf operators (WHO), the RHP and SIOs, a powerful technique now standard in every domain of Toeplitz–WHO–SIO–RHP. This led him to the definition and theory of "local-type operators." These allow the techniques of localization to be applied to the WHO on the cones of $\mathbb{R}^n$ and to the pseudo-differential operators, as well as to the inversion of operators by the *finite section method* (Kozak, Silbermann, and Böttcher; see Chapter 5).

In order to support his family during a difficult period for the country (with a dramatic degradation of living conditions during and after the Second World War, the family was evacuated from Kiev to the Salsk steppes, soon to become part of the battle zones of Stalingrad, etc.), Simonenko – after graduating from a technical school (Technicum) – worked in a factory, and then combined work with university studies by correspondence. He submitted his thesis in 1961 (University of Rostov–Don) and obtained his *Habilitation* in 1967. As professor at the University of Rostov–Don, Simonenko often collaborated with applied mathematicians such as Yudovich, Moiseyev, Vorovich, Gakhov, and others, He founded his own "school," serving as the thesis director for two dozen students.

Lemma 3.2.2 (Douglas and Sarason, 1970) *For every $f \in L^\infty(\mathbb{T})$,*

$$\text{dist}(f, H^\infty + C(\mathbb{T})) = \max_{\lambda \in \mathbb{T}} \text{dist}_\lambda(f, H^\infty),$$

where dist *is the distance in $L^\infty(\mathbb{T})$.*

Proof (Sarason, 1973) Clearly,

$$\text{dist}(f, H^\infty + C(\mathbb{T})) \geq M =: \max_{\lambda \in \mathbb{T}} \text{dist}_\lambda(f, H^\infty).$$

For the converse, let $\epsilon > 0$ and, for every $\lambda \in \mathbb{T}$, $h_\lambda \in H^\infty$ such that $\text{dist}_\lambda(f, h_\lambda) < \epsilon + M$. By property (1) of the local distance, there exists an open arc $A(\lambda)$ such that $\lambda \in A(\lambda)$ and $\text{dist}_\zeta(f, h_\lambda) < 2\epsilon + M$ for every $\zeta \in A(\lambda)$.

Now, let $\alpha = (A(\lambda_j))$, $1 \leq j \leq n$, be a finite subcovering of $\mathbb{T}$ and (a_j) a partition of the unity subordinate to α: $a_j \in C(\mathbb{T})$, $a_j \geq 0$, $\text{supp}(a_j) \subset \overline{A}(\lambda_j)$ and $\sum_j a_j = 1$ (for example, $a_j = b_j / \sum_i b_i$ where $b_j = 0$ outside of $A(\lambda_j)$ and $b_j(\zeta) = \text{dist}(\zeta, \partial A(\lambda_j))$ for $\zeta \in A(\lambda_j)$). We set $g = \sum_j a_j h_{\lambda_j}$. Since $H^\infty + C$ is an algebra (see Lemma 2.2.7), we have $g \in H^\infty + C(\mathbb{T})$, and for every $\lambda \in \mathbb{T}$

$$\text{dist}_\lambda(f, g) = \text{dist}_\lambda\left(\sum_j a_j f, \sum_j a_j h_{\lambda_j}\right)$$

$$\leq \sum_j \text{dist}_\lambda(a_j f, a_j h_{\lambda_j})$$

$$= \sum_{j:\ \lambda \in A(\lambda_j)} \text{dist}_\lambda(a_j f, a_j h_{\lambda_j})$$

$$= \sum_{j:\ \lambda \in A(\lambda_j)} a_j(\lambda) \text{dist}_\lambda(f, h_{\lambda_j})$$

$$< (M + 2\epsilon) \sum_{j:\ \lambda \in A(\lambda_j)} a_j(\lambda)$$

$$\leq M + 2\epsilon.$$

As $\epsilon > 0$ is arbitrarily small, we obtain $\text{dist}(f, H^\infty + C(\mathbb{T})) \leq M$. ∎

3.2.1 Proof of Theorem 3.2.1 (Sarason, 1973)

By Theorem 3.1.3(3), for every $\lambda \in \mathbb{T}$, we have $\epsilon_\lambda =: \text{ess inf}|\varphi_\lambda| > 0$. Let $J(\lambda)$ be an open arc on which $\text{dist}_\zeta(\varphi, \varphi_\lambda) < \epsilon_\lambda/2$ ($\zeta \in J(\lambda)$), and hence $|\varphi(\zeta)| > \epsilon_\lambda - |\varphi(\zeta) - \varphi_\lambda(\zeta)| > \epsilon_\lambda/2$ (for almost every $\zeta \in J(\lambda)$). By selecting a finite subcovering $(J(\lambda_i))$, we obtain $\text{ess inf}_\mathbb{T}|\varphi| > 0$. Since $\varphi(\zeta) = \varphi_\lambda(\zeta) + o(1)$ for $\zeta \to \lambda$, we have

$$\text{dist}_\lambda\left(\frac{\varphi}{|\varphi|}, \frac{\varphi_\lambda}{|\varphi_\lambda|}\right) = \text{dist}_\lambda(\varphi, \varphi_\lambda) = 0,$$

and then, by Lemma 3.2.2, there exists $\lambda \in \mathbb{T}$ such that

$$\text{dist}(\varphi/|\varphi|, H^\infty + C(\mathbb{T})) = \text{dist}_\lambda(\varphi/|\varphi|, H^\infty),$$

and hence

$$\text{dist}\left(\frac{\varphi}{|\varphi|}, H^\infty + C(\mathbb{T})\right) \leq \text{dist}_\lambda\left(\frac{\varphi}{|\varphi|}, \frac{\varphi_\lambda}{|\varphi_\lambda|}\right) + \text{dist}_\lambda(\varphi_\lambda/|\varphi_\lambda|, H^\infty)$$

$$= \text{dist}_\lambda\left(\frac{\varphi_\lambda}{|\varphi_\lambda|}, H^\infty + C(\mathbb{T})\right)$$

$$\leq \text{dist}\left(\frac{\varphi_\lambda}{|\varphi_\lambda|}, H^\infty + C(\mathbb{T})\right) < 1$$

(the last inequality follows from Theorem 3.1.10(2,3)). The same computation but for $\overline{\varphi}$ shows that

$$\text{dist}\left(\frac{\overline{\varphi}}{|\varphi|}, H^\infty + C(\mathbb{T})\right) < 1.$$

Finally, by applying Theorem 3.1.10, we obtain $T_\varphi \in \text{Fred}(H^2)$. ∎

Remark 3.2.3 The proof shows that condition (ii) of Theorem 3.2.1 can be weakened to

$$\mathrm{dist}_\lambda\left(\frac{\varphi_\lambda}{|\varphi_\lambda|}, \frac{\varphi}{|\varphi|}\right) + \mathrm{dist}_\lambda\left(\frac{\varphi_\lambda}{|\varphi_\lambda|}, H^\infty + C(\mathbb{T})\right) < 1 \quad (\forall \lambda \in \mathbb{T}).$$

It is also evident that an analog of Lemma 3.2.2 remains valid for $C(\mathbb{T})$ in place of $H^\infty + C(\mathbb{T})$:

$$\mathrm{dist}(f, C(\mathbb{T})) = \max_{\lambda \in \mathbb{T}} \mathrm{dist}_\lambda(f, C(\mathbb{T})) = \max_{\lambda \in \mathbb{T}} \mathrm{dist}_\lambda(f, \mathbb{C}),$$

where $\mathbb{C}$ is used to denote the set of constant functions.

3.2.2 Examples

(a) If T_φ is locally Fredholm (i.e. $\forall \lambda \in \mathbb{T}$, $\exists V_\lambda$ neighborhood of λ and $\exists T_{\varphi_\lambda} \in$ Fred such that $\varphi = \varphi_\lambda$ on V_λ), then it is Fredholm. (For an independent direct proof see Exercise 3.4.5 below.)

(b) If $\varphi^{\pm 1} \in L^\infty(\mathbb{T})$ and φ is locally sectorial (i.e. $\forall \lambda \in \mathbb{T}$, $\exists V_\lambda$ a neighborhood of λ such that the range $\mathrm{Ran}_{\mathrm{ess}}(\varphi|V_\lambda)$ is contained within an angle of value strictly inferior to π), then $T_\varphi \in$ Fred (see Lemma 3.3.2 below).

(c) If $\varphi^{\pm 1} \in C$ and, for every $\lambda \in \mathbb{T}$, the local oscillation

$$\omega_\lambda(\arg(\varphi)) =: \lim_{\epsilon \to 0} \sup_{u,v \in D(\lambda,\epsilon)} |\arg(\varphi(u)) - \arg(\varphi(v))|$$

is $< \pi$, then $T_\varphi \in$ Fred.

In fact, this is another form of (b); a sufficient condition for (b) and (c) is

$$\omega_\lambda(\varphi) < \pi \cdot \varliminf_{\zeta \to \lambda} |\varphi(\zeta)| \quad (\forall \lambda \in \mathbb{T}).$$

(d) Let $PC(\mathbb{T})$ be the set of *regulated functions* on the circle $\mathbb{T}$, i.e. the uniform closure (hence, in the space $L^\infty(\mathbb{T})$) of the space of "piecewise continuous" functions (it is equivalent to say (exercise!) that $PC(\mathbb{T})$ is the space of φ admitting limits to the left $\varphi(\lambda-)$ and to the right $\varphi(\lambda+)$ at every point of $\mathbb{T}$):

$$PC(\mathbb{T}) = \mathrm{span}_{L^\infty(\mathbb{T})}(\chi_I : I \subset \mathbb{T}, \text{ an arc of } \mathbb{T}).$$

A function $\varphi \in PC(\mathbb{T})$ is always normalized (as an element of $L^\infty(\mathbb{T})$) by the condition $\varphi(\lambda) = \varphi(\lambda+)$ ($\forall \lambda \in \mathbb{T}$). If $\varphi \in PC(\mathbb{T})$, and $0 \notin [\varphi(\lambda-), \varphi(\lambda+)]$ ($\forall \lambda \in \mathbb{T}$), then $T_\varphi \in$ Fred.

Indeed, the last condition implies (b) and (c). ■

3.3 The Principal Criterion of Invertibility

This section contains a complete description of the spectrum $\sigma(T_\varphi)$ of a Toeplitz operator as a function of the symbol φ. The language is at times a bit complicated and we seek alternative criteria for certain particularly important cases.

Lemma 3.3.1 (reduction to unimodular symbols) Let $\varphi \in L^\infty(\mathbb{T})$.

(1) If T_φ is invertible, then $1/\varphi \in L^\infty(\mathbb{T})$ and $\|1/\varphi\|_\infty \leq \|T_\varphi^{-1}\|$.
(2) If $1/\varphi \in L^\infty(\mathbb{T})$, then the following assertions are equivalent.
   (a) T_φ is invertible.
   (b) $T_{\varphi/|\varphi|}$ is invertible.
   (c) $T_{\varphi/[\varphi]}$ is invertible (where $[\varphi]$ is an H^∞ outer function with $|[\varphi]| = |\varphi|$ a.e. on $\mathbb{T}$: see Appendix F).

Moreover, $\|T_\varphi^{-1}\| \leq \|1/\varphi\|_\infty \|T_u^{-1}\|$ where $u = \varphi/|\varphi|$ or $u = \varphi/[\varphi]$.

Proof (1) The property is part of Theorem 3.1.3(3).

(2) Let h be an outer function defined by $|h|^2 = |1/\varphi|$, i.e. $h = [|1/\varphi|^{1/2}]$: see Appendix F. Clearly $h^{\pm 1} \in H^\infty$, and hence T_h is invertible and $T_h^{-1} = T_{1/h}$ (thus $\|T_h^{-1}\| = \|1/h\|_\infty = \|1/\varphi\|_\infty^{1/2}$).

By Lemma 3.1.4(1) and Theorem 2.1.5, $T_{\varphi/|\varphi|} = T_h^* T_\varphi T_h$ and $T_{\varphi/[\varphi]} = T_\varphi T_{1/[\varphi]}$, and the result follows. ∎

Lemma 3.3.2 (the case of a sectorial symbol) Let $\varphi \in L^\infty(\mathbb{T})$ be a sectorial function of angle 2α, i.e. $|\arg(e^{i\theta}\varphi(\zeta))| \leq \alpha$ a.e. on $\mathbb{T}$ where $\alpha < \pi/2$ and $\theta \in [0, 2\pi)$. Then, T_φ is invertible if and only if $1/\varphi \in L^\infty(\mathbb{T})$, and under this condition

$$\|T_\varphi^{-1}\| \leq \frac{\|1/\varphi\|_\infty}{\cos(\alpha)}.$$

Proof Supposing $\|\varphi\|_\infty = 1$, $\theta = 0$ and $1/\varphi \in L^\infty(\mathbb{T})$, we observe that $\mathrm{Ran}_{\mathrm{ess}}(\varphi) \subset \Omega(a, b, \alpha)$, where $b = 1$, $a = \|1/\varphi\|_\infty^{-1}$ and

$$\Omega(a, b, \alpha) = \{z \in \mathbb{C} : a \leq |z| \leq b, |\arg(z)| \leq \alpha\}.$$

Let $\lambda > 0$ and let $\overline{D}(\lambda, R) = \{z \in \mathbb{C} : |z - \lambda| \leq R\}$, $0 < R < \lambda$, be a disk containing the points $a\, e^{\pm i\alpha}$, and hence – for λ large enough – containing all of $\Omega(a, b, \alpha)$; see Figure 3.1.

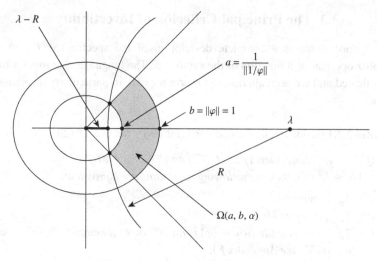

Figure 3.1 The disk $D(\lambda, R)$, $\lambda \to \infty$.

We will then have $|1 - \varphi/\lambda| \leq R/\lambda < 1$, and then $T_\varphi = \lambda(I - (I - T_{\varphi/\lambda}))$, thus

$$\|T_\varphi^{-1}\| = \lambda^{-1}\left\|\sum_{n\geq 0}(I - T_{\varphi/\lambda})^n\right\| \leq \lambda^{-1}\frac{1}{1 - R/\lambda} = \frac{1}{\lambda - R}.$$

However, $\lim_{\lambda \to \infty}(\lambda - R) = \text{Re}(a\,e^{i\alpha}) = \|1/\varphi\|_\infty^{-1}\cos(\alpha)$, and the result follows. ∎

Corollary 3.3.3 (spectral inclusion) *For every $\varphi \in L^\infty(\mathbb{T})$,*

$$\sigma(\varphi) \subset \text{conv}(\text{Ran}_{\text{ess}}(\varphi)),$$

where conv *is the closed convex hull. Moreover, for $\lambda \in \mathbb{C} \setminus \text{conv}(\text{Ran}_{\text{ess}}(\varphi))$,*

$$\|(\lambda I - T_\varphi)^{-1}\| \leq \frac{1}{\text{dist}(\lambda, \text{Ran}_{\text{ess}}(\varphi)) \cdot \cos(\alpha(\lambda, \text{Ran}_{\text{ess}}\varphi))/2)},$$

where $\alpha(\lambda, \text{Ran}_{\text{ess}}(\varphi))$ is the angle through which $\text{Ran}_{\text{ess}}(\varphi)$ is seen from the point λ.

Indeed, if $\lambda \in \mathbb{C} \setminus \text{conv}(\text{Ran}_{\text{ess}}(\varphi))$, the function $\lambda - \varphi$ is sectorial and $\|(\lambda - \varphi)^{-1}\|_\infty^{-1} = \text{dist}(\lambda, \text{Ran}_{\text{ess}}(\varphi))$. We apply Lemma 3.3.2. ∎

Lemma 3.3.4 (unimodular symbols) *Let $u \in L^\infty(\mathbb{T})$, $|u| = 1$ a.e. on $\mathbb{T}$. The following assertions are equivalent.*

(1) *T_u is invertible.*
(2) *$\text{dist}_{L^\infty}(u, H^\infty) < 1$ and $\text{dist}_{L^\infty}(\overline{u}, H^\infty) < 1$.*

3.3 The Principal Criterion of Invertibility

(3) *There exist real functions $a, b \in L^\infty(\mathbb{T})$ and a number $c \in \mathbb{R}$ such that $u = e^{i(a+\tilde{b}+c)}$ and $\|a\|_\infty < \pi/2$ (where $\tilde{b} = 1/i\, \mathbb{H}b$ is the harmonic conjugate of b).*

Moreover, in the case of invertibility, $\mathrm{dist}_{L^\infty}(u, H^\infty) = \mathrm{dist}_{L^\infty}(\bar{u}, H^\infty) =: d$ *and*

$$\|T_u^{-1}\| = \frac{1}{(1-d^2)^{1/2}},$$

$$\|T_u^{-1}\| \leq \frac{1 + e^{L(b)}}{2\, e^{L(b)/2} \cos(\|a\|_\infty)} = \frac{\cosh(L(b)/2)}{\cos(\|a\|_\infty)},$$

where a and b come from the representation of (3) and $L(b) = \overline{M} - \underline{M}$ is the "width of b," with $\overline{M} = \mathrm{ess\,sup}_\mathbb{T} b$, $\underline{M} = \mathrm{ess\,inf}_\mathbb{T} b$.

Proof (1) $\Leftrightarrow$ (2) Recall that according to Appendix D.3, an operator $T: X \to Y$ is invertible if and only if there exists $\epsilon, \delta > 0$ such that

$$\|Tx\| \geq \epsilon\|x\| \quad \text{and} \quad \|T^*y\| \geq \delta\|y\| \quad (\forall x \in X, y \in Y^*),$$

and if T is invertible then $\|T^{-1}\| = 1/\epsilon = 1/\delta$ (with the best possible values of ϵ and δ). In the case $T = T_u$, we have $T_u^* = T_{\bar{u}}$ and, for every $f \in H^2$,

$$\|f\|_2^2 = \|uf\|_{L^2}^2 = \|T_u f\|_2^2 + \|H_u f\|_2^2, \quad \|f\|_2^2 = \|T_u^* f\|_2^2 + \|H_{\bar{u}} f\|_2^2,$$

where $H_u = P_- M_u: H^2 \to H_-^2$ is the Hankel operator with symbol u. The largest possible values for ϵ and δ are

$$\epsilon^2 = \inf_{\|f\|_2 = 1} \|T_u f\|_2^2 = 1 - \|H_u\|^2 = 1 - \mathrm{dist}_{L^\infty}(u, H^\infty)^2,$$

$$\delta^2 = \inf_{\|f\|_2 = 1} \|T_{\bar{u}} f\|_2^2 = 1 - \|H_{\bar{u}}\|^2 = 1 - \mathrm{dist}_{L^\infty}(\bar{u}, H^\infty)^2,$$

hence (1) $\Leftrightarrow$ (2).

(2) $\Rightarrow$ (3) Let $g, h \in H^\infty$ be such that $\|u - h\|_\infty = \mathrm{dist}_{L^\infty}(u, H^\infty) =: d < 1$ and $\|\bar{u} - g\|_\infty = \mathrm{dist}_{L^\infty}(\bar{u}, H^\infty) = d < 1$ (the distances are attained (see Theorem 2.2.4) and equal according to (1) $\Leftrightarrow$ (2)). This gives $|1 - \bar{u}h| \leq d$ and $|1 - ug| \leq d$ a.e. on $\mathbb{T}$, thus (for almost every $\zeta \in \mathbb{T}$) $\bar{u}(\zeta)h(\zeta) \in \overline{D}(1, d)$ and $u(\zeta)g(\zeta) \in \overline{D}(1, d)$ (see Figure 3.2), which implies

$$h(\zeta)g(\zeta) \in \overline{D}(1, d) \cdot \overline{D}(1, d) = \{zw: z, w \in \overline{D}(1, d)\} =: \omega.$$

Clearly (see Figure 3.3 for Ω),

$$\omega \subset \Omega(r, s, \alpha) = \{z \in \mathbb{C}: r \leq |z| \leq s, |\arg(z)| \leq \alpha\},$$
$$r = (1-d)^2, \quad s = (1+d)^2, \quad \alpha = 2\arcsin(d),$$

where $\arcsin(d) < \pi/2$, and $\mathbb{C} \setminus \Omega$ is connected.

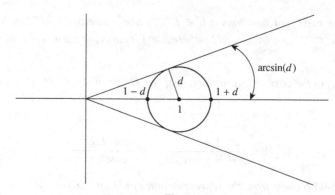

Figure 3.2 The disk $\overline{D}(1,d)$ of Lemma 3.3.4.

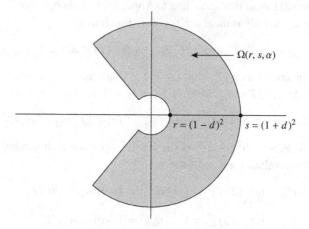

Figure 3.3 The domain $\Omega(r, s, \alpha)$ of Lemma 3.3.4.

The following Lemma 3.3.5 implies $h(z)g(z) \in \Omega = \Omega(r, s, \alpha)$ for every $z \in \mathbb{D}$, giving

$$\frac{(1-d)^2}{\|g\|_\infty} \leq |h(z)| \leq (1+d)^2 \|g\|_\infty \quad (z \in \mathbb{D}),$$

and thus $b =: \log|h| \in L^\infty(\mathbb{T})$ and $h = e^{b+i\tilde{b}+ic}$ where $c \in \mathbb{R}$. Finally,

$$u = u\frac{\overline{h}}{|h|} \cdot \frac{h}{|h|} = e^{ia} e^{i\tilde{b}+ic},$$

where $a = \arg(u\overline{h})$ and, since $\overline{u}h \in \overline{D}(1,d)$,

$$\|a\|_\infty \leq \arcsin(d) < \pi/2.$$

3.3 The Principal Criterion of Invertibility

(3) ⇒ (2) Let a, b, c be the functions of (3), and $h = e^{b+i\tilde{b}+ic}$. Then, $h^{\pm 1} \in H^\infty$ and $\arg(\bar{u}h) = -a$, $\|a\|_\infty < \pi/2$, and thus the values of $\bar{u}h$ are located in the domain $\Omega = \Omega(r, s, \alpha)$,

$$\Omega(r, s, \alpha) = \{z \in \mathbb{C} : r \leq |z| \leq s, |\arg(z)| \leq \alpha\},$$

where $r = \|h^{-1}\|_\infty^{-1}$, $s = \|h\|_\infty$, $\alpha = \|a\|_\infty < \pi/2$. Similarly to Lemma 3.3.2, let $\overline{D}(\lambda, R)$ ($\lambda > 0$) be the disk whose boundary contains the points $re^{\pm i\alpha}$ and $se^{\pm i\alpha}$, and hence the domain Ω (see Figure 3.4). Then, $|\bar{u}h - \lambda| \leq R$ (a.e. on $\mathbb{T}$), hence

$$\mathrm{dist}_{L^\infty}(u, H^\infty) \leq \|(h/\lambda) - u\|_\infty = \|(\bar{u}h/\lambda) - 1\|_\infty \leq R/\lambda.$$

By the definition of $\overline{D}(\lambda, R)$, we have (see also Figure 3.4)

$$(\lambda - r\cos(\alpha))^2 + r^2\sin^2(\alpha) = (\lambda - s\cos(\alpha))^2 + s^2\sin^2(\alpha) = R^2,$$

hence $r^2 - 2r\lambda\cos(\alpha) = s^2 - 2s\lambda\cos(\alpha)$,

$$\lambda = \frac{r+s}{2\cos(\alpha)},$$

and

$$\mathrm{dist}_{L^\infty}(u, H^\infty)^2 \leq (R/\lambda)^2 = (r/\lambda)^2 - 2r\cos(\alpha)/\lambda + 1$$
$$= 1 - \frac{4\cos^2(\alpha)}{1 + (s/r)} + \frac{4\cos^2(\alpha)}{(1 + (s/r))^2} = 1 - \frac{4(s/r)\cos^2(\alpha)}{(1 + (s/r))^2}.$$

Since $0 \leq \alpha < \pi/2$, $\mathrm{dist}_{L^\infty}(u, H^\infty)^2 < 1$. In the same manner (with $\bar{u}$, h^{-1}, etc.), $\mathrm{dist}_{L^\infty}(\bar{u}, H^\infty) < 1$. Assertion (2) thus follows.

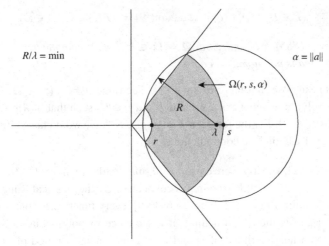

Figure 3.4 The disk $\overline{D}(\lambda, R)$ and the domain $\Omega(r, s, \alpha)$ of Lemma 3.3.4.

To conclude the above estimate, we denote

$$\overline{M}(b) = \operatorname{ess\,sup}_{\mathbb{T}} b, \quad \underline{M}(b) = \operatorname{ess\,inf}_{\mathbb{T}} b \quad \text{and} \quad L(b) = \overline{M}(b) - \underline{M}(b),$$

the "width of b." Then,

$$s/r = \|h\|_\infty / \|h^{-1}\|_\infty^{-1} = e^{L(b)},$$

$$\operatorname{dist}_{L^\infty}(u, H^\infty)^2 \leq 1 - \frac{4\,e^{L(b)} \cos^2(\|a\|_\infty)}{(1 + e^{L(b)})^2},$$

$$\operatorname{dist}_{L^\infty}(u, H^\infty) < 1 - \frac{2\,e^{L(b)} \cos^2(\|a\|_\infty)}{(1 + e^{L(b)})^2},$$

and similarly for $\operatorname{dist}_{L^\infty}(\overline{u}, H^\infty)$. Finally,

$$\|T_u^{-1}\| = \frac{1}{(1 - d^2)^{1/2}} \leq \frac{1 + e^{L(b)}}{2\,e^{L(b)/2} \cos(\|a\|_\infty)}. \qquad \blacksquare$$

Remark The upper bound of $\|T_u^{-1}\|$ given in Lemma 3.3.4 behaves as const. $\cdot\, e^{L(b)/2}$ when $L(b) \to \infty$, and as $\operatorname{const.}/((\pi/2) - \|a\|_\infty)$ when $\|a\|_\infty \to \pi/2$.

Lemma 3.3.5 (polynomial convexity)

(1) *Let $\Omega = \overline{\Omega}$ be a bounded subset of $\mathbb{C}$ that is polynomially convex in the following sense: for $\lambda \in \mathbb{C}$, if for every complex polynomial $p \in \mathcal{P}_a$, $|p(\lambda)| \leq \max_{z \in \Omega} |p(z)|$, then $\lambda \in \Omega$. Then, the following "maximum principle" holds:*

$$(F \in H^\infty(\mathbb{D}), F(\zeta) \in \Omega \text{ a.e. on } \mathbb{T}) \Rightarrow (F(z) \in \Omega, \forall z \in \mathbb{D}).$$

(2) *(Runge, 1885) A bounded set $\Omega = \overline{\Omega} \subset \mathbb{C}$ whose complement $\mathbb{C} \setminus \Omega$ is connected is polynomially convex.*

Proof (1) Suppose that there exists $z \in \mathbb{D}$ such that $\lambda =: F(z) \notin \Omega$. Then, using the polynomial convexity, we could find $p \in \mathcal{P}_a$ such that $|p(\lambda)| > 1$ and $|p(z)| \leq 1$ ($\forall z \in \Omega$). For the function $f = p \circ F \in H^\infty$, we would have $|f(z)| > 1$ and $|f(\zeta)| \leq 1$ a.e. on $\mathbb{T}$, a contradiction.

(2) Let $\lambda \in \mathbb{C} \setminus \Omega$. Clearly there exists $\epsilon > 0$ sufficiently small that $\Omega \cap \overline{D}(\lambda, \epsilon) = \emptyset$ and the set $\mathbb{C} \setminus (\Omega \cup \overline{D}(\lambda, \epsilon))$ remains connected. By the classical Runge theorem (see [Rudin, 1986], Theorems 13.6–13.9), every function F holomorphic on $\Omega \cup \overline{D}(\lambda, \epsilon)$ is the uniform limit of a sequence of polynomials. Setting $F = 1/2$ on a neighborhood of Ω and $F = 2$ on a neighborhood of $\overline{D}(\lambda, \epsilon)$,

3.3 The Principal Criterion of Invertibility

and approximating F by polynomials, we find $p \in \mathcal{P}_a$ such that $|p(\lambda)| > 1$ and $|p(z)| \leq 1$ ($\forall z \in \Omega$). Hence, Ω is polynomially convex. ∎

Carl Runge (1856–1927), a German mathematician and physicist, was the inventor of the *Runge–Kutta method* (the most widely used method for the numerical solution of differential equations), of *Runge's theorem* (on polynomial and rational approximation, which "belongs to every analyst's bag of tricks" according to a contemporary expert), and of the *Runge–Lenz vector* in physics. He submitted his thesis under the direction of Karl Weierstrass (in differential geometry, 1880) and his *Habilitation* thesis under the direction of Leopold Kronecker (numerical solution of algebraic equations, 1883). Professor at the University of Hanover in 1886 (where he spent 18 years), he then obtained the chair of Applied Mathematics at Göttingen (until his retirement in 1923). A friend of Max Planck, Runge made experimental spectroscopy his second field of interest, and there obtained and published many results (in particular, on the classification of spectral lines of helium).

Runge grew up in a family of eight children, bilingual in English and German (his mother Fanny Tolmé was descended from Huguenots who had settled in England, so he also had French roots), but he lost his father when he was eight. In 1885 Runge asked for the hand of Aimée, daughter of the famous physiologist Emil du Bois-Reymond (brother of the mathematician Paul du Bois-Reymond), but he had to wait for her father's permission until he had obtained the position of professor in 1887. One of Runge's sons, Wilhelm, became an electrical engineer and participated in the development of radar. His daughter Iris wrote a biography of her father and became one of the very first female mathematicians to be employed in industry. Another of their four daughters, Nina, married Richard Courant (a close collaborator of Hilbert in the creation of spectral theory). Throughout his life Runge was an active sportsman (in the English tradition), but he died suddenly of a heart attack.

3.3.1 Wiener–Hopf Factorization

Let $\varphi \in L^1(\mathbb{T})$. φ is said to admit a *Wiener–Hopf factorization* if there exist functions φ_1, φ_2 such that

$$\varphi = \varphi_1 \overline{\varphi_2}, \quad \varphi_i^{\pm 1} \in H^2$$

and if the mapping

$$A: f \to \frac{1}{\varphi_1} P_+ \frac{1}{\overline{\varphi_2}} f \quad (f \in \mathcal{P}_a)$$

can be extended to a bounded operator $H^2 \to H^2$.

The link between Wiener–Hopf factorizations and the invertibility of Toeplitz operators was established by Wiener and Hopf [1931], but the final form of this link awaited the works of Widom [1960a] and then of Lacey, Nazarov, Treil, and Volberg (with the discovery of a criterion of continuity of an operator A; see the references and comments in §2.4, §3.3.3, and §3.5).

Theorem 3.3.6 (principal criterion (Widom, 1960; Devinatz, 1964)) *Let $\varphi \in L^\infty(\mathbb{T})$. The following assertions are equivalent.*

(1) T_φ *is invertible.*
(2) $1/\varphi \in L^\infty(\mathbb{T})$ *and* $\text{dist}_{L^\infty}(\varphi/|\varphi|, H^\infty) < 1$, $\text{dist}_{L^\infty}(\overline{\varphi}/|\varphi|, H^\infty) < 1$.
(3) $1/\varphi \in L^\infty(\mathbb{T})$ *and there exist real functions* $a, b \in L^\infty(\mathbb{T})$ *and a number* $c \in \mathbb{R}$ *such that* $\varphi/|\varphi| = e^{i(a+\tilde{b}+c)}$, *where* $\|a\|_\infty < \pi/2$ *and* $\tilde{b} = \mathbb{H}b$ *is the harmonic conjugate of* b.
(4) $\varphi = g e^{b+ia}$, *where* $g^{\pm 1} \in H^\infty$, $a, b \in L^\infty(\mathbb{T})$ *are real and* $\|a\|_\infty < \pi/2$.
(5) $1/\varphi \in L^\infty(\mathbb{T})$ *and* $\varphi = fg$ *where* $g^{\pm 1} \in H^\infty$ *and* f *is sectorial.*
(6) $1/\varphi \in L^\infty(\mathbb{T})$ *and there exists an outer function* $h \in H^2$ *such that* $\varphi/|\varphi|=h/\overline{h}$ *and* $|h|^2$ *satisfies the Helson–Szegő condition* ($|h|^2 = e^{\tilde{a}+b}$ *with* a, b *real functions in* L^∞, $\|a\|_\infty < \pi/2$), *or the Muckenhoupt condition* (A_2) *(equivalent to Helson–Szegő; see [Nikolski, 2019] and Appendix F.3 below).*
(7) φ *admits a Wiener–Hopf factorization* $\varphi = \varphi_1 \overline{\varphi_2}$ *(and in this case,* $T_\varphi^{-1} f = (1/\varphi_1) P_+ (1/\overline{\varphi_2}) f$, $\forall f \in \mathcal{P}_a$).

Moreover, we can add to the list (1)–(7) the assertions (2'), (3') and (6') obtained by replacing $|\varphi|$ in (2), (3), (6) with $[\varphi]$ (the outer function with the same modulus as φ: see Appendix F).

The bounds of the norm $\|T_\varphi^{-1}\|$ resulting from the proof are summarized in §3.3.2.

Proof By Lemmas 3.3.1 and 3.3.4, (1) $\Leftrightarrow$ (2) $\Leftrightarrow$ (3) $\Leftrightarrow$ (2') $\Leftrightarrow$ (3').

(3) $\Rightarrow$ (4) With the notation of (3') it suffices to set $g = [\varphi] e^{b+i\tilde{b}+c}$; then, $\varphi = g e^{ia-b}$ is a representation as in (4).

3.3 The Principal Criterion of Invertibility

(4) $\Rightarrow$ (5) Clear.

(5) $\Rightarrow$ (1) We have $T_\varphi = T_f T_g$, where T_g and T_f are invertible (by Lemma 3.3.2).

(6) $\Rightarrow$ (3) With the notation of (6), we have $h = e^{(\bar{a}+b)/2+i(\bar{b}-a)/2+ic}$ (for a real constant c). The function on the right is an outer function (see Appendix F) whose modulus coincides with $|h|$. This gives $\varphi/|\varphi| = h/\bar{h} = e^{i(\bar{b}-a+2c)}$, hence (3).

(3) $\Rightarrow$ (6) In a similar manner, we set $h = e^{(-\bar{a}+b)/2+i(\bar{b}+a)/2+ic/2}$; since $\|a\|_\infty < \pi/2$, h is an outer function of H^2 (see Appendix F), and then $h/\bar{h} = e^{i(a+\bar{b}+c)} = \varphi/|\varphi|$.

(7) $\Rightarrow$ (1) According to §3.3.1, the mapping

$$A: f \to \frac{1}{\varphi_1} P_+ \frac{1}{\bar{\varphi}_2} f \quad (f \in \mathcal{P}_a)$$

is bounded on L^2, and for every $f \in \mathcal{P}_a$,

$$T_\varphi A f = P_+ \bar{\varphi}_2 \varphi_1 \frac{1}{\varphi_1} P_+ \frac{1}{\bar{\varphi}_2} f = P_+ \bar{\varphi}_2 P_+ \frac{1}{\bar{\varphi}_2} f = P_+ \bar{\varphi}_2 (P_- + P_+) \frac{1}{\bar{\varphi}_2} f = f,$$

and similarly for $A T_\varphi f = f$. Hence, T_φ is invertible and the extension of A coincides with T_φ^{-1}.

(1) $\Rightarrow$ (7) Given (1) $\Leftrightarrow$ (3'), φ can be written as $\varphi = [\varphi] e^{i(a+\bar{b}+c)}$ where $\|a\|_\infty < \pi/2$ and $\tilde{b} = 1/i \, \mathbb{H} b$, $b \in L^\infty$. Setting

$$\varphi_1 = [\varphi] e^{ic+(ia-\bar{a})/2} e^{(b+i\tilde{b})/2}, \quad \varphi_2 = e^{(-ia+\bar{a})/2} e^{-(b+i\tilde{b})/2},$$

we obtain $[\varphi]^{\pm 1} e^{\pm(b+i\tilde{b})/2} \in H^\infty$ and $e^{\pm(ia-\bar{a})/2} \in H^2$ (by Kolmogorov's theorem, in Appendix F.3), and thus $\varphi_1^{\pm 1} \in H^2$, and by a similar reasoning, $\varphi_2^{\pm 1} \in H^2$. Clearly $\varphi = \varphi_1 \bar{\varphi}_2$ and, as in the implication (7) $\Rightarrow$ (1), T_φ^{-1} coincides with A on the polynomials. However, T_φ^{-1} is bounded, hence φ admits a Wiener–Hopf factorization. ∎

3.3.2 Upper Bounds for $\|T_\varphi^{-1}\|$

The following estimates are consequences of Lemmas 3.3.1, 3.3.2, and 3.3.4, and of the proof above.

Under the condition of Theorem 3.3.6(2), $d = \text{dist}_{L^\infty}(\varphi/|\varphi|, H^\infty)$,

$$\|T_\varphi^{-1}\| \leq \left\| \frac{1}{\varphi} \right\|_\infty \frac{1}{(1-d^2)^{1/2}}.$$

Under the condition of Theorem 3.3.6(3),

$$L(b) = \overline{M}(b) - \underline{M}(b) = \operatorname{ess\,sup}_{\mathbb{T}} b - \operatorname{ess\,sup}_{\mathbb{T}} b$$
$$= \|T_\varphi^{-1}\| \leq \left\|\frac{1}{\varphi}\right\|_\infty \frac{\cosh(L(b)/2)}{\cos(\|a\|_\infty)}.$$

Under the condition of Theorem 3.3.6(4),

$$\|T_\varphi^{-1}\| \leq \left\|\frac{1}{g}\right\|_\infty \frac{e^{-\underline{M}(b)}}{\cos(\|a\|_\infty)}.$$

Under the condition of Theorem 3.3.6(5), $\varphi = fg$, $\alpha = \alpha(0, \operatorname{Ran}_{\mathrm{ess}}(f))$ is the angle through which $\operatorname{Ran}_{\mathrm{ess}}(f)$ can be seen from the point 0,

$$\|T_\varphi^{-1}\| \leq \left\|\frac{1}{g}\right\|_\infty \frac{1}{\cos(\alpha/2)} \left\|\frac{1}{f}\right\|_\infty.$$

Under the condition of Theorem 3.3.6(6), the same estimate as in §3.3.1(3).
Under the condition of Theorem 3.3.6(7), see the comments in §2.4 and §3.3.3.

3.3.3 A Comment on Wiener–Hopf Factorization

(a) *Let $\varphi \in L^1(\mathbb{T})$; if a Wiener–Hopf factorization $\varphi = \varphi_1 \overline{\varphi}_2$ exists, it is unique (up to a constant factor), the functions φ_j are outer ($j = 1, 2$), and $\|1/\varphi\|_\infty \leq \|A\|$ where $A = (1/\varphi_1) P_+ (1/\overline{\varphi}_2)$.*

Indeed, if $\varphi = \Phi_1 \overline{\Phi}_2$ is another Wiener–Hopf factorization, then $\varphi_1/\Phi_1 = \overline{\Phi}_2/\overline{\varphi}_2 \in H^1$ and $\Phi_2/\varphi_2 \in H^1$, hence $\varphi_1/\Phi_1 = \mathrm{const.}$ (see Appendix F). It is also clear that each function f satisfying $f, 1/f \in H^p$, $p > 0$, is outer (see Appendix F).

For an upper bound of φ, we write

$$\|A(z^n x)\|_2 = \left\|\frac{1}{\varphi_1} P_+(z^n x/\overline{\varphi}_2)\right\|_2 \leq \|A\| \cdot \|x\|_2,$$

and by passing to the limit $\|(1/\varphi)x\|_2 \leq \|A\| \cdot \|x\|_2$ ($\forall x \in \mathcal{P}_a$) we conclude by applying Lemma 2.1.4. ∎

(b) *Let $f^{\pm 1}, g^{\pm 1} \in H^2(\mathbb{T})$; then the mapping $A: x \mapsto fP_+\overline{g}x = T_f T_{\overline{g}} x$, $x \in \mathcal{P}_a$ can be extended to a bounded operator $H^2 \to H^2$ if and only if the Riesz projection P_+ is continuous as the mapping*

$$P_+: L^2\left(\mathbb{T}, \frac{1}{|g|^2} dm\right) \to L^2(\mathbb{T}, |f|^2 dm).$$

Moreover, $\|A\| = \|P_+\|$.

3.3 The Principal Criterion of Invertibility

Indeed, if A is bounded on $\mathcal{P}_a$, it is bounded on $\mathcal{P}$: for every $x \in \mathcal{P}$,

$$\|fP_+\bar{g}x\|_2 = \|fP_+\bar{g}(P_+ + P_-)x\|_2 = \|fP_+\bar{g}P_+x\|_2$$

$$\leq \|A\| \cdot \|P_+x\|_2 \leq \|A\| \cdot \|x\|_2,$$

and the property follows. Clearly the set $\bar{g}\mathcal{P}$ is dense in the space $L^2(\mathbb{T}, (1/|g|^2)\, dm)$, and denoting $\bar{g}x = h$ ($x \in \mathcal{P}$), we obtain

$$\|P_+h\|_{L^2(|f|^2\, dm)} = \|fP_+\bar{g}x\|_2 \leq \|A\| \cdot \|x\|_2 = \|A\| \cdot \|h\|_{L^2(1/|g|^2\, dm)},$$

hence P_+ can be extended by continuity to a bounded operator $L^2(\mathbb{T}, (1/|g|^2)\, dm) \to L^2(\mathbb{T}, |f|^2\, dm)$, and $\|P_+\| \leq \|A\|$. The reasoning can easily be reversed to obtain the converse. ∎

(c) *Let T_φ be a bounded and invertible Toeplitz operator. Then the Wiener–Hopf factorization of φ is of the form*

$$\varphi = \varphi_1 \bar{\varphi}_2 = f\frac{\bar{g}}{g}$$

where $\varphi_1 = f/g$, $\varphi_2 = g$ and $f^{\pm 1} \in H^\infty$, $g^{\pm 1} \in H^2$, $|g|^2 \in (A_2)$, where (A_2) denotes the Muckenhoupt condition (see Appendix F.3), and

$$\|T_\varphi^{-1}\| = \|T_{1/\varphi_1} T_{1/\bar{\varphi}_2}\| = \|P_+\|_{L^2(|g|^2\, dm) \to L^2(|g|^2/|\varphi|^2\, dm)},$$

$$\frac{1}{\|\varphi\|_\infty}\|P_+\|_{L^2(|g|^2\, dm)} \leq \|T_\varphi^{-1}\| \leq \|1/\varphi\|_\infty \|P_+\|_{L^2(|g|^2\, dm)},$$

$$\|P_+\|_{L^2(|g|^2\, dm)} =: \|P_+\|_{L^2(|g|^2\, dm) \to L^2(|g|^2\, dm)}.$$

Indeed, $\varphi = \varphi_1 \bar{\varphi}_2$ and

$$0 < \|T_\varphi^{-1}\|^{-1} \leq |\varphi| = |\varphi_1 \varphi_2| \leq \|\varphi\|_\infty < \infty \text{ a.e. on } \mathbb{T},$$

and since the φ_j are outer, $\varphi_1 = f/g$ where $f = [\varphi] \in (H^\infty)^{-1}$, $g = \varphi_2$. By (b),

$$P_+: L^2(\mathbb{T}, |\varphi_2|^2\, dm) \to L^2\left(\mathbb{T}, \frac{1}{|\varphi_1|^2}\, dm\right)$$

is bounded, and as $f^{\pm 1} \in H^\infty$, id: $L^2(\mathbb{T}, (1/|\varphi_1|^2)\, dm) \to L^2(\mathbb{T}, |\varphi_2|^2\, dm)$ is also bounded (by the norm $\|\varphi\|_\infty$, hence the left inequality of the statement). Thus by Appendix F, $|g|^2 = |\varphi_2|^2 \in (A_2)$. ∎

(d) Remark. Observe that point (c) is perfectly consistent with Theorem 3.3.6(6). The situation of (c) becomes more complicated as soon as we speak of the inversion of an *unbounded Toeplitz operator* T_φ: $\varphi \in L^2(\mathbb{T})$

and $T_\varphi x = P_+(\varphi x)$ is defined only for $x \in \mathcal{P}_a$. We can see that a Wiener–Hopf factorization $\varphi = \varphi_1 \overline{\varphi}_2$ remains equivalent to the invertibility of T_φ, but the problem of the continuity of $A = T_{1/\varphi_1} T_{1/\overline{\varphi}_2}$ can no longer be reduced to the condition (A_2): it is equivalent to the famous *problem of "two-weight boundedness"* of the projection

$$P_+ : L^2(\mathbb{T}, v\, dm) \to L^2(\mathbb{T}, w\, dm)$$

where $v = |\varphi_2|$, $w = 1/|\varphi_1|^2$. The condition $w \leq cv$, of course, remains necessary, and completed with $v \in (A_2)$, it becomes sufficient, but these naive arguments are very far from the true depths of the problem. See the comments and references in §2.4.

Theorem 3.3.6 will be heavily used in what follows; for now, we will apply it to the real symbols (self-adjoint T_φ), and then to a reinforcement of Corollary 3.1.8 by replacing $C(\mathbb{T})$ with $\mathrm{QC}(\mathbb{T})$ (the Sarason algebra: see Exercise 2.3.5(c)). Later, we will obtain a more complete version of this reinforcement (see Exercise 3.4.3(f, g)).

3.3.4 First Consequences of the Principal Criterion

Lemma 3.3.7 (symbols taking on two values) *Let $\varphi = \alpha \chi_E + \beta \chi_{\mathbb{T} \setminus E}$ where $E \subset \mathbb{T}$ is a Borel set, $\alpha, \beta \in \mathbb{C} \setminus \{0\}$ and $0 < mE < 1$ (m the normalized Lebesgue measure). The following assertions are equivalent.*

(1) T_φ *is invertible.*
(2) φ *is sectorial (see Lemma 3.3.2).*
(3) $|\alpha - \beta| < |\alpha| + |\beta|$.

Proof For simple geometric reasons, (2) $\Leftrightarrow$ (3), and (2) $\Rightarrow$ (1) by Lemma 3.3.2.

Let us show (1) $\Rightarrow$ (3). Supposing the contrary, we obtain $|\alpha - \beta| = |\alpha| + |\beta|$, whence the function $\varphi/|\varphi|$ takes on two diametrically opposed unimodular values: $-\alpha = \beta \in \mathbb{T}$, $\varphi/|\varphi| =: \alpha \chi_E + \beta \chi_{\mathbb{T} \setminus E}$. Moreover, the invertibility of T_φ and Theorem 3.3.6 imply that there exists $h \in H^\infty$ such that $\|\varphi/|\varphi| - h\|_\infty =: r < 1$, meaning that on the set E the function h has its values in the closed disk $\overline{D}(\alpha, r)$, and on $\mathbb{T} \setminus E$ in $\overline{D}(\beta, r)$. But $\overline{D}(\alpha, r) \cap \overline{D}(\beta, r) = \emptyset$, and hence by the classical Runge theorem (see [Rudin, 1986], Theorems 13.6–13.9) there exists a sequence of polynomials (p_n) converging uniformly to 1 on $\overline{D}(\alpha, r)$ and to 0 on $\overline{D}(\beta, r)$. Consequently, $\lim_n \|p_n \circ h - \chi_E\|_\infty = 0$ and, since $p_n \circ h \in H^\infty$, we obtain $\chi_E \in H^\infty$, which is absurd (this contradicts the uniqueness theorem of Appendix F.2). ∎

Theorem 3.3.8 (self-adjoint operators T_φ (Hartman and Wintner, 1954)) *Let T_φ be a self-adjoint Toeplitz operator (i.e. φ is real-valued). Then,*

$$\sigma(T_\varphi) = \text{conv}(\text{Ran}_{\text{ess}}(\varphi)) = [\text{ess inf}(\varphi), \text{ess sup}(\varphi)].$$

Proof The inclusion $\sigma(T_\varphi) \subset \text{conv}(\text{Ran}_{\text{ess}}(\varphi)) = [\text{ess inf}(\varphi), \text{ess sup}(\varphi)]$ is in Corollary 3.3.3. For the converse, we show that $T_\varphi - \lambda I$ is not invertible if ess $\inf(\varphi) < \lambda < $ ess $\sup(\varphi)$. Replacing φ with $\varphi - \lambda$ if necessary, we can assume $\lambda = 0$. Then, the function $\varphi/|\varphi|$ takes on the values ± 1, and by Lemma 3.3.7, $0 \in \sigma(T_\varphi)$. ∎

Another proof of Theorem 3.3.8 is in Exercise 3.4.11. Another corollary of Theorem 3.3.6 is the following reinforcement of Corollary 3.1.8.

Corollary 3.3.9 *Let $\varphi = e^a$, where $a \in \text{QC}(\mathbb{T})$. Then T_φ is invertible.*

Indeed, recall that

$$\text{QC}(\mathbb{T}) = (H^\infty + C) \cap (\overline{H}^\infty + C) = L^\infty \cap \text{VMO}(\mathbb{T})$$

(see Exercise 2.3.5(c)) and hence $\varphi^{\pm 1} \in L^\infty(\mathbb{T})$ and $\varphi/|\varphi| = e^{i(\text{Im}(a))}$, $\text{Im}(a) = u + \tilde{v}$ where $u, v \in C(\mathbb{T})$ are real functions. If p is a trigonometric polynomial such that $\|u - p\|_\infty < \pi/2$, then $\varphi/|\varphi| = e^{i(u-p)+i(-\tilde{p}+v)^\sim+ic}$ where $c \in \mathbb{R}$, and by Theorem 3.3.6(3), the operator T_φ is invertible. ∎

Remark It is not true that every function $\varphi^{\pm 1} \in \text{QC}$ admits a logarithm in QC (for example, $\varphi = z^n$ ($n \neq 0$) or $\varphi = e^{i\tilde{u}}$ if $u \in C(\mathbb{T})$ and $\tilde{u}$ is unbounded; see Sarason's theorem in Exercise 3.4.3(g)). For the Fredholm property, the situation is simpler (see Exercise 3.4.3(f)): for $\varphi \in \text{QC}$, $T_\varphi \in \text{Fred} \Leftrightarrow 1/\varphi \in \text{QC}$.

3.4 Exercises

3.4.0 Basic Exercises: Integral and Multiplication Operators

(a) Estimates for integral operators with kernels on the L^p spaces. Let (Ω_j, μ_j), $j = 1, 2$ be two separable metric spaces, equipped with locally finite Borel measures, $1 < p < \infty$, k a measurable function on $\Omega_2 \times \Omega_1$, and J_k the operator (whose domain of definition will be made precise) defined by

$$J_k f(x) = \int_{\Omega_1} k(x, y) f(y) \, d\mu_1(y), \quad x \in \Omega_2.$$

(1) *The case where $p = 1$ or $p = \infty$. Show that if*

$$A =: \operatorname{ess\,sup}\left\{\int_{\Omega_2} |k(x,y)|\, d\mu_2(x) : y \in \Omega_1\right\} < \infty,$$

then J_k is well-defined and continuous as a mapping $L^1(\Omega_1, \mu_1) \to L^1(\Omega_2, \mu_2)$ and $\|J_k\| = A$, and if

$$B =: \operatorname{ess\,sup}\left\{\int_{\Omega_1} |k(x,y)|\, d\mu_1(y) : x \in \Omega_2\right\} < \infty,$$

then J_k is well-defined and continuous $L^\infty(\Omega_1, \mu_1) \to L^\infty(\Omega_2, \mu_2)$ and $\|J_k\| = B$.

SOLUTION: For $p = \infty$ and for every $f \in L^\infty(\Omega_1, \mu_1)$,

$$\operatorname{ess\,sup}_{x\in\Omega_2}\int_{\Omega_1} |k(x,y) f(y)|\, d\mu_1(y) \leq \|f\|_\infty B,$$

hence J_k is well-defined and $\|J_k\| \leq B$.

For the converse, given $\epsilon > 0$, by the definition of B there is a set E of $x \in \Omega_2$, with positive measure μ_2, such that

$$\int_{\Omega_1} |k(x,y)|\, d\mu_1(y) > B - \epsilon, \forall x \in E.$$

By Luzin's theorem, every measurable mapping is "almost continuous"; we apply this to the measurable mapping $F: E \to L^1(\Omega_1, \mu_1)$ defined by $F(x) = k(x, \cdot)$, $x \in E$: a subset $E' \subset E$ of positive measure can be found such that the restriction $F|E'$ is continuous. Thus there exists an $x_0 \in E'$ and $\delta > 0$ so that $\|F(x) - F(x_0)\|_{L^1(\Omega_1)} < \epsilon$ for every $x \in E' \cap B(x_0, \delta)$ and $\mu_2(E' \cap B(x_0, \delta)) > 0$ (separability of Ω_2) where $B(x_0, \delta) = \{x \in \Omega_2 : d(x, x_0) < \delta\}$ is a ball of radius δ, and d is the distance in Ω_2. With all this, there exists a function $f \in L^\infty(\Omega_1, \mu_1)$ such that $\|f\|_\infty = 1$ and

$$\int_{\Omega_1} k(x_0, y) f(y)\, d\mu_1(y) = \int_{\Omega_1} |k(x_0, y)|\, d\mu_1(y) > B - \epsilon.$$

Hence, for every $x \in E' \cap B(x_0, \delta)$,

$$\left|\int_{\Omega_1} k(x,y) f(y)\, d\mu_1(y)\right| \geq \int_{\Omega_1} k(x_0,y) f(y)\, d\mu_1(y) - \|F(x) - F(x_0)\|_{L^1(\Omega_1)} > B - 2\epsilon.$$

As $\mu_2(E' \cap B(x_0, \delta)) > 0$, we obtain $\|J_k f\|_\infty > B - 2\epsilon$, and finally $\|J_k\| \geq B - 2\epsilon$. Since ϵ is arbitrary, we obtain $\|J_k\| \geq B$, and thus $\|J_k\| = B$.

For $p = 1$, we reason by duality, knowing that $(L^1(\Omega, \mu))^* = L^\infty(\Omega, \mu)$, $J_k^* f(y) = \int_{\Omega_2} k(x,y) f(x)\, d\mu_2(x)$ (duality is bilinear) and $\|J_k\| = \|J_k^*\|$.

NOTE In particular, for every *convolution operator* on a locally compact commutative group G,

$$Jf = f * \varphi = \int_G \varphi(x - y) f(y)\, dy,$$

we have $\|J\| = \|\varphi\|_1$ (where the norm of J is in the spaces $L^1(G, dx)$ or $L^\infty(G, dx)$, and dx is the Haar measure on G); for example, $G = \mathbb{T}$ and $G = \mathbb{R}$ are frequently used in this book.

Another remark, somewhat unexpected, is that $\|J_k\| = \|J_{|k|}\|$, always on these spaces. For example, for an "ϵ-approximation" $J_{k,\epsilon}$ of the Cauchy singular integral operator, §4.1.3 below, or of the Hilbert operator, §4.2.7, we obtain $\|J_{k,\epsilon}\| = \|J_{|k|,\epsilon}\|$ in $L^1(\mathbb{T})$ or $L^\infty(\mathbb{T})$ where

$$J_{k,\epsilon} = \int_{|z-\zeta|\geq\epsilon} \frac{f(\zeta)\,d\zeta}{z-\zeta}, \quad J_{|k|,\epsilon} = \int_{|z-\zeta|\geq\epsilon} \frac{f(\zeta)\,dm(\zeta)}{|z-\zeta|},$$

which strongly contrasts with the behavior of these operators on the space $L^2(\mathbb{T})$ where the norms $\|J_{k,\epsilon}\|$ are uniformly bounded, whereas $\|J_{|k|,\epsilon}\|$ behaves as $\log(1/\epsilon)$ when $\epsilon \to 0$: see the comments in §4.6.

It is also worth noticing that, for a positive kernel $k \geq 0$, the equality for the norm of J_k follows directly from Fubini's theorem: if $f \geq 0$, $f \in L^1(\Omega_1, \mu_1)$, then

$$\|J_k f\|_1 = \int_{\Omega_2}\int_{\Omega_1} k(x, y) f(y)\,d\mu_1(y)\,d\mu_2(x) = \int_{\Omega_1} f(y) \int_{\Omega_2} k(x, y)\,d\mu_2(x)\,d\mu_1(y),$$

and by maximizing over $\|f\|_1 \leq 1$, we obtain $\|J_k\| = A$.

(2) *The case $1 < p < \infty$. The Schur test (1911) in the spaces L^p. Let (Ω_j, μ_j), $j = 1, 2$, be two measure spaces, $1 < p < \infty$, and k a measurable function on $\Omega_2 \times \Omega_1$ such that for certain $\varphi > 0$ and $\psi > 0$,*

$$\int_{\Omega_1} |k(x, y)|\varphi(y)\,d\mu_1(y) \leq (a\psi(x))^{p'/p}, \quad \int_{\Omega_2} |k(x, y)|\psi(x)\,d\mu_2(x) \leq (b\varphi(y))^{p/p'}.$$

Show that the operator J_k,

$$J_k f(x) = \int_{\Omega_1} k(x, y) f(y)\,d\mu_1(y), \quad x \in \Omega_2,$$

is bounded $L^p(\Omega_1, \mu_1) \to L^p(\Omega_2, \mu_2)$ and $\|J_k\| \leq a^{1/p} b^{1/p'}$.

Hint Use Hölder's inequality $\|ab\|_1 \leq \|a\|_p \|b\|_{p'}$, where $1/p + 1/p' = 1$, first multiplying by $\varphi(y)^{1/p} 1/(\varphi(y)^{1/p'})$ under the integral sign.

SOLUTION: Estimate $\int_{\Omega_2} |J_k f(x)|^p \, d\mu_2(x)$ following the Hint:

$$\left(\int_{\Omega_1} |k(x,y)| \cdot |f(y)| \, d\mu_1(y) \right)^p$$

$$= \left(\int_{\Omega_1} |k(x,y)\varphi(y)|^{1/p'} \cdot |k(x,y)|^{1/p} \varphi(y)^{-1/p'} |f(y)| \, d\mu_1(y) \right)^p$$

$$\leq a\psi(x) \int_{\Omega_1} |k(x,y)| \varphi(y)^{-p/p'} \cdot |f(y)|^p \, d\mu_1(y),$$

hence

$$\|J_k f\|_p^p \leq a \int_{\Omega_2} \int_{\Omega_1} |f(y)|^p |k(x,y)| \varphi(y)^{-p/p'} \psi(x) \, d\mu_1(x) \, d\mu_2(x)$$

$$= a \int_{\Omega_1} |f(y)|^p \varphi(y)^{-p/p'} \int_{\Omega_2} |k(x,y)| \psi(x) \, d\mu_2(x) \, d\mu_1(y)$$

$$\leq a b^{p/p'} \int_{\Omega_1} |f(y)|^p \, d\mu_1(y).$$

(3) *The Schur test is sharp on positive kernels.* Let $k(x,y) = k(y,x) \geq 0$ be a symmetric positive kernel on $\Omega \times \Omega$ where (Ω, μ) is a measure space, such that the operator J_k is bounded on the space $L^2(\Omega, \mu)$, $J_k: L^2(\Omega, \mu) \to L^2(\Omega, \mu)$. Show that if $\epsilon > 0$, there exists a positive function φ such that

$$\int_\Omega k(x,y) \varphi(y) \, d\mu(y) \leq (\|J_k\| + \epsilon) \varphi(x) \quad \text{a.e. on } \Omega.$$

Hint Consider the resolvent $R(\lambda, J_k)$ where $\lambda = \|J_k\| + \epsilon$.

SOLUTION: Following the Hint, let $\varphi = \sum_{n\geq 0} \lambda^{-n} J_k^n 1$ (absolutely convergent series). Clearly, $\varphi \geq 1$ and $J_k \varphi = \sum_{n\geq 0} \lambda^{-n} J_k^{n+1} 1 = \lambda(\varphi - 1) \leq \lambda \varphi$.

(4) An application: the Laplace transform on $L^p(\mathbb{R}_+)$, $1 < p < \infty$. Let

$$Lf(\lambda) = \int_{\mathbb{R}_+} f(t) e^{-\lambda t^{p-1}} \, dt, \quad \lambda > 0.$$

Show that $L: L^p(\mathbb{R}_+) \to L^p(\mathbb{R}_+)$ (which for $p = 2$ coincides with the classical Laplace transform) is bounded for $1 < p < \infty$ and

$$\|L\| \leq \min_{0 < \alpha < 1} \Gamma\left(\frac{1-\alpha}{p-1}\right)^{1/p'} \Gamma(\alpha)^{1/p} \frac{1}{(p-1)^{1/p'}}$$

where $1/p + 1/p' = 1$ and Γ is the Euler gamma function. In particular,

$$\|L\| \leq \Gamma\left(\frac{1}{2(p-1)}\right)^{1/p'} \pi^{1/2p}(p-1)^{p-1},$$

giving, among others,

$$\|L\|_{L^2 \to L^2} \leq \sqrt{\pi} \quad \text{and} \quad \|L\|_{L^{4/3} \to L^{4/3}} \leq \left(\frac{1}{2}\right)^{1/4}\left(\frac{1}{3}\right)^{1/3}\sqrt{\pi}.$$

Hint Apply (2) above with $\varphi(t) = t^{-\alpha}$.

SOLUTION: Following the Hint, let $\varphi(t) = t^{-\alpha}$, $0 < \alpha < 1$, and $a = p - 1 = p/p'$. Then

$$L\varphi(\lambda) = \int_{\mathbb{R}_+} t^{-\alpha} e^{-\lambda t^a} dt = \lambda^{(\alpha-1)/a} \int_{\mathbb{R}_+} y^{-\alpha/a} e^{-y} \frac{y^{1/a-1}}{a} dy$$

$$= \frac{1}{a\lambda^{(1-\alpha)/a}} \Gamma\left(\frac{1-\alpha}{a}\right)$$

(just by the definition of Γ), giving

$$L\varphi(\lambda) = (A\psi(\lambda))^{p'/p}, \quad A = \left(a^{-1}\Gamma\left(\frac{1-\alpha}{a}\right)\right)^{p/p'}, \quad \psi(\lambda) = \lambda^{-(1-\alpha)}.$$

Similarly,

$$L^*\psi(t) = \int_{\mathbb{R}_+} \lambda^{-(1-\alpha)} e^{-\lambda t^a} d\lambda = \frac{1}{t^{\alpha a}}\Gamma(\alpha) = (B\varphi(t))^{p/p'},$$

where $B = \Gamma(\alpha)^{p'/p}$. By (2), the operator L is continuous on $L^p(\mathbb{R}_+)$, and we have

$$\|L\| \leq A^{1/p} B^{1/p'} = \left(a^{-1}\Gamma\left(\frac{1-\alpha}{a}\right)\right)^{1/p'} \Gamma(\alpha)^{1/p}$$

$$= \Gamma\left(\frac{1-\alpha}{p-1}\right)^{1/p'} \Gamma(\alpha)^{1/p} \frac{1}{(p-1)^{1/p'}},$$

for every α, $0 < \alpha < 1$.

The known values of Γ, $\Gamma(1/2) = \sqrt{\pi}$ and $\Gamma(3/2) = \sqrt{\pi}/2$, provide the upper bounds for L^2 and $L^{4/3}$.

NOTE It is easy to see that the classical Laplace transform

$$Lf(\lambda) = \int_{\mathbb{R}_+} f(t) e^{-\lambda t} dt, \quad \lambda > 0$$

is *bounded on the space* $L^p(\mathbb{R}_+)$ *if and only if* $p = 2$. Effectively, for $p = 2$, we already know that $\|L\| \leq \sqrt{\pi}$. To show that it is not bounded for the other p, $1 \leq p \leq \infty$, remark first that for $p = 1, \infty$ it is immediate by (1) above: $\sup_\lambda \int_{\mathbb{R}_+} e^{-\lambda t} dt = \infty$. For $1 < p < \infty$, it suffices to consider $1 < p < 2$ (by duality, L will have the same norm on $L^p(\mathbb{R}_+)$ and $L^{p'}(\mathbb{R}_+) = (L^p(\mathbb{R}_+))^*)$ and the family of functions $f_\alpha(t) = e^{-t}t^{-\alpha}$, for which

$$\|f_\alpha\|_p^p = p^{p\alpha-1}\Gamma(1-p\alpha) \quad \text{and} \quad Lf_\alpha(\lambda) = (\lambda+1)^{\alpha-1}\Gamma(1-\alpha).$$

Supposing $p(1-\alpha) > 1$, we get $\|Lf_\alpha\|_p^p = \Gamma(1-\alpha)^p(p(1-\alpha)-1)^{-1}$. When $p(1-\alpha) \to 1$ (hence $1-p\alpha \to 2-p > 0$ and $1-\alpha \to 1/p > 0$), we obtain

$$\sup_\alpha \|f_\alpha\|_p^p < \infty \quad \text{and} \quad \sup_\alpha \|Lf_\alpha\|_p^p = \infty. \qquad \blacksquare$$

(b) On the Fredholm character of multiplication operators. *The general Fredholm theory is presented in Appendix E below, where, in particular, the multiplication operators M_φ: $L^p \to L^p$ are treated: see Appendix E.8(6). We present here a few additional facts linked to the contents of this Chapter 3.*

(1) *The multipliers of a function space are bounded.* Let (Ω, μ) be a measure space and $\Sigma(\Omega, \mu)$ the space of measurable functions on Ω, and let X be a Banach space continuously embedded in Σ; Σ is equipped with the convergence in measure. Suppose for example (as we do below) that there exists a function w integrable and μ-a.e. positive ($w > 0$), then this convergence is metrizable by

$$\rho(f, g) = \int_\Omega \frac{|f-g|}{1+|f-g|} w\, d\mu.$$

To avoid secondary problems, suppose, in addition, that Ω is an essential domain of X: if $E \subset \Omega$ is measurable and $f = 0$ a.e. on E for every $f \in X$, then $\mu E = 0$. Let Mult(X) denote the set of multipliers φ of X, $f \in X \Rightarrow \varphi f \in X$.

Show that every multiplier φ of X ($f \in X \Rightarrow \varphi f \in X$) is bounded (Mult$(X) \subset L^\infty(\Omega, \mu)$), and in addition, the mapping M_φ: $X \to X$, $M_\varphi f = \varphi f$, is bounded and $\|\varphi\|_{L^\infty(\Omega,\mu)} \leq \|M_\varphi\|$.

Hint First, apply the closed graph theorem, Appendix D.3.

SOLUTION: Following the Hint, we show that M_φ is closed, and hence bounded. Effectively, if $\lim_n \|f_n\|_X = 0$ and $\lim_n \|M_\varphi f_n - g\|_X = 0$ where $g \in X$, we obtain by the continuous embedding $X \subset \Sigma$ that $\lim_n f_n = 0$ in measure, thus $g = 0$, hence M_φ is a closed operator.

To conclude, let $R > \|M_\varphi\|$. Since, for every $f \in X$, we have $\|\varphi^n f\|_X \leq \|M_\varphi\|^n \|f\|_X$, giving $\lim_n \|R^{-n}\varphi^n f\|_X = 0$ and hence $\lim_n (R^{-n}\varphi^n f) = 0$ in measure, in particular $\lim_n \mu\{x \in \Omega: R^{-1}|\varphi(x)| \cdot |f(x)|^{1/n} > 1\} = 0$. Thus $R^{-1}|\varphi| \leq 1$ a.e. on the set $\{x \in \Omega: |f(x)| > 0\}$. Since Ω is an essential domain of X, this implies that $R^{-1}|\varphi| \leq 1$ a.e. on Ω (for every $R > \|M_\varphi\|$), and the result follows.

(2) *When is a Fredholm multiplier invertible?* Suppose $1 \in X$ and $\varphi \in \text{Mult}(X)$, and either $\text{Ker} M_\varphi^* = \{0\}$ or $\dim \text{Ker} M_\varphi^* = \infty$. Show that

$$M_\varphi \in \text{Fred} \Leftrightarrow M_\varphi \text{ invertible} \Leftrightarrow \frac{1}{\varphi} \in \text{Mult}(X).$$

SOLUTION: If $1/\varphi \in \text{Mult}(X)$, then clearly M_φ is invertible and $M_\varphi^{-1} = M_{1/\varphi}$ (and of course, $M_\varphi \in \text{Fred}$). Conversely, suppose that M_φ is Fredholm. Then, $M_\varphi X$ is closed and $\text{codim}(M_\varphi X) = \dim \text{Ker} M_\varphi^* < \infty$, thus $\text{Ker} M_\varphi^* = \{0\}$, and hence M_φ is invertible. In particular, there exists $f \in X$ such that $M_\varphi f = 1$, $\varphi f = 1$, therefore $\varphi \neq 0$ a.e. on Ω. The identity $M_\varphi^{-1}(\varphi f) = f$ ($\forall f \in X$) shows that $M_\varphi^{-1}(h) = h/\varphi \in X$ ($\forall h \in X$), hence $1/\varphi \in \text{Mult}(X)$.

NOTE In general, $M_\varphi \in \text{Fred}$ implies neither $1/\varphi \in \text{Mult}(X)$, nor $1/\varphi \in X$ (example: $X = H^2(\mathbb{T})$, $\varphi = z$, see Appendix E.9(7) for details).

(3) *Multipliers of $L^p(\Omega,\mu)$, where μ is a continuous measure (without singletons of positive measure).* Show that $\text{Mult}(L^p) = L^\infty(\Omega,\mu)$ and

$$M_\varphi \in \text{Fred} \Leftrightarrow M_\varphi \text{ invertible} \Leftrightarrow 1/\varphi \in L^\infty(\Omega,\mu).$$

SOLUTION: This is a special case of Appendix E.8(6).

(4) *Multipliers of the Hardy space $H^2(\mathbb{T})$.* Show that

(i) $\text{Mult}(H^2) = H^\infty \ (= H^2 \cap L^\infty(\mathbb{T}))$;
(ii) $M_\varphi \in \text{Fred}(H^2) \Leftrightarrow \varphi = BF$ where B is a finite Blaschke product and $F \in (H^\infty)^{-1}$ *(see also Appendix E.8(7))*;
(iii) *if $M_\varphi \in \text{Fred}(H^2)$, then*

$$\text{ind} M_\varphi = -\deg(B) = -\lim_{r \to 1}(\text{wind}(\varphi_r)),$$

where $\text{wind}(\varphi_r)$ *denotes the Cauchy index of* $\varphi_r(z) = \varphi(rz)$ *(see Definitions 3.1.2). If $\varphi \in H^\infty \cap C(\mathbb{T})$ (disk algebra) then* $\lim_{r \to 1}(\text{ind}(\varphi_r)) = \text{ind}(\varphi(\mathbb{T}))$.

SOLUTION: Practically everything has already been shown elsewhere in the book: for (i) see Exercise 2.3.0(c(4)), and for (ii) Appendix E.8(7). For (iii), note first that $FH^2 = H^2$ and hence $\text{codim}(M_\varphi H^2) = \text{codim}(BH^2) = \dim(BH^2)^\perp = \deg(B)$; the last equality is proved in Exercise 2.3.0 (see (c(3)) and its Remark). There only remains the formula with "wind," which follows from $\text{wind}(B_r F_r) = \text{wind}(B_r) + \text{wind}(F_r)$ (a classical property of the Cauchy index), and then from $\text{wind}(F_r) = 0$ (F is invertible in H^∞, hence F_r is homotopic to the constant $F(0)$). Finally, for r sufficiently close to 1, $\text{wind}(B_r) = \deg(B)$: it suffices to show that $\text{wind}(b_\lambda) = 1$ for every $\lambda \in \mathbb{D}$, where $b_\lambda = (\lambda - z)(1 - \bar{\lambda} z)^{-1}$ is an elementary Blaschke factor, which is evident by wind$((\lambda-z)(1-\bar{\lambda}z)^{-1})$ = wind$(\lambda-z)$ + wind$(1-\bar{\lambda}z)^{-1}$ = wind$(\lambda-z)$ = 1 (note that this last reasoning could be replaced by a reference to the classical argument principle).

(5) *Multipliers of the Bergman space* $L_a^2(\mathbb{D})$,

$$L_a^2(\mathbb{D}) =: L^2(\mathbb{D}, dx\,dy) \bigcap \text{Hol}(\mathbb{D})$$

$$= \left\{ f = \sum_{k \geq 0} \widehat{f}(k) z^k : \|f\|^2 = \frac{1}{\pi} \int_\mathbb{D} |f(z)|^2 dx\,dy = \sum_{k \geq 0} \frac{|\widehat{f}(k)|^2}{k+1} < \infty \right\}.$$

Show the analogs of properties (i)–(iii) of (4) above:

(i) Mult$(L_a^2) = H^\infty$;

(ii) $M_\varphi \in \text{Fred}(L_a^2) \Leftrightarrow \varphi = BF$ where B is a finite Blaschke product and $F \in (H^\infty)^{-1}$;

(iii) *if* $M_\varphi \in \text{Fred}(L_a^2)$, *then*

$$\text{ind}\,M_\varphi = -\deg(B) = -\lim_{r \to 1}(\text{wind}(B_r))$$

where wind(φ_r) *denotes the Cauchy index of* $\varphi_r(z) = \varphi(rz)$.

SOLUTION: (i) By (1) above, Mult$(L_a^2) \subset L^\infty(\mathbb{D}, dx\,dy)$ with the inequality $\|\varphi\|_\infty \leq \|\varphi\|_{\text{Mult}}$, and also Mult$(L_a^2) \subset L_a^2$ (since $1 \in L_a^2$), hence Mult$(L_a^2) \subset L_a^2 \cap L^\infty(\mathbb{D}) = H^\infty$. The converse, $H^\infty \subset \text{Mult}(L_a^2)$ with the inequality $\|\varphi\|_{\text{Mult}} \leq \|\varphi\|_\infty$ is evident by the first formula for the norm in L_a^2.

To justify (ii), note that if $\varphi = BF$ as in (ii), then $FL_a^2 = L_a^2$, and hence $M_\varphi L_a^2 = BL_a^2$. Furthermore, for every $\lambda \in \mathbb{D}$, the evaluation $\psi_\lambda: f \mapsto f(\lambda)$ ($\lambda \in \mathbb{D}$) is a continuous functional on L_a^2: by the mean value theorem, for every $f \in L_a^2$ and $r = 1 - |\lambda|$, we have

$$|f(\lambda)| = \left| \frac{1}{\pi r^2} \int_{D(\lambda,r)} f(z)\,dx\,dy \right| \leq \left(\frac{1}{\pi r^2} \int_{D(\lambda,r)} |f(z)|^2\,dx\,dy \right)^{1/2}$$

$$\leq r^{-1}\left(\frac{1}{\pi} \int_\mathbb{D} |f(z)|^2\,dx\,dy \right)^{1/2} = \frac{1}{1-|\lambda|} \|f\|_2.$$

Hence, for every r, $0 < r < 1$, the norm $\|f\|_2$ on L_a^2 is equivalent to

$$\left(\int_{r < |z| < 1} |f(z)|^2\,dx\,dy \right)^{1/2}.$$

It follows that the range $M_\varphi L_a^2 = BFL_a^2 = BL_a^2$ is closed in L_a^2; the codimension of BL_a^2 is bounded above by induction, codim$(b_\lambda L_a^2) = 1$, where b_λ is an elementary Blaschke factor (since, for every $f \in L_a^2$, $(f - f(\lambda))/b_\lambda \in L_a^2$), and by induction codim$(BL_a^2) \leq n = \deg(B)$, where $B = \prod_{k=1}^n b_{\lambda_k}$. As Ker$M_\varphi = \{0\}$, we obtain $M_\varphi \in \text{Fred}$ (and ind$(M_\varphi) \geq -\deg(B)$).

For the converse, if $M_\varphi \in$ Fred, we know that the subspace $M_\varphi L_a^2 = \varphi L_a^2$ is closed and of finite codimension. As it is M_z-invariant, it is of the form BL_a^2 with a finite Blaschke product, such that $\deg(B) = \operatorname{codim}(\varphi L_a^2)$ (justification to follow). Hence, $\varphi \in BL_a^2$, $\varphi = BF$ where $F \in L_a^2$, and $BFL_a^2 = BL_a^2$ implies $FL_a^2 = L_a^2$, and by (i), $F \in (H^\infty)^{-1}$.

It only remains to justify the statement about the M_z-invariant subspaces $E \subset L_a^2$ of finite codimension, codim $E < \infty$. For the orthogonal complement $E^\perp$, we have $M_z^* E^\perp \subset E^\perp$, $\dim E^\perp < \infty$. Hence, $E^\perp$ is generated by its Jordan subspaces, $\operatorname{Ker}(M^*|E^\perp - \overline{\lambda}I)^k$, $1 \leq k \leq k(\lambda)$, $\overline{\lambda} \in \sigma(M^*|E^\perp)$. However, $\dim \operatorname{Ker}(M_z^* - \overline{\lambda}I) = \operatorname{codim}((z - \lambda)L_a^2) = 1$, and then $\dim \operatorname{Ker}(M_z^* - \overline{\lambda}I)^k = \operatorname{codim}((z - \lambda)^k L_a^2) \leq k$ for every $k \geq 1$, and thus $\operatorname{Ker}(M_z^*|E^\perp - \overline{\lambda}I)^k = \operatorname{Ker}(M_z^* - \overline{\lambda}I)^k$, $1 \leq k \leq k(\lambda)$ (already showing $\sum_{\lambda \in \sigma} k(\lambda) = \dim E^\perp$). It follows that

$$E^\perp = \operatorname{span}(\operatorname{Ker}(M_z^* - \overline{\lambda}I)^k : 1 \leq k \leq k(\lambda), \lambda \in \sigma(M^*|E^\perp))$$

and hence

$$E = \bigcap_\lambda (z - \lambda)^k L_a^2 = BL_a^2, \quad \text{where } B = \prod_{k=1}^n b_{\lambda_k}, \quad \{\overline{\lambda}_k\} = \sigma(M^*|E^\perp).$$

This completes the proof of (ii); the equality $\dim E^\perp = \deg(B)$ (and thus the formula for the index of (iii)) is also validated).

3.4.1 Spectral Inclusions

(a) Continuous, point, and essential spectra. Let $T \in L(H)$, where H is a Hilbert space. *Show that*

$$\sigma_c(T) \subset \sigma_{\text{ess}}(T), \quad \partial \sigma(T) \subset \sigma_c(T) \cup \sigma_p(T),$$

where

$$\sigma_p(T) = \{\lambda \in \mathbb{C}: \ker(\lambda I - T) \neq \{0\}\} \text{ is the point spectrum,}$$

$$\sigma_c(T) = \{\lambda \in \mathbb{C}: \exists (x_n)_{n \geq 1} \subset H, \text{ non-(pre)compact}, \|x_n\| = 1,$$

$$\lim_n \|(\lambda I - T)x_n\| = 0\} \text{ is the continuous spectrum.}$$

SOLUTION: Let $\lambda \in \sigma_c(T)$. If $\lambda \notin \sigma_{\text{ess}}(T)$, there exists a left regularizer $A \in L(H)$, $A(\lambda I - T) = I + K$, $K \in \mathfrak{S}_\infty(H)$ (see Theorem E.7.2), and thus if (x_n) is a sequence as in the definition of $\sigma_c(T)$, then $\lim_n \|x_n + Kx_n\| = 0$, and since $(Kx_n)_n$ is precompact, so is $(x_n)_n$, which is a contradiction. Hence, $\sigma_c(T) \subset \sigma_{\text{ess}}(T)$.

For the second inclusion, let $R_\lambda(T) = (\lambda I - T)^{-1}$ be the *resolvent* of T (at the point $\lambda \in \mathbb{C} \setminus \sigma(T)$), for which we know the inequality $\|R_\lambda(T)\| \geq 1/\operatorname{dist}(\lambda, \sigma(T))$

(see Appendix C). If $\lambda \in \partial\sigma(T)$, there exists $\lambda_n \in \mathbb{C} \setminus \sigma(T)$, $\lim_n \lambda_n = \lambda$, hence $\lim_n \|R_{\lambda_n}\| = \infty$, meaning that there exists $x_n \in H$ satisfying $\|x_n\| = 1$ and $\lim_n \|(\lambda_n I - T)x_n\| = 0$, thus $\lim_n \|(\lambda I - T)x_n\| = 0$. Then, if (x_n) contains a convergent subsequence, we have $\lambda \in \sigma_p(T)$, but if not, $\lambda \in \sigma_c(T)$ by definition of the continuous spectrum.

(b) The case of Toeplitz T_φ. Let $\varphi \in L^\infty(\mathbb{T})$, $\varphi \neq \text{const}$. Show that

$$\sigma_p(T_\varphi) \cap \partial(\text{conv}(\text{Ran}_{\text{ess}}(\varphi))) = \emptyset.$$

In particular, if $\overline{\varphi} = \varphi$, then $\sigma_p(T_\varphi) = \emptyset$.

SOLUTION: Let $\lambda \in \partial(\text{conv}(\text{Ran}_{\text{ess}}(\varphi)))$. If necessary replacing φ with $e^{i\theta}(\varphi - \lambda)$, we can suppose that $\lambda = 0$ and $\text{Im}(\varphi) \geq 0$. If $f \in H^2$, $f \neq 0$, such that $T_\varphi f = P_+(\varphi f) = 0$, then $0 = (T_\varphi f, f) = \int_\mathbb{T} \varphi |f|^2 \, dm$ and thus $\int_\mathbb{T} \text{Im}(\varphi) |f|^2 \, dm = 0$. As $|f|^2 > 0$ a.e. on $\mathbb{T}$ (see Appendix F), we obtain $\text{Im}(\varphi) = 0$, and then $0 = (T_\varphi f, z^n f) = \int_\mathbb{T} \varphi |f|^2 \overline{z}^n \, dm$ for every $n \geq 0$. By complex conjugation it follows that $\varphi |f|^2 = 0$, thus $\varphi = 0$. This contradiction shows that $0 \notin \sigma_p(T_\varphi)$, and the result follows.

The case of a real symbol is also proved by the last three lines of the reasoning.

(c) The case of self-adjoint Toeplitz. Let $\overline{\varphi} = \varphi \in L^\infty(\mathbb{T})$. Show that

$$\sigma_{\text{ess}}(T_\varphi) = \sigma(T) = [\text{ess inf}(\varphi), \text{ess sup}(\varphi)].$$

SOLUTION: By (a) and (b), $\partial\sigma(T_\varphi) \subset \sigma_{\text{ess}}(T_\varphi)$. However,

$\sigma_{\text{ess}}(T_\varphi) \subset \sigma(T_\varphi)$ (evident),

$\sigma(T_\varphi) = [\text{ess inf}(\varphi), \text{ess sup}(\varphi)]$ (by Theorem 3.3.8),

$\partial[\text{ess inf}(\varphi), \text{ess sup}(\varphi)] = [\text{ess inf}(\varphi), \text{ess sup}(\varphi)]$ (the boundary in $\mathbb{C}$),

and the result follows.

3.4.2 Holomorphic Symbols $\varphi \in H^\infty$

Let $\varphi \in H^\infty$. Show that

(a) T_φ is invertible $\Leftrightarrow 1/\varphi \in H^\infty$,
(b) $\sigma(T_\varphi) = \text{clos}(\varphi(\mathbb{D}))$,
(c) $T_\varphi \in \text{Fred}(H^2) \Leftrightarrow (1/\varphi \in L^\infty(\mathbb{T})$ *and the inner part φ_{in} of φ is a finite Blaschke product) (see Appendix F for the terminology).*

Hint For (c) see Appendix E.9(7).

3.4.3 Fredholm Theory for the Algebra alg $\mathcal{T}_{H^\infty+C(\mathbb{T})}$

(a) *Suppose $\bar{f} \in H^\infty + C(\mathbb{T})$ or $g \in H^\infty + C(\mathbb{T})$ ("or" not exclusive). Show that $[T_f, T_g] = T_f T_g - T_{fg} \in \mathfrak{S}_\infty(H^2)$.*

SOLUTION: $T_f T_g - T_{fg} = -H_{\bar{f}}^* H_g$ (by Lemma 2.2.9), and $H_g \in \mathfrak{S}_\infty(H^2) \Leftrightarrow g \in H^\infty + C(\mathbb{T})$ (by Theorem 2.2.8).

(b) *If for every $g \in L^\infty(\mathbb{T})$ (respectively, for every $f \in L^\infty(\mathbb{T})$) $[T_f, T_g] = -H_{\bar{f}}^* H_g \in \mathfrak{S}_\infty(H^2)$, then $\bar{f} \in H^\infty + C(\mathbb{T})$ (respectively, $g \in H^\infty + C(\mathbb{T})$).*

SOLUTION: With $g = \bar{f}$, we have $[T_f, T_{\bar{f}}] = -H_{\bar{f}}^* H_{\bar{f}} \in \mathfrak{S}_\infty(H^2)$, therefore $H_{\bar{f}} \in \mathfrak{S}_\infty(H^2)$ and hence $\bar{f} \in H^\infty + C(\mathbb{T})$ (by Theorem 2.2.8).

(c) *Let $T \in \text{alg}\mathcal{T}_{H^\infty+C(\mathbb{T})}$. Show that $\text{Sym}(T) \in H^\infty + C(\mathbb{T})$ (definition of Sym in Theorem 3.1.3) and $T - T_{\text{Sym}(T)} \in \mathfrak{S}_\infty(H^2)$.*

SOLUTION: For a product $T = \prod_{j=1}^n T_{\varphi_j}$ ($\varphi_j \in H^\infty + C(\mathbb{T})$), we have $T - T_\varphi \in \mathfrak{S}_\infty(H^2)$, $\varphi = \Pi_j \varphi_j$, by (a) and an induction on n. We then move to an aggregate operator $T = \sum_i \prod_j T_{\varphi_{ij}}$ with $\varphi_{ij} \in H^\infty + C(\mathbb{T})$ and $\text{Sym}(T) =: \sum_i \prod_j \varphi_{ij}$, and then to the limit to arrive at an arbitrary $T \in \text{alg}\mathcal{T}_{H^\infty+C(\mathbb{T})}$.

(d) (Douglas, 1969) *Show that*

$$\text{alg}\mathcal{T}_{H^\infty+C(\mathbb{T})} = \mathcal{T}_{H^\infty+C(\mathbb{T})} + \mathfrak{S}_\infty(H^2) \quad \text{(direct sum)},$$

Sym: $T \mapsto \text{Sym}(T)$ *is a contractive surjective homomorphism of Banach algebras whose kernel is $\mathfrak{S}_\infty(H^2)$; its quotient mapping*

$$\widehat{\text{Sym}}: \text{alg}\mathcal{T}_{H^\infty+C(\mathbb{T})}/\mathfrak{S}_\infty(H^2) \to H^\infty + C(\mathbb{T})$$

is an isometric algebra isomorphism.

Hint Using (c) above, repeat the reasoning of Theorem 3.1.5.

(e) (Douglas, 1969) *Let $T \in \text{alg}\mathcal{T}_{H^\infty+C(\mathbb{T})}$, $\varphi = \text{Sym}(T)$.*

(1) $T \in \text{Fred}(H^2) \Leftrightarrow T_\varphi \in \text{Fred}(H^2) \Leftrightarrow \varphi \in (H^\infty + C(\mathbb{T}))^{-1} \Leftrightarrow (1/\varphi \in L^\infty(\mathbb{T})$ *and* $\varphi/[\varphi] \in \text{QC}(\mathbb{T})$, *where $[\varphi]$ is the outer function of modulus $|\varphi|$).*

(2) *If $T \in \text{Fred}(H^2)$, then $T_{1/\varphi}$ is a regularizer of T.*

(3)* *If $T \in \text{Fred}(H^2)$, then $\text{ind}(T) = -\text{wind}(\text{Sym}(T))$.*

SOLUTION: (1) $\Rightarrow$ (2) If $1/\varphi \in H^\infty + C(\mathbb{T})$, then, by (c), $TT_{1/\varphi} = I + K_1$, $T_{1/\varphi} T = I + K_2$ where $K_j \in \mathfrak{S}_\infty(H^2)$.

For (1), clearly $1/\varphi \in H^\infty + C(\mathbb{T})$ implies $T \in$ Fred (the same reasoning as above); the converse is more delicate than the corresponding part of Theorem 3.1.6 (because $C(\mathbb{T})$ is a symmetric sub-algebra (C^* sub-algebra) of $L^\infty(\mathbb{T})$, but $H^\infty + C(\mathbb{T})$ is not), however, by (d) above, we know that $T \in$ Fred $\Rightarrow T_\varphi \in$ Fred $\Rightarrow 1/\varphi \in L^\infty$ (see Theorem 3.1.3(3)), and thus $\varphi = [\varphi]u$, $[\varphi]$ is invertible in H^∞ and $|u| = 1$ a.e. on $\mathbb{T}$, $u \in H^\infty + C(\mathbb{T})$. It suffices to show that u is invertible in $H^\infty + C(\mathbb{T})$ ($\Leftrightarrow u \in$ QC). By Theorem 3.1.10(3), dist($\bar{u}, H^\infty + C$) < 1, so there exists $h \in H^\infty + C$ such that $\|\bar{u} - h\|_\infty < 1$, thus $\|1 - uh\|_\infty < 1$, and hence $uh \in H^\infty + C(\mathbb{T})$ is an invertible element. Then, u (and h) is invertible in $H^\infty + C(\mathbb{T})$. The other equivalences of (1) are also clear.

For point (3)*, as well as for the question *of the definition of* wind(φ), $\varphi^{\pm 1} \in H^\infty + C(\mathbb{T})$, see §3.4 and (especially) [Douglas, 1972], [Nikolski, 1986], and [Böttcher and Silbermann, 1990].

Remark Another means to determine ind(T), $T \in$ Fred $\cap$ alg$\mathcal{T}_{H^\infty + C(\mathbb{T})}$, is Sarason's theorem, in (g) below.

(f) Special case. *If $\varphi \in$ QC($\mathbb{T}$), then $T_\varphi \in$ Fred $\Leftrightarrow 1/\varphi \in$ QC($\mathbb{T}$).*

Hint Immediate by (e).

(g) (Sarason, 1973) *Let $u \in L^\infty(\mathbb{T})$, $|u| = 1$ a.e. on $\mathbb{T}$. Show that*

$$u \in \text{QC}(\mathbb{T}) \Leftrightarrow u = z^n e^{i(a+\bar{b})}, \quad \text{where } n \in \mathbb{Z}, a, b \in C(\mathbb{T}) \text{ (real functions)}.$$

Moreover, $u \in$ QC($\mathbb{T}$) $\Rightarrow T_u \in$ Fred and ind(T) = $-n$, hence

$$T_u \text{ invertible} \Leftrightarrow u = e^{i(a+\bar{b})}.$$

Solution: For the implication $\Leftarrow$, $u = z^n e^{b+ib} \cdot e^{-b+ia} \in H^\infty \cdot C(\mathbb{T}) \subset H^\infty + C(\mathbb{T})$.

For $\Rightarrow$, observe that $u \in$ QC $\subset H^\infty + C(\mathbb{T})$ and hence there exists $k \in \mathbb{Z}$ and $h \in H^\infty$ such that $\|u - z^k h\|_\infty < 1$; as u is invertible in $H^\infty + C(\mathbb{T})$ ($1/u = \bar{u} \in H^\infty + C(\mathbb{T})$ and $\|1/u\|_\infty = 1$), so is the element $z^k h$. This implies that h, and then h_{in} and h_{out} (inner and outer parts of h: see Appendix F) are invertible in $H^\infty + C(\mathbb{T})$. The inner function h_{in} is a finite Blaschke product. Indeed, there exists $g \in H^\infty$ such that $\|\bar{h}_{in} - z^m g\|_\infty < 1$, hence $\|\bar{z}^m - h_{in} g\|_\infty < 1$, which implies

$$\inf\{|h_{in}(\zeta)g(\zeta)|: 1 - \delta < |\zeta| < 1\} > 0$$

for $\delta > 0$ sufficiently small, and hence h_{in} is a finite Blaschke product: see Appendix F.2. The inequality $\|1 - \bar{u}z^k h\|_\infty < 1$ guarantees the existence of $\log(\bar{u}z^k h) \in H^\infty + C(\mathbb{T})$, hence

$$\bar{u}z^k h = e^{f+g} \quad \text{where } f \in C(\mathbb{T}), \ g \in H^\infty,$$

and with $l =: \log |h| \in L^\infty$ we have $u = z^k h_{in} \exp(l+i\tilde{l}-f-g)$ where $l-\mathrm{Re}(g)-\mathrm{Re}(f) = 0$ (since $|u| = 1$), and thus

$$u = z^k h_{in} \exp(i\tilde{l} - i\,\mathrm{Im}(g) - i\,\mathrm{Im}(f)) = z^k h_{in} \exp(i\tilde{b} + ia_1),$$

where $b = l - \mathrm{Re}(g) = \mathrm{Re}(f) \in C(\mathbb{T})$, $a_1 = -\mathrm{Im}(f) \in C(\mathbb{T})$. It only remains to remark that h_{in} is a product of $d < \infty$ factors of type $(z - \lambda)/(1 - \bar{\lambda}z)$, $|\lambda| < 1$, so $\mathrm{wind}(z^{-d}h_{in}) = 0$ and $z^{-d}h_{in} = e^{ia_2}$, $a_2 \in C(\mathbb{T})$, hence $u = z^n e^{i(a+\tilde{b})}$ with $n = k + d$, $a = a_1 + a_2$.

The formula for $\mathrm{ind}(T_u)$ is clear: $u \in (H^\infty + C(\mathbb{T}))^{-1} \Rightarrow T_u \in \mathrm{Fred}$ (by (e)), and $u_0 =: e^{i(a+\tilde{b})}$ is homotopic to 1 by $t \mapsto e^{it(a+\tilde{b})}$ ($0 \le t \le 1$), hence T_{u_0} is invertible and $\mathrm{ind}(T_u) = \mathrm{ind}(T_{z^n}) = -n$.

3.4.4 $H^\infty + C(\mathbb{T})$ is the Minimal Algebra Containing H^∞ (Hoffman and Singer, 1960)

Let $B \subset L^\infty(\mathbb{T})$ be a closed sub-algebra containing H^∞. If $B \ne H^\infty$, then $B \supset H^\infty + C(\mathbb{T})$.

Show the theorem while supposing $(B \cap A) \setminus H^\infty \ne \emptyset$ where $A = \mathrm{clos}_{L^\infty}(H^\infty + \overline{H^\infty})$.

Remark In particular, to validate the supplementary hypothesis, it suffices to have $f \in B \cap C(\mathbb{T})$, $f \notin H^\infty$. Unfortunately, $A \ne L^\infty(\mathbb{T})$: see [Nikolski, 2019], Exercise 2.8.4(i).

SOLUTION: [Cohen, 1961] Let $f \in (B \cap A) \setminus H^\infty$. If necessary multiplying f by λz^n, we can suppose that $\widehat{f}(-1) = 1$. Then there exists $a, b \in H^\infty$ and $c \in L^\infty(\mathbb{T})$ such that $zf = 1 + za + \overline{zb} + c$ and $\|c\|_\infty < 1/2$. Denote $M = \|zb - \overline{zb}\|_\infty$ and observe that, for every $\epsilon > 0$, $\|1 + \epsilon(zb - \overline{zb})\|_\infty \le 1 + \epsilon^2 M^2/2$ (since $\mathrm{Re}(zb - \overline{zb}) = 0$). We rewrite this inequality as

$$\|1 + \epsilon zb - \epsilon(zf - 1 - za - c)\|_\infty = \|1 + \epsilon + \epsilon zb - \epsilon(zf - za - c)\|_\infty \le 1 + \epsilon^2 M^2/2,$$

and then

$$\|1 + \epsilon + \epsilon zb - \epsilon(zf - za)\|_\infty \le \epsilon \|c\|_\infty + 1 + \epsilon^2 M^2/2.$$

Consequently, if $\epsilon M^2 < 1$, then $\|1 - zF\|_\infty < 1$ where $F =: \epsilon(f - a - b)/(1 + \epsilon) \in B$. It follows that zF is invertible in B, and thus so is z, hence $\bar{z} \in B$. As B is a closed sub-algebra, we obtain $H^\infty + C(\mathbb{T}) \subset B$.

3.4.5 Fredholm Theory for the Algebra $\text{alg}\mathcal{T}_{PC(\mathbb{T})}$

See Example 3.2.2(d) for the definition of $PC(\mathbb{T})$.

(a) (Simonenko, 1960; Devinatz, 1964) *Let $\varphi \in PC(\mathbb{T})$ and let φ_* be a continuous parametrization of the range $\varphi(\mathbb{T})$ completed by segments $[\varphi(\lambda-), \varphi(\lambda+)]$ ($\lambda \in \mathbb{T}$; for every $\epsilon > 0$, there is only a finite set of λ such that $|\varphi(\lambda-) - \varphi(\lambda+)| > \epsilon$). Then,*

(1) $T_\varphi \in \text{Fred} \Leftrightarrow \varphi_*(s) \neq 0$ ($\forall s$).
(2) $\sigma_{\text{ess}}(T_\varphi) = \text{Ran}(\varphi_*)$.
(3)* *If $T_\varphi \in \text{Fred}$ then $\text{ind}(T_\varphi) = -\text{wind}(\varphi_*)$.*

Remark We can consider φ_* as the mapping

$$\varphi_* : \mathbb{T}_* =: \mathbb{T} \times [0,1] \to \mathbb{C}$$

defined by $\varphi_*(\lambda, t) = \varphi(\lambda-)t + \varphi(\lambda+)(1-t)$.

SOLUTION: For (3), see [Gohberg and Krupnik, 1973], [Nikolski, 1986], and [Böttcher and Silbermann, 1990]. (1) $\Rightarrow$ (2) is evident.

For $\Leftarrow$ of (1) see Example 3.2.2(d).

For $\Rightarrow$ of (1), we use (3). Suppose $T_\varphi \in \text{Fred}$ and $\varphi_*(s) = 0$ for a certain s. As $1/\varphi \in L^\infty$, there exists a finite set of $\lambda \in \mathbb{T}$ such that $0 = \varphi_*(s) \in (\varphi(\lambda-), \varphi(\lambda+))$ (an open interval). Select $\epsilon > 0$ small enough so that the circle $C = \{|z| = \epsilon\}$ encounters twice each of these intervals and consider the operators $T_\varphi - zI, z \in C$. It follows from (3) and the "$\Leftarrow$" part of (1) that when z crosses one of the above intervals the index $\text{ind}(T - zI)$ changes. But this contradicts the stability of the index (Corollary E.7.4(3)), and hence $\varphi_*(s) \neq 0$ ($\forall s$).

(b)* (Gohberg and Krupnik, 1969)

(1) *The set $\mathfrak{S}_\infty(H^2)$ is the closed bilateral ideal of the algebra $\text{alg}\mathcal{T}_{PC(\mathbb{T})}$ generated by the commutators $[T_f, T_g]$, $f, g \in PC(\mathbb{T})$.*

(2) *Let $(\varphi_{ij}) \subset PC(\mathbb{T})$ be a finite family, $T = \sum_i \prod_j T_{\varphi_{ij}}$ and*

$$\text{Sym}_*(T) =: \sum_i \prod_j (\varphi_{ij})_*,$$

where φ_* is defined in the Remark of (a). Then, $\text{Sym}_* : T \mapsto \text{Sym}_*(T)$ can be extended to a contracting surjective homomorphism of Banach algebras $\text{alg}\mathcal{T}_{PC(\mathbb{T})} \to C(\mathbb{T}_*)$ (where $\mathbb{T}_*$ is equipped with an "exotic topology" induced by the set of functions $\varphi_*, \varphi \in PC(\mathbb{T}))$ whose kernel is $\mathfrak{S}_\infty(H^2)$. Its quotient mapping

$$\widetilde{\text{Sym}}_* : \text{alg}\mathcal{T}_{PC(\mathbb{T})}/\mathfrak{S}_\infty(H^2) \to C(\mathbb{T}_*)$$

is an isometric algebra isomorphism.

(3) *An operator $T \in \mathrm{alg} \mathcal{T}_{PC(\mathbb{T})}$ is Fredholm if and only if* $\mathrm{Sym}_*(T)$ *does not vanish on* $\mathbb{T}_*$, *and* $\mathrm{ind}(T) = -\mathrm{wind}(\mathrm{Sym}_*(T))$ *(the latter is well-defined for the "aggregate" operators $T = \sum_i \prod_j T_{\varphi_{ij}}$ whose family (φ_{ij}) has only a finite set of discontinuities, and then can be extended by continuity).*

3.4.6 A Simplified Local Principle (Simonenko, 1960)

Let $\varphi \in L^\infty(\mathbb{T})$ and suppose that for every $\lambda \in \mathbb{T}$ there exists $\varphi_\lambda \in L^\infty(\mathbb{T})$ such that

(i) $T_{\varphi_\lambda} \in \mathrm{Fred}(H^2)$,
(ii) $\varphi = \varphi_\lambda$ on an open arc V_λ containing λ.

Show (without using Theorem 3.2.1) that $T_\varphi \in \mathrm{Fred}(H^2)$.

SOLUTION: Let $\alpha = (V_{\lambda(i)})$ be a finite subcovering of $\mathbb{T}$ and let (a_i) be a partition of unity subordinate to α ($a_i \in C(\mathbb{T})$); denote by R_i a regularizer of the operator $T_{\varphi_{\lambda(i)}}$, and let $R = \sum_i R_i T_{a_i}$. Then R is a left regularizer of T_φ (A_i, B_i, etc. are compact operators):

$$RT_\varphi = \sum_i R_i T_{a_i} T_\varphi = \sum_i R_i (T_{a_i \varphi} + A_i) = \sum_i R_i (T_{a_i \varphi_{\lambda(i)}} + A_i)$$
$$= \sum_i R_i (T_{\varphi_{\lambda(i)}} T_{a_i} + B_i) = \sum_i (I + C_i) T_{a_i} + D_i = \sum_i T_{a_i} + E = I + E$$

($E \in \mathfrak{S}_\infty(H^2)$). Similarly, $\sum_i T_{a_i} R_i$ is a right regularizer; hence $T_\varphi \in \mathrm{Fred}$.

3.4.7 Fred(H^2) and Local Sectoriality

Let $\varphi^{\pm 1} \in L^\infty(\mathbb{T})$.
(a) *(Reminder) If φ is (locally) sectorial, T_φ is invertible (resp. in $\mathrm{Fred}(H^2)$) (see Example 3.2.2(b)).*

Show that $T_\varphi \in \mathrm{Fred}(H^2) \Leftrightarrow (\varphi = z^n fg$ where $g^{\pm 1} \in H^\infty$ and f is sectorial): see Theorem 3.3.6(5).

SOLUTION: As $T_{z^n} \in \mathrm{Fred}$, it follows that T_φ and $T_\varphi T_{z^n}$ are simultaneously Fredholm (see Appendix E.9(8)), and since $[T_\varphi, T_{z^n}] = T_\varphi T_{z^n} - T_{\varphi z^n} \in \mathfrak{S}_\infty$ (see Exercise 3.4.3(a)), we have $\mathrm{ind}(T_{\varphi z^n}) = \mathrm{ind} T_\varphi + \mathrm{ind} T_{z^n}$ (see Corollary E.7.4(5)); then apply Theorem 3.3.6(5).

(b) **Symbols taking on two values.** *Suppose that $\mathrm{Ran}_{\mathrm{ess}}(\varphi)$ consists of two values $\lambda, \mu \in \mathbb{C}$, so that $\varphi = \lambda \chi_E + \mu \chi_{\mathbb{T} \setminus E}$ for a Borel set $E \subset \mathbb{T}$. Show that $T_\varphi \in \mathrm{Fred} \Leftrightarrow (T_\varphi$ invertible$) \Leftrightarrow \varphi$ is (locally) sectorial.*

144 H^2 Theory of Toeplitz Operators

SOLUTION: Clearly "sectorial" and "locally sectorial" are equivalent for a φ with two values, and φ is not sectorial if and only if $0 \in (\lambda, \mu)$ (the open interval); thus, if φ is sectorial, then T_φ is invertible (see Lemma 3.3.2), and conversely, if $T_\varphi \in$ Fred, then φ is locally sectorial (see Exercise 3.4.1(c)).

(c) Symbols taking on three values. *Suppose that* $\text{Ran}_{\text{ess}}(\varphi)$ *consists of three values* $\lambda_j \in \mathbb{C}$ *(j = 1, 2, 3), so that* $\varphi = \sum_j \lambda_j \chi_{E_j}$ *(where* $E_j \subset \mathbb{T}$ *are Borel sets,* $E_j \cap E_i = \emptyset$ *for* $j \neq i$, $\bigcup E_i = \mathbb{T}$*), Show that:*

(1) *if* E_j *are intervals (arcs), then* $T_\varphi \in \text{Fred}(H^2)$ *if and only if* φ *is locally sectorial* $(\Leftrightarrow 0 \notin (\lambda_1, \lambda_2) \cup (\lambda_1, \lambda_3) \cup (\lambda_2, \lambda_3))$;
(2) *there exists* φ *not locally sectorial and taking on the values* $1, i, -i$ *such that* T_φ *is invertible.*

Hint For (2), use Lemma 3.3.4(3).

SOLUTION: (1) follows from Example 3.2.2(d) or Exercise 3.4.5(a).

For (2), use Lemma 3.3.4(3) and search for φ of the form

$$\varphi = e^{i(a+\tilde{b}+c)},$$

where $a, b \in L^\infty(\mathbb{T})$ are real functions, $c \in \mathbb{R}$ such that $\|a\|_\infty < \pi/2$, and $\tilde{b}$ is the harmonic conjugate of b. Define the domain Ω (an "infinitely oscillating serpent") as

$$\Omega = \{x + iy: 0 < x < \pi/2; |y - \sin(1/x)| < \epsilon(x)\},$$

where $x \mapsto \epsilon(x) \in]0, \frac{1}{2}[$ is a function that decreases sufficiently rapidly as $x \to 0$ (in order for the "wavelets" of the serpent Ω not to be superposed), and then $\omega: \mathbb{D} \to \Omega$ a conformal mapping (which exists according to Riemann's theorem [Rudin, 1986], Theorem 14.8) and set $b = \text{Re}(\omega)$ and $a(\zeta) = \pi/2 - \text{Im}(\omega(\zeta))$ if $\text{Im}(\omega(\zeta)) > 1/2$, $a(\zeta) = -\pi/2 - \text{Im}(\omega(\zeta))$ if $\text{Im}(\omega(\zeta)) < -1/2$, and $a(\zeta) = -\text{Im}(\omega(\zeta))$ if $|\text{Im}(\omega(\zeta))| \leq 1/2$.

It is easy to see that $\|a\|_\infty \leq \pi/2 - 1/2 < \pi/2$, and hence T_φ is invertible, whereas φ is not locally sectorial (in a neighborhood of an accumulation point of the $\omega^{-1}(\Omega_n)$, where $\Omega_n = \{x + iy: x + iy \in \Omega, x \leq 1/n\}$).

(d) Local sectoriality and sectoriality (Simonenko; Sarason). *Show that* $\varphi \in L^\infty(\mathbb{T})^{-1}$ *is locally sectorial (see Example 3.2.2(b) for the definition) if and only if* $\varphi = c \cdot s$, *where* $c \in C(\mathbb{T})$ *and* $s \in L^\infty(\mathbb{T})$ *is sectorial.*

SOLUTION: It is easy to see that φ is locally sectorial if and only if, for every $\lambda \in \mathbb{T}$, $\text{dist}_\lambda(u, \mathbb{C}) < 1$ where $u = \varphi/|\varphi|$; by using Lemma 3.2.2 and Remark 3.2.3, we can find $c \in C(\mathbb{T})$ such that $\|u - c\|_\infty < 1$, which implies $1/c \in C(\mathbb{T})$ and $\|1 - (c/u)\|_\infty < 1$, and thus c/u is sectorial (as is u/c); hence $\varphi = c(|\varphi|u/c)$ where $|\varphi|u/c$ itself is also sectorial.

The converse is evident.

3.4.8 Multipliers Preserving Fred(H^2)

Let $f \in (H^\infty + C)^{-1}$. Show that $T_\varphi \in \text{Fred}(H^2) \Leftrightarrow T_{\varphi f} \in \text{Fred}(H^2)$ ($\forall \varphi \in L^\infty(\mathbb{T})$).

SOLUTION: Since $T_f \in \text{Fred}$ (see Exercise 3.4.3(e)), we have $T_\varphi \in \text{Fred}(H^2) \Leftrightarrow T_\varphi T_f \in \text{Fred}(H^2)$ (see Appendix E.9(8)), and then $T_\varphi T_f = T_{\varphi f} + K$, where $K \in \mathfrak{S}_\infty(H^2)$ (by Exercise 3.4.3(a)), hence $T_\varphi T_f \in \text{Fred}(H^2) \Leftrightarrow T_{\varphi f} \in \text{Fred}(H^2)$.

3.4.9 The Toeplitz Algebra alg$\mathcal{T}_{L^\infty(\mathbb{T})}$: A Necessary Condition

Show that if $A \in \text{alg}\mathcal{T}_{L^\infty(\mathbb{T})}$, then

$$[A, T_\varphi] \in \mathfrak{S}_\infty(H^2) \quad \text{for every } \varphi \in \text{QC}(\mathbb{T}),$$

and consequently $\text{alg}\mathcal{T}_{L^\infty(\mathbb{T})} \neq L(H^2)$ (see also Exercise 3.4.10 below).

SOLUTION: For every $f \in L^\infty(\mathbb{T})$, we have $[T_f, T_\varphi] = [T_f, T_\varphi) - [T_\varphi, T_f) \in \mathfrak{S}_\infty(H^2)$ (by Exercise 3.4.3(a)), and then, by induction,

$$[T_f T_g, T_\varphi] = T_f[T_g, T_\varphi] + [T_f, T_\varphi]T_g \in \mathfrak{S}_\infty(H^2),$$

implying $[A, T_\varphi] \in \mathfrak{S}_\infty(H^2)$ for every aggregate operator $A = \sum_i \prod_j T_{f_{ij}}$. Finally, by passing to the limit, $[A, T_\varphi] \in \mathfrak{S}_\infty(H^2)$ for every $A \in \text{alg}\mathcal{T}_{L^\infty(\mathbb{T})}$.

To exhibit an operator $A \in L(H^2) \setminus \text{alg}\mathcal{T}_{L^\infty(\mathbb{T})}$, take $A = P$, the orthogonal projection onto the subspace span$_{H^2}(z^{2k}: k = 0, 1, \ldots)$ and $\varphi = z \in C(\mathbb{T}) \subset \text{QC}$; then, $[A, T_z]z^{2k+1} = Pz^{2k+2} = z^{2k+2}$ ($\forall k \geq 0$), hence $[A, T_z] \notin \mathfrak{S}_\infty(H^2)$ and $A \notin \text{alg}\mathcal{T}_{L^\infty(\mathbb{T})}$.

3.4.10 Hankel Operators from the Toeplitz Algebra alg$\mathcal{T}_{L^\infty(\mathbb{T})}$

Recall that $\Gamma_\varphi = JH_\varphi: H^2 \to H^2$, where $Jf = \bar{z}f(\bar{z})$ is an involution of $L^2(\mathbb{T})$ exchanging H^2 and H^2_-: see Lemma 2.2.1 and Example 2.2.2.

Show that:

(a) for every $f, g \in L^\infty(\mathbb{T})$, $\Gamma_f \Gamma_g \in \text{alg}\mathcal{T}_{L^\infty(\mathbb{T})}$,
(b) if $\Gamma_f \geq 0$ (positive semi-definite), then $\Gamma_f \in \text{alg}\mathcal{T}_{L^\infty(\mathbb{T})}$,
(c) (Axler, 1980 (see Notes and Remarks)) There exists $f \in L^\infty(\mathbb{T})$ such that $\Gamma_f \notin \text{alg}\mathcal{T}_{L^\infty(\mathbb{T})}$ (for example, $f = (z-i)/|z-i|$).

SOLUTION: Since Γ_f has a matrix $(\widehat{f}(-j-k-1))_{j,k\geq 0}$, then $\Gamma_f^* = \Gamma_{f_*}$ where

$$f_*(\zeta) = \overline{f(\zeta)}, \quad \zeta \in \mathbb{T},$$

and thus $\Gamma_f \Gamma_g = \Gamma_{f_*}^* \Gamma_g = T_{\overline{f}_*} T_g - T_{\overline{f}_* g} \in \text{alg}\mathcal{T}_{L^\infty(\mathbb{T})}$, hence the assertion (a).

For (b), we have $\Gamma_f^2 \in \text{alg}\mathcal{T}_{L^\infty(\mathbb{T})}$ (by (a)), and since the algebra $A = \text{alg}\mathcal{T}_{L^\infty(\mathbb{T})}$ is self-adjoint (a C^* algebra: $T \in A \Rightarrow T^* \in A$), we have $\Gamma_f = (\Gamma_f^2)^{1/2} \in \text{alg}\mathcal{T}_{L^\infty(\mathbb{T})}$.

For (c), we seek an operator Γ_f for which the necessary condition of Exercise 3.4.9 is not satisfied. Suppose $[\Gamma_f, T_\varphi] \in \mathfrak{S}_\infty(H^2)$ for every $\varphi \in QC(\mathbb{T})$, in particular (with $\varphi = z$), $J[\Gamma_f, T_z] \in \mathfrak{S}_\infty(H^2, H^2_-)$. Using $JP_+ J = P_-$, $J^2 = \text{id}$ and $J(ab) = a(\overline{z})Jb$, for $x \in H^2$ we obtain

$$J[\Gamma_f, T_z]x = J\{JP_- fP_+ zx - P_+ zJP_- fx\}$$

$$= P_-\{fzx - JzJP_- fx\}$$

$$= P_-\{fzx - \overline{z}P_- fx\}.$$

However, $x \mapsto P_- \overline{z} P_+ fx$ is an operator of rank 1, so we must have $H_{zf-\overline{z}f} \in \mathfrak{S}_\infty(H^2, H^2_-)$, implying $f(z - \overline{z}) \in H^\infty + C(\mathbb{T})$ (Hartman's Theorem 2.2.8(2)). There exist numerous functions $f \in L^\infty(\mathbb{T})$ such that $f(z - \overline{z}) \notin H^\infty + C(\mathbb{T})$. For example, $f = (z - i)/|z - i|$ (by Lindelöf's theorem (see Appendix F.3), if a function $f \in H^\infty$ admits limits to the left $f(\zeta_-)$ and to the right $f(\zeta_+)$ at some point $\zeta \in \mathbb{T}$, then $f(\zeta_-) = f(\zeta_+)$. In our case, $f(i_-) = -1$, $f(i_+) = 1$).

Remark More generally, in order for $\Gamma_f \in \text{alg}\mathcal{T}_{L^\infty(\mathbb{T})}$, we must have $f(\varphi - \varphi(\overline{z})) \in H^\infty + C(\mathbb{T})$ for every $\varphi \in QC(\mathbb{T})$.

3.4.11 On the Equation $T_\varphi f = 1$ (Another Proof of Theorem 3.3.8)

Suppose $\varphi \in L^\infty(\mathbb{T})$, $\overline{\varphi} = \varphi$ and that there exists $f \in H^2(\mathbb{T})$ such that $T_\varphi f = 1$. Show that $\text{sign}(\varphi) = \text{const}$.

Consequently, $\sigma(T_\varphi) = [\text{ess inf}(\varphi), \text{ess sup}(\varphi)]$.

SOLUTION: $P_+(\varphi f) = 1$ means that $\varphi f = 1 + h$ where $h \in H^2$, or equivalently $\varphi \overline{f} = 1 + \overline{h} \in H^2$, and then $\varphi |f|^2 = (1 + \overline{h})f \in H^1$. However, a real function of H^1 is a constant (see Appendix F), hence $\text{sign}(\varphi) = \text{const}$.

By applying the property shown to $\varphi - \lambda$ ($\lambda \in \mathbb{R}$), we obtain $\sigma(T_\varphi) \supset [\text{ess inf}(\varphi), \text{ess sup}(\varphi)]$, whereas the converse $\sigma(T_\varphi) \subset [\text{ess inf}(\varphi), \text{ess sup}(\varphi)]$ is evident by Corollary 3.3.3.

3.4.12 Is There a Regularizer of T_φ in $\mathcal{T}_{L^\infty(\mathbb{T})}$ and/or in $\text{alg}\mathcal{T}_{L^\infty(\mathbb{T})}$?

We shall see that the answers to these two questions are quite different.

Let $T_\varphi \in \mathcal{T}_{L^\infty} \cap \text{Fred}(H^2)$.

3.4 Exercises

(a) *Show that if there exists $T \in \mathrm{alg}\mathcal{T}_{L^\infty(\mathbb{T})}$ such that $TT_\varphi - I \in \mathfrak{S}_\infty(H^2)$, then $\mathrm{Sym}(T) = 1/\varphi$. In particular, if $T \in \mathcal{T}_{L^\infty(\mathbb{T})}$, then $T = T_{1/\varphi}$.*

SOLUTION: By Theorem 3.1.3, $1 = \mathrm{Sym}(TT_\varphi) = \mathrm{Sym}(T)\varphi$.

(b) *Show that $T_{1/\varphi}$ is a regularizer of T_φ if and only if*

$$\mathrm{alg}(H^\infty, \overline{\varphi}) \cap \mathrm{alg}(H^\infty, 1/\varphi) \subset H^\infty + C(\mathbb{T}) \quad \text{and}$$
$$\mathrm{alg}(H^\infty, \varphi) \cap \mathrm{alg}(H^\infty, 1/\overline{\varphi}) \subset H^\infty + C(\mathbb{T}),$$

where the algebra $\mathrm{alg}(H^\infty, F)$ is defined in §3.5 below (see part (viii) of the comments on Exercise 3.4.3).

Hint Use the Axler–Chang–Sarason–Volberg theorem from part (viii) of the comments on Exercise 3.4.3 in §3.5.

SOLUTION: This is immediate by the theorem referenced in the Hint, since $T_\varphi T_{1/\varphi} - I = [T_\varphi, T_{1/\varphi}]$ and $T_{1/\varphi} T_\varphi - I = [T_{1/\varphi}, T_\varphi]$.

(c) *Suppose that $|\varphi| = 1$ a.e. on $\mathbb{T}$. Show that $T_{1/\varphi}$ is a regularizer of T_φ if and only if $\varphi \in \mathrm{QC}(\mathbb{T})$.*

SOLUTION: By (b) and $\overline{\varphi} = 1/\varphi$, the desired property is equivalent to $\mathrm{alg}(H^\infty, \overline{\varphi}) \subset H^\infty + C(\mathbb{T})$ and $\mathrm{alg}(H^\infty, \varphi) \subset H^\infty + C(\mathbb{T})$, or to $\overline{\varphi}, \varphi \in H^\infty + C(\mathbb{T})$, which is the definition of $\varphi \in \mathrm{QC}(\mathbb{T})$.

(d) *Suppose that φ is real. Show that $T_{1/\varphi}$ is a regularizer of T_φ if and only if $\varphi \in \mathrm{QC}(\mathbb{T})$.*

SOLUTION: Since $\overline{\varphi} = \varphi$, the two inclusions of (b) are reduced to only one:

$$\mathrm{alg}(H^\infty, \varphi) \cap \mathrm{alg}(H^\infty, 1/\varphi) \subset H^\infty + C(\mathbb{T}).$$

Moreover, as $T_\varphi \in \mathrm{Fred}$, the symbol φ does not change sign (see Exercise 3.4.1(c)) – say $\varphi \geq 0$ (in fact $\varphi \geq \delta > 0$) – and thus there exists $\lambda > 0$ and $\epsilon > 0$ such that $0 < \epsilon \leq \lambda/\varphi \leq 1 - \epsilon$, hence $\varphi/\lambda = \sum_{n \geq 0}(1 - (\lambda/\varphi))^n$ and therefore $\varphi \in \mathrm{alg}(H^\infty, 1 - (\lambda/\varphi)) = \mathrm{alg}(H^\infty, 1/\varphi)$. Thus the above inclusion is equivalent to $\mathrm{alg}(H^\infty, \varphi) \subset H^\infty + C(\mathbb{T})$ and hence to $\varphi \in H^\infty + C(\mathbb{T})$, or equivalently $\varphi \in \mathrm{QC}(\mathbb{T})$.

(e) *(Reminder) Suppose $\varphi \in H^\infty + C(\mathbb{T})$. Show that $T_{1/\varphi}$ is a regularizer of T_φ.*

SOLUTION: This is part of Exercise 3.4.3(e).

Remark For $\varphi \in H^\infty$, the result also follows from Exercise 3.4.2(c), as $T_\varphi \in \mathrm{Fred}(H^2) \Rightarrow \varphi = [\varphi]B$ where B is a finite Blaschke product and $[\varphi]^{\pm 1} \in H^\infty$, hence

$$\text{alg}(H^\infty, \overline{\varphi}) \cap \text{alg}(H^\infty, 1/\varphi) \subset \text{alg}(H^\infty, 1/B) \subset H^\infty + C(\mathbb{T}).$$

The other inclusion of (b) is also evident because $\text{alg}(H^\infty, \varphi) = H^\infty$.

(f) *Give an example of $T_\varphi \in \text{Fred}(H^2)$ (and even T_φ invertible) that does not have Toeplitz regularizers (in $\mathcal{T}_{L^\infty(\mathbb{T})}$).*

Hint Use (c) or (d).

SOLUTION: By (c), it suffices to find T_φ invertible, $|\varphi| = 1$ a.e., and $\varphi \notin QC(\mathbb{T})$. There are many such φ. For example, every φ taking on two different unimodular values $\varphi = \lambda \chi_E + \mu \chi_{\mathbb{T} \setminus E}$ ($E \subset \mathbb{T}$ a Borel set, $m(E) > 0$, $m(\mathbb{T}\setminus E) > 0$, $|\lambda| = |\mu| = 1$, $\lambda \neq \pm \mu$): indeed, T_φ is invertible by sectoriality of φ (see Exercise 3.4.7(b) above). However, $\varphi \notin H^\infty + C(\mathbb{T})$. If we suppose the contrary, we obtain $f =: \chi_E \in H^\infty + C(\mathbb{T})$, hence $[T_f, T_f] = T_f T_f - T_{f^2} = -H_{\overline{f}}^* H_f \in \mathfrak{S}_\infty$, which is impossible, as $T_f^2 - T_f$ is a self-adjoint operator such that $\sigma(T_f^2 - T_f) = h(\sigma(T_f)) = h([0,1]) = [-1/4, 0]$, where $h(x) = x^2 - x$ (using the spectral mapping theorem: see Appendix C or D).

By (d), it is even easier: let

$$\varphi = 1 + \frac{1}{4}(V_1 + V_{-1})$$

where

$$V_s = \exp\left(-s\frac{1+z}{1-z}\right).$$

Then φ is real, $\varphi \geq 1/2$ on $\mathbb{T}$, thus T_φ is invertible (see Theorem 3.3.8). However, $\varphi \notin H^\infty + C$ because $V_1 \in H^\infty$ and $V_{-1} = \overline{V}_1 \notin H^\infty + C$ given the non-compactness of the Hankel operator $H_{\overline{V}_1} = P_- \overline{V}_1 | H^2$ which is, in fact, isometric on an infinite-dimensional subspace $K_{V_1} =: H^2 \ominus V_1 H^2$ ($f \in H^2$ is in K_{V_1} if and only if $\overline{V}_1 f \in H^2_-$).

Remark In fact, the above argument using (c), as well as the formula $\|H_f\|_{\text{ess}} = \text{dist}(f, H^\infty + C)$, shows that $\text{dist}(\chi_E, H^\infty + C) = 1/2$.

Here again, the impossibility of the equation $\chi_E = h + c$ where $h \in H^\infty$, $c \in C(\mathbb{T})$ is linked to a theorem of Lindelöf (see Appendix F.3, as well as Nikolski [2002a], Appendix B.5.3.5): the one-sided limits at a point of a function in H^∞ (if they exist) are equal.

(g) *Show that the algebra $\text{alg}\mathcal{T}_{L^\infty(\mathbb{T})}$ is stable for inversion and for the regularizers: (1) if $T \in \text{alg}\mathcal{T}_{L^\infty(\mathbb{T})}$ is invertible, then $T^{-1} \in \text{alg}\mathcal{T}_{L^\infty(\mathbb{T})}$, and (2) if $T \in \text{alg}\mathcal{T}_{L^\infty(\mathbb{T})} \cap \text{Fred}(H^2)$, then the regularizers of T are also in $\text{alg}\mathcal{T}_{L^\infty(\mathbb{T})}$.*

Hint Show these properties for a C^* algebra containing $\mathfrak{S}_\infty$.

SOLUTION: Effectively, $A = \text{alg}\mathcal{T}_{L^\infty(\mathbb{T})}$ is a C^* algebra (see Appendix C), since $T_\varphi^* = T_{\bar\varphi}$.

Proof of (1) If $a \in A$ is invertible in $L(H^2)$ then so is the positive self-adjoint operator a^*a, and thus with an appropriate λ such that $0 < \lambda < \infty$, we have $\|I - \lambda a^*a\| < 1$, hence $(\lambda a^*a)^{-1} = \sum_{n\geq 0}(I - \lambda a^*a)^n \in A$, and then $b = (a^*a)^{-1}a^* \in A$ is a left inverse of a $((a^*a)^{-1}a^*a = I)$, but since a is invertible, then $a^{-1} = (a^*a)^{-1}a^*$. ∎

Proof of (2) Since $\mathfrak{S}_\infty \subset A = \text{alg}\mathcal{T}_{L^\infty(\mathbb{T})}$ and the quotient algebra $A/\mathfrak{S}_\infty$ is a C^* sub-algebra of the Calkin algebra $\mathcal{K} = L(H^2)/\mathfrak{S}_\infty$, we obtain the result by applying (1) ($T \in \text{Fred}(H^2) \Leftrightarrow \Pi(T)$ is invertible in $\mathcal{K}$: see Appendix E). ∎

3.4.13 Fredholm Theory for Almost Periodic Symbols

Let
$$V_s = \exp\left(-s\frac{1+z}{1-z}\right) \quad (s \in \mathbb{R})$$

and let
$$\text{AP}_1(\mathbb{T}) = \text{span}_{L^\infty(\mathbb{T})}(V_s : s \in \mathbb{R})$$

be the algebra of functions "almost periodic at the point $\zeta = 1$." We could equally consider and apply to Toeplitz operators the functions "almost periodic at an arbitrary point" $\zeta \in \mathbb{T}$, by replacing V_s with

$$\exp\left(-s\frac{\zeta + z}{\zeta - z}\right).$$

Clearly
$$\text{AP}_1(\mathbb{T}) = \text{AP}(\mathbb{R}) \circ \Omega = \{f \circ \Omega : f \in \text{AP}(\mathbb{R})\},$$

where
$$\Omega(z) = i\frac{1+z}{1-z}$$

(a conformal mapping $\mathbb{D} \to \mathbb{C}_+ = \{w \in \mathbb{C}: \text{Im}(w) > 0\}$) and $\text{AP}(\mathbb{R})$ is the Bohr algebra of classical almost periodic functions, uniform limits of finite trigonometric sums:

$$\text{AP}(\mathbb{R}) = \text{span}_{L^\infty(\mathbb{R})}(e_s : s \in \mathbb{R}), \quad e_s(x) = e^{isx} \quad (x \in \mathbb{R}).$$

We refer to Appendix C for basic notions on the algebra $\text{AP}(\mathbb{R})$, in particular for *Bohr's theorem of mean motion* (or of *mean rotation*): *for every* $f \in \text{AP}(\mathbb{R})$ *satisfying* $\inf_\mathbb{R} |f| > 0$, *the limit*

$$w_f = \lim_{T \to \infty} \frac{1}{2T}(\arg(f(T)) - \arg(f(-T)))$$

exists (arg is a continuous branch of the argument) and $f(x) = e^{iw_f x} e^{g(x)}$ *where* $g \in AP(\mathbb{R})$.

(a) *Show that* $AP_1(\mathbb{T}) + C(\mathbb{T})$ *is a closed sub-algebra of* $L^\infty(\mathbb{T})$ *and every* $\varphi \in AP_1(\mathbb{T}) + C(\mathbb{T})$ *admits a unique representation* $\varphi = \varphi_1 + c$ *where* $\varphi_1 \in AP_1(\mathbb{T})$, $c \in C(\mathbb{T})$ *and* $c(1) = 0$. *Moreover,* $\|\varphi_1\|_\infty \leq \|\varphi\|_\infty$.

SOLUTION: Clearly every function of $AP_1(\mathbb{T})$ is continuous on $\mathbb{T}\setminus\{1\}$ and there exists a representation $\varphi = \varphi_1 + c$. Therefore $AP_1(\mathbb{T}) + C(\mathbb{T})$ is an algebra. Moreover, $\varphi_1 = f \circ \Omega$ where $f \in AP(\mathbb{R})$, and we know (see Appendix C) that $\sup_\mathbb{R} |f| = \overline{\lim}_{x \to \infty} |f(x)|$ (and $\inf_\mathbb{R} |f| = \underline{\lim}_{x \to \infty} |f(x)|$), hence $\|\varphi_1\|_\infty \leq \|\varphi\|_\infty$. From this estimate follows the uniqueness of the representation and the closed nature of the algebra $AP_1(\mathbb{T}) + C(\mathbb{T})$.

(b) *Let* $\varphi \in AP_1(\mathbb{T}) + C(\mathbb{T})$, $\inf_\mathbb{T} |\varphi| > 0$ *and* $\varphi = \varphi_1 + c$ *the representation of (a). Show that* $\varphi/\varphi_1 \in C(\mathbb{T})$ *and hence* $\varphi = \varphi_1 \varphi_0$ *where* $\varphi_1 \in AP_1(\mathbb{T})$, $\varphi_0 \in C(\mathbb{T})$.

SOLUTION: Indeed, $\lim_{z \to 1} \varphi(z)/\varphi_1(z) = 1$ (since $\inf_\mathbb{T} |\varphi_1| = \underline{\lim}_{z \to 1} |\varphi_1(z)| = \underline{\lim}_{z \to 1} |\varphi(z)| \geq \inf_\mathbb{T} |\varphi| > 0$: see Solution (a)), which implies $\varphi/\varphi_1 \in C(\mathbb{T})$.

(c) *Let* $\varphi \in AP_1(\mathbb{T}) + C(\mathbb{T})$. *Show that the following assertions are equivalent.*

(i) T_φ *is Fredholm.*
(ii) $\inf_\mathbb{T} |\varphi| > 0$ *and* $w_f = 0$ *where* $\varphi = f \circ \Omega$.

Moreover, if $T_\varphi \in$ Fred, *then* $\mathrm{ind}(T_\varphi) = -\mathrm{wind}(\varphi_0)$ *where* φ_0 *comes from the factorization* $\varphi = \varphi_1 \varphi_0$ *of (b) above. In particular, for* $\varphi \in AP_1(\mathbb{T})$, $T_\varphi \in$ Fred $\Leftrightarrow T_\varphi$ *is invertible.*

SOLUTION: Clearly the condition $\inf_\mathbb{T} |\varphi| > 0$ is necessary, thus there exists a factorization $\varphi = \varphi_1 \varphi_0$ of (b) and, thanks to $\varphi_0 \in C(\mathbb{T})$, $T_\varphi = T_{\varphi_1} T_{\varphi_0} + K$ where $K \in \mathfrak{S}_\infty$. Since $\inf |\varphi_0| > 0$, we have $T_{\varphi_0} \in$ Fred.

Next, writing $\varphi_j = f_j \circ \Omega$ ($j = 0, 1$), we have $w_f = w_{f_1} =: s$, and by Bohr's theorem (above) $\varphi_1 = V_s e^\psi$ where V_s is defined previously and $\psi \in AP_1(\mathbb{T})$. Since $V_s \in H^\infty$ for $s \geq 0$ and $\overline{V}_s \in H^\infty$ for $s \leq 0$, we have either $T_{\varphi_1} = T_{e^\psi} T_{V_s}$, or $T_{\varphi_1} = T_{V_s} T_{e^\psi}$. We conclude the proof by showing that T_{e^ψ} is invertible, and using that T_{V_s} is Fredholm if and only if $s = 0$. For T_{e^ψ}, since $\psi \in AP_1(\mathbb{T})$, there exist finite sums $p = \sum_{s \geq 0} a_s V_s$ and $q = \sum_{s < 0} b_s V_s$ such that $\|\psi - p - q\|_\infty < 1/2$ and

$$T_{e^\psi} = T_{e^q} T_{e^{\psi-p-q}} T_{e^p},$$

where the three operators of the product are invertible (for the middle one, set $a = \psi - p - q$, $\|I - T_{e^a}\| = \|1 - e^a\|_\infty \leq \|a\|_\infty e^{\|a\|_\infty} < 2^{-1} e^{1/2} < 1$).

3.4 Exercises

Remark Note that $w_f = 0$ does not mean that φ is continuous at the point $z = 1$.

3.4.14 Fredholm Operators T_φ with Matrix-Valued Symbols

We re-use the notation of Exercise 2.3.11 and Appendix F.6.

(a) (i) *Show that for all* $\psi \in H^\infty(\mathbb{T}, L(\mathbb{C}^N))$, $\varphi \in L^\infty(\mathbb{T}, L(\mathbb{C}^N))$, *we have* $T_\varphi T_\psi - T_{\varphi\psi} = 0$ *and* $T_{\psi^*} T_\varphi - T_{\psi^*\varphi} = 0$, *where* $\psi^*(\zeta) = \psi(\zeta)^*$ *for every* $\zeta \in \mathbb{T}$.

(ii) *For* $n \in \mathbb{Z}$, $a \in L(\mathbb{C}^N)$ *and* $\varphi \in L^\infty(\mathbb{T}, L(\mathbb{C}^N))$, *we have* $\mathrm{rank}(T_\varphi T_{z^n a} - T_{\varphi z^n a}) < \infty$ *and* $\mathrm{rank}(T_{z^n a} T_\varphi - T_{z^n a \varphi}) < \infty$.

Hint Repeat the proof of Lemma 3.1.4(2): for example, as in the scalar case, for $n < 0$,
$$(T_\varphi T_{z^n a} - T_{\varphi z^n a})|(z^{-n} H^2(\mathbb{T}, \mathbb{C}^N)) = 0,$$
and the result follows.

(b) *Show that*
$$[T_\varphi, T_\psi], [T_\psi, T_\varphi] \in \mathfrak{S}_\infty(H^2(\mathbb{T}, \mathbb{C}^N)), \quad \forall \varphi \in L^\infty(\mathbb{T}, L(\mathbb{C}^N)), \; \forall \psi \in C(\mathbb{T}, L(\mathbb{C}^N)).$$

Hint Using (a), repeat the proof of the first statement of Theorem 3.1.5.

(c) *Let* $\varphi \in L^\infty(\mathbb{T}, L(\mathbb{C}^N))$. *Show that*
$$T_\varphi \in \mathrm{Fred}(H^2(\mathbb{T}, \mathbb{C}^N)) \Rightarrow \varphi \in L^\infty(\mathbb{T}, L(\mathbb{C}^N))^{-1} \Leftrightarrow \det(\varphi) \in L^\infty(\mathbb{T})^{-1}$$

(i.e. $\varphi^{-1} \in L^\infty(\mathbb{T}, L(\mathbb{C}^N))$ *where* $\varphi^{-1}(\zeta) = \varphi(\zeta)^{-1}$*), and moreover*
$$\|\varphi^{-1}\|_\infty \leq \|\Pi(T_\varphi)^{-1}\|, \quad \mathrm{Ran}_{\mathrm{ess}}(\sigma(\varphi(\cdot))) \subset \sigma_{\mathrm{ess}}(T_\varphi)$$

where
$$\mathrm{Ran}_{\mathrm{ess}}(\sigma(\varphi(\cdot))) = \{\lambda \in \mathbb{C} : \forall \epsilon > 0, \, m\{\zeta \in \mathbb{T} : \mathrm{dist}(\lambda, \sigma(\varphi(\zeta))) < \epsilon\} > 0\}$$
$$= \{\lambda \in \mathbb{C} : \mathrm{ess\,inf} |\det(\lambda I - \varphi)| = 0\}.$$

Hint Adapt the proof of Theorem 3.1.3(3).

(d) (**Simonenko, 1961; Gohberg, 1964; Douglas, 1968**) *Let* $\varphi \in C(\mathbb{T}, L(\mathbb{C}^N))$. *Show that*

(i) $T_\varphi \in \mathrm{Fred} \Leftrightarrow \varphi \in C(\mathbb{T}, L(\mathbb{C}^N))^{-1}$.
(ii) *If* $T_\varphi \in \mathrm{Fred}$, *then* $T_{\varphi^{-1}}$ *is a regularizer of* T_φ ($\varphi^{-1}(\zeta) = \varphi(\zeta)^{-1}$).

Hint Using (b) and (c), repeat the proof of Theorem 3.1.6(1).

Remark The analog of the index formula of Theorem 3.1.6(3) also remains valid for $\varphi \in C(\mathbb{T}, L(\mathbb{C}^N))$, $T_\varphi \in \text{Fred}(H^2(\mathbb{T}, \mathbb{C}^N)) \Rightarrow \text{ind}(T_\varphi) = -\text{wind}(\det(\varphi(\cdot)))$, but the known proofs are more laborious. For these, see the specialized monograph by [Krupnik, 1984], §1.3.

(e)* (**Douglas, 1968**) *Let $\varphi \in H^\infty(\mathbb{T}, L(\mathbb{C}^N)) + C(\mathbb{T}, L(\mathbb{C}^N))$. Then,*

(i) $T_\varphi \in \text{Fred}(H^2) \Leftrightarrow \varphi \in (H^\infty + C)^{-1}$,

(ii) *if $T_\varphi \in \text{Fred}(H^2)$, then $T_{\varphi^{-1}}$ is a regularizer of T_φ, and $\text{ind}(T_\varphi) = -\text{wind}(\det(\varphi(\cdot)))$ (the latter is defined with the aid of a harmonic extension of $\det(\varphi(\cdot))$).*

Remark For more comments and references concerning Toeplitz–Fredholm operators with matrix-valued symbols, see §3.5 and §4.6.

(f)* **Birkhoff–Wiener–Hopf matrix factorization.** By definition, a *right Wiener–Hopf factorization* of a function $\varphi \in L^\infty(\mathbb{T}, L(\mathbb{C}^N))$ is a representation

$$\varphi = \varphi_- D \varphi_+,$$

where $\varphi_-^{*\pm 1} \in H^2(\mathbb{T}, L(\mathbb{C}^N))$ ($\varphi_-^*(z) = \varphi_-(z)^*$ for every $z \in \mathbb{T}$), $\varphi_+^{\pm 1} \in H^2(\mathbb{T}, L(\mathbb{C}^N))$, D is a diagonal matrix

$$D = \text{diag}(z^{k_j})_{j=1}^N$$

where $k_j \in \mathbb{Z}$ (these are called the partial indices of T_φ), and the mapping

$$A_\varphi : f \mapsto \varphi_+^{-1} P_+ D^{-1} \varphi_-^{-1} f \quad (f \in \mathcal{P}_+(\mathbb{C}^N))$$

can be extended to a bounded operator on $H^2(\mathbb{T}, L(\mathbb{C}^N))$. A "left" factorization is defined symmetrically: $\varphi = \varphi_+ D \varphi_-$.

Theorem (Simonenko, 1968). *Let $\varphi \in L^\infty(\mathbb{T}, L(\mathbb{C}^N))$. The following assertions are equivalent.*

(i) $T_\varphi \in \text{Fred}(H^2(\mathbb{T}, L(\mathbb{C}^N)))$.

(ii) *φ possesses a right Wiener–Hopf factorization $\varphi = \varphi_- D \varphi_+$.*

In the case where (i)–(ii) are satisfied, the mapping A_φ (from the definition) is a regularizer of T_φ,

$$\dim \text{Ker}(T_\varphi) = \sum_{k_j < 0} |k_j|, \quad \dim \text{Ker}(T_\varphi^*) = \sum_{k_j > 0} k_j$$

and $\text{ind}(T_\varphi) = -\sum_j k_j$, and hence T_φ is invertible if and only if $k_j = 0$, $\forall j$ (in this case, the factorization is said to be canonical).

(g) *Show the implication (ii) $\Rightarrow$ (i) of Theorem (f).*

SOLUTION: For every diagonal matrix $D = \mathrm{diag}(z^{k_j})_{j=1}^{N}$ with monomial diagonal elements, the operators $P_+ - DP_+D^{-1}$, $P_- - DP_-D^{-1}$ and $P_+D^{-1}P_-$ are of finite rank, and therefore, for every polynomial $f \in \mathcal{P}_+(\mathbb{C}^N)$,

$$A_\varphi T_\varphi f = \varphi_+^{-1} P_+ D^{-1} \varphi_-^{-1} P_+ \varphi_- D \varphi_+ f$$
$$= \varphi_+^{-1}(P_+ D^{-1} \varphi_-^{-1} \varphi_- D \varphi_+ f + Kf) = f + Kf,$$

where $\mathrm{rank}(K) < \infty$, and similarly for $T_\varphi A_\varphi$. Hence, T_φ possesses regularizers, i.e. $T_\varphi \in \mathrm{Fred}$ (see Appendix E).

3.4.15 "Truncated" Toeplitz Operators

The compressed operators $PM_\varphi|(\mathrm{Ran}P)$, where P is an orthogonal projection onto a subspace of $H^2(\mathbb{T})$, are called "truncated Toeplitz operators." The best-known are the Wiener–Hopf operators on a finite interval $(0, a)$, which, up to a unitary transformation (see §4.4 below), correspond to $P = P_\theta = P_+ - \theta P_+ \bar\theta$, where

$$\theta = \theta_a = \exp\left(-a\frac{1+z}{1-z}\right), \quad a > 0.$$

More generally, we consider the *model subspaces*

$$K_\theta =: H^2 \ominus \theta H^2$$

where θ is an inner function (see Appendix F on these subspaces), whose orthogonal projection is P_θ, and we let

$$T_\varphi^\theta = P_\theta M_\varphi|K_\theta$$

denote a *truncated Toeplitz operator with a symbol* $\varphi \in L^2(\mathbb{T})$ defined by $T_\varphi^\theta f = P_\theta(\varphi f)$ for $f \in K_\theta \cap H^\infty$. Note that $K_\theta \cap H^\infty$ is a vector subspace dense in K_θ since the functions

$$k_\lambda =: P_\theta\left(\frac{1}{1-\bar\lambda z}\right) = \frac{1 - \overline{\theta(\lambda)}\theta}{1-\bar\lambda z}, \quad |\lambda| < 1$$

belong to $K_\theta \cap H^\infty$, and if $f \in K_\theta$, $f \perp k_\lambda$ then $0 = (f, k_\lambda) = f(\lambda)$, $\forall \lambda \in \mathbb{D}$, therefore $f = 0$. Hence, if $T_\varphi^\theta: K_\theta \cap H^\infty \to K_\theta$ is bounded, it uniquely extends to a bounded operator on K_θ.

(a) Let $\theta = z^{n+1}$, $n \geq 0$. Show that $K_\theta = \mathrm{Vect}(z^k: 0 \leq k \leq n) = \mathcal{P}_n$, the subspace of polynomials of degree $\leq n$, and that the matrix of T_φ^θ in the base $(z^k)_0^n$ is the matrix $T_{\varphi,n}$ (the matrix of T_φ truncated to the size $[0, n] \times [0, n]$).

Hint Evident.

(b) (Sarason, 2007) *Let θ be a non-constant inner function and $\varphi \in L^2(\mathbb{T})$. Show that*

(i) *$\varphi \mapsto T_\varphi^\theta$ is linear, and $T_\varphi^\theta = T_\psi^\theta$ if and only if $\varphi - \psi \in \bar{\theta}zH_-^2 \oplus \theta H^2$,*

(ii) *$\|T_\varphi^\theta\| \leq \inf\{\|\varphi - h\|_\infty : h \in \bar{\theta}zH_-^2 + \theta H^2\}$.*

SOLUTION: Clearly (i) $\Rightarrow$ (ii). For (i), first let $\varphi \in \bar{\theta}zH_-^2 \oplus \theta H^2$, then $\varphi = a + b$ where $a \in \bar{\theta}zH_-^2$, $b \in \theta H^2$, thus $a = \bar{\theta}zA$ with $A \in H_-^2$ and $b = \theta B$ with $B \in H^2$. Then $\forall f \in K_\theta \cap H^\infty = \{f \in H^\infty : \bar{\theta}f \in H_-^2 \cap L^\infty\}$, we have $\varphi f = af + bf$ where $af = Az\bar{\theta}f \in Az(H_-^2 \cap L^\infty) \subset H_-^2$ and $bf \in \theta H^2$, hence $P_\theta(\varphi f) = 0$.

For the converse, write $\varphi = a + \bar{b}$ where $a, b \in H^2$, and suppose that $T_\varphi^\theta = 0$. As $a \in H^2$, the operator T_a^θ commutes with every T_ψ^θ, $\psi \in H^\infty$ (see point (ii) of (c) below, which is independent of our current statement). Since

$$T_a^\theta = T_{\bar{\bar{b}}}^\theta = (T_{-b}^\theta)^*,$$

it also commutes with every $T_{\bar{\psi}}^\theta = (T_\psi^\theta)^*$, $\psi \in H^\infty$, and hence with $I - T_z^\theta T_{\bar{z}}^\theta =: P$. Clearly $P = P_\theta(I - SS^*)|K_\theta$ (where S and S^* are respectively the shift and backward shift operators on H^2), therefore $Pf = f(0)P_\theta 1$ for every f. Denote $k_0 = P_\theta(1) = 1 - \overline{\theta(0)}\theta$, $k_0 \in K_\theta$, and thus $Pf = f(0)k_0$. By commutation,

$$PT_a^\theta k_0 = T_a^\theta P k_0$$

and $k_0(0) \neq 0$, and we obtain $T_a^\theta k_0 = ck_0$ with a certain $c \in \mathbb{C}$, i.e. $T_{a-c}^\theta k_0 = 0$. The next observation (evident but important) is that

$$\psi \in H^2, \; T_\psi^\theta k_0 = 0 \Rightarrow \psi \in \theta H^2.$$

Indeed, $0 = P_\theta(\psi k_0) = P_\theta(\psi(1 - \overline{\theta(0)}\theta))$ implies $\psi \in \theta H^2$, thus $T_{a-c}^\theta = 0$, and consequently $a - c \in \theta H^2$. But then $0 = (T_{a-c}^\theta)^* = T_{-(b+\bar{c})}^\theta$, and in turn $b + \bar{c} \in \theta H^2$, which proves $\varphi = (\bar{b} + c) + (a - c) \in \bar{\theta}zH_-^2 \oplus \theta H^2$.

Remark In the special case where $\theta(0) = 0$ (thus, $k_0 = 1$), we could skip the previous argument and conclude directly: if $T_\varphi^\theta = 0$, then $0 = P_\theta \varphi k_0 = P_\theta \varphi$ and $P_+ \varphi \in \theta H^2$. By conjugation, $T_{\bar{\varphi}}^\theta = (T_\varphi^\theta)^* = 0$, therefore $P_+ \bar{\varphi} \in \theta H^2$, and the result follows. ∎

(c) Analytic truncated Toeplitz operators (Sarason, 1967). *Let θ be an inner function, $\varphi \in H^2$, and let $T_\varphi^\theta f = P_\theta(\varphi f)$ for $f \in K_\theta \cap H^\infty$ (a truncated Toeplitz operator with analytic symbol). Show that*

(i) *if T_φ^θ is bounded, there exists $\psi \in H^\infty$ such that $T_\varphi^\theta = T_\psi^\theta$, and*

$$\|T_\varphi^\theta\| = \|T_\psi^\theta\| = \inf\{\|\psi - h\|_\infty : h \in \theta H^\infty\} = \mathrm{dist}_{L^\infty(\mathbb{T})}(\psi, \theta H^\infty),$$

(ii) *the mapping* $\psi \mapsto T_\psi^\theta$ *is a representation (linear and multiplicative) of* H^∞, *and hence* $T_\psi^\theta = \psi(T_z^\theta)$ *for* $\psi \in W_a(\mathbb{T})$ *(the Wiener algebra)*.

Hint Use the Hankel operators and Theorem 2.2.4.

SOLUTION: For (i), note that

$$\bar{\theta} T_\varphi^\theta f = \bar{\theta} P_\theta(\varphi f) = P_-(\bar{\theta}\varphi f)$$

for $f \in K_\theta \cap H^\infty$, which shows that the Hankel operator $H_{\bar{\theta}\varphi}: f \mapsto P_-(\bar{\theta}\varphi f)$ is bounded (being 0 on θH^2) and $\|T_\varphi^\theta\| = \|H_{\bar{\theta}\varphi}\|$. The rest follows from Theorem 2.2.4.

For (ii), for every $f \in K_\theta$, we have $T_\varphi^\theta T_\psi^\theta f = P_\theta(\varphi\theta P_-\bar{\theta}(\psi f)) = P_\theta(\varphi\theta\bar{\theta}(\psi f))$ (since $\theta P_+\bar{\theta}(\psi f) \in \theta H^2$ and thus $P_\theta(\varphi\theta P_+\bar{\theta}(\psi f)) = 0$), giving $T_\varphi^\theta T_\psi^\theta = T_{\varphi\psi}^\theta$. In particular, $T_{z^n}^\theta = (T_z^\theta)^n$ ($n \geq 0$) and hence $\psi(T_z^\theta) = \sum_{n \geq 0} \widehat{\psi}(n)(T_z^\theta)^n = T_\psi^\theta$ (if $\sum_n |\widehat{\psi}(n)| < \infty$).

(d) *Let θ be an inner function, $\varphi \in L^2(\mathbb{T})$ and $T_\varphi^\theta f = P_\theta(\varphi f)$ for $f \in K_\theta \cap H^\infty$. Show that*

(i) $M_\theta^* M_\theta = I$, $M_\theta M_\theta^* = I - P_\theta$, *where* $M_\theta: H^2 \longrightarrow H^2$, $M_\theta f = \theta f$,
(ii) $T_\varphi^\theta \in \text{Fred}(K_\theta) \Leftrightarrow T_\psi \in \text{Fred}(H^2 \oplus H^2)$, *where*

$$\psi = \begin{pmatrix} \bar{\theta} & 0 \\ \varphi & \theta \end{pmatrix}$$

(see Exercises 2.3.11 and 3.4.14 on the matrix-valued Toeplitz operators).

SOLUTION: For (i), this is a direct verification using $M_\theta^* = P_+ M_{\bar{\theta}}|H^2$. For (ii), in the principal special case

$$\theta = \theta_a = \exp\left(-a\frac{1+z}{1-z}\right),$$

this is Theorem 4.4.5 below. For the case of a general inner function, we can repeat the proof of Theorem 4.4.5 by using the same notion of Fredholm equivalence of operators $A \simeq B$ (see Theorem 4.4.5) and by modifying steps (i)–(iv) of Theorem 4.4.5. We begin with the equivalence $T_\varphi^\theta \simeq P_\theta T_\varphi P_\theta + P_\theta': H^2 \longrightarrow H^2$ where $P_\theta' = I - P_\theta$, then $P_\theta T_\varphi P_\theta + P_\theta' \simeq T_\varphi P_\theta + P_\theta'$, then use $A - BC \simeq I \oplus (A - BC): H^2 \oplus H^2 \longrightarrow H^2 \oplus H^2$, and finally apply Lemma 4.4.6 to $A = T_\varphi$, $B = (T_\varphi - I)M_\theta$, $C = M_\theta^*$; see §4.4 for all the details.

3.5 Notes and Remarks

In this chapter, we have presented an introduction to the *spectral theory of Toeplitz operators* on the space $L^2(\mathbb{T})$. It is well known that at the beginning of its existence, the spectral theory was presented under the form of "integral equations of the second kind," i.e. equations

$$\lambda x - Tx = y$$

where $\lambda \in \mathbb{C}$ is a "spectral parameter," the "second member" $y \in X$ is a known element, the "first member" $x \in X$ is an element to be found, and $T \colon X \to X$ is a linear integral mapping "with kernel k" on a space X, $Tx(s) = \int k(s,t)x(t)\,dt$. The founders of the theory include Carl Neumann [1877], Henri Poincaré [1895], Vito Volterra [1896], Erik Ivar Fredholm [1900, 1903], and David Hilbert [1904, 1912]; they were followed by the "second wave" with Erhard Schmidt, Frigyes Riesz, Torsten Carleman, Fritz Noether, and many others (here we will not discuss the non-linear integral equations known by Liouville (1838) and Fuchs (1870) in the framework of ODEs). Poincaré's role was particularly important: by reducing the problem of a vibrating membrane to an integral equation, he introduced the spectral parameter λ in order to find the eigenfunctions (corresponding to $y = 0$), thus inspiring the work by Fredholm and later Hilbert.

Clearly the λ for which the equation admits a unique solution x for every "second member" y form the set $\mathbb{C} \setminus \sigma(T)$; finding the spectrum $\sigma(T)$ is the first of the principal goals of spectral theory (which has occupied us in this chapter for the case of Toeplitz operators). Incidentally, according to Dieudonné [1981], the term "spectrum of a transform" was introduced by Hilbert [1904] following Wirtinger [1897] (where it came from a result on the Hill operator interpreted in terms of optical spectra); see also [Harrell, 2004]. There is a curious historical detail in Dieudonné's book: he cites the participants in the Hilbert seminar, who recalled that the master began the year 1904 by declaring "This year, we will look for a self-adjoint transformation A whose spectrum $\sigma(A)$ coincides with $i(Z - 1/2)$ where Z is the set of zeros of the Riemann (Euler) ζ function." To date (2018), no one has accomplished this program of Hilbert (equivalent to the Riemann hypothesis on the zeros of the ζ function).

To finish with the positioning of this chapter with respect to general spectral theory, note that the second goal of the theory – and the main one – is to find the eigenfunctions $Tx_\lambda - \lambda x_\lambda = 0$, $\lambda \in \sigma(T)$ (perhaps belonging to an enlarged space) and to establish a sort of "spectral resolution" of the identity,

$$x = \sum_\lambda \langle x, x'_\lambda \rangle x_\lambda, \quad \text{or} \quad x = \int \langle x, x'_\lambda \rangle x_\lambda \, d\mu(\lambda),$$

where x'_λ are the eigenfunctions of the adjoint operator T^*. This type of problem is beyond the scope of this book; see nonetheless a comment below concerning Theorem 3.3.8, as well as the very last paragraph of this §3.5.

What we know today as the spectral theory of Toeplitz operators is the confluence of three more or less independent streams of twentieth-century analysis:

3.5 Notes and Remarks

(1) the *Riemann–Hilbert boundary problems* and the associated singular integral operators (the "RH problem," again coming from Riemann's famous thesis (1851), and then Hilbert's "Problem 21" in his list of problems for the twentieth century (1900)),

(2) the *Wiener–Hopf equations* (originating in [Wiener and Hopf, 1931]: see Chapter 4 for details),

(3) the *Toeplitz matrices themselves*.

We have known since Hilbert that (1) reduces (on the space L^2) to the singular integral operators $T = \varphi P_+ + \psi P_-$. Then, (1) and (2) were identified by Rappoport [1948a], then professor at Odessa in Krein's seminar (incidentally, in the same paper, probably for the first time, the analytical index of a Wiener–Hopf operator W_k was calculated with the aid of wind($\mathcal{F}k$)); we will return to the RH problems and the WH equations in Chapter 4. As for (3), it seems that the problem of the spectrum $\sigma(T_\varphi)$ of a "true Toeplitz operator" (or of a Wiener–Hopf discrete operator: see Example 2.1.7(2) and Exercise 2.3.3) was also first tackled by Rappoport [1948b]. The unitary equivalence between theories (2) and (3) was only established in 1965 [Rosenblum, 1965], and the simplest and most natural argument was found by Devinatz [1967] (it is reproduced in §4.2 and §4.3 below); see also the presentation by Rosenblum and Rovnyak [1985]. It was with these identifications that the theory of Toeplitz (and Hankel) operators crossed a "red line" in its evolution: there was a "before," when it wandered a bit in search of methods for spectral problems, relying on "artisanal" and makeshift techniques, and an "after," once the appropriate techniques were identified – those of Hardy spaces and the associated harmonic analysis. During the 15 years following the fusion in question (around the beginning of the 1980s) all the principal problems of the theory were resolved. We can only note that the spectral theory of Ha-plitz operators/matrices, part of which is presented in this chapter, is a real triumph of the techniques of Hardy spaces. Moreover, the interaction between Hardy spaces and Toeplitz/Hankel operators was never one-way. It is true that the Ha-plitz theory is an avid consumer of the techniques of the H^p spaces, but it also acted as a stimulus for the creation of new subjects of research: it suffices to recall the theory of "Douglas algebras" A, $H^\infty \subset A \subset L^\infty$ (see below), the problem of weak factorizations (§2.4), two-weight estimations (§2.4), etc.

This "Ha-plitzian revolution" of the years 1960–1980 carries three striking colors on its flag, as it were.

(1) The "essentialization" (or "Fredholmization") of the theory, consisting of first creating a "soft" theory modulo $\mathfrak{S}_\infty$ in the Calkin algebra $L(H^2)/\mathfrak{S}_\infty$,

and then raising it back up to $L(H^2)$. The history of this idea began with Fredholm, Noether, Mikhlin...

(2) The *algebraization and Abelianization*. Instead of treating each operator T_φ separately, we generate the algebra of "aggregate operators," then the quotient by the commutator ideal to pass to an Abelian algebra. We can then use the symbolic calculus by profiting from small algebraic tricks (such as the formulas for $[T_f, T_g]$, etc.) as well as the full power of the methods of Banach algebras. Among the pioneers of this technique were (Poincaré, 1910), Bertrand (1921), Mikhlin (1936, 1948) (all three in the context of singular integral operators), Krein (1958), Krein (1958), Gohberg (1960), and Douglas, 1968).

(3) The *localization*. The "local principles" in a broad sense (the properties of $Tf(x)$ depend above all on the behavior of f around of x) were observed long ago for integral operators of different types (beginning with the Fourier transform, then the singular integral operators, equivalent to Toeplitz), but the first general and effective theorems were only proved in the 1960s (Simonenko, 1960).

Numerous different forms of localization have been found over time, in particular those of Douglas (from 1970 onwards), (from 1970 onwards), Gohberg and Krupnik (1973), and Sarason (1973). In the presentation of this chapter we have mainly relied on the language and techniques of Douglas and Sarason.

For all of these important results, we will give detailed references below.

Formal credits, subject by subject.

Lemma 3.1.1 and Theorem 3.1.3. Essential (Fredholm) spectrum.

The Toeplitz algebras alg$\mathcal{T}_X$ emerged in the works of Gohberg [1964], Gohberg and Feldman [1967], and Coburn [1967] (for the case $X = C(\mathbb{T})$), and in a series of publications by Gohberg, with applications to the spectrum and essential spectrum of singular integral and Wiener–Hopf operators; see more references in the book by Gohberg and Feldman [1967]. The systematic use of the algebras alg$\mathcal{T}_X$ (for $X = H^\infty + C$, QC, etc.) began with Douglas (1968); see [Douglas, 1972, 1973] for surveys and the original references. An advanced presentation of this algebraic approach, including the case of alg$\mathcal{T}_{L^\infty(\mathbb{T})}$, can be found in [Nikolski, 1986], Appendix 4, pages 299–399. However, the idea of introducing the aggregate Toeplitz operators dates back to Brown and Halmos [1963].

A brief word on the classical terminology of Definitions 3.1.2, in particular on the "topological index" or the "index of a curve" $\zeta \mapsto \varphi(\zeta)$ (with respect to the origin $z = 0$), denoted wind(φ). The concept of the index of a curve was already implicitly present in Cauchy's residue calculus (1826), but the question of knowing when the term "index" was formally introduced remains to be explained (see [Lindelöf, 1905] for the initial history of Cauchy's formula). In any case, in the very first (founding) article of Noether [1921] where (*de facto*) the equation "ind(T_φ) = $-$wind(φ)" appears, the author speaks especially of "*Cauchysches Integral* $\frac{1}{2\pi i}\int_\gamma \frac{dz}{z}$" and uses the word "Index" (for its value) only in quotes.

See also the comments on Exercises 3.4.3 and 3.4.5 below.

Lemma 3.1.4. [Coburn, 1967, 1969].

Theorem 3.1.5. [Gohberg, 1964; Coburn, 1967, 1969].

Theorem 3.1.6. [Gohberg, 1964; Douglas, 1968a]. This Fredholm theory of the algebra alg$\mathcal{T}_{C(\mathbb{T})}$, at the time of its appearance, was already known in a more general form, for the operators of an abstract C^*-algebra generated by an isometry $V: H \to H$ of a Hilbert space: see [Gohberg and Feldman, 1967] and [Douglas, 1972] (which includes historical information).

Theorem 3.1.10. (Douglas and Sarason [1970] (except for the expression of the norm of the essential inverse)). By comparing this criterion of being Fredholm with those of invertibility in Lemma 3.3.4 and Theorem 3.3.6, we can note that the passage from an operator $T_\varphi \in L(H^2)$ to the equivalence class $T_\varphi + \mathfrak{S}_\infty$ in the Calkin algebra $L(H^2)/\mathfrak{S}_\infty$, at the level of symbols, corresponds to replacing dist(φ, H^∞) with dist($\varphi, H^\infty + C(\mathbb{T})$), which is of course explained by the complementarity formulas of Lemma 2.2.9 and the norms $\|H_\varphi\|$ (Nehari, Theorem 2.2.4) and $\|H_\varphi\|_{\text{ess}}$ (AAK, Theorem 2.2.8).

An important property that we have not treated is the *connectedness of* $\sigma_{\text{ess}}(T_\varphi)$ ($\forall \varphi \in L^\infty$) proved by Douglas [1972].

3.2. Localization.

Theorem 3.2.1. [Simonenko, 1960, 1964, 1965a,b].

The idea of local descriptions of commutative (!) Banach algebras, and in particular, the local nature of the property of invertibility, dates back at least as far as Wiener [1933] (with his famous "$1/f$ theorem," and then the general Tauberien theorem), but in fact the principle of "locality" was known in harmonic analysis as far back as Riemann. The idea of algebraization (the passage from $\mathcal{T}_X$ to alg$\mathcal{T}_X$, $X \subset L^\infty(\mathbb{T})$) went well with the localizations of the algebra X over the space of maximal ideals of X and resulted in the small masterpiece of Theorem 3.2.1 between operators and harmonic analysis on $\mathbb{T}$. The inventions of Douglas and Sarason [1970] and Sarason [1973b] are important for the subject (the definition of local distances, their properties, in particular, Lemma 3.2.2 and its applications to the algebra $H^\infty + C$).

In addition, Douglas [1972, 1973] proposed another localization technique that works in an abstract C^*-algebra and is called "localization on the center of a C^*-algebra." (It should be noted that, even in a more general setting and without relations with the Toeplitz algebras, similar techniques appeared earlier in [Allan, 1968].) The localization for such an algebra A consists of considering its *center*

$$Z =: \{b \in A : ab = ba \ \forall a \in A\}$$

and a sub-C^*-algebra, $X \subset Z$. Letting J_λ denote the bilateral ideal of A generated by a maximal ideal λ of X, we consider the quotient algebra A/J_λ and the canonical projection

$$P_\lambda \colon A \to A/J_\lambda.$$

The theorem (see [Douglas, 1973, 1980]) states that an element $a \in A$ is invertible if and only if every "local element" $P_\lambda(a)$ is invertible in A/J_λ. In applications, it is often the case that all the fibres A/J_λ are identical. For example, for $A = \mathrm{alg}\mathcal{T}_{L^\infty(\mathbb{T})}/\mathfrak{S}_\infty$, we can show that $Z = \mathcal{T}_{\mathrm{QC}(\mathbb{T})} + \mathfrak{S}_\infty \approx \mathrm{QC}(\mathbb{T})$; a sub-algebra of $\mathrm{QC}(\mathbb{T})$ with the simplest space of maximal ideals is $X = C(\mathbb{T})$, and thus, by regarding a natural extension of projections,

$$P_\lambda \colon \mathrm{alg}\mathcal{T}_{L^\infty(\mathbb{T})} \to A = \mathrm{alg}\mathcal{T}_{L^\infty(\mathbb{T})}/\mathfrak{S}_\infty \to A/J_\lambda,$$

we obtain that *an operator $T \in \mathrm{alg}\mathcal{T}_{L^\infty(\mathbb{T})}$ is Fredholm if and only if every local operator $P_\lambda(T)$ is invertible*. This last invertibility can be expressed in terms of the Poisson extension of the symbol $\mathrm{Sym}(T)$; see [Douglas, 1972, 1973] and [Power, 1982] for details and more applications.

Of course, the techniques of C^*-algebras are not obligatory for a localization, and in more or less the same style as for the algebras $C(\mathbb{T})$ and $H^\infty + C(\mathbb{T})$, a localization and a criterion to be Fredholm can be obtained for other algebras of symbols such as $\mathrm{PC}(\mathbb{T})$, $\mathrm{QC}(\mathbb{T})$,

$$\mathrm{AP}_\lambda(\mathbb{T}) =: \mathrm{span}_{L^\infty}\left(\exp\left(s\frac{z+\lambda}{z-\lambda}\right) \colon s \in \mathbb{R} \right)$$

(functions almost periodic "at the point $\lambda \in \mathbb{T}$") and their combinations (between themselves and with H^∞). Two examples of such generalizations are given in Exercises 3.4.3 and 3.4.5; an overview of these subjects and an outline of the (vast) literature can be found in [Böttcher and Silbermann, 1990] and [Nikolski, 1986, 2002a].

See also the comments for Exercises 3.4.6–3.4.8 below.

3.3. The spectrum. The *invertibility of a Toeplitz operator* T_φ is the result of a refined interplay between the local properties of the symbol φ and its topological nature. In this handbook we attempt to polish the computations a little more, in particular by giving *precise numerical estimates of the inverses* $\|T_\varphi^{-1}\|$ as explicitly as possible as a function of the conditions on the symbol φ. Nevertheless, we must admit that, in general, we succeeded in separating the local/global effects (and thus obtained simplified criteria of invertibility) for symbols that are quite simple and special, belonging to $C(\mathbb{T})$, $PC(\mathbb{T})$, $QC(\mathbb{T})$, $AP_\lambda(\mathbb{T})$, H^∞, or to certain combinations of these classes. For a generic symbol, we are reduced to applying the principal criterion of Theorem 3.3.6, which is not so easy to apply! Note also that a property common to all the symbols $\varphi \in L^\infty$: $\sigma(T_\varphi)$ is always connected [Widom, 1964].

Lemma 3.3.2–Corollary 3.3.3. The results are due to Brown and Halmos [1963] (with the exception of the estimates), but the term "sectorial" (as well as "locally sectorial") was introduced by Douglas and Widom [1970].

Lemma 3.3.4. This decisive step towards Theorem 3.3.6 is due to Widom [1960a,b] (as above, with the exception of the computation of the norm), but his techniques were strongly inspired by the famous paper by Helson and Szegő [1960] (which, now a classic, is widely disseminated [Nikolski, 2019; Böttcher and Silbermann, 1990; Nikolski, 1986; Douglas, 1972]).

Lemma 3.3.5 is a classic element of complex analysis, see [Rudin, 1986]; Runge's paper is [Runge, 1885].

§3.3.1, §3.3.3. The *Wiener–Hopf factorization* appeared in [Wiener and Hopf, 1931], an article that played (and continues to play) an exceptional role in both pure and applied mathematics and whose value would be difficult to overestimate. It is briefly described in §1.1.3, §1.3.2, and §4.6. We can add that in [Wiener and Hopf, 1931] and in the literature of the years 1940–1950 on the Riemann–Hilbert problem, the Wiener–Hopf equations, and the singular integral equations, the terminology associated with factorization and the contents of this term varied considerably. In the beginning, it was only applied to functions φ holomorphic in a neighborhood of the contour ($\mathbb{T}$, in our case), thus requiring the same for the factors $\varphi = \varphi_1 \overline{\varphi}_2$ (or $\varphi = \varphi_1 z^n \overline{\varphi}_2$, for the case of non-invertible Fredholm operators): φ_1 holomorphic in $|z| < 1 + \epsilon$, and $\overline{\varphi}_2$ in $|z| > 1 - \epsilon$ [Plemelj, 1908b; Gakhov, 1963]. Next, with φ in a Banach algebra (or just in a functional class), the factors were sought in the same algebra (class); this approach is well presented in [Krein, 1958], [Gohberg and

Krupnik, 1973], and [Clancey and Gohberg, 1981] (under the name "canonical factorization"). For symbols "without any qualities" (just $\varphi \in L^\infty$) this language is already insufficient, and it was modified several times until the definition in §3.3.1 ([Widom, 1960a]; [Simonenko, 1968]; [Gohberg and Krupnik, 1973]; [Spitkovsky, 1976]; [Clancey and Gohberg, 1981]; [Litvinchuk and Spitkovsky, 1987]; [Böttcher, Karlovich, and Spitkovsky, 2002]; [Böttcher and Karlovich, 1997]), always under different names, such as "generalized factorization," "Φ-factorization" (Φ for Fredholm) – or even without a name at all.

We can also mention that an analytic/co-analytic factorization $\varphi = \varphi_1 \overline{\varphi}_1$ of positive functions φ with only the restriction $\log(\varphi) \in L^1(\mathbb{T})$, and with factors $\varphi_1, \overline{\varphi}_1$ of the same class of integrability as φ, was discovered by Szegő [1921b], and then developed by Helson, Lowdenslager (and others) for the case of matrix- or operator-valued functions: see [Helson, 1964], or [Nikolski, 1986] for details. This factorization is widely used in the analysis of stochastic processes: see [Rozanov, 1963] and [Dym and McKean, 1976].

The equivalence described in §3.3.3(d) brings this classic subject once again (2013) to the center of "hard analysis," this time of the twenty-first century! See the explanations and references in §2.4.

Theorem 3.3.6 [Widom, 1960b; Devinatz, 1964]. The latter article contains several improvements to the arguments of Widom, as well as an approach also valid for "abstract" Hardy spaces (generated by an arbitrary Dirichlet algebra). We can reiterate that these remarkable results still remain "in the shadow of the ideas of Helson–Szegő" [Helson and Szegő, 1960]. They were rediscovered in the 1960s, especially by the Russian school (Simonenko, Krein, Gohberg, Spitkovsky: see the references (selected! they represent a small cape of an enormous continent) mentioned in the preceding point), significantly reworked, and adapted to more general situations (Toeplitz operators on sophisticated curves in the plane $\mathbb{C}$, vector-valued operators, in weighted functional spaces, etc.). However, we have no chance of flying over this continent of knowledge without drastically augmenting the size of these comments. We count on the curiosity of the reader to lead her/him to seek out these powerful theories, perhaps by following the vague indications given above.

Just one exception: since the Wiener–Hopf factorization has already been sufficiently discussed, we note how it can be adapted to the spaces $H^p(\mathbb{T})$, $p \neq 2$: $\varphi = \varphi_1 z^n \overline{\varphi}_2$ and $(1/\varphi_1) P_+ (1/\overline{\varphi}_2) : H^p(\mathbb{T}) \longrightarrow H^p(\mathbb{T})$ is bounded, where $\varphi_1 \in H^{p'}(\mathbb{T})$, $1/\varphi_1 \in H^p(\mathbb{T})$, $\varphi_2 \in H^p(\mathbb{T})$, $1/\varphi_2 \in H^{p'}(\mathbb{T})$ and $(1/p) + (1/p') = 1$, $1 < p < \infty$; see for example [Widom, 1960c] and [Gohberg and Krupnik, 1973].

3.5 Notes and Remarks

§3.3.2. The upper bounds of §3.3.2 are the results of a rewrite (as in the text) of the classical proofs of Lemma 3.3.4 and Theorem 3.3.6 to adapt them to more clearly track the numerical constants.

Theorem 3.3.8 [Hartman and Wintner, 1954] (with an alternative proof given in Exercise 3.4.11). In fact, the structure of self-adjoint Toeplitz operators is known in more detail. First of all, Rosenblum [1960] showed that the spectrum (i.e. the spectral measure) of a T_φ, $\overline{\varphi} = \varphi \neq$ const. is *absolutely continuous* (with respect to the Lebesgue measure). Then Ismagilov [1963] (announced without a proof) and Rosenblum [1965] described the spectral measure of T_φ as a function of φ. Even better, [Rosenblum, 1965] provided an explicit construction of the diagonalization of a self-adjoint T_φ. Rosenblum's approach (in an improved form), as well as comments and references to other publications on the question, are also presented in [Rosenblum and Rovnyak, 1985]. We quote below only a couple of these remarkable results.

(1) ([Ismagilov, 1963]; [Rosenblum, 1965]) Let $\varphi \in L^\infty(\mathbb{T})$, $\overline{\varphi} = \varphi$ and E_k ($k = 1, 2, \ldots$) the set of $x \in \sigma(T_\varphi)$ ($=$ [ess inf(φ), ess sup(φ)]: see Theorem 3.3.8) such that the Lebesgue set $\{\zeta \in \mathbb{T}: \varphi(\zeta) < x\}$ consists exactly of k intervals (arcs) mod(0), We set $E_\infty = \mathbb{T} \setminus \bigcup_{k \geq 1} E_k$ and define the "diagonal" operators A_k and B_k by

$$A_k f(x) = x f(x), \quad x \in E_k, \quad A_k: L^2(E_k, dx) \to L^2(E_k, dx), \quad B_k = \sum_{j=1}^{k} \oplus A_k,$$

an orthogonal sum of k copies of A_k defined by

$$\sum_{j=1}^{k} \oplus L^2(E_k, dx) = L^2(E_k, \mathbb{C}^k, dx).$$

Then, T_φ is unitarily equivalent to the sum

$$B_\infty \oplus \sum_{k \geq 1} \oplus B_k,$$

and hence, the spectral multiplicity of T_φ is $\mu(T_\varphi) =: \sup\{k: |E_k| > 0\}$.

(2) *Explicit diagonalization of a self-adjoint Toeplitz operator T_φ with simple spectrum (of multiplicity $\mu(T_\varphi) = 1$).* With the notation of (1), suppose $E_1 = \sigma(T_\varphi)$ (mod (0)) and $x \mapsto a(x), b(x)$ are two measurable functions such that

$$\{e^{it}: \varphi(e^{it}) \geq x\} = [a(x), b(x)]$$

(modulo an m-negligible set) and $0 < b(x) - a(x) < 2\pi$ (for every $x \in \sigma(T_\varphi)$). Let U be the mapping

$$Uf(z) = \int_{\sigma(T_\varphi)} f(x) u(z,x) \, d\rho(x) \quad (z \in \mathbb{D}),$$

where $d\rho(x) = (1/\pi) \sin((b(x) - a(x))/2) \, dx$,

$$u(z,x) = \frac{1}{E(z,x)} \left(\frac{k_z(e^{-ia(x)}) - k_z(e^{-ib(x)})}{z(e^{-ia(x)} - e^{-ib(x)})} \right)^{1/2},$$

with $k_z(\zeta) = 1/(1 - z\bar{\zeta})$ (Szegő kernel at the point $\bar{z}$) and $E(\cdot, x) = [|\varphi - x|^{1/2}]$, an outer function satisfying $|E(\cdot, x)| = |\varphi - x|^{1/2}$ a.e. on $\mathbb{T}$. To be sure of the existence of E for almost every x it suffices to see that

$$\int_\mathbb{T} \int_{|x| < 2\|\varphi\|_\infty} \log |\varphi - x| \, dx \, dm > -\infty.$$

Then, U is a unitary mapping

$$U : L^2(\sigma(T_\varphi), d\rho) \to H^2(\mathbb{D}),$$

diagonalizing T_φ:

$$(U^{-1} T_\varphi U) f(x) = x f(x), \quad f \in L^2(\sigma(T_\varphi), d\rho).$$

Exercise 3.4.0. Basic exercises. The integral operators ("with kernel $k(x,y)$") are treated in (almost) all texts on functional analysis/theory of operators: see Appendix D for a few references. The presentation in Exercise 3.4.0(a) follows that of Nikolski [2002a], where among others can be found comments and references on the "Schur test." The study of the Fredholm character of the multiplication operators (the "multipliers") is part of the "folklore"; see also the comments in Appendices E.8(6) and E.9(7), and for the multipliers of the Bergman space [Zhu, 2007].

Exercise 3.4.1. For part (a) see [Weyl, 1910a]; see also the comments in Appendix E.10. For parts (b) and (c) see [Hartman and Wintner, 1954].

Exercise 3.4.3. Semi-commutators, Douglas algebras, harmonic extensions. Historically, during the years 1970–1980, the contents of Exercise 3.4.3 were at the heart of important developments in the theory of operators and functional algebras on the circle $\mathbb{T}$. As a result, a large part of the modern theory of Hardy spaces was inspired by these developments; see for example the reference monograph by Garnett [1981], where the technically more advanced chapters are devoted to the fruits of this progress. We present here a few examples linked with propositions (a) and (b) of §3.4.3, but first we comment on the rest of the exercise.

3.5 Notes and Remarks

The theorem in (g) comes from [Sarason, 1973a]. The results in (a)–(f) are due to Douglas [1969b], including the principal result, i.e. theorem (e), as well as the *definition of* wind(φ) *for* $\varphi \in H^\infty + C(\mathbb{T})$ *and the criterion of invertibility* in this algebra; see also [Douglas, 1972]. As always (see Appendix F), we let $f(\zeta) = P_\zeta * f$ ($\zeta \in \mathbb{D}$) denote the harmonic extension of a function $f \in L^1(\mathbb{T})$ in the disk $\mathbb{D}$. Then:

(i) A function φ is in $(H^\infty + C)^{-1}$ if and only if there exists ρ, $0 < \rho < 1$ such that

$$\delta_\rho(\varphi) =: \inf\{|\varphi(\zeta)|: \rho \leq |\zeta| < 1\} > 0.$$

Further,

(ii) in the case where $\delta_\rho(\varphi) > 0$, the wind$(\varphi(rz))$, $\rho < r < 1$ does not depend on r and we set wind$(\varphi) = $ wind$(\varphi(rz))$, $\rho < r < 1$.

This implies, in addition, two important consequences.

(iii) If $\varphi \in H^\infty + C(\mathbb{T})$, then

$$\sigma_{\text{ess}}(T_\varphi) = \cap_{\rho < 1} \text{clos}\{\varphi(\zeta): \rho < |\zeta| < 1\}$$

(which easily implies that $\sigma_{\text{ess}}(T_\varphi)$ *and* $\sigma(T_\varphi)$ *are connected*).

(iv) If $\varphi \in H^\infty + C(\mathbb{T})$ and $\delta_0(\varphi) = \inf_{\zeta \in \mathbb{D}} |\varphi(\zeta)| > 0$, then T_φ is invertible.

(The last fact clearly follows from the above and from the observation that all $\varphi(rz)$, $0 \leq r < 1$ have the same Cauchy index since they are homotopic in $\mathbb{C} \setminus \{0\}$.)

The properties (i)–(iv) gave rise to the question of the links between the harmonic extension $\varphi(\zeta)$ ($\zeta \in \mathbb{D}$) and the invertibility of T_φ. The following facts have been observed without completely solving the problem; see [Nikolski, 1986], §165.50 for a discussion, proofs, and references.

(v) (Example of Douglas [1973]) There exists T_φ invertible ($\varphi \in L^\infty(\mathbb{T})$) such that $\delta_\rho(\varphi) = 0$ ($\forall \rho, 0 < \rho < 1$). Douglas questioned if there exists a $\delta > 0$ such that $\delta \leq |\varphi(\zeta)| \leq 1$ ($\forall \zeta \in \mathbb{D}$) implies the invertibility of T_φ.

(vi) (Example of Wolff [1983]) There exists $\varphi \in L^\infty(\mathbb{T})$) with $\delta_0(\varphi) > 0$ but T_φ is non-invertible.

(vii) (Tolokonnikov [1981]; for the improved version below see [Nikolski, 1986]) If $\|\varphi\|_\infty \leq 1$ and

$$\delta_0(\varphi) > a =: \left(\frac{4e}{1+4e}\right)^{1/2} \approx 0.956962$$

(which means $a < \delta_0(\varphi) \le |\varphi(\zeta)| \le 1$ ($\forall \zeta \in \mathbb{D}$); we can see that $a < 23/24 \approx 0.958333$), then T_φ is invertible and

$$\|T_\varphi^{-1}\| \le ((1+4e)\delta_0(\varphi)^2 - 4e)^{-1/2} \le c(\delta_0(\varphi) - a)^{-1/2},$$

where $c = 1/(2((1+4e)e)^{1/4}) \approx 0.2098 < 0.21$, or $\|T_\varphi^{-1}\| \le (24\delta_0(\varphi) - 23)^{-1/2}$.

Hence, if we define a "Douglas–Wolff–Tolokonnikov constant" as

$$D =: \inf\{\delta : \delta \le |\varphi(\zeta)| \le 1 (\forall \zeta \in \mathbb{D}) \Rightarrow T_\varphi \in L(H^2)^{-1}\},$$

theorems (vi) and (vii) state that

$$0 < D \le a = \left(\frac{4e}{1+4e}\right)^{1/2}.$$

It is curious to note that the version of RKT for the Hankel operators discussed in §2.4 (following Treil) gives exactly the same bound a for D as in (vii). Perhaps it is sharp, $D = a$?

We give a few comments on the development of the "easy parts" (a) and (b) of §3.4.3. Point (a) gives a sufficient condition $\overline{f} \in H^\infty + C$ for the inclusion $[T_f, T_g] \in \mathfrak{S}_\infty(H^2)$, and (b) shows that over a class of symbols g containing $g = f$ the condition is also necessary. A "real" necessary and sufficient condition is more profound and is given in the following theorem, which uses the language of sub-algebras of $L^\infty(\mathbb{T})$. Given subsets $A, B \subset L^\infty(\mathbb{T})$, we denote

$$\mathrm{alg}(A,B) = \mathrm{clos}_{L^\infty}\left(\sum_i \prod_j f_{ij}g_{ij} : f_{ij} \in A, g_{ij} \in B\right)$$

the minimal closed sub-algebra of $L^\infty(\mathbb{T})$ containing A, B.

(viii) ([Axler, Chang, and Sarason, 1978]; [Volberg, 1982]) The following assertions are equivalent.

1. $[T_f, T_g) \in \mathfrak{S}_\infty(H^2)$.
2. $\mathrm{alg}(H^\infty, \overline{f}) \cap \mathrm{alg}(H^\infty, g) \subset H^\infty + C(\mathbb{T})$.

In the decisive special case where f and $\overline{g}$ are *inner functions* (of H^∞), assertions 1–2 are also equivalent to the following.

3. $\lim_{|\zeta| \to 1} \max\{|f(\zeta)|, |\overline{g}(\zeta)|\} = 1$.
4. $P_f P_{\overline{g}} \in \mathfrak{S}_\infty(H^2)$ where $P_\theta = \theta P_{-\overline{\theta}}$ is the orthogonal projection onto the subspace $K_\theta = H^2 \ominus \theta H^2$.
5. $T_{\overline{f}}|K_{\overline{g}} = U + A$ where U is a unitary operator on $K_{\overline{g}}$ and $A \in \mathfrak{S}_\infty(K_{\overline{g}})$.

Clearly, if the spectra of f and $\overline{g}$ (see Appendix F) are disjoint, $\sigma(f) \cap \sigma(\overline{g}) = \emptyset$, then 3 is satisfied, which implies the other properties. The proof of

(viii) is quite involved and requires considerable effort concerning uniform sub-algebras of $L^\infty(\mathbb{T})$; in particular, it uses the famous Chang–Marshall theorem. A closed sub-algebra $\mathcal{D} \subset L^\infty(\mathbb{T})$ is said to be a *Douglas algebra* if $H^\infty \subset \mathcal{D}$. Given a set $A \subset L^\infty(\mathbb{T})$, let $\overline{A} = \{\overline{f} : f \in A\}$ denote the conjugate of A.

(ix) ([Chang, 1976]; [Marshall, 1976]) The following assertions are equivalent.

1. $\mathcal{D}$ is a Douglas algebra.
2. There exists a family A of inner functions such that $\mathcal{D} = \mathrm{alg}(H^\infty, \overline{A})$.

If $\mathcal{D}$ is a Douglas algebra, then $\mathcal{D} = H^\infty + C_\mathcal{D}$, where $C_\mathcal{D} = \mathrm{alg}(B, \overline{B})$ and $B = \{f \text{ inner}: f, \overline{f} \in \mathcal{D}\}$.

In particular, $\mathcal{D} = L^\infty(\mathbb{T})$ is a Douglas algebra, and hence $\mathcal{D} = \mathrm{alg}(H^\infty, \overline{A})$, where A is the family of (all) inner functions of H^∞ (this is an older theorem of Douglas and Rudin [1969]). Other than the original publications, proofs of (ix) can be found in [Garnett, 1981], [Sarason, 1978], [Nikolski, 1986], and proofs of (viii) and (almost all of) (ix) in [Nikolski, 1986].

Exercise 3.4.4. Minimality of the algebra $H^\infty + C(\mathbb{T})$. In fact, the elementary reasoning of Cohen, given in the solution, was invented to simplify the proof of the "maximality theorem" of Wermer [1953] (if A is a Banach algebra such that $C_a(\mathbb{T}) \subset A \subset C(\mathbb{T})$ and $A \neq C_a(\mathbb{T})$, then $A = C(\mathbb{T})$). The proof by Hoffman and Singer of the complete version of Exercise 3.4.4 (see [Hoffman and Singer, 1960], [Hoffman, 1962], or [Garnett, 1981]) requires some knowledge of the properties (quite refined) of uniform algebras.

Exercise 3.4.5. The *symbols of $PC(\mathbb{T})$ ("piecewise continuous")*. Theorem (a) is due to Simonenko [1960] (using his local principle (§3.2) and the language of singular operators $aI + b\mathbb{S}$: see §4.1, §4.5), and then in the Toeplitz form by [Devinatz, 1964] (and [Widom, 1960a] for the case where $\arg(\varphi)$ is piecewise continuous). Theorem (b) is by Gohberg and Krupnik [1969]; the principal point is to identify the commutator ideal $\mathrm{Com}\mathcal{T}_{PC(\mathbb{T})}$ of the algebra $\mathrm{alg}\mathcal{T}_{PC(\mathbb{T})}$ with $\mathfrak{S}_\infty(H^2)$ (part (1) of the theorem). Note that according to these authors, for $f, g \in PC(\mathbb{T})$, we always have $[T_f, T_g] \in \mathfrak{S}_\infty(H^2)$, but if f, g share a common point of discontinuity, then $[T_f, T_g] \notin \mathfrak{S}_\infty(H^2)$.

Sarason [1977a] found analogs of Exercise 3.4.5(a,b) for $T \in \mathrm{alg}\mathcal{T}_{PQC(\mathbb{T})}$, where $PQC(\mathbb{T}) = \mathrm{alg}_{L^\infty}(PC(\mathbb{T}), QC(\mathbb{T}))$ ($T \in$ Fred is equivalent to $\varphi_\#$ does not vanish, where $\varphi = \mathrm{Sym}(T)$ and $\varphi_\#$ is an "integral mean" of φ_*), and also gave another necessary and sufficient condition in terms of the harmonic extension:

$T \in \mathrm{Fred} \Leftrightarrow \inf\{|\mathrm{Sym}T(\zeta)| : \rho < |\zeta| < 1\} > 0$ for a certain ρ, $0 < \rho < 1$.

Another important fact for the Toeplitz operators on a Hardy space H^p, $1 < p < \infty$, was observed for symbols φ in PC($\mathbb{T}$): in general, the spectrum $\sigma_{\text{ess}}(T_\varphi\colon H^p \to H^p)$ depends on p (which is not the case for $\varphi \in C(\mathbb{T})$). In fact, we always have $\sigma_{\text{ess}}(T_\varphi) = \text{Ran}(\varphi_*)$, but in the definition of φ_* we must replace the segments joining the jumps of φ with circular arcs whose size (i.e. the center and the radius) depends on p. We do not give the exact statements, but refer the reader to [Gohberg and Krupnik, 1973] and [Böttcher and Silbermann, 1990] for all the details.

Later on, many other algebras of symbols were analyzed using the same scheme: see [Douglas, 1973], [Power, 1982], and [Nikolski, 1986] for overviews of the subject. In particular, Douglas [1972, 1973] established a direct decomposition $\text{alg}\mathcal{T}_{L^\infty(\mathbb{T})} = \text{Com}\mathcal{T}_{L^\infty(\mathbb{T})} + \mathcal{T}_{L^\infty(\mathbb{T})}$ (similar to Theorem 3.1.5, Exercise 3.4.4(e), Exercise 3.4.5(b)).

Exercise 3.4.6 ([Simonenko, 1960])

Exercise 3.4.7. The discussion of Exercise 3.4.7(a–c) is taken from [Nikolski, 1986], and proposition (d) from [Böttcher and Silbermann, 1999], Chapter 1; see also [Böttcher and Silbermann, 1990], Sections 4.71–4.78.

Exercise 3.4.8. For a discussion of multipliers f preserving Fred ($T_\varphi \in$ Fred $\Rightarrow$ $T_{\varphi f} \in$ Fred) and invertibility (T_φ invertible $\Rightarrow T_{\varphi f}$ invertible), see [Spitkovsky, 1980], [Nikolski, 1986], and the references therein. In particular, f preserves invertibility if and only if T_{f^α} is invertible, $\forall \alpha > 0$.

Exercises 3.4.9–3.4.10. These results appeared in [Power, 1980], including the observation by Axler in Exercise 3.4.10(c). Moreover, Power used the links of Hankel operators Γ_φ with $\text{alg}\mathcal{T}_{L^\infty(\mathbb{T})}$ to describe $\sigma_{\text{ess}}(\Gamma_\varphi)$ for several classes of symbols: see [Power, 1980] and [Nikolski, 2002a].

Exercise 3.4.11. This is an original argument of Hartman and Wintner [1954]. The similar equation $(T_\varphi)_n f = (1, 0, \ldots, 0) \in \mathbb{C}^n$ will be crucial for the $n \times n$ Toeplitz matrices: see Chapter 5 below.

Exercise 3.4.12. Regularizers in $\mathcal{T}_{L^\infty(\mathbb{T})}$ and/or in $\text{alg}\mathcal{T}_{L^\infty(\mathbb{T})}$. This material, as well as the other discussions on regularizers scattered later throughout the text, is new. Note that, for example, the problem of compactness of semicommutators $T_\varphi T_\psi - T_{\varphi\psi}$ (resolved by the Axler–Chang–Sarason–Volberg theorem: see the comments on Exercise 3.4.3 above) was implicitly motivated by the partial case $\psi = 1/\varphi$, which corresponds exactly to the problem of Toeplitz regularizers.

Exercise 3.4.13. Fredholm theory for almost periodic symbols. The theorems in Exercise 3.4.13 (a) and (c) were found independently in [Gohberg and Feldman, 1968] and [Coburn and Douglas, 1969]. It is worth noting that it is exactly here that we find an argument showing traces of the original Wiener–Hopf technique [Wiener and Hopf, 1931]: if $\varphi = \varphi_1 \varphi_0 \bar{\varphi}_2$ where $\varphi_{1,2}^{\pm 1} \in H^\infty$ and $\|1 - \varphi_0\|_\infty < 1$, then T_φ is invertible on H^2.

Similar results are known for the symbols in the algebra $AP_\sigma(\mathbb{T}) + C(\mathbb{T})$ of functions "with almost periodic behavior on the points of a set σ" (for every $\sigma \subset \mathbb{T}$, including $\sigma = \mathbb{T}$), where

$$AP_\sigma(\mathbb{T}) = \mathrm{span}_{L^\infty(\mathbb{T})}(V_s(\bar{\zeta}z): \zeta \in \sigma, s \in \mathbb{R});$$

see [Gohberg and Krupnik, 1973], [Douglas, 1973], and [Nikolski, 1986]. These results led Douglas [1969b] to define the "Douglas algebras" and triggered the intensive study of the sub-algebras of L^∞ (see the comments on Exercise 3.4.3 above). One of the symbol algebras most useful for applications was introduced by Sarason [1977b]: this is the algebra $SAP_1(\mathbb{T}) + C(\mathbb{T})$. The transfer of $SAP_1(\mathbb{T}) + C(\mathbb{T})$ to the real axis is defined by

$$SAP(\mathbb{R}) + C_0(\mathbb{R}) = \{f: f = hf_1 + (1-h)f_2 + g; f_j \in AP(\mathbb{R}), g \in C_0(\mathbb{R})\}$$

where $h \in C(\mathbb{R})$ such that $\lim_{x \to +\infty} h(x) = 1$, $\lim_{x \to -\infty} h(x) = 0$; $SAP(\mathbb{R})$ is made up of the functions with different almost periodic behavior at $\pm\infty$.

The vector-valued versions of the theory of the preceding paragraph represent a very active field of research (2018) and are of great value for applications: see [Gohberg, Goldberg, and Kaashoek, 1990], [Böttcher, Karlovich, and Spitkovsky, 2002], [Gohberg, Kaashoek, and Spitkovsky, 2003], and [Böttcher and Spitkovsky, 2013]. In particular, the spectral theory of "truncated Toeplitz operators" in Exercise 3.4.15 (and hence of "truncated Wiener–Hopf operators," §4.4) can be reduced to operators of this type (see Exercise 3.4.15(c) and Theorem 4.4.5); note that the principal technique of the subject – the factorization of symbols (of Birkhoff–Wiener–Hopf) – becomes a very delicate subject for matrix-valued symbols with almost periodic behavior (see §4.4.5, Exercise 4.5.5 and the comments on §4.6).

Exercise 3.4.14. Fredholm operators T_φ with matrix-valued symbols. History of the Birkhoff–Wiener–Hopf factorization. The theory of vector-valued Toeplitz operators,

$$T_\varphi: H^2(\mathbb{T}, \mathbb{C}^N) \longrightarrow H^2(\mathbb{T}, \mathbb{C}^N),$$

was developed in parallel with the "scalar theory" (presented in this chapter) and not long after this model case, since it is not just a simple generalization, but the generic case – the matricial T_φ are responsible for the treatment (at the demand of numerous applications) of *systems of Toeplitz equations*

$$\sum_j T_{\varphi_{ij}} f_j = g_i,$$

and their equivalents for the half-line – the *systems of Wiener–Hopf equations* (and then, the systems of SIE, and the *vectorial Riemann–Hilbert Problems (RHP)*): see Exercise 4.5.2 below. To underline the importance of these links and equivalences, remember that in particular, it is exactly the vectorial RHPs (and the equivalent systems of SIE) that deal with the problem of Fuchsian groups (the 21st Hilbert problem): see §4.6 for more details.

The discovery of the factorization $\varphi = \varphi_- D \varphi_+$ of Exercise 3.4.14(f) was, with hindsight, a major event for a large part of mathematics: today, an immense domain of pure and applied mathematics depends on this technique, far beyond operator theory or even beyond the whole of analysis (see the explanations below). The very first factorization theorem was proved by Birkhoff [1909] in the context of the solution of Hilbert's Problem 21 (in [Plemelj, 1908b], the method of factorization only appears in an implicit form); see §4.6 for a few explanations.

George David Birkhoff (1884–1944), an American mathematician, famous for *Birkhoff factorization*, the proof of the "last geometric theorem of Poincaré," and the Birkhoff ergodic theorem, was one of the founders of algebraic graph theory. He justified the Schwarzschild geometry (with its "black holes") as the unique symmetric solution of the Einstein field equations. He worked on the four-color problem, proposed *Birkhoff's axioms* as a foundation of Euclidean geometry (different from these of Hilbert) and created a mathematical theory of aesthetics (with a formula $M = O/C$ where M, O, C are, respectively, measures of the aesthetics, the order (or the symmetry), and the complexity of an object). By common consensus, Birkhoff was the most influential American mathematician of the first third of the twentieth century.

The principal contribution attributed to Birkhoff is without doubt the creation of a theory of dynamical systems, especially his famous ergodic theorem (1931) which resolved a fundamental problem of statistical mechanics. It is curious to note that, by contrast, the discovery of the *Birkhoff factorization* (1909), while fundamental (as we have seen in §3.5 and §4.6) for an impressive number of important subjects (integrable systems, diffusion theory, systems of ODE, Toeplitz and Wiener–Hopf operators, etc.), remains completely overshadowed by Birkhoff's other creations: none of Birkhoff's biographies even mention the factorization method, which, by itself, would be sufficient to make him a great figure in mathematics, and for posterity.

There is another aspect to Birkhoff's personality: Einstein described him as "one of the academic world's worst anti-Semites." Throughout his life, Birkhoff declared that Einstein's theory of general relativity was a "useless theory." It is well known and documented (see *History in Sum: 150 Years of Mathematics at Harvard* by Steve Nadis and Shing-Tung Yau, Harvard University Press, 2013) that Birkhoff dominated the Mathematics Department at Harvard for 30 years (1912–1944), preventing the recruitment of mathematicians of Jewish origin, including such geniuses as Norbert Wiener (rejected by Harvard on multiple occasions) and Oscar Zariski. That said, Saunders Mac Lane defended his memory (1994), affirming that Birkhoff "shared the rather diffuse and varied forms of anti-Semitism common to many (most?) of his contemporaries. We should not, fifty years later, make a scapegoat of him."

Birkhoff was a student at the University of Chicago (with E. H. Moore), then at Harvard (with Maxime Bôcher). However, his doctoral thesis on ODEs and the Riemann–Hilbert problem (Chicago, 1907) was an independent work inspired especially by his extensive study of the research of Henri Poincaré. Birkhoff remained a professor at Harvard for most of his life (1912–1944). In 1924–1926 he was President of the American Mathematical Society. He received several professional distinctions, and a lunar crater was named "Birkhoff" in his honor. Birkhoff was married and had three children (his son Garrett Birkhoff also became a great mathematician). He died in his sleep, due to a weak heart.

Before sketching the subsequent development of Birkhoff factorizations, we note the special case of functions φ with positive definite values $\varphi(z) \geq \mathbb{O}$ ($z \in \mathbb{T}$). Here, in the scalar case, the factorization, in the form of a Hermitian square $\varphi = \overline{\varphi}_+\varphi_+$ and under the weakened hypothesis $\log|\varphi| \in L^1(\mathbb{T})$, has been known since Szegő [1921b]. It was generalized to the matricial (and operator) case by Lowdenslager and Helson (and also Masani–Wiener and Rozanov) $\varphi = \varphi_+^*\varphi_+$ under the condition $\log|\det(\varphi)| \in L^1(\mathbb{T})$: see [Helson, 1964]. The two factorizations, of Birkhoff–Wiener–Hopf and of Szegő–Lowdenslager (coincident for the case $\varphi(z) \geq cI$, $c > 0$ ($z \in \mathbb{T}$)), are indispensable tools in the theory and applications of stationary random process, as well as for Lévy processes (with independent increments). The influence of the factorization method on these theories took on such importance that, for example, when entering the words "factorization method" into an Internet search engine, you will first obtain several pages of references to stochastic processes, whereas the references to operators (also quite numerous!) are literally lost behind. See, for example, the special edition of the *Journal of Engineering Mathematics* (**59:4**, 2007) devoted to the 75th anniversary of the publication of Wiener and Hopf [1931].

Coming back to the Birkhoff factorization, the second notable appearance of the method is in Wiener and Hopf's famous article [1931], where it is applied to the integral equations now known as "Wiener–Hopf equations": see Chapter 4 for their theory. Notwithstanding all the equivalences known today (see Chapter 4 for these equivalences) between the RHP (Riemann–Hilbert Problems), the SIO (singular integral operators), the WHO (Wiener–Hopf operators) and the TO (Toeplitz operators), it seems that Wiener and Hopf's article was not linked back to the previous result of Birkhoff [1909]. This article exercised an enormous influence on the diffusion of the factorization method to several domains of pure and applied mathematics: signal processing, stochastic processes, inverse problems, non-linear integrable systems, orthogonal polynomials, etc. Here are a few glimpses of some of these subjects (an immense literature is devoted to each one!): [Lawrie and Abrahams, 2007] and [Noble, 1998] (with applications to diffraction, scattering, elasticity, electromagnetism, the theory of crystals, etc.); [Linton and McIver, 2001], especially Chapter 5 (applications to the theory of diffusion, wave propagation, etc.); [Deift, 2000] (inverse problems, random matrices, orthogonal polynomials); [Its, 2003] (factorization and integrable systems, with a panorama of other applications, including an overview of Hilbert's Problem 21).

The subsequent steps of the Birkhoff–Wiener–Hopf factorization theory (for matrix-valued functions) are linked to the milestone articles of Gohberg and Krein [1958] and Simonenko [1960, 1965a, 1965b, 1968] as well as to the texts by Clancey and Gohberg [1981], Litvinchuk and Spitkovsky [1987], and Böttcher, Karlovich, and Spitkovsky [2002]; see also the survey articles by Böttcher and Spitkovsky [2013] and Gohberg, Kaashoek, and Spitkovsky [2003], and the textbook by Gohberg, Goldberg, and Kaashoek [1990]. Note the great influence of the techniques of Banach algebras to the problem of the factorization (beginning with the pioneering results of Gohberg (1950s) and Douglas (1960s)). These even caused the basic definition of factorization to be modified: if A is a functional Banach algebra generated by polynomials, the A-factorization is defined as $a = a_- D a_+$ where $a_+ \in A_+ =: \mathrm{clos}_A \mathcal{P}_a$, $a_- \in A_- =: \{f^* : f \in A_+\}$, and D is a "monomial factor." One can show the following.

(1) ([Gohberg, 1964]; see also [Gohberg and Feldman, 1967]) Every invertible element $a \in A$ admits an A-factorization if and only if $A = A_- \dotplus A_+$ (a direct sum up to a constant). The result remains valid for the matricial case $\varphi = (\varphi_{ij})_1^N$, $\varphi_{ij} \in A$.

(2) ([Litvinchuk and Spitkovsky, 1987]; see also [Böttcher and Silbermann, 1990]) The assertions of (1) are equivalent to the fact that $T_a \in \mathrm{Fred}(A_+^N \longrightarrow A_+^N)$ for every invertible a, where $A_+^N = A_+ \times \cdots \times A_+$.

In particular, for the almost periodic matrix-valued functions

$$A = \mathrm{AP}_1(\mathbb{T}, L(\mathbb{C}^N)) =: \{(\varphi_{ij})_{1 \le i, j \le N} : \varphi_{ij} \in \mathrm{AP}_1(\mathbb{T})\},$$

$$\mathrm{AP}_1(\mathbb{T}) = \mathrm{alg}_{L^\infty(\mathbb{T})}\left(\theta_s =: \exp\left(-s\frac{1+z}{1-z}\right) : s \in \mathbb{R}\right),$$

the role of D is played by the matrices $D = \mathrm{diag}(\theta_{s_j})_{j=1}^N$, and the principal result of Gohberg and Krein [1958] states that an operator $T_\varphi : H^2(\mathbb{T}, \mathbb{C}^N) \to H^2(\mathbb{T}, \mathbb{C}^N)$, $\varphi \in A$, is Fredholm if and only if it is invertible, and if and only if φ possesses a *canonical AP-factorization* (i.e. with $D = \mathrm{id}$); see [Böttcher, Karlovich, and Spitkovsky, 2002] for details.

For *Exercise 3.4.14(a,b)*, see [Gohberg and Krein, 1958], [Simonenko, 1961], [Gohberg, 1964], and [Douglas, 1968b], but in fact the essence of these properties was already known to Mikhlin [1948, 1962].

Exercise 3.4.14(c) is from [Simonenko, 1968] and [Douglas, 1968b].

Exercise 3.4.14(d) is one of the first striking results of the vectorial theory in "natural generality"; it is from [Simonenko, 1961, 1968], [Gohberg, 1964], and [Douglas, 1968b]. The Fredholm criterion for the Toeplitz operators T_φ with "piecewise continuous" symbols $\varphi \in \mathrm{PC}(\mathbb{T}, L(\mathbb{C}^N))$ and the operators of the algebra $\mathrm{alg} \mathcal{T}_{\mathrm{PC}(\mathbb{T}, L(\mathbb{C}^N))}$ (under the form $\det(\varphi_*) \neq 0$ similar to Exercise 3.4.5(a,b), and quite different from $(\det(\varphi))_* \neq 0$!)) was obtained in [Gohberg and Krupnik, 1969] (and for the scalar case in [Widom, 1960a]).

For Exercise 3.4.14(e) see [Douglas, 1968b, 1973]. For similar results for other classes of matricial symbols, such as $AP + C$, $SAP + C$, etc., see [Gohberg and Feldman, 1967], [Douglas, 1973], and [Böttcher, Karlovich, and Spitkovsky, 2002].

The results of Exercise 3.4.14(f) – the ultimate form of the factorization method – are due to Simonenko [1968], including the definitive form of the definition of the "(Birkhoff)–Wiener–Hopf factorization" (often called a "generalized factorization").

The spectral theory of Toeplitz operators (and the SIO) on the weighted spaces $L^p(\gamma, w(s) ds)$ on a curve $\gamma \subset \mathbb{C}$ (in the place of $\mathbb{T}$) is rich, well-developed and full of new phenomena, but it is completely beyond the elementary scope of this handbook; see [Gohberg and Krupnik, 1973] (the scalar case), [Litvinchuk and Spitkovsky, 1987], [Böttcher and Silbermann, 1990], [Böttcher and Karlovich, 1997], and [Böttcher, Karlovich, and Spitkovsky, 2002].

Exercise 3.4.15. "Truncated" Toeplitz operators. The study of the special case

$$\theta = \theta_a = \exp\left(-a\frac{1+z}{1-z}\right), \quad a > 0,$$

of the T_φ^θ corresponding to Wiener–Hopf integral operators on a finite interval (or on the Paley–Wiener space) began as early as the time of Fredholm and Hilbert; see the comments about this case in §4.6 (comment on §4.4). Note also the precursor article of Widom [1960c] where the spectral theory of the operators $W: L^p(E) \longrightarrow L^p(E)$,

$$Wf(x) = a(x)f(x) + b(x)\int_E \frac{f(y)}{y - x} dy, \quad x \in E \subset \mathbb{R},$$

is constructed, which is particularly complete in the case when E is a finite union of intervals (a single interval corresponds to the case of truncated Wiener–Hopf operators: see §4.4).

The general study of operators T_φ^θ, under the name of *skew Toeplitz operators*, began with applications to problems of optimal control [Bercovici, Foias, and Tannenbaum, 1988, 1998], but a new impetus (and the name *truncated Toeplitz*) was given in a survey article [Sarason, 2007]. Today (2019), the research on "truncated Toeplitz operators" is widespread and promising, but the theory has not yet attained the maturity of classical Toeplitz operator theory. Even a basic question (posed by [Sarason, 2007]) on the existence (and a description) of bounded symbols $\psi \in L^\infty(\mathbb{T})$ for a bounded truncated Toeplitz $T_\varphi^\theta = T_\psi^\theta$ is not exactly totally transparent, notwithstanding the significant progress made in the past few years. We refer to the survey article by

3.5 Notes and Remarks

Baranov, Bessonov, and Kapustin [2011], where a description of the T_φ^θ admitting a bounded symbol is given in the language of "weak factorizations" of functions $f \in K_\theta^1$, $f = \sum_{k\geq 1} g_k h_k$ ($g_k, h_k \in K_\theta^2$, $\sum_k \|g_k\|_2 \|h_k\|_2 < \infty$) and of an embedding theorem of Carleson type. In particular, such a symbol always exists in the case of singular functions $\theta_a = \exp(-a((1+z)/(1-z)))$: this result goes back to [Rochberg, 1987]. The very first example of a bounded T_φ^θ without any bounded symbol (and hence providing a negative response to Sarason's question) was given by Baranov (2010; the exact reference can be found in [Baranov, Bessonov, and Kapustin, 2011]). See also [Garcia and Ross, 2012] for a sketch of the state of the subject in 2012.

Exercise 3.4.15(b) is taken from [Sarason, 2007]. It is a direct analog of a result of Rochberg [1987] (just cited) where the inequality is transformed into an equivalence: see §4.4 (Theorem 4.4.4) and §4.6 for details. Note that by the Remark on Exercise 3.4.15(b), the proof becomes truly trivial for the case where $\theta(0) = 0$. One could suppose that the general case could be reduced to this special case by using the Crofoot transform C_θ [Crofoot, 1994],

$$C_\theta f = f \frac{(1 - |\theta(0)|^2)^{1/2}}{1 - \overline{\theta(0)}\theta},$$

which is a unitary mapping $K_\theta \longrightarrow K_\Theta$ where

$$\Theta = \frac{\theta - \theta(0)}{1 - \overline{\theta(0)}\theta}$$

(already satisfying $\Theta(0) = 0$).

Exercise 3.4.15(c) was proved by Sarason [1967], who initiated the powerful *commutatant lifting method* (a special case is Exercise 3.4.15(c)). On the commutatant lifting method and its innumerable consequences, see [Szőkefalvi-Nagy and Foias, 1967] and [Nikolski, 2002b, 1986]. By comparing (c) and (b), we can note that for $\varphi \in H^\infty$

$$\mathrm{dist}_{L^\infty(\mathbb{T})}(\varphi, \theta H^\infty) = \|T_\varphi^\theta\| \leq \mathrm{dist}_{L^\infty(\mathbb{T})}(\varphi, \theta H^\infty + \overline{\theta} z H_-^\infty),$$

and therefore $\mathrm{dist}_{L^\infty(\mathbb{T})}(\varphi, \theta H^\infty) = \mathrm{dist}_{L^\infty(\mathbb{T})}(\varphi, \theta H^\infty + \overline{\theta} z H_-^\infty)$, but it is easy to find an example of $\varphi \in L^\infty$ where

$$\mathrm{dist}_{L^\infty(\mathbb{T})}(\varphi, \theta H^\infty) > \mathrm{dist}_{L^\infty(\mathbb{T})}(\varphi, \theta H^\infty + H_-^\infty)$$

(see §5.2.1(a)). The nature of this phenomenon (i.e. of the following property: if $F =: \varphi - \theta h$ ($h \in H^\infty$) is an extremal element in the sense where $\|\varphi - \theta h\|_\infty = \mathrm{dist}_{L^\infty(\mathbb{T})}(\varphi, \theta H^\infty)$ then $\|F\|_\infty = \mathrm{dist}_{L^\infty(\mathbb{T})}(F, \overline{\theta} z H_-^\infty)$) is not entirely clear. We can suppose that it is linked to a remarkable property known for Hankel operators (which, by the proof of (c), are responsible for the norm $\|T_\varphi^\theta\|$) – the kernel of a Hankel operator $\mathrm{Ker}H_a = \{f \in H^2 : H_a f = 0\}$ is invariant for M_z, and hence is either trivial $\{0\}$, or occupies a "dominant portion" of H^2 (for every θ inner, θH^2 is "much larger" than $H^2 \ominus \theta H^2$) – and the latter property is always the case for $a = \bar{\theta}\varphi$ with $\varphi \in H^\infty$. More generally, for the extremal problems in H^∞, see the works of Adamyan, Arov, and Krein (beginning with [Adamyan, Arov, and Krein, 1968]) and their presentations in [Garnett, 1981], [Nikolski, 1986], and [Peller, 2003].

Exercise 3.4.15(d) is a generalization of Theorem 4.4.5 by Krupnik–Feldman (below). Given the existence of many compact or finite rank operators T_φ^θ (and even on a finite-dimensional space; $\dim K_\theta < \infty$ if and only if θ is a finite Blaschke product), we can expect to have some worries in applying Exercise 3.4.15(d) to obtain a criterion to belong to the Fredholm class, in particular, to construct a Birkhoff–Wiener–Hopf factorization of a matricial symbol (even of size 2×2); see also the comments in §4.4, §4.6.

***

To conclude, note an important application of the criterion of invertibility in Theorem 3.3.6 to the problem of *bases of reproducing kernels* of the model spaces $K_\theta = H^2 \ominus \theta H^2$ where θ is an inner function (see Appendix F). The reproducing kernel of K_θ is

$$k_\lambda^\theta = P_\theta k_\lambda(z) = \frac{1 - \overline{\theta(\lambda)}\theta(z)}{1 - \bar{\lambda}z},$$

where $|\lambda| < 1$ and P_θ is the orthogonal projection of H^2 on K_θ. Recall that by definition a Riesz *sequence (respectively, a basis)* (x_j) in a Hilbert space is an isomorphic range $V: H \to H$ of an orthogonal sequence (respectively, of a basis) of H. The conditions to be a Riesz sequence (a basis) for sequences $x_j = k_{\lambda_j}^\theta$, $\lambda_j \in \mathbb{D}$, can be expressed in terms of the invertibility of Toeplitz operators in the following manner.

(i) If $(k_{\lambda_j}^\theta)_j$ is a Riesz sequence, then (λ_j) is a Blaschke sequence (see Appendix F) and the *Carleson condition (C)*

$$\inf_j \left|\frac{B}{b_{\lambda_j}}(\lambda_j)\right| > 0$$

is satisfied, where $B = \prod_j b_{\lambda_j}$ is the Blaschke product and

$$b_\lambda(z) = \frac{|\lambda|}{\lambda} \times \frac{\lambda - z}{1 - \bar{\lambda}z}$$

(see Appendix F).

(ii) If (C) is satisfied and the operator $T_{\theta \bar{B}}$ is an isomorphism on its range (respectively, invertible), then $(k^{\theta}_{\lambda_j})_j$ is a Riesz sequence (respectively, basis).

(iii) If the points λ_j do not approach the spectrum of θ, so that $\sup_j |\theta(\lambda_j)| < 1$, then the conditions of (ii) are also necessary. *In particular, $(k^{\theta}_{\lambda_j})_j$ is a Riesz basis of K_θ if and only if the condition (C) holds and $T_{\theta \bar{B}}$ is invertible.*

In the special case

$$\theta = \exp\left(-a\frac{1+z}{1-z}\right), \quad a > 0,$$

and after the unitary transform of the space $H^2(\mathbb{D})$ to the space $H^2(\mathbb{C}_+)$ followed with Fourier transform (see §4.2.1 below), the kernel k^{θ}_λ is transformed into exponentials

$$e^{i\mu x}, \mathrm{Im}(\mu) > 0 \text{ in the space } L^2(0, a)$$

on the finite interval $(0, a)$. We thus arrive at the classical problem of exponential bases

$$F(x) = \sum_j a_j e^{i\mu_j x}$$

in the space $L^2(0, a)$. The multitude of conditions of Theorem 3.3.6, equivalent to the invertibility of $T_{\theta \bar{B}}$, allows the deduction of necessary and sufficient conditions for bases $(e^{i\mu_j x})$ as a function of the geometric distribution of the frequencies (μ_j). We refer to [Nikolski, 2002b] for the exact statements, proofs, and references.

4
Applications: Riemann–Hilbert, Wiener–Hopf, Singular Integral Operators (SIO)

Topics

- Riemann–Hilbert problem, Hilbert transform on $\mathbb{T}$ and $\mathbb{R}$.
- Lebesgue points, Sokhotsky–Plemelj formulas.
- Wiener–Hopf operators.
- Singular integral equations, finite difference equations, Laguerre functions and polynomials, Wiener–Hopf operators on an interval.
- Systems of equations and matricial Wiener–Hopf operators, Hankel operators on $L^2(\mathbb{R}_+)$.

Biographies Nikolaï Luzin, Norbert Wiener, Eberhard Hopf, Edmond Laguerre, Josip Plemelj.

In this chapter we establish the links of Toeplitz operators with the subjects mentioned in the title. In particular, we express the Wiener–Hopf integral operators in terms of Toeplitz matrices over the basis of dilated Laguerre functions.

4.1 The Riemann–Hilbert Problem and the SIO

Using the definitions of §1.1, we speak of a *Riemann–Hilbert problem (RHP)* when there is an equation

$$f_+ + g f_- = h$$

with respect to unknown functions $f_\pm$, where h and g are given functions defined on the circle $\mathbb{T}$; we seek $f_+ \in \text{Hol}(\mathbb{D})$ (a function defined and holomorphic in $\mathbb{D}$) and $f_- \in \text{Hol}_0(\mathbb{D}_-) = \{a \in \text{Hol}(\mathbb{D}_-) : a(\infty) = 0\}$, $\mathbb{D}_- = \mathbb{C} \setminus \overline{\mathbb{D}}$, so that $f_\pm$ possess values on the boundary $\mathbb{T}$, and hence the equation indeed makes sense. Suppose that $h \in L^2(\mathbb{T})$ and $f_\pm \in H_\pm^2(\mathbb{T})$ ($H_+^2(\mathbb{T}) = H^2(\mathbb{T})$, $H_-^2(\mathbb{T}) = L^2(\mathbb{T}) \ominus H^2(\mathbb{T})$). Then, $f = f_+ + f_- \in L^2(\mathbb{T})$ and $f_\pm = P_\pm f$, $P_- = I - P_+$ and P_+ is the Riesz projection: see Appendix F. The equation thus takes the form

$$Rf = h, \text{ where } R = P_+ + gP_-;$$

R is known as the *RHP operator*. By convention, the *RHP* is said to be *normally solvable* if the operator R is Fredholm $L^2(\mathbb{T}) \to L^2(\mathbb{T})$ and $\mathrm{ind}(R)$, $\dim \ker(R)$ and $\dim \mathrm{coker}(R)$ are known. In fact (and following Hilbert), we will study a slightly more general expression,

$$R_{a,b} = aP_+ + bP_-,$$

where a, b are two given functions.

4.1.1 The RHP and Toeplitz Operators

Theorem 4.1.1

(1) Let $a, b \in L^2(\mathbb{T})$. Then, $R_{a,b}$ is bounded $L^2(\mathbb{T}) \to L^2(\mathbb{T})$ if and only if $a, b \in L^\infty(\mathbb{T})$.

(2) Let $a, b \in L^\infty(\mathbb{T})$.

$$R_{a,b} \in \mathrm{Fred}(L^2(\mathbb{T})) \Leftrightarrow (1/a, 1/b \in L^\infty(\mathbb{T}) \text{ and } T_{a/b} \in \mathrm{Fred}(H^2)),$$

where $T_{a/b}$ is the Toeplitz operator with symbol a/b. Moreover, if $R_{a,b}$ is Fredholm, it is invertible, either to the left, or to the right (or to both), and $\mathrm{ind}(R_{a,b}) = \mathrm{ind}(T_{a/b})$,

$$\|1/a\|_\infty, \|1/b\|_\infty \leq \|\Pi(R_{a,b})^{-1}\|,$$

$$\|\Pi(R_{a,b})^{-1}\| \leq \|1/b\|_\infty (1 + \|\Pi(T_{a/b})^{-1}\|)\left(1 + \mathrm{dist}_{L^\infty}\left(\frac{a}{b}, H^\infty\right)\right),$$

where Π stands for the canonical projection on the Calkin algebra defined in Appendices D.6 and E.7.1.

Proof (1) If $R_{a,b} \in L(L^2)$, then $P_+ R_{a,b}|H^2 = T_a \in L(H^2)$, $P_- R_{a,b}|H_-^2 = T_b \in L(H_-^2)$, and by Theorem 2.1.5, $a, b \in L^\infty(\mathbb{T})$. The converse is evident.

(2) If $R_{a,b} \in \mathrm{Fred}(L^2(\mathbb{T}))$, then by a slight modification of the reasoning of Theorem 3.1.3(3) (using $z^n f$, first with $n \to \infty$, and then with $n \to -\infty$), we obtain $\|1/a\|_\infty, \|1/b\|_\infty \leq \|\Pi(R_{a,b})^{-1}\|$.

On the other hand, if $1/a, 1/b \in L^\infty(\mathbb{T})$, then

$$R_{a,b} = b\left(\frac{a}{b}P_+ + P_-\right) = b\left(P_+ \frac{a}{b}P_+ + P_-\right)\left(P_- \frac{a}{b}P_+ + I\right),$$

where b and $P_-(a/b)P_+ + I$ are invertible $((P_-(a/b)P_+ + I)^{-1} = I - P_-(a/b)P_+)$, and $P_+(a/b)P_+ + P_-$ is a "block-diagonal" operator with respect to the decomposition $L^2 = H^2 \oplus H^2_-$. Hence, $P_+(a/b)P_+ = T_{a/b}$ is Fredholm if and only if $R_{a,b}$ is Fredholm, and

$$\|\Pi(R_{a,b})^{-1}\| \le \left\|\frac{1}{b}\right\|_\infty (1 + \|\Pi(T_{a/b})^{-1}\|)\left(1 + \mathrm{dist}_{L^\infty}\left(\frac{a}{b}, H^\infty\right)\right),$$

since

$$\left\|I - P_-\frac{a}{b}P_+\right\| \le 1 + \left\|P_-\frac{a}{b}P_+\right\| = 1 + \mathrm{dist}_{L^\infty}\left(\frac{a}{b}, H^\infty\right)$$

by Nehari's Theorem 2.2.4.

The index formula follows from the same representation of $R_{a,b}$, and the one-sided invertibility – the fact analogous to that for Toeplitz operators, since either $\ker(T_{a/b}) = \{0\}$, or $\mathrm{coker}(T_{a/b}) = \ker(T^*_{a/b}) = \{0\}$ (see Lemma 3.1.4(5)). ∎

Remark A symmetric version of the criterion consists of starting with $R_{a,b} = a(P_+ + (b/a)P_-)$ and finishing with the equivalence

$$R_{a,b} \in \mathrm{Fred} \Leftrightarrow P_-\frac{b}{a}\bigg|H^2_- \in \mathrm{Fred}(H^2_-) \Leftrightarrow T_{(b/a)_*} \in \mathrm{Fred},$$

where $(b/a)_*(z) = b(\bar{z})/a(\bar{z})$ (in fact, $P_-(b/a)|H^2_- = JT_{(b/a)_*}J$ where $Jf(z) = \bar{z}f(\bar{z})$).

If $\mathrm{wind}(a/b)$ is well-defined (as, for example, in the cases in Chapter 3, $a/b \in H^\infty + C(\mathbb{T})$ or $a/b \in PC(\mathbb{T})$, etc.), then, of course, $\mathrm{wind}(b/a)_* = \mathrm{wind}(a/b)$.

4.1.2 The Hilbert Transform $\mathbb{H}$ and SIOs

The analytic properties of the harmonic conjugation operator (also known as the *Hilbert transform*)

$$\mathbb{H}: u \mapsto i\tilde{u}$$

are the subject of Exercise 2.8.4 and §4.5 of [Nikolski, 2019]. Their links with the Hankel operators are treated in Exercises 2.3.4 and 2.3.5 above. We now give a classical interpretation of $\mathbb{H}$ as a singular integral operator (SIO), limiting ourselves to the space $L^2(\mathbb{T})$. We begin with the a.e. existence of the "principal values" of singular integrals (Luzin's Theorem 4.1.2). For this, certain preliminaries are required.

4.1 The Riemann–Hilbert Problem and the SIO

Recall (see Appendix F.3) that $\mathbb{H}$ is well-defined on $L^2(\mathbb{T})$ by $\mathbb{H}z^k = (k/|k|)z^k$ for $k \in \mathbb{Z} \setminus \{0\}$ and $\mathbb{H}1 = 0$. Moreover,

$$\mathbb{H} = P_+ - P_- - (\cdot, 1)_{L^2} 1$$

and hence, for every $f \in L^2(\mathbb{T})$, we have $f + \mathbb{H}f \in H^2$ (the sum can be extended to a holomorphic function in the disk $\mathbb{D}$: see Appendix F). As above (see Exercise 2.3.4(b) or Appendix F), given a function $f \in L^1(\mathbb{T})$, let $f(z)$ ($z \in \mathbb{D}$) denote its harmonic *Poisson extension* in the disk (see [Nikolski, 2019]):

$$f(z) = f * P(z) = \int \frac{1 - |z|^2}{|1 - \bar{\zeta}z|^2} f(\zeta)\, dm(\zeta), \quad z \in \mathbb{D}.$$

Recall also that in complex analysis, given a real harmonic function u in $\mathbb{D}$, the *harmonic conjugate of u* is the function $v = \tilde{u}$ such that $u + iv$ is holomorphic and $v(0) = 0$; it is unique since a real function holomorphic in $\mathbb{D}$ is a constant.

The following properties then hold.

(1) $\mathbb{H}u$ and the harmonic conjugate of u. *For every real function $u \in L^2(\mathbb{T})$, there exists a unique real function $v \in L^2(\mathbb{T})$ such that $u + iv \in H^2$ and $\widehat{v}(0) = 0$. We have $v = -i\mathbb{H}u$, and $z \mapsto v(z)$ is the harmonic conjugate of $z \mapsto u(z)$.*

Indeed, $u + \mathbb{H}u \in H^2$ and $-i\mathbb{H}u$ is real (recall that $v \in L^1(\mathbb{T})$ is real if and only if $\widehat{v}(-k) = \overline{(\widehat{v}(k))}$ for every $k \in \mathbb{Z}$). The uniqueness follows from the fact that a real function of H^2 is a constant, and with $v(0) = \widehat{v}(0) = 0$, then $v = 0$. ∎

(2) The conjugate Poisson kernel. *The Poisson kernel is*

$$P(z) = \frac{1 - |z|^2}{|1 - z|^2} = \operatorname{Re} \frac{1 + z}{1 - z}$$

and its conjugate is

$$Q =: \tilde{P} = \operatorname{Im} \frac{1 + z}{1 - z} = \frac{z - \bar{z}}{|1 - z|^2}.$$

For a real function $u \in L^2(\mathbb{T})$, its harmonic conjugate is

$$\tilde{u}(r e^{i\theta}) = u * Q_r(\theta) = \int_0^{2\pi} Q_r(\theta - t) u(e^{it}) \frac{dt}{2\pi},$$

$$Q_r(\theta) =: Q(r e^{i\theta}) = \frac{2r \sin(\theta)}{1 - 2r\cos(\theta) + r^2}.$$

Moreover,

$$\operatorname{ctg} \frac{\theta}{2} - Q_r(\theta) = \operatorname{ctg} \frac{\theta}{2} \cdot \frac{(1-r)^2}{1 - 2r\cos(\theta) + r^2} = \operatorname{ctg} \frac{\theta}{2} \cdot \frac{1-r}{1+r} P(r e^{i\theta}),$$

where P is the Poisson kernel.

Indeed,
$$u(z) = u * P(z) = \int_{\mathbb{T}} P(\bar{\zeta}z)u(\zeta)\,dm(\zeta) = \int_{\mathbb{T}} \operatorname{Re} \frac{1+\bar{\zeta}z}{1-\bar{\zeta}z} u(\zeta)\,dm(\zeta),$$

and the integral
$$f(z) =: \int_{\mathbb{T}} \frac{1+\bar{\zeta}z}{1-\bar{\zeta}z} u(\zeta)\,dm(\zeta), \quad z \in \mathbb{D},$$

is holomorphic, $f = u + i\tilde{u}$ and thus
$$\tilde{u}(z) = \int_{\mathbb{T}} \operatorname{Im} \frac{1+\bar{\zeta}z}{1-\bar{\zeta}z} u(\zeta)\,dm(\zeta).$$

The identity for the difference can be verified with a direct calculation. ∎

(3) Lebesgue points of a function $u \in L^1$. Recall that, according to a theorem of Lebesgue (see [Rudin, 1986], Theorem 7.6), given $u \in L^1(I)$ ($I \subset \mathbb{R}$ is an interval), almost every point $x \in I$ is a Lebesgue point of u, i.e.

$$\lim_{\epsilon \to 0} \frac{1}{2\epsilon} \int_{|t|<\epsilon} |u(x+t) - u(x)|\,dt = 0.$$

(4) Improper integrals in the sense of the p.v. (principal value). Let $I \subset \mathbb{R}$ be an open interval, $x \in I$ and $u \in L^1_{\text{loc}}(I \setminus \{x\})$ (i.e. u is integrable over every compact subset of $I \setminus \{x\}$). The *principal value (p.v.) improper integral of u* is defined as

$$\text{p.v.} \int_J u(t)\,dt =: \lim_{\epsilon \to 0} \int_{J,\,|x-t|>\epsilon} u(t)\,dt$$

(if it exists), where J is an open interval, $x \in J \subset \bar{J} \subset I$. The existence does not depend on the choice of J; if $u \in L^1(I)$, then for every $J \subset I$, p.v.$\int_J u(t)\,dt = \int_J u(t)\,dt$.

Theorem 4.1.2 (Luzin, 1913)

(1) *If $u \in L^2(\mathbb{T})$, then the integrals*

$$\mathbb{X}u(e^{i\theta}) =: \text{p.v.} \int_{-\pi}^{\pi} u(e^{it}) \operatorname{ctg} \frac{\theta - t}{2} \cdot \frac{dt}{2\pi}, \quad \text{and}$$

$$\mathbb{Y}u(e^{i\theta}) =: \text{p.v.} \int_{-\pi}^{\pi} \frac{u(e^{it})}{\theta - t} \cdot \frac{dt}{\pi}$$

exist a.e. on $\mathbb{T}$, and $\mathbb{X}u = \tilde{u}$.

(2) *The operator*

$$\mathbb{X} \colon L^2(\mathbb{T}) \to L^2(\mathbb{T})$$

4.1 The Riemann–Hilbert Problem and the SIO

(also called the Hilbert transform) is a contraction (unitary on the subspace $\{f \in L^2(\mathbb{T}): \widehat{f}(0) = 0\}$) and $\mathbb{Y} - \mathbb{X}$ is a compact Hilbert–Schmidt operator on $L^2(\mathbb{T})$ (see Appendix D).

Proof We know that $u + i\tilde{u} \in H^2$ and hence, by Fatou's theorem (see Appendix F.1), the limit $\lim_{r \to 1} \tilde{u}(r\, e^{i\theta})$ exists a.e. on $\mathbb{T}$. For every Lebesgue point $e^{i\theta}$ of u (see §4.1.2(3)), we show that

$$\lim_{r \to 1}(\tilde{u}(r\, e^{i\theta}) - X_{1-r}(e^{i\theta})) = 0,$$

where

$$X_\epsilon(e^{i\theta}) = \int_{\epsilon < |\theta - t| < \pi} u(e^{it}) \operatorname{ctg} \frac{\theta - t}{2} \times \frac{dt}{2\pi}.$$

Indeed, by using §4.1.2(2) and the fact that $Q_r(t)$ and $\operatorname{ctg}(t/2)$ are odd functions, at every Lebesgue point $e^{i\theta}$ of u we obtain

$$\tilde{u}(r\, e^{i\theta}) - X_{1-r}(e^{i\theta})$$
$$= \int_{1-r < |t| < \pi} u(e^{i(\theta - t)})\left(Q_r(t) - \operatorname{ctg}\frac{t}{2}\right)\frac{dt}{2\pi} + \int_{|t| < 1-r} u(e^{i(\theta - t)}) Q_r(t) \frac{dt}{2\pi}$$
$$= \int_{1-r < |t| < \pi} (u(e^{i(\theta - t)}) - u(e^{i\theta}))\left(Q_r(t) - \operatorname{ctg}\frac{t}{2}\right)\frac{dt}{2\pi}$$
$$+ \int_{|t| < 1-r} (u(e^{i(\theta - t)}) - u(e^{i\theta})) Q_r(t) \frac{dt}{2\pi}$$
$$=: I_1 + I_2,$$

$$|I_1| \le \int_{1-r < |t| < \pi} |u(e^{i(\theta - t)}) - u(e^{i\theta})| \operatorname{ctg}\frac{|t|}{2} \times \frac{1-r}{1+r} P(r\, e^{it}) \frac{dt}{2\pi}$$
$$\le \operatorname{ctg}\frac{1-r}{2} \times \frac{1-r}{1+r} \int_{|t| < \pi} |u(e^{i(\theta - t)}) - u(e^{i\theta})| P(r\, e^{it}) \frac{dt}{2\pi}$$
$$\le 2 \int_{|t| < \pi} |u(e^{i(\theta - t)}) - u(e^{i\theta})| P(r\, e^{it}) \frac{dt}{2\pi} = o(1)$$

when $r \to 1$ (since $e^{i\theta}$ is a Lebesgue point: see §4.1.2(3)), and

$$|I_2| \le \int_{|t| < 1-r} |u(e^{i(\theta - t)}) - u(e^{i\theta})| \times |Q_r(t)| \frac{dt}{2\pi}$$
$$\le \int_{|t| < 1-r} |u(e^{i(\theta - t)}) - u(e^{i\theta})| \times \frac{2r(1-r)}{(1-r)^2} \frac{dt}{2\pi}$$
$$\le \frac{1}{\pi(1-r)} \int_{|t| < 1-r} |u(e^{i(\theta - t)}) - u(e^{i\theta})|\, dt = o(1)$$

as $r \to 1$ (again since $e^{i\theta}$ is a Lebesgue point). Since the limit $\lim_{r \to 1} \tilde{u}(re^{i\theta}) = \tilde{u}(e^{i\theta})$ exists (for the same reason), we obtain the convergence of the singular integral $\mathbb{X}u(e^{i\theta})$ and the equation $\mathbb{X}u(e^{i\theta}) = \tilde{u}(e^{i\theta})$.

For the difference $\mathbb{Y} - \mathbb{X}$, we have $(\mathbb{Y} - \mathbb{X})u(e^{i\theta}) = \int K(\theta, t) u(e^{it}) \, dt / 2\pi$ where $K(\theta, t) = k(\theta - t)$, $k(t) =: 2/t - \operatorname{ctg}(t/2) \leq k(\pi) = 2/\pi$ (since $k' > 0$), and thus $K \in L^2([-\pi, \pi]^2)$, and the result follows (see Appendix D). ∎

Nikolaï N. Luzin (in Russian Николай Николаевич Лузин, 1883–1950) was a Russian/Soviet mathematician, founder of harmonic analysis in Russia/USSR, creator of the "Luzitania" (an allusion to the ancient Roman province Lusitania corresponding to Portugal) – a research group based around his seminar at the University of Moscow where he discovered and illuminated the talents of his many students, including Suslin, Aleksandrov, Khinchin, Uryson, Menshov, Kolmogorov, Bari, Shnirelman, Novikov, Lusternik, Lyapunov, Keldysh, Lavrentiev, and Kronrod.

Luzin began his studies at the University of Moscow with Nikolai Bugaev and (especially) Dmitri Egorov, and then continued in Paris (1906–1908), but this soon degenerated into a crisis of self-confidence and the questioning of science as a profession (leading to suicide attempts). The second European visit was more productive, first at Göttingen (1910–1912, under the direction of Edmund Landau), then in Paris (1912–1914, at Hadamard's seminar, where Luzin forged a close friendship with Émile Borel and Henri Lebesgue). His thesis "The integral and trigonometric series," completed on the eve of the First World War and submitted in Moscow (1915), is a marker of the birth of harmonic analysis in Russia. Luzin's famous seminar, a veritable incubator of talent, operated at the University of Moscow between 1915 and 1925. Luzin developed a personal style of research collaboration with his students (a "written version" of this style can already be distinguished in his thesis), totally informal, and full of conjectures, comparisons of ideas, discussions, and perspectives – in brief, such a collaboration produced a true detailed plan of future research. This style was neither appreciated nor well understood by everyone at the time, but its effectiveness was nonetheless evident. In particular, some of Luzin's ideas had a long-lasting influence on the development of twentieth-century harmonic analysis (such as his plea for research in "real analysis" techniques to treat singular integrals, or his famous conjecture (1915) on the a.e. convergence of L^2 Fourier series, proved by Carleson 50 years later). It is difficult to overestimate Luzin's impact on the creation of the twentieth-century Russian mathematical school and the culmination of its "golden age" a few decades later.

A dozen mathematical objects bear Luzin's name, including *Luzin's theorem* ("every measurable function is nearly continuous" according to Littlewood), the *Denjoy–Luzin theorem* (on absolutely convergent trigonometric series), the *Luzin–Privalov uniqueness theorem* (with its construction of "Luzin's ice-cream cone"), the *Luzin–Suslin analytic sets*, etc. His research monographs (his thesis mentioned above, as well as his *Leçons sur les ensembles analytiques et leurs applications* (Paris, Gauthier–Villars, 1930)) have kept their mathematical value to this day.

Luzin's career and the running of the "Luzitania" was abruptly interrupted by a serious conflict between Luzin and a certain number of his students, entered in the annals as "the affair of academician Nikolai Nikolaevich Luzin." What in other circumstances would have been a standard conflict of generations, or a manifestation of an "Œdipus complex" according to Freud, was transformed under Stalin's regime of the USSR into a manhunt, cruel and unjust for the victim and humiliating for all involved. With almost total unanimity of the observers (see the references below), this was one of the most shameful events of Russian intellectual life of all periods. In the years 1931–1936, aiming for Luzin's removal from his positions, a group of his students, with the support of Stalin's senior officials such as Kolman and Mekhlis (the chief editor of *Pravda*, the principal mouthpiece of the regime), published several accusations against Luzin on "acts of sabotage" against "proletarian Soviet mathematics" and on his nature as a "servant of Fascist occidental sciences," evoking, among others, his friendly relations with foreign colleagues (Borel, Denjoy, Lebesgue, Montel, and Sierpinski). In this period between Stalin's "Great Turn" (at the beginning of the 1930s) and his "Great Purge" (the mass repression against the elite of the USSR, culminating in the loss of at least a million lives, at the end of the 1930s) such denunciations, in almost 100% of cases, were immediately transformed into affairs managed by the NKVD/KGB (the secret police), with the life of the accused at stake. For example, in similar circumstances, a close friend of Luzin, the philosopher Pavel Florensky (who, incidentally, figured equally on the sidelines of the "Luzin affair"), perished in the bowels of the NKVD. The most active roles in this group of students were played by Aleksandrov, Khinchin, Shnirelman, Kolmogorov, Sobolev, Gelfond, and Pontryagin, the stars of Soviet/Russian mathematics. (*O tempora! O mores!*... It seems that this pestilential regime managed to pervert the whole of Russian society – a society that, in other times, knew Tolstoy, Chekhov, and Berdiaev. However, there remained a few courageous individuals – Sergei Bernstein, and then Nina Bari and Dmitrii Menshov – who publicly supported Luzin). According to a severe critique by Semën Kutateladze, the Luzin affair was a signal from the Russian scientific community to Stalin and his *oprichniks*: "We will crush anyone, just give us a sign" (see "The tragedy of mathematics in Russia," arXiv:math/0702632v7, or a more complete Russian version, www.math.nsc.ru/LBRT/g2/english/ssk/case.html#lessons). But this time, no sign was given (on the contrary, there are witnesses according to whom Stalin himself suggested lowering the tone of the accusations): Luzin was not arrested but "only" relieved of his positions, and forced to cease his activities and hide for many years in an engineering institute. The affair itself, as well as the ostracism that hit Luzin (he lived 14 years branded as an "enemy under a Soviet mask"), was kept secret for 50 years (!) by all the participants in this trial. Today the full details of the affair (at some points horribly shameful) have been published, including numerous letters of support to Luzin from French mathematicians: see Youschkevitch and Dugac, " 'L'affaire' de l'académicien Luzin" (*Gazette des mathématiciens* **38** (1988), pages 30–35); Demidov and Ford, *N. N. Luzin and the Affair of the "National Fascist Center"* (Academic Press, 1996); Demidov and Levshin, eds., *The Case of Academician Nikolai Nikolaevich Luzin* (AMS, 2016: reports of the commission investigating the case); Lorentz, "Mathematics and politics in the Soviet Union from 1928 to 1953" (*J. Approx. Theory* **116** (2002), pages 169–223); and http://en.wikipedia.org/wiki/Nikolai_Luzin.

4.1.3 Comment: Operators and Singular Integral Equations

The only point in the proof of Theorem 4.1.2 where we used $u \in L^2(\mathbb{T})$ is the fact that $\tilde{u}$ admits boundary limits a.e. on $\mathbb{T}$ (since $u+i\tilde{u} \in H^2$: see Appendix F). But the same property of $\tilde{u}$ can be justified by only supposing that $u \in L^1(\mathbb{T})$:

such a u is a combination $u = u_1 - u_2 + iu_3 - iu_4$ of integrable functions $u_j \geq 0$, where the $\tilde{u}_j$ admit boundary limits a.e., since this is the case for the functions $f_j \in H^\infty$ where $f_j = e^{-(u_j + i\tilde{u}_j)}$, and given the continuity of log. Hence, taking into account this remark on the proof of §4.1.2, we have proved the following theorem.

Theorem 4.1.3 (Privalov, 1919) *Let $u \in L^1(\mathbb{T})$. Then the integrals*

$$\mathbb{X}u(e^{i\theta}) =: \text{p.v.} \int_{-\pi}^{\pi} u(e^{it}) \text{ctg}\,\frac{\theta - t}{2} \cdot \frac{dt}{2\pi} \quad \text{and}$$

$$\mathbb{Y}u(e^{i\theta}) =: \text{p.v.} \int_{-\pi}^{\pi} \frac{u(e^{it})}{\theta - t} \cdot \frac{dt}{\pi}$$

exist a.e. on $\mathbb{T}$, and $\mathbb{X}u = \tilde{u}$.

The *classical singular integral equations* are

$$(aI + b\mathbb{X})f = g, \quad \text{or } (aI + b\mathbb{Y})f = g, \quad \text{or } (aI + b\mathbb{S})f = g.$$

The first two, parametrized over the interval $(-\pi, \pi)$, are called having the *Hilbert kernel* and the *Cauchy kernel* respectively; the third involves the principal value of the *Cauchy complex singular integral*,

$$\mathbb{S}f(z) = \text{p.v.} \int_{\mathbb{T}} \frac{f(\zeta)}{\zeta - z} \frac{d\zeta}{\pi i}, \quad z \in \mathbb{T}.$$

In each case, a and b are given functions, $g \in L^2(\mathbb{T})$ is known, and $f \in L^2(\mathbb{T})$ is to be found. It is easy to see (as in Theorem 4.1.2(2)) that

$$\mathbb{S} - i\mathbb{X} \in \mathfrak{S}_\infty, \quad \mathbb{S} - i\mathbb{Y} \in \mathfrak{S}_\infty, \quad \mathbb{S} - \mathbb{H} \in \mathfrak{S}_\infty$$

(in fact, the last difference is only of rank 1: see Exercise 4.5.1 below), and hence the Fredholm theory of the SIE of the three types is always the same. In particular, Theorem 4.1.1, Theorem 4.1.2, and the preceding comments immediately lead to the following criteria.

Corollary 4.1.4 *Let $a, b \in L^\infty(\mathbb{T})$.*

(1) *The operator*

$$A = aI + b\mathbb{X} \quad \text{and/or} \quad B = aI + b\mathbb{Y}$$

is Fredholm if and only if $(a \pm ib)^{-1} \in L^\infty(\mathbb{T})$ and the Toeplitz operator T_c, $c = (a - ib)/(a + ib)$, is Fredholm,

$$\text{ind}(A) = \text{ind}(B) = \text{ind}(T_c) = \text{wind}\frac{a + ib}{a - ib}$$

(in the case of coefficients a, b in $H^\infty + C(\mathbb{T})$ or $PC(\mathbb{T})$).

(2) *The operator*

$$C = aI + bS$$

is Fredholm if and only if $(a \pm b)^{-1} \in L^\infty(\mathbb{T})$ *and the Toeplitz operator* T_c, $c = (a + b)/(a - b)$, *is Fredholm,*

$$\operatorname{ind}(C) = \operatorname{ind}(T_c) = \operatorname{wind}\frac{a-b}{a+b}$$

(in the case of coefficients a, b in $H^\infty + C(\mathbb{T})$ *or* $PC(\mathbb{T})$).

(1) Indeed, $\mathbb{X} = -i\mathbb{H} = iP_- - iP_+ + iK$, $\operatorname{rank}(K) = 1$, hence $A = (a - ib)P_+ + (a + ib)P_- + L$, $\operatorname{rank}(L) = 1$, and the result follows from Theorem 4.1.1 and Corollary E.7.4(5). ∎

(2) Indeed, by §4.1.3,

$$C = aI + b(P_+ - P_-) + K = (a+b)P_+ + (a-b)P_- + K$$

where $K \in \mathfrak{S}_\infty$, and we conclude as in (1). ∎

See also the Sokhotsky–Plemelj formulas in Exercise 4.5.1 below.

For the convenience of the reader, we summarize below the relations between operators introduced earlier: $P_\pm$ (defined in Appendix F.3; §2.1.1), $\mathbb{H}$ (defined in Appendix F.2; Exercise 2.3.4, §4.1.2), $\mathbb{S}$ (defined in §4.1.3), $\mathbb{X}$ and $\mathbb{Y}$ (defined in §4.1.2, §4.1.3):

- $\mathbb{H} = P_+ - P_- - (\cdot, 1)_{L^2} 1$ (Hilbert transformation, §4.1.2),
- $\mathbb{S} = P_+ - P_-$ (Exercise 4.5.1 below),
- $\mathbb{X} = -i\mathbb{H}$ (harmonic conjugation, §4.1.2(1)),
- $\mathbb{Y} = \mathbb{X} + K$, $K \in \mathfrak{S}_2$ (Theorem 4.1.2).

4.2 Toeplitz on $H^2(\mathbb{C}_+)$ and Wiener–Hopf on $L^2(\mathbb{R}_+)$

In this section we transfer the theory of Toeplitz operators on $H^2(\mathbb{D})$ to the space $H^2(\mathbb{C}_+)$ on the upper half-plane (with the aid of a conformal mapping), and then to the space $L^2(\mathbb{R}_+)$ (with the aid of the Fourier transform). In the latter case, we will arrive at the Wiener–Hopf integral operators, which turn out to have Toeplitz matrices with respect to the orthonormal basis of *Laguerre functions* (dilated by a factor 2) $l_n(x) = (1/n!) e^{x/2} (x^n e^{-x})^{(n)}$, $n = 0, 1, \ldots$ (see §4.3).

4.2.1 On the Space $H^2(\mathbb{C}_+)$: The Paley–Wiener Theorem

First recall a few correspondences known between the spaces $H^2(\mathbb{D})$, $H^2(\mathbb{C}_+)$ and $L^2(\mathbb{R}_+)$; in a more complete form, these can be found in [Nikolski, 2019] or in Appendix F below. Let

$$\mathbb{C}_+ = \{z \in \mathbb{C}: \operatorname{Im}(z) > 0\}$$

and let $\omega \colon \mathbb{C}_+ \to \mathbb{D}$ be the conformal mapping

$$\omega(z) = \frac{z-i}{z+i}.$$

The restriction of ω to the boundary $\partial \mathbb{C}_+ = \mathbb{R}$ is a bijection $\mathbb{R} \to \mathbb{T} \setminus \{1\}$ whose Jacobian (derivative) is

$$|J(x)| = \frac{2}{1+x^2}, \quad x \in \mathbb{R}.$$

Hence, the mapping U,

$$Uf(x) = \frac{1}{\sqrt{\pi}(x+i)} \cdot f\!\left(\frac{x-i}{x+i}\right), \quad x \in \mathbb{R},$$

is a unitary isomorphism between the spaces $L^2(\mathbb{T})$ and $L^2(\mathbb{R})$,

$$U \colon L^2(\mathbb{T}) \to L^2(\mathbb{R}).$$

In particular,

$$\int_\mathbb{T} |f|^2 \, dm = \frac{1}{\pi} \int_\mathbb{R} \left|f\!\left(\frac{x-i}{x+i}\right)\right|^2 \frac{dx}{x^2+1} \quad \forall f \in L^2(\mathbb{T})),$$

and if $f \in L^2(\mathbb{T})$ and $\varphi \in L^\infty(\mathbb{T})$, then $U(\varphi f) = (Uf) \cdot \varphi \circ \omega$. As above, given a function $\Phi \in L^\infty(\mathbb{R})$, we denote M_Φ *the multiplication operator* $M_\Phi \colon L^2(\mathbb{R}) \to L^2(\mathbb{R})$,

$$M_\Phi f = \Phi f \quad (f \in L^2(\mathbb{R})).$$

In what follows, given a Borel set $E \subset \mathbb{R}$, we consider the space $L^2(E, dx)$ as a subspace of $L^2(\mathbb{R})$ by extending every function $f \in L^2(E)$ by zero in the complement $\mathbb{R} \setminus E$. We define the subspace $H^2(\mathbb{C}_+)$ of $L^2(\mathbb{R})$ by

$$H^2(\mathbb{C}_+) = U H^2(\mathbb{D}),$$

and recall that $H^2(\mathbb{C}_+)$ is generated in $L^2(\mathbb{R})$ by the Cauchy kernels (the ranges under U of the reproducing kernels of $H^2(\mathbb{D})$, up to a numerical factor)

$$H^2(\mathbb{C}_+) = \operatorname{span}_{L^2(\mathbb{R})}\!\left(\frac{1}{x-\bar{\mu}} \colon \operatorname{Im}(\mu) > 0\right).$$

The standard orthonormal basis $(z^n)_{n\in\mathbb{Z}}$ of $L^2(\mathbb{T})$ is transformed into the "Laguerre basis" (in applications, it is also called a "Laguerre sampling system," or the *Laguerre fractions*)

$$\Lambda_n(x) = Uz^n(x) = \frac{1}{\sqrt{\pi}(x+i)} \cdot \left(\frac{x-i}{x+i}\right)^n, \quad n \in \mathbb{Z}.$$

The unitary mapping $\mathcal{F}_P \colon L^2(\mathbb{R}) \to L^2(\mathbb{R})$ given by the *Fourier–Plancherel transform* $\mathcal{F}_P =: (2\pi)^{-1/2}\mathcal{F}$, where

$$\mathcal{F}f(z) = \int_{\mathbb{R}} f(x)\, e^{-ixz}\, dx, \quad \mathcal{F}^{-1}f(z) = \frac{1}{2\pi}\int_{\mathbb{R}} f(x)\, e^{ixz}\, dx, \quad \mathcal{F}_P^{-1} = (2\pi)^{1/2}\mathcal{F}^{-1},$$

is also extremely important. In particular,

$$\mathcal{F}_P^{-1} L^2(\mathbb{R}_+) = H^2(\mathbb{C}_+) \quad \text{(Paley–Wiener theorem)};$$

see [Nikolski, 2019], as well as Appendix F.

Here is some notation used below:

$$\mathcal{F}f = \widehat{f}, \quad 2\pi\mathcal{F}^{-1}f = f^{\vee}.$$

Theorem 4.2.1 *Let $T \in L(H^2(\mathbb{C}_+))$. The following assertions are equivalent.*

(1) $T = P_+ M_\Phi | H^2(\mathbb{C}_+)$ *where* $\Phi \in L^\infty(\mathbb{R})$ *and P_+ is the orthogonal projection of $L^2(\mathbb{R})$ onto $H^2(\mathbb{C}_+)$.*
(2) $M_\omega^* T M_\omega = T$.
(3) *For every $s > 0$, $M_{e_s}^* T M_{e_s} = T$, where $e_s(x) = e^{isx}$ ($x \in \mathbb{R}$).*
(4) $T = U T_\varphi U^{-1}$ *where T_φ is a Toeplitz operator on $H^2(\mathbb{D})$.*

If (1)–(4) hold, then $\Phi = \varphi \circ \omega$; we denote $T = T_\Phi$ and call it the Toeplitz operator *on $H^2(\mathbb{C}_+)$ (with symbol Φ).*

Proof First, we transfer propositions (1)–(3) to the space $H^2(\mathbb{D})$ with the aid of the mapping U, temporarily letting $P_+^{\mathbb{D}}$ denote the Riesz projection on $L^2(\mathbb{T})$:

$$(1) \Leftrightarrow U^{-1}TU = U^{-1}P_+ M_\Phi U = U^{-1}P_+ UU^{-1}M_\Phi U = P_+^{\mathbb{D}} M_\varphi | H^2(\mathbb{D}),$$

where $\varphi = \Phi \circ \omega^{-1}$,

$$\begin{aligned}(2) \Leftrightarrow U^{-1}M_\omega^* T M_\omega U &= (U^{-1}M_\omega^* U) U^{-1} T U (U^{-1} M_\omega U)\\ &= (U^{-1}M_\omega U)^* U^{-1} T U (U^{-1} M_\omega U) = U^{-1}TU,\end{aligned}$$

$$\begin{aligned}(3) \Leftrightarrow U^{-1} M_{e_s}^* T M_{e_s} U &= U^{-1} M_{e_s}^* U U^{-1} T U U^{-1} M_{e_s} U\\ &= (U^{-1} M_{e_s} U)^* U^{-1} T U (U^{-1} M_{e_s} U)\\ &= U^{-1}TU \quad (\forall s > 0).\end{aligned}$$

However, $U^{-1}M_\omega U = M_z$ (given the equation $U(\varphi f) = (Uf) \cdot \varphi \circ \omega$) and $U^{-1}M_{e_s}U = M_{e_s \circ \omega^{-1}}$, where

$$e_s \circ \omega^{-1} =: V_s = \exp\left(-s\frac{1+z}{1-z}\right),$$

and hence by applying Theorem 2.1.5 we obtain (1) $\Leftrightarrow$ (4), and by Lemma 2.1.2 (or §2.1.6) we obtain (2) $\Leftrightarrow$ (4).

For (3) $\Leftrightarrow$ (4), it only remains to justify that the equations (in $H^2(\mathbb{D})$) $M_z^* A M_z = A$ and $M_{V_s}^* A M_{V_s} = A$ ($\forall s > 0$) are equivalent. However, $M_z^* A M_z = A$ signifies that A is a Toeplitz operator $A = T_\varphi$, and thus, for every $h \in H^\infty$, $(M_h)^* A M_h = T_{\bar{h}} T_\varphi T_h = T_{|h|^2 \varphi}$ (see Lemma 3.1.4(1)); and since $h = V_s$ is an inner function (see Appendix F), thus unimodular ($|V_s| = 1$ a.e. on $\mathbb{T}$), we have $M_{V_s}^* A M_{V_s} = A$ ($\forall s > 0$).

For the converse (i.e. for (3) $\Rightarrow$ (4)), observe that $M_{V_s}^* A M_{V_s} = A$ ($\forall s > 0$) is equivalent to

$$(AM_{V_s}f, M_{V_s}g) = (Af, g) \quad (\forall f, g \in \mathcal{P}_a)$$

(the set of analytic polynomials $\mathcal{P}_a$ is dense in $H^2(\mathbb{D})$), and that for $f = (1-z)F$, $g = (1-z)G$, where $F, G \in \mathcal{P}_a$, we have

$$M_{V_s}f = f - s\frac{1+z}{1-z}f + o(s), \quad M_{V_s}g = g - s\frac{1+z}{1-z}g + o(s),$$

where $\lim_{s \to 0} \|o(s)/s\|_2 = 0$. By replacing f, g and $M_{V_s}f, M_{V_s}g$ in our equation with these expressions, we obtain

$$-s(A(1+z)F, (1-z)G) - s(A(1-z)F, (1+z)G) + o(s) = 0 \quad \text{(as } s \to 0),$$

and hence $(AF + AzF, G - zG) + (AF - AzF, G + zG) = 0$, therefore $2(AF, G) - 2(AzF, zG) = 0$ for every $F, G \in \mathcal{P}_a$. We arrive at $M_z^* A M_z = A$. ∎

4.2.2 Pseudo-Measures and Wiener–Hopf Operators

In contrast with the discrete group $\mathbb{Z}$ (see Example 2.1.7, parts (2) and (4)), the characterization of Wiener–Hopf operators requires an elementary familiarity with the language of the Schwartz tempered distributions; everything needed is (largely) covered in [Rudin, 1991].

Recall that the space $\mathcal{S}'(\mathbb{R})$ of *tempered distributions* is the dual of the space $\mathcal{S}(\mathbb{R})$ of C^∞ functions rapidly decreasing at infinity, that the *Fourier transform* $\mathcal{F}$ is well-defined on $\mathcal{S}'(\mathbb{R})$ (by duality) and that $\mathcal{F}(\mathcal{S}'(\mathbb{R})) = \mathcal{S}'(\mathbb{R})$. For every p, $1 \leq p \leq \infty$, we have $L^p(\mathbb{R}) \subset \mathcal{S}'(\mathbb{R})$, as well as $\mathcal{M}(\mathbb{R}) \subset \mathcal{S}'(\mathbb{R})$ (the space of complex Borel measures on $\mathbb{R}$); a distribution $S \in \mathcal{S}'(\mathbb{R})$ is said to be a *pseudo-measure* if

$$\mathcal{F}(S) \in L^\infty(\mathbb{R})$$

(note that when S is a measure $S = \mu \in \mathcal{M}(\mathbb{R})$, we have the classical Fourier transform $\mathcal{F}(\mu)(x) = \widehat{\mu}(x) =: \int_\mathbb{R} e^{-itx} d\mu(t)$, $x \in \mathbb{R}$). The set of pseudo-measures is denoted by

$$\mathcal{P}M(\mathbb{R}).$$

The *convolution* $u * S$ ($u \in \mathcal{S}(\mathbb{R})$) is well-defined by $u * S(x) = \langle \tau_x u_*, S \rangle$, where τ_x is the *translation* on $\mathbb{R}$,

$$\tau_x f(y) = f(y - x), \quad y \in \mathbb{R},$$

$u_*(t) = u(-t)$ ($t \in \mathbb{R}$), and $\langle \cdot, \cdot \rangle$ is a form of duality between $\mathcal{S}(\mathbb{R})$ and $\mathcal{S}'(\mathbb{R})$ (in particular, if $u \in \mathcal{S}(\mathbb{R})$ and μ is a measure, then $\langle u, \mu \rangle = \int_\mathbb{R} u \, d\mu$). The mapping $u \mapsto u * S$ ($u \in \mathcal{S}(\mathbb{R})$) can be extended to a bounded mapping $L^2(\mathbb{R}) \to L^2(\mathbb{R})$ if and only if $S \in \mathcal{P}M(\mathbb{R})$ (this is an analog on $\mathbb{R}$ of Lemma 2.1.4; in fact, we will not use this in what follows).

Definition A *Wiener–Hopf operator,* $W: L^2(\mathbb{R}_+) \to L^2(\mathbb{R}_+)$, $\mathbb{R}_+ = (0, \infty)$, is a bounded mapping defined by a pseudo-measure $k \in \mathcal{P}M(\mathbb{R})$ according to the formula

$$W_k f = P^+(k * f) \quad (f \in \mathcal{S}(\mathbb{R}_+)),$$

where $P^+ g = \chi_{\mathbb{R}_+} g$, the orthogonal projection of $L^2(\mathbb{R})$ onto $L^2(\mathbb{R}_+)$, and $\mathcal{S}(\mathbb{R}_+) = \{f \in \mathcal{S}(\mathbb{R}): \text{supp}(f) \subset \mathbb{R}_+\}$.

Norbert Wiener (1894–1964) was an American mathematician, inventor of cybernetics (1948) and communications theory (a credit shared with Vladimir Kotelnikov and Claude Shannon), creator of the theory of stochastic processes and of generalized harmonic analysis (1930, with the Wiener measure and Brownian motion), of *Banach spaces* (1923, independently of Stefan Banach), and of Tauberian theory. He was the author of innovative works in mathematical physics, potential theory, and optimal prediction of random processes (with applications to the automatic correction of the firing of anti-aircraft guns, a priority shared with Andrey Kolmogorov).

He spent most of his career at the Massachusetts Institute of Technology (MIT), which he, along with Claude Shannon, Vannevar Bush, and others, transformed from an ordinary provincial establishment into one of the best technical universities in the world. An admirer of Leibniz, Lebesgue, and Hadamard, Norbert Wiener was one of the geniuses of the twentieth-century who changed the face of mathematics and the sciences. The reader can find a remarkable overview of Wiener's scientific impact (as well as a biographical essay by Norman Levinson) in a special issue of the *Bulletin of the American Mathematical Society* (**72:1** (1966)).

After receiving his Bachelor's degree at 14, Wiener studied at Master's level in zoology at Harvard and philosophy at Cornell, and then in mathematics at Harvard. After submitting his thesis in 1912 (at the age of 17!), he went to Europe for his post-doc (successively under the direction of Bertrand Russell, G. H. Hardy, J. E. Littlewood, Edmund Landau, and David Hilbert). On his return to the United States, Wiener was refused a position at Harvard due to the anti-Semitic atmosphere of the establishment (and due to Birkhoff's opposition in particular). In contrast with many other high-level scientists, Wiener refused to participate in the Manhattan Project, or to direct a laboratory of scientific computations (in California) under the Department of War. He published an article destined for the public at large on the responsibility of scientists engaged in weapons of destruction (in *The Atlantic Monthly*, 1947); this resulted in a period of ostracism and isolation, beginning with sudden removal of his name from the already printed program of a major conference at Harvard on the electronic computing machines ENIAC/MANIAC. A committed pacifist, after the Second World War he systematically refused any government financing of his research and never participated in military projects.

Several dozen mathematical objects (and not the least!) bear Wiener's name (the Wiener process, Wiener equation, Wiener's Tauberian theorem (WTT), Wiener–Hopf factorization, the Paley–Wiener theorem (and space), Wiener algebra, etc.). Norbert Wiener was also the author of several texts on philosophy, reflection and history, such as *Cybernetics* (Hermann, Paris, 1948), *Cybernetics and Society* (Boston, 1950), *Ex-prodigy: My Childhood and Youth* (MIT, 1953), *God & Golem, Inc.* (MIT,

1964), and *I am a Mathematician: The Later Life of a Prodigy* (London, 1965). We conclude with a few quotations by and about Wiener (taken especially from the last-mentioned book):

[with an invitation to Göttingen, 1926] I was in a very exulted mood ... and felt that I had now got from under the pressure and the indifference of Birkhoff and Veblen.　　　　　　　　　　　　　　　(Chapter 5 of the MIT Press edition, 1956)
　　It is the battle for learning which is significant, and not the victory. Every victory that is absolute is followed at once by the Twilight of the gods, in which the very concept of victory is dissolved in the moment of its attainment.
　　A. Khinchin and A. Kolmogorov ... have long been involved in the same field in which I was working. For more than twenty years, we have been on one another's heels, either they had proved a theorem which I was about to prove, or I had been ahead of them by the narrowest of margins.
　　I'm tired of being Norbert Wiener's daughter; I want to be Peggy Wiener!
　　　　　　　　　　　　　　　　　　　　　　　(Reported by Wiener himself)
　　Wiener always had a feeling of insecurity.　(Stanisław Ulam, *Adventure of a Mathematician*, New York (1983))

Eberhard Hopf (1902–1983), a German–American mathematician, is famous as one of the founders of ergodic theory and of the theory of bifurcations, as well as his fundamental results on PDEs, the Wiener–Hopf equations, fluid dynamics, and differential geometry. (He should not be confused with the famous topologist Heinz Hopf.) Hopf studied at the University of Berlin, where he submitted his thesis in 1926 (with Erhard Schmidt and Issai Schur), and then his *Habilitation* in mathematical astronomy (1929), which already contains his famous "maximum principle" for elliptic PDEs. In 1930, Hopf received a grant from the Rockefeller Foundation to work with George Birkhoff (Harvard) in classical mechanics (according to Hopf himself, Birkhoff only accepted his candidature after assuring himself that Hopf was not Jewish). Then, with Wiener's aid, he obtained a position at MIT (1931–1936). There, he wrote his influential book *Mathematical Problems of Radiative Equilibrium* (1934, re-issued in 1964)

and collaborated with Wiener on the founding article "Über eine Klasse singulären Integralgleichungen" [Wiener and Hopf, 1931]. In 1936, under pressure from his family (his wife Ilse and daughter Barbara), Hopf accepted a position as professor at the University of Leipzig (just vacated by Leon Lichtenstein, a cousin of Wiener's father Leo – it's a small world! – driven out by the Nazis already in power). On his arrival, Hopf published his other reference text, *Ergodentheorie*, again written at MIT (and only containing 81 pages!). Between 1942 and 1944 he worked for the Deutsche Forschungsanstalt für Segelflug on aeronautical subjects at the demand of the Luftwaffe. Hopf was certainly not a supporter of the Nazi regime, but his collaboration cost him much of his reputation: there was a certain isolation and incomprehension from his colleagues on his return to the United States (1947), even a "deliberate forgetting" of some of his results (for example, his name disappeared from *Wiener–Hopf filters*, transformed into *Wiener filters*, etc.). In 1971 Hopf became the Gibbs Lecturer of the American Mathematical Society (one of the highest professional distinctions in the United States), and then received the Leroy P. Steele Prize (1981). He served as editor of the *Indiana University Mathematical Journal* for 30 years (1951–1981).

Theorem 4.2.2 *Let $W \in L(L^2(\mathbb{R}_+))$. The following assertions are equivalent.*

(1) $W = W_k$ where $k \in \mathcal{P}M(\mathbb{R})$.
(2) $\mathcal{F}_P^{-1} W \mathcal{F}_P$ is a Toeplitz operator T_Φ on $H^2(\mathbb{C}_+)$.
(3) For every $t > 0$, $\tau_t^* W \tau_t = W$.

If (1)–(3) hold, then $\Phi = 2\pi \mathcal{F}^{-1} k$; Φ is called the symbol *of the Wiener–Hopf operator W.*

Proof (1) $\Leftrightarrow$ (2) For every $f \in \mathcal{S}(\mathbb{R}_+)$, we have

$$\mathcal{F}_P^{-1} W_k \mathcal{F}_P(\mathcal{F}_P^{-1} f) = \mathcal{F}_P^{-1} W_k f = \mathcal{F}_P^{-1} P^+ \mathcal{F}_P \mathcal{F}_P^{-1}(k * f)$$
$$= \mathcal{F}_P^{-1} P^+ \mathcal{F}_P(\Phi \cdot \mathcal{F}_P^{-1} f) = P_+(\Phi \cdot \mathcal{F}_P^{-1} f) = T_\Phi(\mathcal{F}_P^{-1} f),$$

where $\Phi = 2\pi \mathcal{F}^{-1} k$ (in the sense of distributions). As the set of these f is a vector subspace dense in $L^2(\mathbb{R}_+)$, the equivalence is shown.

(2) $\Leftrightarrow$ (3) by Theorem 4.2.1, since the equation $\tau_s^* W \tau_s = W$ ($\forall s > 0$) is equivalent to $\mathcal{F}_P^{-1} \tau_s^* W \tau_s \mathcal{F}_P = \mathcal{F}_P^{-1} W \mathcal{F}_P$ ($\forall s > 0$) and

$$\mathcal{F}_P^{-1}\tau_s^* W \tau_s \mathcal{F}_P = \mathcal{F}_P^{-1}\tau_s^* \mathcal{F}_P(\mathcal{F}_P^{-1} W \mathcal{F}_P)\mathcal{F}_P^{-1}\tau_s \mathcal{F}_P = M_{e_s}^*(\mathcal{F}_P^{-1} W \mathcal{F}_P)M_{e_s}$$

(the equality $\tau_s \mathcal{F}_P = \mathcal{F}_P M_{e_s}$ is well known and very easy to verify; $e_s(x) = e^{isx}$). ∎

Corollary 4.2.3 *A Wiener–Hopf operator*

$$W_k \colon L^2(\mathbb{R}_+) \to L^2(\mathbb{R}_+) \quad (k \in \mathcal{PM}(\mathbb{R}))$$

with symbol $\Phi = 2\pi \mathcal{F}^{-1} k$ *is unitarily equivalent to a Toeplitz operator* $T_\varphi \colon H^2(\mathbb{D}) \to H^2(\mathbb{D})$,

$$W_k = \mathcal{U} T_\varphi \mathcal{U}^{-1} \quad \text{where } \mathcal{U} = \mathcal{F}_P U, \quad \varphi = \Phi \circ \omega^{-1},$$

and

$$Uf(x) = \frac{1}{\sqrt{\pi}(x+i)} \cdot f(\omega(x)) \quad (x \in \mathbb{R}), \quad \omega^{-1}(z) = i\frac{1+z}{1-z}.$$

Indeed, this is a direct consequence of Theorems 4.2.1 and 4.2.2. ∎

4.2.3 Transfer of Spectral Theory to Wiener–Hopf Operators

Given the preceding corollary, several properties of Wiener–Hopf operators are automatic consequences of the results of Chapters 2 and 3; an abbreviated list is given below.

(1) $\|W_k\| = \|W_k\|_{\text{ess}} = \|\Phi\|_{L^\infty(\mathbb{R})}$ where $\Phi = 2\pi \mathcal{F}^{-1} k$ is the symbol of W_k (in particular, $W_k \in \mathfrak{S}_\infty \Rightarrow k = 0$), and similarly for all the powers $\|W_k^n\| = \|W_k^n\|_{\text{ess}} = \|\Phi\|_{L^\infty(\mathbb{R})}^n$, $n \geq 0$ (in particular, $\sigma(W_k) = \{0\} \Rightarrow k = 0$).

(2) If $W_k \in \text{Fred}(L^2(\mathbb{R}_+))$, then $1/\Phi \in L^\infty(\mathbb{R})$ and $\|1/\Phi\|_{L^\infty(\mathbb{R})} \leq \|\Pi(W_k)^{-1}\|$,

$$\text{Ran}_{\text{ess}}(\Phi) \subset \sigma_{\text{ess}}(W_k) \subset \sigma(W_k) \subset \text{conv}(\text{Ran}_{\text{ess}}(\Phi)).$$

(3) If $k \neq 0$, then either $\ker W_k = \{0\}$, or $\ker W_k^* = \{0\}$, and consequently W_k is invertible if $W_k \in \text{Fred}$ and $\text{ind}(W_k) = 0$.

(4) If $\Phi \in C(\overline{\mathbb{R}})$, where $\overline{\mathbb{R}} = \mathbb{R} \cup \{\infty\}$, then $W_k \in \text{Fred} \Leftrightarrow \Phi(x) \neq 0 \, \forall x \in \overline{\mathbb{R}}$ and

$$\text{ind}(W_k) = -\text{wind}(\Phi) = \lim_{T \to \infty} \frac{1}{2T}(\arg(\Phi(T)) - \arg(\Phi(-T))).$$

Moreover, $\sigma_{\text{ess}}(W_k) = \Phi(\overline{\mathbb{R}})$ and the regularizer of W_k is W_l with symbol $1/\Phi$ (the pseudo-measure l is the convolutive inverse of k).

(5) $W_k \in \text{Fred} \Leftrightarrow (1/\Phi \in L^\infty(\mathbb{R})$ and

$$\text{dist}_{L^\infty(\mathbb{R})}\left(\frac{\Phi}{|\Phi|}, H^\infty(\mathbb{C}_+) + C(\overline{\mathbb{R}})\right) < 1, \quad \text{dist}_{L^\infty(\mathbb{R})}\left(\frac{\overline{\Phi}}{|\Phi|}, H^\infty(\mathbb{C}_+) + C(\overline{\mathbb{R}})\right) < 1.$$

4.2 Toeplitz on $H^2(\mathbb{C}_+)$ and Wiener–Hopf on $L^2(\mathbb{R}_+)$

In fact, if $W_k \in$ Fred, the distances in question are equal, say d, and $\|\Pi(W_k)^{-1}\| = (1-d^2)^{-1/2}$.

(6) If $\Phi^{\pm 1} \in L^\infty(\mathbb{R})$ and Φ is locally sectorial on $\overline{\mathbb{R}}$, then $W_k \in$ Fred.

(7) W_k is invertible if and only if
   (i) $\Phi = fg$ where $f^{\pm 1} \in H^\infty(\mathbb{C}_+)$, $g^{\pm 1} \in L^\infty(\mathbb{R})$ and g is sectorial, or
   (ii) $1/\Phi \in L^\infty(\mathbb{R})$ and $\Phi/|\Phi| = e^{i(a+\tilde{b}+c)}$ where a,b,c are real ($c \in \mathbb{R}$), $\|a\|_\infty < \pi/2$, or
   (iii) Φ admits a Wiener–Hopf factorization $\Phi = \Phi_1\Phi_2$ (see §3.3.1 for the definition). See also Theorem 3.3.6 for other conditions equivalent to the invertibility.

(8) If W_k is self-adjoint ($\Leftrightarrow \Phi$ is real $\Leftrightarrow k(t) = \overline{k(-t)}$, $t \in \mathbb{R}$), then $\sigma(W_k) = \sigma_{\text{ess}}(W_k) = [\text{ess inf}(\Phi), \text{ess sup}(\Phi)]$.

(9) If $\Phi \in H^\infty(\mathbb{C}_+) + C(\overline{\mathbb{R}})$, then $W_k \in$ Fred $\Leftrightarrow 1/\Phi \in H^\infty(\mathbb{C}_+) + C(\overline{\mathbb{R}})$; the regularizer of W_k is W_l with symbol $1/\Phi$. For a formula for the index ind(W_k), see Exercise 3.4.3.

(10) In the case where $\Phi \in PC(\overline{\mathbb{R}})$, the essential spectrum $\sigma_{\text{ess}}(W_k)$ and the spectrum $\sigma(W_k)$ can be written in the same manner as in Exercise 3.4.5 (with the aid of the extended symbol Φ_*).

4.2.4 Classical Wiener–Hopf Equations and Operators

Let $k \in L^1(\mathbb{R})$ and $\lambda \in \mathbb{C}$. The equation in question is

$$\lambda f(x) - \int_0^\infty k(x-y)f(y)\,dy = g(x), \quad x \in \mathbb{R}_+.$$

Here $g \in L^2(\mathbb{R}_+)$ is given (arbitrary), and we seek $f \in L^2(\mathbb{R}_+)$. The Fredholm theory of the corresponding operator W_k, $k = \lambda\delta_0 - k\,dx$ (by writing a pseudo-measure k as a measure; δ_0 is the unit Dirac delta mass at the origin) is in §4.2.3(4), since $\Phi = \lambda + k^\vee \in C(\overline{\mathbb{R}})$, where to abbreviate $k^\vee = 2\pi\mathcal{F}^{-1}k$,

$$k^\vee(x) = \int_\mathbb{R} k(y)\,e^{ixy}\,dy.$$

In particular, for $\lambda = 0$ we have $\Phi(\infty) = 0$ (Riemann–Lebesgue lemma), hence the equation is not Fredholm and we cannot describe the case of resolvability (and thus the range $W_k(L^2(\mathbb{R}_+))$, which is not closed) in a satisfactory manner.

4.2.5 Finite Difference Operators

Let $k = \mu \in \mathcal{M}(\mathbb{R})$ (a measure of finite variation) and $\mu = \mu_d + a\,dx$, $a \in L^1(\mathbb{R})$, where μ_d is a discrete measure,

$$\mu_d = \sum_{s \in \sigma} c_s \delta_s,$$

and $\sigma \subset \mathbb{R}$ is set at most countable, $c_s \in \mathbb{C}$ and $\sum_s |c_s| < \infty$. The operator W_μ becomes

$$W_\mu f(x) = \sum_{s \in \sigma} c_s f(x-s) + \int_0^\infty a(x-y) f(y)\, dy \quad (x \in \mathbb{R}_+),$$

and its symbol

$$\Phi(x) = \mu_d^\vee + a^\vee = \sum_{s \in \sigma} c_s e^{isx} + a^\vee(x)$$

is in $AP(\mathbb{R}) + C(\overline{\mathbb{R}})$.

If the principal part μ_d is "causal," $\mathrm{supp}(\mu_d) \subset \overline{\mathbb{R}}_+$ (see also §4.2.6 below), then $\Phi \in H^\infty(\mathbb{C}_+) + C(\overline{\mathbb{R}})$, and we can treat W_k according to §4.2.3(9); in particular, if $W_k \in \mathrm{Fred}$, there exists a Wiener–Hopf regularizer W_l with symbol $1/\Phi$.

If μ_d *is not causal*, we apply Exercise 3.4.13, and hence for $W_k \in \mathrm{Fred}$ we must have $1/\Phi \in L^\infty(\mathbb{R})$ and the mean rotation zero $w_\Phi = 0$. We then calculate the index according to the rule in Exercise 3.4.13(c):

$$\mathrm{ind}(W_k) = -\mathrm{wind}(\Phi/\mu_d^\vee).$$

However, this time, even in the simplest case, when the symbol Φ is a trigonometric polynomial, the regularizer is not necessarily a Wiener–Hopf operator. By Exercise 3.4.12(f), this anomaly already occurs for

$$\Phi = 1 + \frac{1}{4}(e^{ix} + e^{-ix}) = 1 + \frac{1}{2}\cos(x),$$

hence for $\mu = \delta_0 + \frac{1}{4}(\delta_1 + \delta_{-1})$ and for the equation (normally solvable, since $w_\Phi = 0$)

$$f(x) + \frac{1}{4}(f(x-1) + f(x+1)) = g(x), \quad x \in \mathbb{R}_+.$$

4.2.6 Operators W_μ with Causal Measures μ

A *pseudo-measure* $S \in \mathcal{PM}(\mathbb{R})$ *is said to be causal* if $\mathrm{supp}(S) \subset \overline{\mathbb{R}}_+$, which is equivalent to saying $S^\vee =: 2\pi \mathcal{F}^{-1}(S) \in H^\infty(\mathbb{C}_+)$. For operators W_k with a causal kernel perturbed by a classical term $a \in L^1(\mathbb{R})$, $k = S + a\,dx$, we can again apply the theory of Exercise 3.4.3 (and in particular, ensure the existence of a Wiener–Hopf regularizer), but in the general case – even when S is a true measure $S = \mu \in \mathcal{M}(\mathbb{R})$ – the techniques employed above are insufficient to assess the spectral nature of W_k (other than, of course, to apply the general criteria of §4.2.3(5) and §4.2.3(7)). See §4.6 for comments and supplementary information.

4.2.7 The Hilbert SIO on $L^2(\mathbb{R}_+)$

Another famous example of a singular integral operator (SIO) is the convolution with a pseudo-measure $k = p.v. (i/\pi t)$ (*the Hilbert transform on* $\mathbb{R}$), in the sense where, for every $f \in \mathcal{S}(\mathbb{R})$,

$$\langle f, k \rangle = \frac{i}{\pi} \text{p.v.} \int_{\mathbb{R}} \frac{f(t)}{t} dt =: \lim_{\epsilon \to 0} \frac{i}{\pi} \int_{\epsilon < |t| < 1/\epsilon} \frac{f(t)}{t} dt.$$

The limit exists and indeed defines a linear functional on $\mathcal{S}(\mathbb{R})$ since

$$\lim_{\epsilon \to 0} \int_{\epsilon < |t| < 1/\epsilon} \frac{f(t)}{t} dt = \lim_{\epsilon \to 0} \left(\int_{\epsilon < |t| < 1} \frac{f(t) - f(0)}{t} dt + \int_{1 < |t| < 1/\epsilon} \frac{f(t)}{t} dt \right)$$

and the last two integrals converge absolutely. The symbol of the convolution operator

$$Sf = k * f$$

is $\Phi = 2\pi \mathcal{F}^{-1} k$ defined by the duality

$$\langle \mathcal{F}^{-1} k, f \rangle = \langle k, \mathcal{F}^{-1} f \rangle \quad (\forall f \in \mathcal{S}(\mathbb{R})),$$

where

$$\langle k, 2\pi \mathcal{F}^{-1} f \rangle = \lim_{\epsilon \to 0} \frac{i}{\pi} \int_{\epsilon < |t| < 1/\epsilon} \frac{f^{\vee}(t)}{t} dt = \lim_{\epsilon \to 0} \frac{i}{\pi} \int_{\epsilon < |t| < 1/\epsilon} \frac{1}{t} \int_{\mathbb{R}} e^{ity} f(y) dy\, dt$$

$$= \lim_{\epsilon \to 0} \int_{\mathbb{R}} f(y) \frac{i}{\pi} \int_{\epsilon < |t| < 1/\epsilon} \frac{e^{ity}}{t} dt\, dy,$$

and

$$\frac{i}{\pi} \int_{\epsilon < |t| < 1/\epsilon} \frac{e^{ity}}{t} dt = \frac{i}{\pi} \int_{\epsilon}^{1/\epsilon} \frac{e^{ity} - e^{-ity}}{t} dt = \frac{-2}{\pi} \int_{\epsilon}^{1/\epsilon} \frac{\sin(ty)}{t} dt$$

$$= \frac{-2\text{sign}(y)}{\pi} \int_{\epsilon|y|}^{|y|/\epsilon} \frac{\sin(t)}{t} dt,$$

which converges (still remaining dominated) to

$$\frac{-2\text{sign}(y)}{\pi} \int_0^{\infty} \frac{\sin(t)}{t} dt = -\text{sign}(y).$$

Therefore
$$\Phi = 2\pi \mathcal{F}^{-1}k(y) = k^\vee(y) = -\text{sign}(y), \quad y \in \mathbb{R},$$
hence k is a pseudo-measure and the operator
$$\mathbb{S}f = k * f(x) = \frac{1}{\pi i}\,\text{p.v.} \int_{\mathbb{R}} \frac{f(t)}{t-x}\,dt \quad (f \in \mathcal{S}(\mathbb{R}),\ x \in \mathbb{R})$$
can be extended to a unitary operator on $L^2(\mathbb{R})$ (since $|\Phi| = 1$ a.e.). Its Wiener–Hopf counterpart is a compression of $\mathbb{S}$ onto $L^2(\mathbb{R}_+)$:
$$W_k = P^+ \mathbb{S}|L^2(\mathbb{R}_+).$$
This is a self-adjoint operator (with real symbol Φ), hence $\sigma(W_k) = \sigma_{\text{ess}}(W_k) = [-1, 1]$ (whereas on the space $L^2(\mathbb{R})$, $\sigma(\mathbb{S}) = \{-1, 1\}$).

4.3 The Matrix of W_k in the Laguerre ONB

The unitary equivalence of Corollary 4.2.3 between the Wiener–Hopf operators W_k and the Toeplitz operators T_φ, $W_k = \mathcal{U}T_\varphi\mathcal{U}^{-1}$ where $\mathcal{U} = \mathcal{F}_P U$, shows that the matrices $(T_\varphi z^j, z^k)$ and $(W_k \mathcal{L}_j, \mathcal{L}_k)$, where $\mathcal{L}_k = \mathcal{U}z^k$ ($k \geq 0$), coincide, and are Toeplitz matrices. Let us calculate the orthonormal basis $(\mathcal{L}_k)_{k \geq 0}$ of $L^2(\mathbb{R}_+)$. Recall the definition of the *Laguerre functions* l_k,
$$l_k(x) = \frac{1}{k!}\, e^{x/2}(x^k e^{-x})^{(k)}, \quad x \in \mathbb{R}_+,$$
$k = 0, 1, \ldots$ (we consider the restriction of l_k to $\mathbb{R}_+$).

Edmond N. Laguerre (1834–1886) was a French mathematician, known for his work in analysis and geometry, where he published around 140 first-class articles. In particular, he discovered the Laguerre polynomials and equation, as well as algorithms to localize roots of polynomials. Unfortunately, his innovative results in geometry were superseded by the works of Arthur Cayley and Felix Klein and by the theory of Lie groups. However, his monograph *Recherches sur la géométrie de direction. Méthode de transformation: anticaustiques* (1885) was re-issued in 1986 (Kessinger Publishing), as was *Œuvres de Laguerre* (1972; the original edition of 1905 was under the direction of Hermite, Poincaré, and Rouché). Today, Laguerre is cited especially as the discoverer of the polynomials bearing his name (see §4.3 and §4.6). In the *Complete Dictionary of Scientific Biography* (2008, www.encyclopedia.com), the entry for Laguerre (by Michael Bernkopf) concludes with:

In his short working life, actually less than twenty-two years, he produced a quantity of first-class papers. Why, then, is his name so little known and his work so seldom cited? Because as brilliant as Laguerre was, he worked only on details – significant details, yet nevertheless details ... The result is that his work has mostly come down as various interesting special cases of more general theories discovered by others.

All the biographies of Laguerre emphasize that his poor health was a curse that afflicted him from childhood throughout all his life. This explains in particular how a young prodigy, who published his first article on projective geometry (and one of the most important!) when only 19, was only ranked 46th in his class at the École Polytechnique. At 20, on leaving the École Polytechnique, he took a modest position in an artillery armaments factory where he worked for 10 years (!). He then returned to the École Polytechnique as a tutor, and then examiner, and was finally elected to the Académie des Sciences just before his premature death.

Lemma 4.3.1

(i) $(Uz^k)_{k \in \mathbb{Z}}$ is an orthonormal basis (ONB) of $L^2(\mathbb{R})$ (the Laguerre fractions) and $(Uz^k)_{k \in \mathbb{Z}_+}$ is an ONB of $H^2(\mathbb{C}_+)$,

$$Uz^k(x) =: \Lambda_k(x) =: \frac{1}{\sqrt{\pi}(x+i)} \cdot \left(\frac{x-i}{x+i}\right)^k \quad (x \in \mathbb{R}).$$

(ii) $\mathcal{F}_p^{-1} l_k(x) = i\sqrt{2} \Lambda_k(2x) (x \in \mathbb{R})$, where $(l_k)_{k \in \mathbb{Z}_+}$ is an ONB of $L^2(\mathbb{R}_+)$ (the Laguerre functions, as defined previously).

(iii) $\mathcal{L}_k(x) =: \mathcal{U}z^k(x) = -i\sqrt{2} \cdot l_k(2x) (x \in \mathbb{R}_+)$, where $(\mathcal{L}_k)_{k \in \mathbb{Z}_+}$ is an ONB of $L^2(\mathbb{R}_+)$.

Proof (i) is clear, since U is unitary.

(ii) Since $l_k \in L^2(\mathbb{R}) \cap L^1(\mathbb{R})$, $\mathcal{F}_P^{-1} l_k$ is the classical Fourier transform,

$$\mathcal{F}_P^{-1} l_k(t) = \frac{1}{\sqrt{2\pi}} \int_0^\infty \frac{1}{k!} e^{itx} e^{x/2} (x^k e^{-x})^{(k)} dx,$$

where

$$\int_0^\infty e^{(it+1/2)x} d(x^k e^{-x})^{(k-1)} = \int_0^\infty e^{(it+1/2)x} d(x^k e^{-x})^{(k-1)}$$

$$= -(it + 1/2) \int_0^\infty e^{(it+1/2)x} (x^k e^{-x})^{(k-1)} dx = \ldots$$

$$= (-1)^k (it + 1/2)^k \int_0^\infty e^{(it+1/2)x} x^k e^{-x} dx$$

$$= (-1)^k (it + 1/2)^k \int_0^\infty (it - 1/2)^{-1} x^k d(e^{(it-1/2)x})$$

$$= (-1)^k (it + 1/2)^k (-1)^k k! (it - 1/2)^{-k} \int_0^\infty e^{(it-1/2)x} dx$$

$$= -k! \left(\frac{it + 1/2}{it - 1/2}\right)^k \frac{1}{it - 1/2},$$

giving

$$\mathcal{F}_P^{-1} l_k(t) = \frac{1}{\sqrt{2\pi}} \left(\frac{2t - i}{2t + i}\right)^k \frac{2i}{2t + i} = i\sqrt{2} \Lambda_k(2t).$$

The ONB property follows from the unitary character (on $L^2(\mathbb{R})$) of the dilation $f(x) \mapsto \sqrt{\lambda} f(\lambda x)$, $\lambda > 0$.

(iii) By (i) and (ii),

$$U z^k(x) = \frac{1}{i\sqrt{2}} (\mathcal{F}_P^{-1} l_k)(x/2),$$

and since

$$(\mathcal{F}_P f(\lambda \cdot))(t) = \frac{1}{\lambda} (\mathcal{F}_P f)(t/\lambda) \quad \lambda > 0,$$

we have

$$\mathcal{U} z^k(x) = \mathcal{F}_P U z^k(x) = -i\sqrt{2} \cdot l_k(2x). \qquad \blacksquare$$

Corollary 4.3.2

(i) *Every operator $T_\Phi = P_+ \Phi | H^2(\mathbb{C}_+)$, $\Phi \in L^\infty(\mathbb{R})$ is a Toeplitz operator with respect to the ONB $(\Lambda_k)_{k \in \mathbb{Z}_+}$ of Laguerre fractions.*

(ii) *Every operator W_k, $k \in \mathcal{P}M(\mathbb{R})$ is a Toeplitz operator with respect to the ONB $(\mathcal{L}_k)_{k \in \mathbb{Z}_+}$ of dilated Laguerre functions.*

4.4 Wiener–Hopf Operators on a Finite Interval

Here, we briefly consider the operators $W_k^a \colon L^2(0,a) \to L^2(0,a)$,

$$W_k^a f(x) = \int_0^a k(x-y)f(y)\,dy \quad (x \in (0,a)),$$

where $a \in (0, \infty)$. These "truncated" (or "finite") Wiener–Hopf operators – in fact, the compressions of the Wiener–Hopf operators to $L^2(0,a)$ – are analogs of the finite Toeplitz matrices (Chapter 5). At first sight, it seems that the analysis of the W_k^a should be simpler than that of the W_k (they act on a finite interval, where the Lebesgue measure is finite, etc.), but this is a misleading impression: they are more complicated and their study is far from complete. This begins with the determination of the distributions k for which W_k^a is bounded (the answer is known, and we can say "naturally expected," but much less elementary than the corresponding theorem of §4.2.3(1) for the W_k: see Theorem 4.4.4 below). Of course, if the operator W_k is well-defined (which is not *a priori* necessary for the existence of W_k^a), then

$$W_k^a = P^a W_k | L^2(0,a),$$

where $P_a f = \chi_{(0,a)} f$ is the orthogonal projection onto $L^2(0,a)$. We begin with a series of simple observations, the first of which is the non-uniqueness of the "symbol" $\Phi = 2\pi \mathcal{F}^{-1} k(y)$.

4.4.1 Determination of the Symbol

Proposition 4.4.1 *Let $a > 0$ and $k \in \mathcal{S}'(\mathbb{R})$. The following assertions are equivalent.*

(i) *For every $f \in \mathcal{S}(0,a) =: \{f \in \mathcal{S}(\mathbb{R}) \colon \mathrm{supp}(f) \subset (0,a)\}$, we have $k * f = 0$ on $(0,a)$.*

(ii) $\mathrm{supp}(k) \subset \mathbb{R} \setminus (-a,a)$.

Moreover, if $k \in L^2(\mathbb{R})$, (i)–(ii) are equivalent to (iii):

(iii) $\Phi = 2\pi \mathcal{F}^{-1} k \in e^{iax} H^2(\mathbb{C}_+) + e^{-iax} H^2(\mathbb{C}_-)$.

Proof Indeed, for (i) $\Rightarrow$ (ii), we have $\langle \tau_x k_*, f \rangle = 0$ ($\forall f \in \mathcal{S}(0,a)$ and $\forall x \in (0,a)$), therefore $\mathrm{supp}(\tau_x k_*) \subset \mathbb{R} \setminus (0,a)$ ($\forall x \in (0,a)$), meaning (by the definition of $k_*(t) = k(-t)$) $x - \mathrm{supp}(k) \subset \mathbb{R} \setminus (0,a)$ ($\forall x \in (0,a)$), hence $\mathrm{supp}(k) \subset \mathbb{R} \setminus (-a,a)$.

To add (iii), we simply remark that

$$\mathcal{F}^{-1} L^2(a, \infty) = e^{iax} H^2(\mathbb{C}_+) \quad \text{and} \quad \mathcal{F}^{-1} L^2(-\infty, -a) = e^{-iax} H^2(\mathbb{C}_-). \quad \blacksquare$$

Proposition 4.4.2 *If $a > 0$ and $k' = k + k_0$, $\mathrm{supp}(k_0) \subset \mathbb{R} \setminus (-a, a)$, then $W_k^a = W_{k'}^a$. Moreover, W_k^a uniquely determines the restriction $k|(-a, a)$ (in the sense of distributions).*

Proof This is immediate by Proposition 4.4.1. ∎

4.4.2 W_k^a of Rank 1

Proposition 4.4.3 *Let*

$$e_\lambda(t) = e^{i\lambda t} \quad (t \in \mathbb{R})$$

where $\lambda \in \mathbb{C}$. Then, for every $a > 0$, $W_{e_\lambda}^a$ is a bounded operator of rank 1.

Proof Indeed, $W_{e_\lambda}^a f(x) = \int_0^a e^{i\lambda(x-t)} f(t)\, dt = e^{i\lambda x} \widehat{f}(\lambda)$ for every $f \in L^2(0, a)$. ∎

4.4.3 Bounding the Norm $\|W_k^a\|$ by the Best Extension

(i) If $k \in \mathcal{PM}(\mathbb{R})$, then $\|W_k^a\| \leq \|k^\vee\|_\infty$, and hence

$$\|W_k^a\| \leq \inf\{\|h^\vee\|_\infty : h \in \mathcal{PM}(\mathbb{R}), \quad h = k \quad \text{on } (-a, a)\}.$$

(ii) If $k = p\, dt$, $p \in L^1(-a, a)$, then $W_k^a \in \mathfrak{S}_\infty(L^2(0, a))$.

Indeed, (i) is clear by §4.4.2 and $\|W_k^a\| \leq \|W_k\| = \|k^\vee\|_\infty$. Then (ii) follows from $\|W_k^a - W_l^a\| \leq \|k - l\|_{L^1(-a,a)}$ where $l \in E =: \mathrm{Vect}(e_\lambda : \lambda \in \mathbb{C})$, the fact that $\mathrm{rank}(W_l^a) < \infty$ for every $l \in E$ (see §4.4.2), and $\mathrm{clos}_{L^1(-a,a)}(E) = L^1(-a, a)$ (the last equation is clear by invoking, for example, the Stone–Weierstrass theorem: E is dense in $C[-a, a]$ for the norm $\|\cdot\|_\infty$). ∎

The following theorem states that, in fact, a truncated Wiener–Hopf operator always admits a symbol in $\mathcal{PM}(\mathbb{R})$ and that the inequality of §4.4.3(i) is sharp up to a constant. For a proof, see the original publication [Rochberg, 1987]. On the other hand, the example following the theorem (in §4.4.4) shows that, in general, we cannot choose the symbol $k \in \mathcal{PM}(\mathbb{R})$ with $\mathrm{supp}(k) \subset [-a, a]$.

Theorem 4.4.4 (Rochberg, 1987) *Let $a > 0$ and $k \in \mathcal{S}'(\mathbb{R})$. W_k^a (defined initially on $\mathcal{S}(0, a)$) is bounded $L^2(0, a) \to L^2(0, a)$ if and only if there exists $h \in \mathcal{PM}(\mathbb{R})$ such that $h = k$ on $(-a, a)$ (and hence $W_k^a = W_h^a$). Moreover, the norms $\|W_k^a\|$ and of the best $\mathcal{PM}(\mathbb{R})$ extension are equivalent: there exists $c > 0$ (depending only on a) such that*

$$\inf\{\|h^\vee\|_\infty : h \in \mathcal{PM}(\mathbb{R}), \quad h = k \quad \text{on } (-a, a)\} \leq c\|W_k^a\|.$$

4.4.4 Example: An Operator W_k^a Bounded but Without Symbol $k \in \mathcal{P}M(\mathbb{R})$ With Support in $[-a, a]$

Without loss of generality, let $a = 2\pi$ and let W be a Fourier multiplier, $We_n = \lambda_n e_n$ ($e_n(t) = e^{int}$, $n \in \mathbb{Z}$). Since $(e_n)_{n\in\mathbb{Z}}$ is an orthonormal basis of $L^2(0, 2\pi)$, it is clear that W can be extended to a bounded linear operator if and only if $(\lambda_n)_{n\in\mathbb{Z}} \in l^\infty(\mathbb{Z})$. If this is the case, $S := \sum_n \lambda_n e_n$ is a 2π-periodic distribution and $Wf = f * S|(0, 2\pi)$ (for every $f \in \mathcal{S}(0, 2\pi)$). It is easy to see that $k := S|(-2\pi, 2\pi) \notin \mathcal{P}M(\mathbb{R})$. Indeed, for every $x \in \mathbb{R}$,

$$k^\vee(x) = \langle e_x, k \rangle = \left\langle e_x, \sum_n \lambda_n e_n | (-2\pi, 2\pi) \right\rangle$$

(the series $\sum_n \lambda_n e_n | (-2\pi, 2\pi)$ converges in $\mathcal{S}'(\mathbb{R})$), hence

$$k^\vee(x) = \sum_n \lambda_n \langle e_x, e_n | (-2\pi, 2\pi) \rangle = \sum_n \lambda_n \int_{-2\pi}^{2\pi} e^{i(x+n)y} \, dy$$
$$= 2 \sum_n \lambda_n \frac{\sin(2\pi(x+n))}{x+n} = 2 \sum_n \lambda_n \frac{\sin(2\pi x)}{x+n}.$$

It is easy to find a sequence $(\lambda_n)_{n\in\mathbb{Z}} \in l^\infty(\mathbb{Z})$ such that this sum is not bounded (for example, near $x = 1/4$), and hence $k \notin \mathcal{P}M(\mathbb{R})$. ∎

We conclude with a quite unexpected reduction of the Fredholm property $W_k^a \in$ Fred to the same property for a "complete" Wiener–Hopf operator W_h, but with vectorial values. This last property is discussed in Exercises 4.5.2 and 3.4.14.

Theorem 4.4.5 (Krupnik and Feldman, 1985) *Let $k \in \mathcal{P}M(\mathbb{R})$, $a > 0$ and let*

$$h = \begin{pmatrix} \delta_{-a} & 0 \\ k & \delta_a \end{pmatrix}$$

be a pseudo-measure on $\mathbb{R}$ with values in $L(\mathbb{C}^2)$. The following assertions are equivalent.

(1) $W_k^a \in \text{Fred}(L^2(0, a))$.
(2) $W_h \in \text{Fred}(L^2((0, \infty), \mathbb{C}^2))$; *the symbol (matrix-valued) of W_h is*

$$\Psi = \begin{pmatrix} e^{-iax} & 0 \\ \Phi & e^{iax} \end{pmatrix},$$

where $\Phi = 2\pi \mathcal{F}^{-1} k$ is the symbol of W_k.

Proof For two operators A and B, where the latter is a "transform" of A, $B = B(A)$, possibly acting on another Hilbert space, we write

$$A \simeq B$$

(and say "Fredholm equivalent") if $A \in$ Fred $\Leftrightarrow B \in$ Fred and, if they are Fredholm, $\text{ind}(A) = \text{ind}(B)$. We will show with several elementary steps that $W_k^a \simeq W_h$. Clearly the relation $\simeq$ is transitive.

We use the following notation: $W_{\delta_a} = V_a$, an isometry on $L^2(0, \infty)$, in fact, a translation $V_a f = \tau_a f$, satisfying $V_a^* V_a = I$, $V_a V_a^* = P_a$, an orthogonal projection on $L^2(a, \infty)$,

$$P_a f = \chi_{(a,\infty)} f.$$

The complementary projection is denoted $P^a =: I - P_a$, $P^a f = \chi_{(0,a)} f$. We have the following equivalences.

(i) $W_k^a \simeq P^a W_k P^a + P_a$, an operator $L^2(0, \infty) \to L^2(0, \infty)$.
(ii) $P^a W_k P^a + P_a \simeq W_k P^a + P_a$ since $P^a W_k P^a + P_a = (W_k P^a + P_a)(I - P_a W_k P^a)$ and the operator $I - P_a W_k P^a$ is invertible (with $I + P_a W_k P^a$ as inverse; $P^a P_a = 0$).
(iii) Writing this last operator in the form

$$W_k P^a + P_a = W_k - (W_k - I) P_a =: A - BC,$$

where $A = W_k$, $B = (W_k - I) V_a$, $C = V_a^*$ (all these operators are on $L^2(0, \infty)$), we observe that (the second operator below is on $L^2(0, \infty) \oplus L^2(0, \infty)$)

$$A - BC \simeq \begin{pmatrix} I & 0 \\ 0 & A - BC \end{pmatrix}.$$

(iv) By applying the matrix identity of Lemma 4.4.6 below, we have

$$\begin{pmatrix} I & 0 \\ 0 & A - BC \end{pmatrix} \simeq \begin{pmatrix} I & C \\ B & A \end{pmatrix},$$

where, in our case,

$$\begin{pmatrix} I & C \\ B & A \end{pmatrix} = \begin{pmatrix} I & V_a^* \\ (W_k - I) V_a & W_k \end{pmatrix} = \begin{pmatrix} V_a^* & 0 \\ W_k & V_a \end{pmatrix} \cdot \begin{pmatrix} V_a & I \\ -I & 0 \end{pmatrix} = W_h \cdot \begin{pmatrix} V_a & I \\ -I & 0 \end{pmatrix},$$

and therefore

$$\begin{pmatrix} I & C \\ B & A \end{pmatrix} \simeq W_h. \qquad \blacksquare$$

Lemma 4.4.6 *For any operators A, B, C on a Hilbert space H,*

$$\begin{pmatrix} I & C \\ B & A \end{pmatrix} = \begin{pmatrix} I & 0 \\ B & I \end{pmatrix} \cdot \begin{pmatrix} I & 0 \\ 0 & A - BC \end{pmatrix} \cdot \begin{pmatrix} I & C \\ 0 & I \end{pmatrix}.$$

Proof We use direct matrix multiplication. $\blacksquare$

4.4.5 Example: The Volterra Operator

The best-known truncated Wiener–Hopf operator is, perhaps, the *Volterra operator* $V: L^2(0,a) \to L^2(0,a)$ given by

$$Vf(x) = \int_0^x f(t)\,dt, \quad x \in (0,a).$$

For $f \in L^2(0,a)$, we have (given $f(t) = 0$ for $t < 0$)

$$f * \chi_{(0,a)}(x) = \int_{\mathbb{R}} f(t)\chi_{(0,a)}(x-t)\,dt = \int_{x-a}^x f(t)\,dt = Vf(x) \quad (0 < x < a),$$

and hence $V = W_k^a$, where $k = \chi_{(0,1)}\,dx$ is a truncated Wiener–Hopf operator with symbol

$$\Phi(x) = \int_0^a e^{itx}\,dt = \frac{1}{ix}(e^{iax} - 1) = a\,e^{iax/2}\frac{\sin(ax/2)}{(ax/2)}, \quad x \in \mathbb{R},$$

where $\Phi \in C_0(\mathbb{R})$.

It is easy to see that $V \in \mathfrak{S}_\infty$ and $\sigma(V) = \{0\}$ since the spectral radius is zero: $\lim_n \|V^n\|^{1/n} = 0$ since $V^n f = f * k_n$, $k_n(t) = t^{n-1}/(n-1)!$ (to be verified by induction on n) and thus $\|V^n\| \leq a^n/n!$. Of course, $V \notin \text{Fred}$ and hence, by Theorem 4.4.5, the operator $W_h: L^2(\mathbb{R}_+, \mathbb{C}^2) \to L^2(\mathbb{R}_+, \mathbb{C}^2)$ is not Fredholm. The convolutor h, as well as the function $\Psi = 2\pi\mathcal{F}^{-1}k$, the symbol of W_h, are defined by the formulas of Theorem 4.4.5, giving

$$\Psi \in \text{AP}(\mathbb{R}, L(\mathbb{C}^2)) + C_0(\mathbb{R}, L(\mathbb{C}^2)), \quad \det \Psi(x) = 1 \quad (\forall x \in \mathbb{R}).$$

Nevertheless, the function Ψ *is not factorizable* AP (see §4.6 for the definition), since $W_h \notin \text{Fred}$. ∎

4.5 Exercises

4.5.0 Basic Exercises: From the Hilbert Singular Operator to the Riesz Transforms ("Method of Rotation")

We consider here the singular integral operators on $\mathbb{R}^n$ of the form

$$T_\Omega f(x) = \lim_{\epsilon \to 0} \int_{|y|>\epsilon} \frac{\Omega(y/|y|)}{|y|^n} f(x-y)\,dy,$$

where

$$\Omega \in L^1(S^{n-1}, m) \quad \text{and} \quad \int_{S^{n-1}} \Omega\,dm = 0,$$

where $S^{n-1} = \{x \in \mathbb{R}^n : |x| = 1\}$ is the unit sphere and m the standard surface measure on S^{n-1} induced by the Lebesgue measure in $\mathbb{R}^n$ (for example, in

208   Applications: Riemann–Hilbert, Wiener–Hopf, SIO

spherical coordinates $(\theta_1, \ldots, \theta_{n-1})$, $\theta_j \in (0, \pi)$ $(1 \le j < n - 1)$, $\theta_{n-1} \in (0, 2\pi)$, the density of m is $\prod_{j=1}^{n-2} S\sin^{n-1-j}(\theta_j)$). Such a $\lim_{\epsilon \to 0}$ is called the "principal value integral"; using the abbreviation $y' = y/|y|$ for the projection on the unit sphere, it is denoted by

$$T_\Omega f(x) = \text{p. v.} \int_{\mathbb{R}^n} \frac{\Omega(y')}{|y|^n} f(x - y)\, dy.$$

In particular, for $n = 2$ we have $S^1 = \mathbb{T}$, and for $n = 1$ we have $S^0 = \{-1, 1\}$. In this last case $T_\Omega = c\mathbb{S}$, where $c = \Omega(1)\pi/i$ and $\mathbb{S}$ is the *Hilbert SIO* defined and studied above in §4.2.7. As in §4.2, we will need the language of Schwartz distributions: the class $\mathcal{S}(\mathbb{R}^n)$ of rapidly decreasing C^∞ functions, its dual, the space $\mathcal{S}'(\mathbb{R}^n)$ of tempered distributions, as well as their standard elementary properties. In particular, every distribution $a \in \mathcal{S}'(\mathbb{R}^n)$ operates on $\mathcal{S}(\mathbb{R}^n)$ as a convolution operator, $f \in \mathcal{S}(\mathbb{R}^n) \Rightarrow a * f \in \mathcal{S}(\mathbb{R}^n)$, where $a * f(x) = \langle (f^\vee)_x, a \rangle$, $f^\vee(t) = f(-t)$, $f_x(t) = f(t - x)$ (hence $(f^\vee)_x(y) = f(x - y)$).

(a) Convolution with a tempered distribution. *Let Ω be as above. Show that for every $f \in \mathcal{S}(\mathbb{R}^n)$, the limit*

$$\langle f, a \rangle =: \lim_{\epsilon \to 0} \int_{|y| > \epsilon} \frac{\Omega(y')}{|y|^n} f(y)\, dy$$

exists and defines a distribution a denoted $a = \text{p. v.}\, \Omega/|x|^n \in \mathcal{S}'(\mathbb{R}^n)$.

SOLUTION: Since $\int \Omega\, dm = 0$, for every $\epsilon > 0$, we have

$$\int_{\epsilon < |y| < 1} \frac{\Omega(y')}{|y|^n}\, dy = 0,$$

hence

$$\lim_{\epsilon \to 0} \int_{|y| > \epsilon} \frac{\Omega(y')}{|y|^n} f(y)\, dy = \lim_{\epsilon \to 0} \int_{1 > |y| > \epsilon} \frac{\Omega(y')}{|y|^n} (f(y) - f(0))\, dy + \int_{|y| > 1} \frac{\Omega(y')}{|y|^n} f(y)\, dy$$

$$= \int_{|y| < 1} \frac{\Omega(y')}{|y|^n} (f(y) - f(0))\, dy + \int_{|y| > 1} \frac{\Omega(y')}{|y|^n} f(y)\, dy,$$

and the limit exists because the two last integrals converge absolutely. Clearly they define continuous functionals on $\mathcal{S}'(\mathbb{R}^n)$.

(b) Directional operators on $\mathbb{R}^n$. Let

$$R \colon L^p(\mathbb{R}) \to L^p(\mathbb{R})$$

be a bounded linear operator such that $R\mathcal{S}(\mathbb{R}) \subset C(\mathbb{R})$, and $u \in S^{n-1}$. Given $x \in \mathbb{R}^n$ and a function $f \in \mathcal{S}(\mathbb{R}^n)$, the *trace of f on the line passing through x in the direction u* is defined by the equality

$$f^u(t) = f(x + (t - (x, u))u) = f(tu + \overline{x}), \quad t \in \mathbb{R},$$

where $\bar{x} \perp u$. A *directional operator (in the direction u)* is then defined by

$$R^u f(x) = R f^u((x, u)), \quad x \in \mathbb{R}^n.$$

(1) *Show that $\|R^u f\|_p \le \|R\| \cdot \|f\|_p$ for every f, and hence R^u can be extended by continuity to a mapping $L^p(\mathbb{R}^n) \to L^p(\mathbb{R}^n)$ with $\|R^u\| \le \|R\|$.*

Hint Use Fubini's theorem.

SOLUTION: Let $x = (x, u)u + \bar{x}$ where $\bar{x}$ is a portion of x in the orthogonal complement, $\bar{x} \in \{u\}^\perp$. Then, by Fubini,

$$\int_{\mathbb{R}^n} |R^u f(x)|^p \, dx = \int_{\{u\}^\perp} \int_{\mathbb{R}} |R f^u(tu + \bar{x})|^p \, dt \, d\bar{x}$$

$$\le \int_{\{u\}^\perp} \|R\|^p \int_{\mathbb{R}} |f(tu + \bar{x})|^p \, dt \, d\bar{x} = \|R\|^p \int_{\mathbb{R}^n} |f(x)|^p \, dx.$$

(2) *Deduce that for every $\Omega \in L^1(S^{n-1})$ the operator R_Ω,*

$$R_\Omega f(x) = \int_{S^{n-1}} \Omega(u) R^u f(x) \, dm(u),$$

is bounded $L^p(\mathbb{R}^n) \to L^p(\mathbb{R}^n)$ and $\|R_\Omega\| \le \|\Omega\|_1 \|R\|$.

SOLUTION: Clearly

$$\|R_\Omega\| \le \int_{S^{n-1}} |\Omega(u)| \cdot \|R^u\| \, dm(u) \le \|\Omega\|_1 \cdot \|R\|.$$

(3) *Reminder: the Hilbert singular operator is defined in §4.2.7 by*

$$\mathbb{S} f = \text{p. v.} \frac{i}{\pi x} * f,$$

where the kernel $1/x$ can also be written as

$$\frac{\Omega(x')}{|x|}, \quad \Omega(\pm 1) = \pm 1.$$

In §4.2.7 we showed that $\mathbb{S}$ is bounded (and even unitary) on the space $L^2(\mathbb{R})$, but for its continuity on $L^p(\mathbb{R})$, $1 < p < \infty$, we refer to Nikolski (2019, Chapter 2, Exercise 2.8.4), or Duoandikoetxea (2001).

(c) Singular operators with odd kernels. *Let*

$$T_\Omega f(x) = \text{p. v.} \int_{\mathbb{R}^n} \frac{\Omega(y')}{|y|^n} f(x - y) \, dy,$$

where $\Omega \in L^1(S^{n-1}, m)$ is an odd function. With the notation of (b) above, show that

$$T_\Omega = \frac{\pi}{2i} R_\Omega$$

where $R = \mathbb{S}$ is the Hilbert SIO of §4.2.7, and consequently, $T_\Omega: L^p(\mathbb{R}^n) \to L^p(\mathbb{R}^n)$, $1 < p < \infty$, is bounded, and

$$\|T_\Omega\| \leq \frac{\pi}{2} \|\Omega\|_1 \|\mathbb{S}\|_{L^p \to L^p}.$$

Hint Integrate using polar coordinates $y = ru$ ($r > 0$, $u \in S^{n-1}$), $dy = r^{n-1} \, dr \, dm(u)$.

SOLUTION: For $f \in \mathcal{S}(\mathbb{R}^n)$ and $x \in \mathbb{R}^n$, we have

$T_\Omega f(x)$

$= \lim_{\epsilon \to 0} \int_{|y| > \epsilon} \frac{\Omega(y')}{|y|^n} f(x - y) \, dy$

$= \lim_{\epsilon \to 0} \int_{S^{n-1}} \Omega(u) \int_{r > \epsilon} f(x - ru) \frac{dr}{r} \, dm(u)$ (and since $\Omega(-u) = -\Omega(u)$)

$= \lim_{\epsilon \to 0} \frac{1}{2} \int_{S^{n-1}} \Omega(u) \int_{|r| > \epsilon} f(x - ru) \frac{dr}{r} \, dm(u)$ (and since $\int_{1 > |r| > \epsilon} f(x) \frac{dr}{r} = 0$)

$= \lim_{\epsilon \to 0} \frac{1}{2} \int_{S^{n-1}} \Omega(u) \int_{1 > |r| > \epsilon} (f(x - ru) - f(x)) \frac{dr}{r} \, dm(u)$

$\quad + \frac{1}{2} \int_{S^{n-1}} \Omega(u) \int_{|r| > 1} f(x - ru) \frac{dr}{r} \, dm(u).$

Note that here we can pass to the limit under the integral sign because the inner integrals are bounded on S^n.

Moreover, by §4.2.7,

$$\mathbb{S}^u f(x) = \frac{i}{\pi} \text{p. v.} \int_\mathbb{R} \frac{f(su + \bar{x})}{t - s} \, ds = \frac{i}{\pi} \text{p. v.} \int_\mathbb{R} \frac{f(x - ru)}{r} \, dr,$$

which coincides up to a multiplicative constant (by (a) above) with

$$\lim_{\epsilon \to 0} \frac{1}{2} \int_{1 > |r| > \epsilon} (f(x - ru) - f(x)) \frac{dr}{r} + \frac{1}{2} \int_{|r| > 1} f(x - ru) \frac{dr}{r}.$$

Consequently,

$$R_\Omega f(x) = \frac{\pi}{2i} \int_{S^{n-1}} \Omega(u) \mathbb{S}^u f(x) \, dm(u).$$

NOTE According to Pichorides (1972, see Notes and Remarks), for $1 < p < \infty$, we have

$$\|\mathbb{S}\|_{L^p(\mathbb{R}) \to L^p(\mathbb{R})} = \max(\tan(\pi/2p), \cot(\pi/2p));$$

see Grafakos [2008] for original references and a fairly short proof. If $|\Omega| \leq 1$, then

$$\|\Omega\|_1 \leq m(S^{n-1}) = \frac{2\pi^{n/2}}{\Gamma(n/2)},$$

where Γ is the Euler gamma function.

(d) Example. The Riesz transforms (1949) are the operators $R_j = T_{\Omega_j}$, $j = 1, 2, \ldots, n$, where

$$\Omega_j(y) = c_n y_j/|y|, \quad y \in \mathbb{R}^n,$$

$$R_j f(x) = \text{p.v.} \, c_n \int_{\mathbb{R}^n} \frac{y_j}{|y|^{n+1}} f(x-y) \, dy$$

with the normalizing constant $c_n = \Gamma((n+1)/2)\pi^{-(n+1)/2}$.

Show that the operators $R_j \colon L^p(\mathbb{R}^n) \to L^p(\mathbb{R}^n)$ are bounded for $1 < p < \infty$, and that

$$\|R_j\| \leq \frac{\pi}{2}\|\Omega_j\|_1 \|\mathbb{S}\|_{L^p \to L^p} \leq \|\mathbb{S}\|_{L^p \to L^p}.$$

SOLUTION: The continuity of the R_j and the first inequality follow from (c), and for the computation of

$$\|\Omega_j\|_1 = c_n \int_{S^{n-1}} |y_j| \, dm(y) = c_n \frac{2\pi^{(n-1)/2}}{\Gamma((n+1)/2)} = 2/\pi,$$

see Stein [1993] or Grafakos [2008].

NOTE On the norm $\|\mathbb{S}\|_{L^p(\mathbb{R}) \to L^p(\mathbb{R})}$ see (c) above (in particular, $\|\mathbb{S}\|_{L^2(\mathbb{R}) \to L^2(\mathbb{R})} = 1$). The choice of the normalizing constant is motivated by the fact that R_j is a convolution $R_j f = r_j * f$ where $r_j(y) = \Omega_j(y)/|y|^n$ ($y \in \mathbb{R}^n$) and the Fourier transform $\widehat{r_j}$ is $\widehat{r_j}(x) = ix_j/|x|$ (see Stein [1993], Duoandikoetxea [2001], or Grafakos [2008]), and hence

$$\widehat{(R_j f)}(x) = i\frac{x_j}{|x|}\widehat{f}(x), \quad (x \in \mathbb{R}^n);$$

in particular, $\sum_{j=1}^n R_j^2 = -I$. Many important applications of the Riesz transforms R_j (in particular, to PDEs) can be found in the sources mentioned above.

4.5.1 Sokhotsky–Plemelj Formulas

Let $\mathbb{S}$ be the complex Cauchy singular integral,

$$\mathbb{S}f(z) = \text{p.v.} \int_\mathbb{T} \frac{f(\zeta)}{\zeta - z} \frac{d\zeta}{\pi i}, \quad z \in \mathbb{T},$$

defined a.e. on $\mathbb{T}$ ($f \in L^2(\mathbb{T})$): see §4.1.3 and Corollary 4.1.4. Show that

$$\mathbb{S} = P_+ - P_-,$$

in other words

$$P_+ = \frac{1}{2}(id + \mathbb{S}), \quad P_- = \frac{1}{2}(id - \mathbb{S})$$

(formulas of Sokhotsky (1873) and Plemelj (1908)).

Josip Plemelj (1873–1967), an Austro-Hungarian (under the Habsburgs)/Yugoslav/Slovenian mathematician, worked in complex analysis, differential equations, and potential theory. After submitting his thesis at the University of Vienna (1898), he continued his studies in Berlin and Göttingen (1899–1901) where, under the influence of David Hilbert and Felix Klein, he was one of the first to apply the integral equations (operators) to the harmonic functions of potential theory and to the Riemann–Hilbert problem. For potential theory, his approach is presented in his book *Potentialtheoretische Untersuchungen* (Leipzig, 1911), but today his name is particularly associated with the Sokhotsky–Plemelj formula of Exercise 4.5.1 and then with the partial solution of Hilbert's Problem 21 in [Plemelj, 1908b] (discussed in Chapter 1 (see in particular Hilbert's recollections cited in §1.1.2, page 11) and in §3.5 and the comments on Exercise 3.4.14). For many decades his 1908 result was considered a complete solution of the problem of Fuchsian groups, until the appearance of the counterexamples of Andrei Bolibruch (1989) showing that the hypothesis of diagonalizability of the Fuchsian matrix (as is the case with Plemelj) is essential.

Plemelj played an important role in the development of mathematics in Slovenia (including the creation of a Slovenian mathematical terminology) and in the (re)foundation of the University of Ljubljana (1919), where he became the first Rector. His last book, *Problems in the Sense of*

> *Riemann and Klein* (London, 1964), contains a survey of his mathematical contributions. Plemelj received numerous distinctions in the course of his career, and bequeathed his villa in Bled to the Association of Mathematicians, Physicists and Astronomers of Slovenia, who made it into a museum devoted to his memory.

Hint Follow the method of Theorem 4.1.2.

SOLUTION: (Privalov, 1919) As in Theorem 4.1.2, compare $P_+f(r\zeta)$ and $S_{1-r}f(\zeta)$, where $\zeta \in \mathbb{T}$ is a Lebesgue point of f and

$$S_\epsilon f(\zeta) = \int_{\mathbb{T}\setminus\{|\zeta-u|<\epsilon\}} \frac{f(u)}{u-\zeta}\frac{du}{\pi i};$$

up to performing a rotation, we can assume that $\zeta = 1$. We have

$$P_+f(r) - \frac{1}{2}S_{1-r}f(1)$$

$$= \int_{\mathbb{T}} \frac{f(u)}{u-r}\frac{du}{2\pi i} - \int_{\{|1-u|>1-r\}} \frac{f(u)}{u-1}\frac{du}{2\pi i}$$

$$= \frac{1}{2\pi i}\Bigl(\int_{\{|1-u|>1-r\}} f(u)\Bigl\{\frac{1}{u-r} - \frac{1}{u-1}\Bigr\}du + \int_{|1-u|<1-r} \frac{f(u)}{u-r}du\Bigr)$$

$$=: I_1(f,r) + I_2(f,r).$$

First assume that $f(1) = 0$ (recall that at a Lebesgue point, an integrable function has a well-defined value). Then

$$|I_1| \le \int_{\{|1-u|>1-r\}} |f(u)| \cdot \Bigl|\frac{1}{u-r} - \frac{1}{u-1}\Bigr| dm(u)$$

$$= \int_{\{|1-u|>1-r\}} |f(u)| \cdot \frac{1-r}{|u-r|\cdot|u-1|} dm(u)$$

(taking into account $|u-r| \le |u-1| + 1 - r \le 2|u-1|$)

$$\le \int_{\{|1-u|>1-r\}} |f(u)| \cdot \frac{2(1-r)}{|u-r|^2} dm(u)$$

$$\le \int_{\mathbb{T}} |f(u)| \cdot \frac{2(1-r^2)}{|u-r|^2} dm(u),$$

which tends to $2|f(1)| = 0$ ($\zeta = 1$ is a Lebesgue point of f). For the integral I_2, we have

$$|I_2| \le \frac{1}{1-r}\int_{|1-u|<1-r} |f(u)| dm(u) \to 0 \quad (\text{as } r \to 1)$$

(Lebesgue point). In the general case, we have

$$P_+f(r) - \frac{1}{2}S_{1-r}f(1)$$
$$= P_+(f - f(1))(r) + f(1) - \frac{1}{2}S_{1-r}(f - f(1))(1) - \frac{1}{2}S_{1-r}(f(1))(1)$$
$$=: J_1 + f(1) - J_2 - J_3,$$

where $J_1 - J_2 \to 0$ as $r \to 1$ (by the above) and for J_3 (with suitable $\epsilon = \epsilon(r) > 0$ satisfying $\lim_{r \to 1} \epsilon(r) = 0$),

$$J_3 = f(1)\frac{1}{2\pi i}\int_{\{|1-u|>1-r\}}\frac{du}{u-1} = \frac{f(1)}{2\pi i}\int_{(-\pi,\pi)\setminus(-\epsilon,\epsilon)}\frac{i e^{it} dt}{e^{it} - 1}$$
$$= \frac{f(1)}{2\pi}\int_{(-\pi,\pi)\setminus(-\epsilon,\epsilon)}\frac{1 - e^{it}}{|e^{it} - 1|^2}dt = \frac{f(1)}{\pi}\int_{(0,\pi)\setminus(0,\epsilon)}\frac{1 - \cos(t)}{|e^{it} - 1|^2}dt$$
$$= \frac{f(1)}{\pi}\int_{(0,\pi)\setminus(0,\epsilon)}\frac{1 - \cos(t)}{2(1 - \cos(t))}dt = f(1)\frac{\pi - \epsilon}{2\pi},$$

which tends to $f(1)/2$, and hence

$$\lim_{r \to 1}(P_+f(r) - \frac{1}{2}S_{1-r}f(1)) = f(1)/2,$$

which completes the proof.

Remark A consequence (already known from the theory of Hardy spaces) – in a classical reading: the jump of the complex Cauchy integral

$$Cf(z) =: \frac{1}{2\pi i}\int_{\mathbb{T}}\frac{f(\zeta)\,d\zeta}{\zeta - z}, \quad z \in \mathbb{C}\setminus\mathbb{T},$$

when traversing the boundary of the domain $\mathbb{D}$ from the exterior to the interior is equal to the density of the integral in question,

$$\lim_{\epsilon \to 0}Cf((1-\epsilon)z) - \lim_{\epsilon \to 0}Cf((1+\epsilon)z) = f(z) \quad \text{(a.e. } z \in \mathbb{T}\text{).}$$

In fact, in this form, the fact was already known to Pierre Fatou [1906]; see [Nikolski, 2019] for details.

4.5.2 Systems of Equations and Matrix Wiener–Hopf Operators

Given pseudo-measures $k_{ij} \in \mathcal{PM}(\mathbb{R})$ ($1 \leq i, j \leq N$), we consider the system of equations

$$\sum_{j=1}^{n}W_{k_{ij}}f_j = g_i \quad (1 \leq i \leq N),$$

where $f_j \in L^2(\mathbb{R}_+)$ are functions to be found and $g_i \in L^2(\mathbb{R}_+)$ are given. The system can naturally be written with a *matricial Wiener–Hopf operator* W_k, $k = (k_{ij})_{1 \leq i,j \leq N}$,

$$W_k f = g,$$

where $f = (f_j)^{\text{col}}$, $g = (g_i)^{\text{col}}$ are in the space $L^2(\mathbb{R}_+, \mathbb{C}^N)$ (functions of class L^2 on $\mathbb{R}_+$ with values in $\mathbb{C}^N$). The rewriting in terms of the matrix-valued symbol Φ is

$$W_k f = \mathcal{F}_P^{-1} P_+(\Phi \mathcal{F}_P f),$$

where $\Phi = k^\vee = (k_{ij}^\vee) \in L^\infty(\mathbb{R}, L(\mathbb{C}^N))$ and P_+ is the orthogonal projection of $L^2(\mathbb{R}, \mathbb{C}^N)$ onto $H^2(\mathbb{C}_+, \mathbb{C}^N)$. It is left to the reader to transfer the basic properties of matrix-valued Toeplitz operators (see Exercises 2.3.11 and 3.4.14) to the operators $T_\Phi f = P_+(\Phi f)$ on the space $H^2(\mathbb{C}_+, \mathbb{C}^N)$ (with the aid of a unitary transform $U_N = U \oplus \cdots \oplus U$), but we present a few consequences for the W_k as exercises.

(a) *Let $k = (k_{ij})_{1 \leq i,j \leq N} \in \mathcal{S}'(\mathbb{R}, L(\mathbb{C}^N))$. Show that W_k, where $W_k f = P^+(k * f)$ for $f \in \mathcal{S}(\mathbb{R})$ having support in $\mathbb{R}_+$ (supp$(f) \subset \mathbb{R}_+$), can be extended to a bounded operator $L^2(\mathbb{R}_+, \mathbb{C}^N) \to L^2(\mathbb{R}_+, \mathbb{C}^N)$ if and only if $\Phi \in L^\infty(\mathbb{R}, L(\mathbb{C}^N))$ where $\Phi = k^\vee$ (hence, if and only if $k \in \mathcal{PM}(\mathbb{R}, L(\mathbb{C}^N))$), and then*

$$\|W_k\| = \|\Phi\|_\infty =: \text{ess sup}_{\mathbb{R}} \|\Phi(x)\|_{L(\mathbb{C}^N)}.$$

Hint Follow the correspondences of Theorems 4.2.1 and 4.2.2 and apply the theorem of Exercise 2.3.11(b)) to $T_\varphi = \mathcal{U}^{-1} W_k \mathcal{U}$ (where $\mathcal{U} = \mathcal{F}_P U$, $\varphi = \Phi \circ \omega^{-1}$).

(b) *Let $k \in \mathcal{PM}(\mathbb{R}, L(\mathbb{C}^N))$ such that $\Phi = k^\vee \in C(\overline{\mathbb{R}}, L(\mathbb{C}^N))$. Show that*
(i) $W_k \in \text{Fred}(L^2(\mathbb{R}_+, \mathbb{C}^N)) \Leftrightarrow \Phi^{-1} \in C(\overline{\mathbb{R}}, L(\mathbb{C}^N)) \Leftrightarrow \det \Phi(x) \neq 0 \; (\forall x \in \overline{\mathbb{R}})$,
(ii) *if $W_k \in \text{Fred}$, then W_l, where $l^\vee = \Phi^{-1}$, is a regularizer of W_k ($\Phi^{-1}(\zeta) = (\Phi(\zeta))^{-1}$).*

Hint This is the analog of Exercise 3.4.14(d) with the aid of $T_\varphi = \mathcal{U}^{-1} W_k \mathcal{U}$ where $\mathcal{U} = \mathcal{F}_P U$, $\varphi = \Phi \circ \omega^{-1}$.

Remark Here, the same remark as in Exercise 3.4.14(d) is applicable: the formula of the index $\text{ind}(W_k) = -\text{wind}(\det(\Phi(\cdot)))$ holds, but the known proofs are more laborious: see [Krupnik, 1984], §1.3.

(c) *Deduce from Exercise 3.4.14(e) its Wiener–Hopf analog: if*

$$k \in \mathcal{PM}(\mathbb{R}, L(\mathbb{C}^N))$$

such that

$$\Phi = k^\vee \in H^\infty(\mathbb{C}_+, L(\mathbb{C}^N)) + C(\overline{\mathbb{R}}, L(\mathbb{C}^N)),$$

then

$$W_k \in \text{Fred}(L^2(\mathbb{R}_+, \mathbb{C}^N)) \Leftrightarrow \Phi^{-1} \in H^\infty + C,$$

and if $W_k \in$ Fred, the operator W_l, $l^\vee = \Phi^{-1}$ is a regularizer of W_k, and $\text{ind}(W_k) = -\text{wind}(\det(\Phi(\cdot)))$.

4.5.3 Hankel Operators on $H^2(\mathbb{C}_+)$ and $L^2(\mathbb{R}_+)$

This exercise contains a sketch of the theory of Hankel integral operators. From a naive point of view, a Hankel operator on $L^2(\mathbb{R}_+)$ is an integral operator whose kernel depends on the sum of variables,

$$f \mapsto \int_0^\infty N(x+y) f(y)\, dy, \quad x \in \mathbb{R}_+.$$

In reality we consider kernels that are pseudo-measures in the style of §4.2.

(a) Characterization-definition of Hankel operators. Let $\mathcal{G}: L^2(\mathbb{R}_+) \to L^2(\mathbb{R}_+)$ be a bounded operator, and $\mathcal{J}: L^2(\mathbb{R}) \to L^2(\mathbb{R})$ an involution $\mathcal{J}f(x) = f(-x)$. Show that the following assertions are equivalent.

(i) There exists a pseudo-measure $k \in \mathcal{PM}(\mathbb{R})$ such that $\mathcal{G} = \mathcal{G}_k$, where

$$\mathcal{G}_k f = \mathcal{J}(k * f)|_{\mathbb{R}_+} \quad (f \in L^2(\mathbb{R}_+)).$$

(ii) For every $t > 0$, $\tau_t^* \mathcal{G} = \mathcal{G}\tau_t$, where $\tau_t: L^2(\mathbb{R}_+) \to L^2(\mathbb{R}_+)$, $\tau_t f(x) = f(x-t)$ (recall that every function $f \in L^2(\mathbb{R}_+)$ is zero on $\mathbb{R}_-$).

(iii) There exists a function $\Phi \in L^\infty(\mathbb{R})$ such that $\mathcal{G} = \mathcal{F}_P G_\Phi \mathcal{F}_P^{-1}$ where $G_\Phi = \mathcal{J}P_- M_\Phi|H^2(\mathbb{C}_+)$.

(iv) There exists a function $\varphi \in L^\infty(\mathbb{T})$ such that $\mathcal{G} = \mathcal{U}\Gamma_\varphi \mathcal{U}^{-1}$ where $\mathcal{U} = \mathcal{F}_P U$.

If (i)–(iv) hold, then $\varphi = -\Phi \circ \omega^{-1}$ $(\omega^{-1}(z) = i((1+z)/(1-z)))$, $\Phi = k^\vee$.

Hint Proceed as in Theorems 4.2.1 and 4.2.2.

SOLUTION: For (i) $\Leftrightarrow$ (iii), we have for every $f \in \mathcal{S}(\mathbb{R})$ with support in $\mathbb{R}_+$ (using $P_+\mathcal{J} = \mathcal{J}P_-$),

$$\mathcal{F}_P^{-1} \mathcal{G}_k \mathcal{F}_P(\mathcal{F}_P^{-1}f) = (\mathcal{F}_P^{-1} P^+ \mathcal{F}_P)(\mathcal{F}_P^{-1}\mathcal{J}F_P)\mathcal{F}_P^{-1}(k*f) = P_+\mathcal{J}(\Phi \cdot \mathcal{F}_P^{-1}f)$$

$$= \mathcal{J}P_-(\Phi \cdot \mathcal{F}_P^{-1}f) = G_\Phi(\mathcal{F}_P^{-1}f),$$

where $\Phi = 2\pi \mathcal{F}^{-1}k$ (in the sense of distributions), showing the equivalence. For (iii) $\Leftrightarrow$ (iv), it remains to transfer Γ_φ to G_Φ. Using $U(\varphi f) = (Uf) \cdot \varphi \circ \omega$ mentioned in §4.2.1, a temporary notation $\mathcal{J}^T h = \bar{z}h(\bar{z})$, $h \in L^2(\mathbb{T})$, and the identity (to be verified) $U\mathcal{J}^T = -\mathcal{J}U$, we obtain

$$-U\Gamma_\varphi U^{-1} = -U\mathcal{J}^T U^{-1} U P_-^T M_\varphi U^{-1} = \mathcal{J} U P_-^T U^{-1} U M_\varphi U^{-1}$$
$$= \mathcal{J} P_- M_\Phi = G_\Phi.$$

To link (ii) with (i), (iii), (iv), we apply Lemma 2.2.1 and Theorem 2.2.4 (which characterize $\Gamma = \Gamma_\varphi$ as operators satisfying $M_z^* \Gamma = \Gamma M_z$), then the reasoning of Theorem 4.2.1 (which shows $M_z^* \Gamma = \Gamma M_z \Leftrightarrow M_{V_s}^* \Gamma = \Gamma M_{V_s}$ ($\forall s > 0$), where $V_s = \exp(-s((1+z)/(1-z)))$ and its transform in $H^2(\mathbb{C}_+)$: $G = G_\Phi \Leftrightarrow M_{e_s}^* G = GM_{e_s}$, where $e_s(x) = e^{isx}$ ($x \in \mathbb{R}$)), and finally the equation $\tau_s \mathcal{F}_P = \mathcal{F}_P M_{e_s}$! These computations are the same as those already made in Theorems 4.2.1 and 4.2.2.

(b) *If $k \in \mathcal{PM}(\mathbb{R})$ is locally integrable, then*

$$\mathcal{G}_k f(x) = \int_{\mathbb{R}_+} k(-x-y) f(y)\, dy, \quad x \in \mathbb{R}_+ \quad (\forall f \in \mathcal{S}(\mathbb{R}),\ \mathrm{supp}(f) \subset \mathbb{R}_+).$$

(c) *Show that*

$$\|\mathcal{G}_k\| = \mathrm{dist}_{L^\infty(\mathbb{R})}(k^\vee, H^\infty(\mathbb{C}_+)),$$
$$\|\mathcal{G}_k\|_{ess} = \mathrm{dist}_{L^\infty(\mathbb{R})}(k^\vee, H^\infty(\mathbb{C}_+) + C(\overline{\mathbb{R}})).$$

In particular, $\mathcal{G}_k \in \mathfrak{S}_\infty(L^2(\mathbb{R}_+)) \Leftrightarrow k^\vee \in H^\infty(\mathbb{C}_+) + C(\overline{\mathbb{R}})$.

SOLUTION: This is immediate according to $\mathcal{G} = \mathcal{U}\Gamma_\varphi \mathcal{U}^{-1}$ of (a) and Theorems 2.2.4 and 2.2.8.

(d) *Show that the matrix of every Hankel operator $\mathcal{G}_k$ with respect to the basis of dilated Laguerre functions $(\mathcal{L}_k)_{k \in \mathbb{Z}_+}$ is a Hankel matrix.*

SOLUTION: This is immediate according to $\mathcal{G} = \mathcal{U}\Gamma_\varphi \mathcal{U}^{-1}$ of (a) and to §4.3, especially Lemma 4.3.1.

(e) The Hilbert operator

$$\mathcal{G}_k f(x) = \int_{\mathbb{R}_+} \frac{f(y)}{x+y}\, dy, \quad x \in \mathbb{R}_+.$$

Show that $\mathcal{G}_k$ is bounded on $L^2(\mathbb{R}_+)$, and $\|\mathcal{G}_k\| = \|\mathcal{G}_k\|_{ess} = \pi$.

Hint Use §4.2.7.

SOLUTION: By §4.2.7, for $k = \mathrm{p.v.}(i/x)$ we have $\Phi = 2\pi \mathcal{F}^{-1}k(y) = k^\vee(y) = \pi\, \mathrm{sign}(y)$, $y \in \mathbb{R}$, and by the same Lindelöf theorem (see Appendix F.3) used in Exercise 3.4.10(c), $\mathrm{dist}_{L^\infty(\mathbb{R})}(k^\vee, H^\infty(\mathbb{C}_+) + C(\overline{\mathbb{R}})) = \pi$.

Remark The equality $\|\mathcal{G}_k\| = \pi$ (the important point of (e)) is also an easy consequence of the famous *Schur test* (see Appendix D). Indeed, if $f(x) = x^{-1/2}$, $x > 0$, then

$$\int_{\mathbb{R}_+} \frac{f(y)}{x+y}\, dy = \int_{\mathbb{R}_+} \frac{2}{x+t^2}\, dt = \frac{2}{\sqrt{x}}\left[\arctan\left(\frac{t}{\sqrt{x}}\right)\right]_{t=0}^{t=\infty} = \frac{\pi}{\sqrt{x}},$$

and the result follows. ∎

4.5.4 Laguerre Polynomials

Let

$$l_k(x) = \frac{1}{k!} e^{x/2} (x^k e^{-x})^{(k)}, \quad x \in \mathbb{R}_+ \quad (k = 0, 1, \ldots)$$

be the Laguerre functions.

(a) *Show that* $l_k(x) e^{x/2} =: L_k(x)$ *is a polynomial of degree k and of leading coefficient $(-1)^k/k!$ (kth Laguerre polynomial).*

SOLUTION: The Leibniz formula for $(fg)^{(k)}$.

(b) *Show that the Laguerre functions l_k ($k = 0, 1, \ldots$) form an orthonormal basis in $L^2(\mathbb{R}_+)$.*

SOLUTION: *First solution.* This follows directly from Lemma 4.3.1.

Second solution (independent of U and $\mathcal{F}_P$). First the orthogonality: supposing $n \leq k$, we integrate by parts in the scalar product

$$(l_n, l_k) = \int_0^\infty \frac{1}{k!}(x^k e^{-x})^{(k)} L_n(x)\, dx = -\int_0^\infty \frac{1}{k!}(x^k e^{-x})^{(k-1)} L_n'(x)\, dx = \cdots$$

$$= (-1)^k \int_0^\infty \frac{1}{k!} x^k e^{-x} L_n^{(k)}(x)\, dx = \delta_{nk},$$

which is clear for $n < k$ ($\deg(L_n) = n$), and for $n = k$, it follows (by (a)) that

$$(-1)^k \int_0^\infty \frac{1}{k!} x^k e^{-x} (-1)^k dx = k \int_0^\infty \frac{1}{k!} x^{k-1} e^{-x}\, dx = \cdots = 1.$$

And now the completeness: let $f \in L^2(\mathbb{R}_+)$ be such that $(f, l_k) = 0 \ \forall k \geq 0$. Given (a), we obtain

$$\int_0^\infty f(x) x^n e^{-x/2}\, dx = 0, \quad n = 0, 1, \ldots.$$

Denoting $F(z) = \int_0^\infty f(x) e^{-xz}\, dx$, $\operatorname{Re}(z) > 0$, we obtain a holomorphic function in $\{z: \operatorname{Re}(z) > 0\}$ such that $F^{(n)}(1/2) = 0$ ($\forall n \geq 0$), so $F = 0$, and hence f is orthogonal to every function $e_\lambda(x) = e^{-\lambda x}$, $\operatorname{Re}(\lambda) > 0$. However, the linear hull $\operatorname{Vect}(e_\lambda: \operatorname{Re}(\lambda) > 0)$ is a sub-algebra of $C_0[0, \infty)$ dense in this space (for the uniform norm $\|\cdot\|_\infty$) by the Stone–Weierstrass theorem. Hence $(f, e^{-x}h) = 0$ for every $h \in C_0[0, \infty)$, which of course implies $f = 0$.

(c) *Show that the set of these polynomials is dense in the weighted space $L^2(\mathbb{R}_+, e^{-x}\,dx)$; moreover, the sequence $((-1)^k L_k)_{k \geq 0}$ is the orthogonalization of $(x^k)_{k \geq 0}$ and represents an ONB of the space.*

SOLUTION: The mapping $Vf(x) = f(x)e^{-x/2}$, $x > 0$, is unitary $L^2(\mathbb{R}_+, e^{-x}\,dx) \to L^2(\mathbb{R}_+)$ and $VL_k = l_k$, from which everything follows, except for the property of orthogonalization which is a consequence of (a): $\deg(L_n) = n$, $\forall n \geq 0$.

(d)* *(Laguerre, 1879) The Laguerre polynomial L_n is a solution of the Laguerre equation*

$$xL_n''(x) + (1 - x)L_n'(x) + nL_n(x) = 0,$$

$L(0) = 1$, $L'(0) = -n$. *The sequence (L_n) satisfies the following recurrence equations:*

$$(n+1)L_{n+1}(x) + (x - 2n - 1)L_n(x) + nL_{n-1}(x) = 0, \quad L_0 = 1, \quad L_1(x) = -x + 1,$$

$$L_n(x) = \frac{1}{n!}\left(\frac{d}{dx} - 1\right)^n x^n,$$

$$\frac{d}{dx}L_n = \left(\frac{d}{dx} - 1\right)L_{n-1}.$$

4.5.5 Compact W_k^a Operators

Let $k \in L^1(-a, a)$ such that

$$K^2 =: \int_{-a}^{a} (a - |t|)|k(t)|^2\,dt < \infty.$$

Show that $W_k^a \in \mathfrak{S}_2(L^2(0, a))$ ($\subset \mathfrak{S}_\infty(L^2(0, a))$: see Appendix D) and $\|W_k^a\| \leq \|W_k^a\|_{\mathfrak{S}_2} = K$.

SOLUTION: We know by Appendix D.8 that an integral operator

$$I_h f = \int_\Omega h(x, y) f(y)\,dy$$

with kernel $h \in L^2(\Omega \times \Omega)$ is Hilbert–Schmidt (hence compact), and $\|I_h\|_{\mathfrak{S}_2} = \|h\|_{L^2(\Omega \times \Omega)}$. However, $W_k^a = I_h$ where $h(x, y) = k(x - y)$, thus

$$\|W_k^a\|_{\mathfrak{S}_2}^2 = \int_0^a \int_0^a |k(x - y)|^2\,dx\,dy = \int_{-a}^{a} |k(t)|^2 (a - |t|)\,dt = K^2.$$

4.6 Notes and Remarks

The *Riemann boundary problem*, posed in 1851, consists in finding functions $f_\pm \in \text{Hol}(\gamma_\pm)$, where γ_+ and γ_- denote the interior and exterior of a closed curve $\gamma \subset \mathbb{C}$, such that $f_+ - bf_- = c$ on γ (where b and c are given functions on γ). Then Hilbert [1904, 1912] generalized the problem by transforming it into an operatorial form $aP_+ f + bP_- f = c$ and then reducing it to a singular integral equation. The links of the Riemann–Hilbert problem (the RHP, as was stated in §4.1) with singular integrals were confirmed by the Sokhotsky–Plemelj formulas of Exercise 4.5.1 (found in the *Habilitation* thesis of Yulian Sokhotsky [1873], then rediscovered by Plemelj [1908a,b] for his partial solution of Hilbert's Problem 21). The theory of boundary problems of analytic functions (Riemann–Hilbert problems) was thus developed *in the framework of singular integral equations and operators*, especially due to the work of Poincaré, Birkhoff, Carleman, Noether, Mikhlin, Krein, Rappoport, and Gohberg, and then by the generation of the 1960s (Widom, Sarason, Douglas), whose results are discussed above for particular questions of the theory (see §2.4 and §3.5, as well as Chapter 1).

The work on *Wiener–Hopf equations and operators* (WHO, i.e. convolution operators restricted to the half-line $\mathbb{R}_+$) appears to have arisen independently (see Chapter 1), bringing in inversion of the operators W_k, T_Φ and T_φ with the aid of the factorization of the symbol (invented earlier by Birkhoff [1909] for a vectorial version of the Riemann–Hilbert problem).

The *systems of singular integral equations*, or Wiener–Hopf systems, have even more applications in mathematical physics and in applied sciences, but to stay within the context of an introductory handbook to the subject we limited ourselves to a few exercises – nonetheless quite representative (see Exercises 2.3.11, 3.4.14, 3.4.15, and 4.5.2, as well as Theorem 4.4.5). It must also be added that the corresponding theory of the operators W_k, T_Φ, $R_{a,b}$ (etc.) with matrix-valued symbols is not yet as complete as the scalar version developed in this book; see also the comments below for Corollary 4.1.4 and Exercise 4.5.2.

With Mikhlin's [1936] discovery (see §2.4) of the symbolic calculus of operators of the types mentioned, the classical period of the theory (characterized by the work in spaces of smooth functions (Hölder classes, etc.) and with smooth symbols) was rapidly completed. For an overview see [Muskhelishvili, 1947], [Hirschman and Widder, 1955], [Mikhlin, 1962], [Gakhov, 1963], and [Gakhov and Chersky, 1978].

The new age of the triad in question (RHP–SIO–WHO) began in the 1960s with the appearance of the modern tools of harmonic analysis and functional

4.6 Notes and Remarks

spaces, and was marked by the fundamental works of Krein, Gohberg, Widom, Sarason, Douglas, Simonenko, and many others. The identification of Toeplitz operators (and the RHP, and the SIO) with the Wiener–Hopf operators (§4.2–4.3) is a true sign of maturity of the theory. The publications of Rosenblum [1965] and Devinatz [1967], where this identification is carried out, ended a long-standing misunderstanding: in reality these theories coincide, after having coexisted and developed independently for several decades. It seems that the most surprising point of this history concerns the means necessary for the said identification: they had already been known for over forty years (Fourier–Plancherel, Wiener–Paley, not to mention the Laguerre polynomials)!

Additional comments are distributed below topic by topic.

Formal credits.

Theorem 4.1.1. The *operator approach* to the Riemann–Hilbert problem comes from Hilbert's course on integral equations [Hilbert, 1904, 1912], but for the actual form of Theorem 4.1.1 (with the exception of estimates of the norms), see [Devinatz, 1967] and [Gohberg and Krupnik, 1973], where there are also additional references.

§4.1.2. The expression of the projections $P_\pm$ in the form of singular integrals comes from Hilbert; the kernel $\operatorname{ctg}(\theta - t)/2$ is sometimes called the Hilbert kernel.

Theorem 4.1.2. This theorem of Luzin [1913] is one of the key results of his thesis [Luzin, 1915] prepared during a period of study in Paris just before August 1, 1914, and submitted in 1915 in Moscow. The theorem was extended to the case of functions $u \in L^1(\mathbb{T})$ (see §4.1.3) and even to arbitrary finite measures by Privalov [1919]. In his thesis, Luzin discussed the "hidden reasons" for the almost everywhere convergence of the singular integral

$$\text{p.v.} \int \frac{f(t)}{x-t} dt = \lim_{\epsilon \to 0} \int_\epsilon^\pi \frac{f(x+s) - f(x-s)}{s} ds,$$

speaking of a certain "almost everywhere symmetry" of the values $f(x - s)$, $f(x + s)$ at symmetric points, in contrast with the possible divergence (even almost everywhere) of the integrals

$$\int \left| \frac{f(t)}{x-t} \right| dt, \quad \text{or} \quad \lim_{\epsilon \to 0} \int_\epsilon^\pi \frac{f(x+s) - f(x)}{s} ds,$$

which do not take into account the mutual annihilation of the large values (see [Luzin, 1915], pages 216–223, 458–473, and [Zygmund, 1959], Chapter 7). Dissatisfied with his proof of the convergence (that of Theorem 4.1.2) which depends on "magic tricks of complex analysis," Luzin posed the question of a "pure real variable" proof; the question led to one of the principal axes of twentieth-century harmonic analysis and attracted much attention. Different "real" proofs were obtained by Besicovich, Luzin himself, Privalov, Plessner, and Marcinkiewicz (see the references in the comments in the second edition of [Luzin, 1915]). We can say that the final form of these efforts led to the proof of the existence of the p.v.$\int$ drawn from properties of the *Hardy–Littlewood maximal function* and to the theory of Calderón–Zygmund operators; see [Stein and Weiss, 1971] and [Stein, 1993] for a presentation of a powerful theory that deals with multidimensional analogs of singular integral operators (in a special case, some analogs of the Hilbert operator: the Riesz transforms) and "systems of harmonic conjugates" satisfying a sort of EDP replacing the Cauchy–Riemann equations for u and $\tilde{u}$.

Corollary 4.1.4. The *Fredholm theory of the RHP and SIO* presented in this corollary can be completed by other criteria for being Fredholm, if the coefficients a and b are in classes where the result is already known for the corresponding Toeplitz operators (for example, for $AP_1(\mathbb{T})$, $AP_\zeta(\mathbb{T}) + C(\mathbb{T})$, etc.). See [Böttcher, Karlovich, and Spitkovsky, 2002], [Böttcher and Silbermann, 1990], [Duduchava, 1979], and [Gohberg and Krupnik, 1973] for details and the original references.

The *Fredholm theory of the algebras* $\mathrm{alg}(A, \mathbb{S})$ can be developed in the same way as for the algebras $\mathrm{alg}(\mathcal{T}_X)$ of Chapter 3; here A is a sub-algebra of $L^\infty(\mathbb{T})$ and $\mathrm{alg}(A, \mathbb{S})$ is generated by composite SIOs (or aggregate SIOs)

$$\sum_i \prod_j (a_{ij} + b_{ij}\mathbb{S}).$$

Moreover, it should be noted that the very idea of this algebraization of the theory (see §3.5) appeared precisely in the framework of a "calculus (composition) of SIOs," first of all by Poincaré [1910a] (in his theory of tides) and Bertrand [1921], and then by Mikhlin [1936]. The famous *Poincaré–Bertrand formula* (in fact proved by Hardy [1908] and then by Poincaré ([1910a], Chapter X)), which in a special case consists of

$$\mathbb{S}(b\mathbb{S}a) + \mathbb{S}(a\mathbb{S}b) = ab + \mathbb{S}a \cdot \mathbb{S}b,$$

inspired the semi-commutator formula of Lemma 2.2.9(1), motivated Mikhlin's symbolic calculus, and was extensively used and generalized for numerous applications. For the general case of the formula and references, see [Muskhelishvili, 1947].

The *Fredholm theory of systems of singular integral equations* and vector-valued RHPs (equivalent to the same theory of Wiener–Hopf equations: see

Exercise 4.5.2) was the first subject of the triad RHP–SIO–WHO where the discovery of the principal technique of the theory was made – the Birkhoff factorization of the symbol $\varphi = \varphi_- D\varphi_+$. From this grew all of its ulterior ramifications, such as the Wiener–Hopf factorization, etc. The rapid development of this theory was stimulated by its direct links with several problems of mathematical physics, random processes, orthogonal polynomials, etc., as well as Hilbert's Problem 21 (on systems of ODEs with a given Fuchsian group); see [Its, 2003], [Muskhelishvili, 1947], [Mikhlin, 1962], and [Gakhov, 1963] for the rich history of this part of analysis. See also the comments in §3.5 on Exercise 3.4.14.

§4.2.1. *The passage from $H^2(\mathbb{D})$ to $H^2(\mathbb{C}_+)$ and $L^2(\mathbb{R}_+)$* is well presented in any of the books on harmonic analysis, beginning with [Hoffman, 1962], [Koosis, 1980], [Garnett, 1981], and [Nikolski, 1986, 2002a, 2019]. The question of the normalization of the Fourier transform merits a small comment: in the case of a compact group (such as $\mathbb{T}$) the choice is forced by the normalization of the Haar measure (the Lebesgue measure in the case of $\mathbb{T}$) which already implies the normalization of the basis of the exponentials in $L^2(\mathbb{T})$, the *Fourier–Plancherel theorem* (isometry between $L^2(\mathbb{T})$ and $l^2(\mathbb{Z})$), as well as the crucial calculation rule for convolutions $\mathcal{F}(f*g) = \mathcal{F}f \cdot \mathcal{F}g$. However, for the case of a non-compact group such as $\mathbb{R}^N$, the corresponding normalization $(2\pi)^{-N/2} dx$ of the Haar measure (which applied to $\mathcal{F}$ and to $*$ gives the same results, and moreover is symmetric with respect to the duality between $\mathbb{R}^N$ and $(\mathbb{R}^N)^\wedge$) conflicts with the simultaneous treatment of the $\mathcal{F}f$ ($f \in L^1(\mathbb{R}^N)$) and $\mathcal{F}\mu$ (where μ is a measure on $\mathbb{R}^N$), which is not acceptable for the integral equations. The authority of this latter tradition having won, we remain with the normalization chosen in §4.2.1. A modification of this normalization for the Fourier–Plancherel theorem is used for the isometry of $\mathcal{F}_P$ on the space $L^2(\mathbb{R})$. In the case of locally compact Abelian groups G different from the classical groups $\mathbb{T}^N$, $\mathbb{Z}^N$, $\mathbb{R}^N$, the choice of the normalization is made through the structural theorem reducing the general case to the classical groups: see §30 of [Weil, 1940], or §36C of [Loomis, 1953], or Chapter 2 §4 of [Bourbaki, 1967] for the definition of *associated Haar measures*. On the system of Laguerre fractions (Λ_n), we can mention that it was at first valued more for applications than for pure analysis, see [Lee, 1960], pages 481–501, or [Khrushchev, 1977], §5 for applications to random processes.

Theorems 4.2.1, 4.2.2 and Exercise 4.5.3(a). The treatment of *Wiener–Hopf operators as a transfer of Toeplitz operators* follows Rosenblum [1965] (see also [Rosenblum and Rovnyak, 1985]) and (especially) Devinatz [1967], with the possible exception of the equations of Theorems 4.2.1(c), 4.2.2(c), and

Exercise 4.5.3(a)ii, whose role was outlined in B.4.8 of [Nikolski, 2002a]. The contents of this paragraph represent a major success of Toeplitz theory, but also a confusion in the noble family of analysts (not entirely overcome even today (2019)). After three decades of intense development of these two important theories, regarded, perhaps, as fairly similar but certainly independent, each with its own values and history, etc., they were discovered to be identical, up to performing a few quite banal transformations such as a change of variables and a Fourier transform. Worse, we showed in §4.3 that the matrices of $T_\Phi \colon H^2(\mathbb{C}_+) \to H^2(\mathbb{C}_+)$ and of $W_k \colon L^2(\mathbb{R}_+) \to L^2(\mathbb{R}_+)$ are simply Toeplitz matrices, and this with respect to bases known as early as the mid-nineteenth century! For Toeplitz, in 1912, who held to the point of view of *forms* (bilinear or quadratic), these objects (T_φ, T_Φ, W_k) should evidently have been seen as identical, but it took more than half a century of intensive study to recognize this.

§4.2.3. The spectral theory of Wiener–Hopf operators. In our presentation, everything is automatic by the theory of Toeplitz operators on $H^2(\mathbb{D})$ and the transfer of Theorems 4.2.1 and 4.2.2 and Exercise 4.5.3, whereas a good number of these properties (as well as the properties of the $R_{a,b}$ of §4.1) were historically discovered directly in the language of Wiener–Hopf equations in the period of "parallel existence" of Toeplitz operators, the SIO, and the Wiener–Hopf operators. For references about this period see [Gohberg and Krupnik, 1973], [Böttcher and Silbermann, 1990], and [Böttcher, Karlovich, and Spitkovsky, 2002].

Other details of the theory in $H^2(\mathbb{D})$ can also be transferred just as easily; in particular, the Fredholm theory in the algebras alg$\mathcal{W}_A$ (analogs of the algebras alg$\mathcal{T}_A$), some local principles, the Wiener–Hopf factorization (!). In fact, the use and adaptation of this last technique for the Wiener–Hopf equations was quite simply invented in the very first publication on the theory [Wiener and Hopf, 1931].

We can add that around the time of the publication of [Wiener and Hopf, 1931], the technique of Plemelj–Birkhoff type factorizations $\Phi = \Phi_- D \Phi_+$ (with respect to a contour γ) was already a powerful and reputed tool, which played a role in the partial solutions of RH problems [Plemelj, 1908b] – or *Riemann–Hilbert barriers*, as they are called in the applications – and in the Riemann problem on the Fuchsian groups of systems of ODE (Problem 21 from Hilbert's list of 1900 [Birkhoff, 1909]). At that time, the generality of the results was not yet the priority, and the study was limited to Φ holomorphic on the contour γ and the factors Φ_+, Φ_- holomorphic in int(γ) and ext(γ), respectively. This is perhaps one of the reasons why Wiener and Hopf, the pioneers of the mathematical theory of the equations $\lambda f - W_k f = g$, worked with exponentially decreasing kernels k, $k(t) = O(e^{-a|t|})$ for $|t| \to \infty$, and with functions f, g satisfying $f(t), g(t) = O(e^{ct})$ as $t \to \infty$ ($0 < c < a$), and hence where the symbols to be factorized $\mathcal{F}k$ are holomorphic in a band of $\mathbb{C}$ containing $\mathbb{R}$.

See [Böttcher, Karlovich, and Spitkovsky, 2002], [Gohberg, Kaashoek, and Spitkovsky, 2003], and [Böttcher and Spitkovsky, 2013] for the current state of the spectral theory of Wiener–Hopf operators.

§4.2.4 and §4.2.7. These "classical Wiener–Hopf" operators and equations appeared already in [Fredholm, 1900, 1903] and [Hilbert, 1904, 1912].

§4.2.5–§4.2.6. *The Wiener–Hopf operators W_μ with a (true) measure $\mu \in \mathcal{M}(\mathbb{R})$.* The (finite) difference integral operators and equations (also called "delay equations and operators") are common in several applications, in particular, in signal processing (feedback control circuits). Moreover, in the case of an arbitrary measure $\mu \in \mathcal{M}(\mathbb{R})$, it is not clear that the treatment of the operators $W_\mu \colon L^2(\mathbb{R}_+) \to L^2(\mathbb{R}_+)$ defined by a (true) measure presents any advantage compared to general W_k, with $k \in \mathcal{P}M(\mathbb{R})$. For example, the invertibility of the symbol $k^\vee \in L^\infty(\mathbb{R})^{-1}$ does not imply the invertibility of the measure $\mu \in \mathcal{M}(\mathbb{R})^{-1}$ ("Wiener–Pitt–Schreider phenomena": see [Gelfand, Raikov, and Shilov, 1960] and [Rudin, 1962]). However, even if μ is an invertible measure (hence, if there exists $\mu^{-1} \in \mathcal{M}(\mathbb{R})$ such that $\mu * \mu^{-1} = \delta_0$), the operator $W_{\mu^{-1}}$ in general will not, as we know, be a regularizer of W_μ.

The situation is a little clearer for the Wiener–Hopf operators on the space $L^1(\mathbb{R})$. More precisely, by taking into account the fact that $\mathcal{M}(\mathbb{R})$ is the algebra of convolutors of the space $L^1(\mathbb{R})$ (a classical result of Frigyes Riesz: see [Gelfand, Raikov, and Shilov, 1960] and [Rudin, 1962] for references), Douglas and Taylor [1972] showed that

$$W_\mu \in \mathrm{Fred}(L^1(\mathbb{R})) \Rightarrow \mu \in \mathcal{M}(\mathbb{R})^{-1},$$

and, conversely, if $\mu \in \mathcal{M}(\mathbb{R})^{-1}$ then

$$\mu^\vee = e^{icx}\omega^n(e^{*\nu})^\vee$$

where $c \in \mathbb{R}$, $n \in \mathbb{Z}$,

$$\omega(x) = \frac{x-i}{x+i}$$

and $\nu \in \mathcal{M}(\mathbb{R})$ ($e^{*\nu} = \sum_{k \geq 0}(\nu^{*k})/k!$) (this is a Joseph Taylor's characterization of invertible measures in $\mathcal{M}(\mathbb{R})$: see [Taylor, 1973]) and the operator $W_\mu \colon L^1(\mathbb{R}) \to L^1(\mathbb{R})$ is Fredholm *if and only if* $c = 0$ (the number c plays the role of the mean motion of almost periodic functions); in this case,

$$\mathrm{ind}(W_\mu) = -n.$$

In particular, W_μ is *invertible on* $L^1(\mathbb{R})$ *if and only if* μ is an exponential of an element of $\mathcal{M}(\mathbb{R})$ ($\mu = e^{*\nu}$).

Returning to $W_\mu \colon L^2(\mathbb{R}_+) \to L^2(\mathbb{R}_+)$, we might have had a more transparent Fredholm criterion (or of invertibility) for the symbols of the algebra

$$A = \mathrm{clos}_{L^\infty(\mathbb{R})}(\mu^\vee : \mu \in \mathcal{M}(\mathbb{R})),$$

but this is not the case here either, and there is not even a reasonable description of the algebra A itself. The only *a priori* credible conjecture, according to which A would coincide with the Eberlein algebra WAP($\mathbb{R}$) (of "weakly almost periodic functions"; a function f is (by definition) *weakly almost periodic* if the translates $(\tau_t f)_{t \in \mathbb{R}}$ are weakly precompact in $C(\mathbb{R}) \cap L^\infty(\mathbb{R})$), was contradicted long ago by Rudin [1959].

§4.3 and Exercise 4.5.4. T_Φ *and* W_k *are Toeplitz operators.* As mentioned in the comments on Theorem 4.2.1, this discovery (found in [Rosenblum, 1965] and especially in [Devinatz, 1967]: see also [Rosenblum, 1965]) dethroned the theory of Wiener–Hopf operators from its status as a subject in itself and made it into a simple double for the theory of the operators $T_\varphi \colon H^2(\mathbb{T}) \to H^2(\mathbb{T})$; this double is nonetheless very important in applications, and often suggests and justifies the study of classes of interesting symbols (such as AP, SAP, etc.).

The *Laguerre polynomials and functions* were introduced and studied by Edmond Laguerre (1879; see [Hermite, Poincaré, and Rouché, 1898]) starting from the differential equation of Exercise 4.5.4(d), with the goal of calculating numerically certain integrals of the form $\int_0^\infty f(x) e^{-x} dx$. They were generalized by Sonin in 1880 to a family of polynomials also containing, depending on the values of their parameters, the Hermite polynomials and certain cases of hypergeometric functions.

§4.4. The *truncated Wiener–Hopf operators* $W_k^a = P^a W_k | L^2(0, a)$ are unitarily equivalent (with the aid of the same unitary transforms as in Theorems 4.2.1 and 4.2.2) to the truncated Toeplitz operators

$$T_\varphi^\theta = P_\theta M_\varphi | K_\theta$$

where $\theta = \theta_a = \exp(-a((1+z)/(1-z)))$ and $K_\theta = H^2(\mathbb{D}) \ominus \theta H^2(\mathbb{D})$, or again, to the "Toeplitz operators on the Paley–Wiener spaces," $T = PM_\Phi P$ where P is orthogonal projection onto the *Paley–Wiener space* $\mathcal{F}_P^{-1} L^2(-a/2, a/2)$. The T_φ^θ (truncated Toeplitz operators) are briefly treated in Exercise 3.4.15 and discussed in §3.5.

Before the explosion of studies of truncated Toeplitz operators (initiated by Sarason [2007]; see a brief description in the comments on Exercise 3.4.15 in §3.5), the principal references on the spectral theory of the W_k^a were the works of Widom [1960c] and Rochberg [1987]. [Widom, 1960c] presented a well-developed theory for the compressions $W_k^E = P_E W_k | L^2(E)$ on the spaces $L^2(E)$ for general subsets $E \subset \mathbb{R}$ (more detailed for finite unions of intervals). The properties of Propositions 4.4.1 and §4.4.2, §4.4.3, and Theorems 4.4.4 and 4.4.5 can be found in [Rochberg, 1987]. They have counterparts for T_φ^θ with a general inner function θ, except perhaps for Theorem 4.4.4: the existence of a bounded symbol for a bounded T_φ^θ is equivalent to a "weak factorization" of the functions K_θ^1 (see the comments on Exercise 3.4.15 in §3.5) and a Carleson-type embedding (see [Baranov, Bessonov, and Kapustin, 2011]), which is not guaranteed for every θ, but nonetheless is the case for θ "having one connected component" (in particular, for θ_a corresponding to truncated Wiener–Hopf operators W_k^a; see Exercise 3.4.15 for explanations). As already mentioned, since the operators T_φ^θ are important for the techniques of "model spaces" and "de Branges spaces," they are the object of numerous recent studies. However, neither the theory of these operators, nor their domains of application, are yet clear enough to merit a separate chapter in an introductory handbook such as this one. See [Garcia and Ross, 2012] and [Baranov, Bessonov, and Kapustin, 2011] for an overview of the subject and some references, and also the comments on Exercise 3.4.15 in §3.5.

Returning to the operators W_k^a, we note that determination of their spectrum or Fredholm spectrum by the intermediary of the (very nice) Theorem 4.4.5 (followed by criteria of invertibility of the matrix-valued operators W_k) nonetheless remains problematic, in particular because of the difficulties of the Wiener–Hopf factorization for matrix-valued symbols (see the Remarks in §4.4.5 and Exercise 4.5.5). See [Böttcher, Karlovich, and Spitkovsky, 2002] or [Böttcher and Spitkovsky, 2013] for the current state of the subject.

Theorem 4.4.5 is proved in [Krupnik and Feldman, 1985]; see also Theorem 1.25 and the comments about it in [Böttcher, Karlovich, and Spitkovsky, 2002].

Exercise 4.5.0. Basic exercises. This brief series of exercises represents an introduction for neophytes to a vast domain of modern analysis: the theory of singular integral operators on $\mathbb{R}^n$ and other metric spaces. This theory plays a crucial role for the EDP, as well as in harmonic analysis on the Euclidean spaces. To navigate the flood of existing literature on the subject, see the sources already mentioned in Section 4.5.0, in the comments on Exercise 4.5.0(a–d), among which [Duoandikoetxea, 2001] is particularly well adapted to the needs of beginners. See also [Mikhlin, 1962] and [Volberg, 2003]. The Riesz transforms of Exercise 4.5.0(d) implicitly appeared in [Riesz, 1949], and then with Calderón and Zygmund; see the history of the subject in the sources mentioned in the text of (a)–(d). The method of rotation in (a)–(d) is presented following the approach of [Duoandikoetxea, 2001].

Exercise 4.5.1. The Sokhotsky–Plemelj formulas. See [Sokhotsky, 1873], and then [Plemelj, 1908a] (independently), where they served in a partial solution to Hilbert's Problem 21. In the form given in the Remark of Exercise 4.5.1, as already mentioned, the formulas are consequences of Fatou's theorem [Fatou, 1906] on boundary limits.

Exercise 4.5.2. Matricial Wiener–Hopf equations and operators. The theory, as presented here, is an extract of the vectorial theory of Toeplitz operators, see Exercises 3.4.14 and 2.3.11 and §3.5 (for comments and references). Recall that, historically, the subjects were developed in inverse order: the principal facts (such as the Fredholm criterion, the Birkhoff–Wiener–Hopf matricial factorization, the formulas for the dimensions of the kernels and co-kernels, etc.) were first discovered for the RHP and SIO operators, then for those of Wiener–Hopf, and only later were they rediscovered for the Toeplitz operators. This dramatic story is explained in detail in Chapter 1, §2.4, and §3.5.

Exercise 4.5.2(b,c). See [Gohberg and Krein, 1958], [Gohberg and Feldman, 1967], [Douglas, 1973], and [Böttcher, Karlovich, and Spitkovsky, 2002].

The corresponding result for the almost periodic symbols (of great value for applications), cited in the comments on Exercise 3.4.14 in §3.5, is much more complicated and depends on the AP factorization; see [Böttcher, Karlovich, and Spitkovsky, 2002] for a modern presentation and some references.

Exercise 4.5.3. Hankel operators on $\mathbb{R}_+$. The same remarks as for the Toeplitz operators (above, as well as those in §2.4 and §3.5) can be applied to the Hankel operators: for a long time, they were treated in parallel to the classical theory of Hankel matrices (hence, the Hankel operators on $l^2(\mathbb{Z}_+)$ or on $H^2(\mathbb{T})$) before these two realizations were shown to be identical: see [Devinatz, 1967], as well as [Rosenblum and Rovnyak, 1985]. For the metric properties of Hankel operators, as well as for numerous applications, see basic sources of the subject, e.g. [Peller, 2003], [Power, 1980], and [Nikolski, 2002a].

Exercise 4.5.3(e) can be found already in Hilbert's course [Hilbert, 1904, 1912]. Recall that there is also a discrete version of the Hilbert transform (and the computation of its norm, more complicated than in the continuous case of Exercise 4.5.3(e)) treated in [Nikolski, 2019], §2.8 (followed by numerous historical remarks in §2.9).

Exercise 4.5.4. Laguerre polynomials. Initially (1879), the polynomials L_n were defined by the differential equation of Exercise 4.5.4(d), whereas our definition in Exercise 4.5.4(a) is known as the "Rodrigues formula." The equation led to the "generalized (or associated) Laguerre polynomials" (by Sonin in 1880), among which we find certain other special functions, such as the Hermite polynomials. The generating function of the Laguerre polynomials is an exponential

$$(1-t)^{-1} \exp\left(-\frac{tx}{1-t}\right) = \sum_{n \geq 0} t^n L_n(x).$$

The polynomials L_n have many applications in mathematical physics, signal processing, probability, etc.

5
Toeplitz Matrices: Moments, Spectra, Asymptotics

Topics

- Trigonometric moment problem, positive definite Toeplitz matrices, orthogonal polynomials.
- Arveson's distance formula, the four-block lemma, Julia's lemma.
- Inversion of a Toeplitz matrix by Gohberg–Sementsul and Baxter–Hirschman, inversion by the finite section method.
- Circulants.
- Toeplitz determinants and their asymptotics, equidistribution following Weyl, asymptotic distribution of Toeplitz spectra (following Szegő) and of circulants.
- The "strong Szegő theorem," Krein–Dirichlet algebra.
- The Helton–Howe and Berger–Shaw formulas of det/trace.
- Positive harmonic functions, holomorphic functions of the Schur class
- Inversion of Wiener–Hopf operators by the finite section method, case of a rational symbol.
- Cauchy and Hilbert determinants, Libkind–Widom and Borodin–Okounkov formulas.

Biographies Constantin Carathéodory, Gábor Szegő, Hermann Weyl.

5.1 Positive Definite Toeplitz Matrices, Moment Problems, and Orthogonal Polynomials

To a finite $N \times M$ matrix, $A = (a_{jk})_{0 \leq j < M, 0 \leq k < N}$ we associate (and even identify with) the linear mapping $A: \mathbb{C}^N \to \mathbb{C}^M$ it defines,

$$Ax = A\left(\sum_{0 \leq k < N} x_k e_k\right) = \sum_{0 \leq k < N} x_k A e_k = \left(\sum_{0 \leq k < N} a_{jk} x_k\right)_{0 \leq j < M},$$

5.1 Positive Definiteness, Moment Problems, Orthogonal Polynomials 231

where $(e_k)_{0 \le k < N}$ is the standard orthonormal basis of $\mathbb{C}^N$, $e_k = (\delta_{kl})_{0 \le l < N}$ (δ_{kl} is the Kronecker "delta"), hence $a_{jk} = (Ae_k, e_j)$ (the scalar product in $\mathbb{C}^M$). Recall that according to Hilbert's terminology (see Appendix D) a square matrix $A \colon \mathbb{C}^N \to \mathbb{C}^N$ is *positive definite* (or positive semi-definite), denoted $A \gg 0$, if

$$(Ax, x) = \sum_{0 \le j,k < N} a_{jk} x_k \overline{x}_j \ge 0 \quad (\forall x \in \mathbb{C}^N).$$

Such an A is self-adjoint ($a_{jk} = \overline{a}_{kj}$) and its eigenvalues are positive or zero.

The expression "moment problem" has a multitude of meanings (see Chapter 1 and §5.8 below), but there is one that arose as the very first application of Toeplitz matrices (forms); this is the *trigonometric moment problem* (TMP): given a family of complex numbers $(a_k)_{k \in \sigma}$, find a measure (positive, or satisfying some other condition) $\mu \in \mathcal{M}(\mathbb{T})$ such that

$$a_k = \widehat{\mu}(k) = \int_{\mathbb{T}} z^{-k} d\mu(z), \quad k \in \sigma.$$

In the classical context, the set of indices $\sigma \subset \mathbb{Z}$ can be $\sigma = \mathbb{Z}$, $\sigma = \mathbb{Z}_+$, or a finite interval of $\mathbb{Z}$ (or some other subset of $\mathbb{Z}$). The very first theorem on the TMP was proved by Carathéodory [1907]; soon afterwards (1911) it was modified to use the language of Toeplitz forms.

With an extension of the notation, given a sequence $(a_k)_{k \in \mathbb{Z}} \in l^\infty(\mathbb{Z})$, a distribution (on $\mathbb{T}$) $f = \sum_{k \in \mathbb{Z}} a_k z^k$ and an integer $n \in \mathbb{Z}_+$, an $(n+1) \times (n+1)$ *Toeplitz matrix* is defined by

$$T_{f,n} = (a_{k-j})_{0 \le k,j \le n} \quad \text{or} \quad T_{a,n} = (a_{k-j})_{0 \le k,j \le n}.$$

Theorem 5.1.1 (Carathéodory, 1907, Toeplitz, 1911) *Let $(a_k)_{k \in \mathbb{Z}} \in l^\infty(\mathbb{Z})$ be such that for every k, $a_{-k} = \overline{a}_k$. The following assertions are equivalent.*

(i) *There exists a unique measure $\mu \in \mathcal{M}(\mathbb{T})$, $\mu \ge 0$, such that $a_k = \widehat{\mu}(k)$ ($\forall k \in \mathbb{Z}$).*

(ii) *The Toeplitz matrices $T_{a,n}$ are positive definite: $T_{a,n} \gg 0$ ($\forall n \in \mathbb{Z}_+$), i.e.*

$$(T_{a,n} x, x) = \sum_{0 \le j,k \le n} a_{j-k} x_k \overline{x}_j \ge 0 \quad (\forall x = (x_j)_0^n \in \mathbb{C}^{n+1}, \forall n \in \mathbb{Z}_+).$$

A complex sequence $(a_k)_{k \in \mathbb{Z}}$ satisfying condition (ii) of the theorem is also said to be *positive definite*. As they are important for several applications, the positive definite sequences (and more generally, the positive definite functions: see §5.8) are the object of a large number of studies; we will come back to these in §5.1.2 and §5.8.

Historically, several important techniques were developed for the different proofs of Theorem 5.1.1 and its analogs for other moment problems. In this handbook we present *three-and-a-half proofs* based on quite distinct ideas.

The *first proof* (that of §5.1.1, Stone and Akhiezer) depended on the spectral theorem for unitary operators.

The *second proof* (see Theorem 5.1.3) was suggested by a study of orthogonal polynomials (Szegő).

The *third proof* (Exercise 5.7.1(b) below, Carathéodory and Herglotz) used an integral representation of positive harmonic functions.

The *fourth proof* (realized only halfway: see Exercise 5.7.4) was a calculatory masterpiece by Szegő.

Constantin Carathéodory (1873–1950), a German mathematician (descendant of an internationally influential Greek family of $Καραθεοδωρης$) made significant contributions to a large spectrum of disciplines (calculus of variations, general measure theory, conformal representations, real analysis, thermodynamics, relativity, geometrical optics, etc.). Among other realizations, Carathéodory was the founder of dynamic optimization (by his discovery of "Bellman's equation" 20 years before its second appearance), as well as the theory of "prime ends" of conformal representations, and the axiomatic approach to thermodynamics (where he introduced several new concepts, including the "adiabatic accessibility of states," and clarified the second law of thermodynamics). He is the author of very useful constructions in measure theory and of the "Carathéodory conjecture" in differential geometry.

Carathéodory began his studies in Brussels (where his father was the ambassador of the Ottoman Empire) at the École Royale Militaire, and then at the École d'Applications, and found his first job (1898) in the British colonial services as an engineer working on the construction of dams in Egypt (the Assiut Barrage in Asyut). There, Carathéodory performed new measures on the Cheops pyramid, which became the subject of his first scientific publication. On his return to Europe, Carathéodory had to make a difficult choice between Paris and Berlin for the continuation of his studies; he decided on the latter because in Paris, "he had too many friends that would keep him from concentrating on his studies" (according to witnesses). In Berlin, he suddenly decided to abandon his career and to devote himself to mathematics (according to legend, under the influence of a seminar by Lipót Fejér at the Hermann Schwarz colloquium, 1900). Carathéodory devised his own program of study (!) and submitted his thesis at Göttingen in 1902, in front of Felix Klein and Karl Schwarzschild (an unthinkable path in our uniformizing epoch which has led to the definitive loss of such independence of the spirit under the diktat of the Internet and never-ending communications). He obtained his *Habilitation* at Göttingen in 1904 under the direction of Hermann Minkowski.

As an eminent scientist and member of a Greek family with important influence, in 1919 Carathéodory was invited to organize a new "Ionian University of Smyrna"; he personally toured all of Europe to collect the material and the books necessary for the research library. In 1920, Carathéodory became the first President of the University, but the project was stopped by the war of 1922 between Greece and Turkey, followed by the famous "Great Fire of Smyrna" in September 1922 (Carathéodory participated in the rescue of the library up to the very last moment, and was himself saved by a journalist who transported him in a rowing boat to the Greek warship "Naxos" in Smyrna's harbour).

Between 1924 and 1938, Carathéodory occupied the chair of mathematics at the University of Munich. Carathéodory was not spared the ethical problems linked to relations with the Nazi regime, especially because many of his friends and colleagues were Jewish. According to witnesses, his position was typical of German intellectuals – not approaching the regime, not participating, if possible helping the persons persecuted, but without actively protesting, and even letting the regime use his name for their propaganda.

Carathéodory's best-known students include Rademacher, Finsler, Hamburger, Seidel, and others. Carathéodory wrote about a dozen influential monographs, including *Conformal Representations* (1932),

Geometrische Optik (1937), and *Vorlesungen über reelle Funktionen*, volume 1 (1918), whose volume 2 was burnt at the publishing house Teubner during the bombardment of Leipzig in 1943). A dozen mathematical objects bear Carathéodory's name, including the *Carathéodory metric*, the *Carathéodory theorem of extension to the boundary*, the *Carathéodory–Fejér theorem*, the *Carathéodory conjecture*, etc.

There exists an extensive literature about Carathéodory and the events in which he participated; we cite here only a single very well-researched book by Maria Georgiadou [2004], where dozens of references can be found. Carathéodory's heritage was conserved by the opening of a museum of the Carathéodory family in Nea Vyssa (Greece) and the Karatheodoris museum in Kotomini (Greece); one of the largest conference rooms at the University of Munich bears his name.

In 1909 Carathéodory married Euphrosyne Carathéodory (his aunt, 11 years his junior), the daughter of Alexander Karatheodori Pasha, Governor of Crete, Minister of Foreign Affairs of the Ottoman Empire, and First Translator of His Imperial Majesty the Sultan. A perfect mastery of languages was a family tradition: Alexander was said to be fluent in 16 and Constantin in 10. Marriages within the family were a Carathéodory tradition; they had two children.

For more details on the spectral theorem (of Hilbert–von Neumann) see Appendix D, and for its proof see (for example) §C.1.5.1(k) of [Nikolski, 2002b]; here is a reminder of its statement.

Theorem 5.1.2 (von Neumann, 1929) *Let H be a Hilbert space and $N: H \to H$ a bounded normal operator with simple spectrum, i.e. such that $NN^* = N^*N$ and such that there exists $x \in H$ generating the entire space H:*

$$H = \mathrm{span}_H(N^j N^{*k} x: j, k \in \mathbb{Z}_+).$$

Then there exists a positive finite Borel measure $\mu \in \mathcal{M}(\mathbb{C})$ with compact support and a unitary mapping

$$\mathcal{U}: H \to L^2(\mu)$$

such that $\mathcal{U}x = 1$ and

$$\mathcal{U}N = M_z \mathcal{U},$$

where $M_z: L^2(\mu) \to L^2(\mu)$ is the operator $M_z f = zf$ of multiplication by an "independent variable" $z(\zeta) = \zeta$ ($\forall \zeta$).

The support of μ coincides with the spectrum of N: $\sigma(N) = \mathrm{supp}(\mu)$.

5.1 Positive Definiteness, Moment Problems, Orthogonal Polynomials

5.1.1 Proof of Theorem 5.1.1 (Following Stone, 1932)

(i) $\Rightarrow$ (ii) since, for every $x \in \mathbb{C}^{n+1}$,

$$0 \leq \int_\mathbb{T} \left| \sum_{0 \leq k \leq n} x_k z^k \right|^2 d\mu = \sum_{0 \leq j,k \leq n} x_k \bar{x}_j \int_\mathbb{T} z^{k-j} d\mu = \sum_{0 \leq j,k \leq n} a_{j-k} x_k \bar{x}_j = (T_{a,n} x, x).$$

(ii) $\Rightarrow$ (i) by the spectral theorem; indeed, for two sequences $x = (x_k)_{k \in \mathbb{Z}} \in \mathcal{S}_0(\mathbb{Z})$, $y = (y_k)_{k \in \mathbb{Z}} \in \mathcal{S}_0(\mathbb{Z})$ with finite support, set

$$\langle x, y \rangle = \sum_{j,k} a_{j-k} x_k \bar{y}_j,$$

and observe that $\langle \cdot, \cdot \rangle$ is a semi-inner product on $\mathcal{S}_0(\mathbb{Z})$ (see Appendix B) and that the shift (translation) operator $S(x_k)_{k \in \mathbb{Z}} =: (x_{k-1})_{k \in \mathbb{Z}}$ is unitary $S: \mathcal{S}_0(\mathbb{Z}), \langle \cdot, \cdot \rangle \to \mathcal{S}_0(\mathbb{Z}), \langle \cdot, \cdot \rangle$:

$$\langle Sx, Sy \rangle = \sum_{j,k \in \mathbb{Z}} a_{j-k} x_{k-1} \bar{y}_{j-1} = \sum_{j,k \in \mathbb{Z}} a_{j-k} x_k \bar{y}_j = \langle x, y \rangle.$$

Using (if necessary) standard procedures of the quotient $\mathcal{S}_0(\mathbb{Z})/\mathrm{Ker}\langle \cdot, \cdot \rangle$ and of completion, this provides a Hilbert space H_a on which the operator S can first be projected as $S \mapsto S^\bullet =: S/\mathrm{Ker}\langle \cdot, \cdot \rangle$ and then naturally extended to a unitary operator $S^\bullet: H_a \to H_a$. Clearly, it has a simple spectrum, since

$$\mathrm{Vect}(S^k e_0: k \in \mathbb{Z}) = \mathrm{Vect}(e_k: k \in \mathbb{Z}) = \mathcal{S}_0(\mathbb{Z}),$$

where $e_k = (\delta_{jk})_{j \in \mathbb{Z}}$ is the natural basis of $\mathcal{S}_0(\mathbb{Z})$. By the spectral theorem, there exists a Borel measure $\mu \geq 0$ and a unitary equivalence $\mathcal{U}: H_a \to L^2(\mu)$ such that $\mathcal{U}S^\bullet = M_z \mathcal{U}$ and $\mathcal{U}e_0^\bullet = 1$ (and hence $\mathcal{U}e_k^\bullet = z^k$, $k \in \mathbb{Z}$, where $x^\bullet$ is the range of $x \in \mathcal{S}_0(\mathbb{Z})$ in the quotient space H_a) with $M_z: L^2(\mu) \to L^2(\mu)$; clearly $\mathrm{supp}(\mu) \subset \mathbb{T}$ ($S^\bullet$ is unitary) and

$$a_n = \langle e_0^\bullet, e_n^\bullet \rangle = (1, z^n)_{L^2(\mu)} = \int_\mathbb{T} \bar{z}^n d\mu = \widehat{\mu}(n), \quad \forall n \in \mathbb{Z}. \qquad \blacksquare$$

5.1.2 The Truncated TMP: Extension to a Positive Definite Sequence

Let $a_1, a_2, \ldots, a_n$ be complex numbers. *Is it possible to include them in a positive definite sequence* $(a_k)_{k \in \mathbb{Z}}$? Given Theorem 5.1.1, the response is manifestly positive (see the Proposition below), but what is much less evident is the existence of an extension of this type *independent of Theorem 5.1.1*, constructive, and in a certain sense "minimal": see Carathéodory's Theorem 5.1.3 below. Moreover, this will allow us to obtain another proof of Theorem 5.1.1.

Proposition *Let $c_1, c_2, \ldots, c_n$ be complex numbers. There exists a positive definite sequence $(a_k)_{k \in \mathbb{Z}}$ such that $a_k = c_k$, $1 \le k \le n$ (and hence, by Theorem 5.1.1 a measure $\mu \in \mathcal{M}(\mathbb{T})$, $\mu \ge 0$, such that $a_k = \widehat{\mu}(k)$, $\forall k \in \mathbb{Z}$).*

Proof Set $b_k = c_k$, $1 \le k \le n$, $b_j = 0$ for $j > n$ and $j = 0$, and $b_j = \overline{b}_{|j|}$ for $j < 0$. Then $T_b = (b_{j-k})_{j,k \ge 0}$ is a bounded self-adjoint operator (matrix) on $l^2(\mathbb{Z}_+)$. Its spectrum $\sigma(T_b)$ is real and has a finite lower bound $\lambda = \min\{t : t \in \sigma(T_b)\}$, and so the Toeplitz matrix $T_b - \lambda I$ is positive definite, $T_b - \lambda I \gg 0$. By setting $a_k = b_k$ for $k \ne 0$ and $a_0 = -\lambda$, and applying Theorem 5.1.1, we obtain the result. ∎

5.1.3 Truncated Toeplitz Operators

In preparation for the presentation of Carathéodory's theorem (see Theorem 5.1.3), we begin with a few observations on Toeplitz matrices.

(1) Truncated Toeplitz matrices and operators: notation. As for the case of Hilbert spaces equipped with an infinite orthonormal basis (see §2.1.3–§2.1.4), we can associate with every Toeplitz matrix $T_{a,n} = (a_{k-j})_{0 \le k,j \le n}$ the linear mapping (operator) on the Euclidean space $\mathbb{C}^{n+1} = l^2([0,n] \cap \mathbb{Z})$,

$$T_{a,n} : \mathbb{C}^{n+1} \to \mathbb{C}^{n+1},$$

$$T_{a,n} x = \left(\sum_{0 \le j \le n} a_{k-j} x_j \right)_{0 \le k \le n} \quad (\forall x = (x_j)_0^n \in \mathbb{C}^{n+1}),$$

or even – keeping the same notation – on the space

$$\mathcal{P}_n = \mathrm{Vect}(z^k : k = 0, 1, \ldots, n)$$

of polynomials of degree less than or equal to n,

$$T_{a,n} : \mathcal{P}_n \to \mathcal{P}_n,$$

$$T_{a,n} p = q, \quad \text{where } p = \sum_0^n \widehat{p}(j) z^j, \quad q = \sum_{0 \le k \le n} \widehat{q}(k) z^k; \quad \widehat{q}(k) = \sum_{0 \le j \le n} a_{k-j} \widehat{p}(j).$$

A *symbol of* $T_{a,n}$ is an arbitrary distribution (or formal trigonometric series) $f = \sum_{k \in \mathbb{Z}} \widehat{f}(k) z^k$ such that $\widehat{f}(k) = a_k$, $k = -n, \ldots, n$; we will use the notations

$$T_{a,n} = T_{f,n}$$

as identical. A projection P_n is defined for sequences or formal series as

$$P_n x = P_n(x_j)_{j \in \mathbb{Z}} = (x_j)_{0 \le j \le n}, \quad P_n f = P_n \left(\sum_{k \in \mathbb{Z}} \widehat{f}(k) z^k \right) = \sum_{k=0}^n \widehat{f}(k) z^k,$$

5.1 Positive Definiteness, Moment Problems, Orthogonal Polynomials 237

and $(\cdot,\cdot)$ denotes the standard scalar product in $\mathbb{C}^{n+1}$ or in $\mathcal{P}_n \subset L^2(\mathbb{T})$; for example, $(p,q) = \int_{\mathbb{T}} p\bar{q}\,dm$ ($\forall p,q \in \mathcal{P}_n$).

(2) Truncated Toeplitz is a compression of multiplication. *More precisely,*

$$T_{f,n} = P_n M_f | \mathcal{P}_n,$$

where $M_f p = fp$ ($\forall p \in \mathcal{P}$), and fp is the formal product (a convolution on $\mathbb{Z}$); if $f \in L^1(\mathbb{T})$, then

$$(T_{f,n} p, q) = \int_{\mathbb{T}} f p \bar{q}\,dm \quad (\forall p, q \in \mathcal{P}_n).$$

Indeed, $(T_{f,n} p, q) = (P_n M_f p, q) = (M_f p, q) = \int_{\mathbb{T}} f p \bar{q}\,dm$ ($\forall p, q \in \mathcal{P}_n$). ∎

Remark With the notation of Exercise 3.4.15, we have $\mathcal{P}_n = H^2 \ominus z^{n+1} H^2$ and

$$T_{\varphi,n} = T_\varphi^{z^{n+1}},$$

the special case $\theta = z^{n+1}$ of a truncated Toeplitz operator.

(3) Equations of Toeplitz matrices. The functional equation of Lemma 2.1.2 is transformed into the following series of identities including the "truncated shift" S_n and its adjoint S_n^* on $\mathbb{C}^{n+1}$

$$S_n x = S(x_0, x_1, \ldots, x_n) = (0, x_0, \ldots, x_{n-1}) \quad (\forall x = (x_j)_0^n \in \mathbb{C}^{n+1}),$$

$$S_n^* x = S^*(x_0, x_1, \ldots, x_n) = (x_1, x_2, \ldots, x_n, 0).$$

More precisely, $T = (t_{k,j})_{0 \le j,k \le n}$ is a Toeplitz matrix if and only if, for every k, $0 \le k \le n$,

$$S_n^{*k} T S_n^k = P_{n-k} T P_{n-k}.$$

Indeed, if the equations are satisfied, then $S_n^{*k} T S_n^k z^j = P_{n-k} T z^j$ for $0 \le j \le n-k$ and every k, and hence, for every l, $0 \le l \le n - k$,

$$t_{l,j} = (Tz^j, z^l) = (P_{n-k} T z^j, z^l) = (S_n^{*k} T S_n^k z^j, z^l) = (T z^{j+k}, z^{l+k}) = t_{l+k, j+k}.$$

By setting $a_l = t_{l,0}$, $a_{-l} = t_{0,l}$ ($0 \le l \le n$), we obtain $t_{\alpha\beta} = a_{\alpha-\beta}$ for every $0 \le \alpha$, $\beta \le n$, thus $T = T_{a,n}$.
The converse is even easier. ∎

5.1.4 The Operator Approach to Orthogonal Polynomials (Akhiezer and Krein, 1938)

In the quest for a sharp form of the positive definite extension theorem of §5.1.2, we need certain properties of orthogonal polynomials with respect to a measure μ (future solution of the problem). A key observation is that these polynomials depend especially on the first few moments of μ (which are known), and not on the whole of μ (which is to be found). Another illuminating idea (by Akhiezer and Krein [1938]) is to use the shift operator, as in §5.1.1, but on a finite-dimensional unitary space. The Gram–Schmidt orthogonalization procedure with respect to the same scalar product plays a predominant role. A general form of the orthogonalization procedure in an abstract unitary space (Hilbert space) can be found in Appendix B.

(1) Unitary space associated with a Toeplitz matrix. As before, let

$$T_{f,n} = T_{a,n} = (a_{k-j})_{0 \le k, j \le n}$$

be a Toeplitz matrix and let $< x, y >_a$ be the corresponding bilinear form on $\mathbb{C}^{n+1}$, or on $\mathcal{P}_n$,

$$\langle x, y \rangle_a =: (T_{f,n} x, y)_{\mathbb{C}^{n+1}} = \sum_{j,k} a_{k-j} x_j \bar{y}_k, \quad \langle p, q \rangle_a =: \sum_{j,k} a_{k-j} \widehat{p}(j) \overline{\widehat{q}(k)},$$

so that

$$\langle z^j, z^k \rangle_a = a_{k-j} \quad (0 \le k, j \le n).$$

We can suppose that $f = \sum_{|k| \le n} a_k z^k$ is a polynomial, and hence

$$\langle p, q \rangle_a = \langle p, q \rangle_f = \int_{\mathbb{T}} f p \bar{q} \, dm.$$

In all of §5.1.4 we *suppose that $T_{f,n}$ is non-zero and positive definite*,

$$T_{f,n} \gg 0, \quad \text{i.e.} \quad \langle p, p \rangle_a \ge 0 \quad (\forall p \in \mathcal{P}_n),$$

in which case $\langle \cdot, \cdot \rangle_a$ is a semi-inner product defining a non-null seminorm

$$\|p\|_a = \langle p, p \rangle_a^{1/2}$$

and $T_{f,n}$ is the *Gram matrix* of vectors $(z^j)_{0 \le j \le n}$ in this pre-Hilbert space. In particular, we have the Cauchy inequality

$$|\langle p, q \rangle_a| \le \|p\|_a \|q\|_a \quad (\forall p, q \in \mathcal{P}_n).$$

Clearly $\| \cdot \|_a$ is a norm on $\mathcal{P}_k$ ($0 \le k \le n$) if and only if $D_k > 0$, where D_k is the *principal minor of $T_{a,n}$ of order $k+1$* ($k = 0, \ldots, n$)

$$D_k = \det(a_{l-j})_{0 \le l, j \le k},$$

(or a *Toeplitz determinant*) and also, if and only if the vectors $(z^j)_{0 \le j \le k}$ are independent in $(\mathcal{P}_k, \| \cdot \|_a)$, in other words

$$\dim(\mathcal{P}_k, \| \cdot \|_a) = k + 1.$$

5.1 Positive Definiteness, Moment Problems, Orthogonal Polynomials

In general,

$$\dim(\mathcal{P}_k, \|\cdot\|_a) = \mathrm{rank}(\langle z^j, z^l\rangle_a)_{0\le l,j\le k} = \mathrm{rank}(a_{l-j})_{0\le l,j\le k}.$$

In what follows, we use the notation $(\mathcal{P}_k, \|\cdot\|_a)$ for the Hermitian space equipped with the semi-inner product described above as well as for the Euclidean quotient space $\mathcal{P}_k / \mathrm{Ker}\|\cdot\|_a$.

(2) The shift operator $S_{k,a}$. As before, S is the "shift" (translation) on the family of monomials (z^k): for every $p \in \mathcal{P}$,

$$Sp = zp.$$

Note that for $0 \le k < n$, S is *a linear mapping* $S: \mathcal{P}_k \to \mathcal{P}_n$ *preserving the semi-inner product* $\langle \cdot, \cdot\rangle_a$. For every $p, q \in \mathcal{P}_k$, we have

$$\langle Sp, Sq\rangle_a = (TSp, Sq)_{\mathcal{P}_n} = (S^*TSp, q)_{\mathcal{P}_n} = (P_{n-1}TP_{n-1}p, q)_{\mathcal{P}_n}$$
$$= (Tp, q)_{\mathcal{P}_n} = \langle p, q\rangle_a.$$

In particular, S is well-defined as the mapping

$$S: (\mathcal{P}_k, \|\cdot\|_a) \to (\mathcal{P}_n, \|\cdot\|_a).$$

Letting $P_{k,a}$ denote the *orthogonal projection onto $\mathcal{P}_k$ in the space* $(\mathcal{P}_n, \|\cdot\|_a)$ (in the strict sense: onto the canonical range of $\mathcal{P}_k$ in the quotient space $(\mathcal{P}_n, \|\cdot\|_a)$), we set $S_{k,a} = P_{k,a} S|(\mathcal{P}_k, \|\cdot\|_a)$, thus

$$S_{k,a}f = P_{k,a}(zf) \quad (\forall f \in \mathcal{P}_k),$$

so that $S_{k,a}$ becomes a mapping

$$S_{k,a}: (\mathcal{P}_k, \|\cdot\|_a) \to (\mathcal{P}_k, \|\cdot\|_a).$$

(3) $S_{k,a}$ is a contraction. For $0 \le k < n$, $S: (\mathcal{P}_k, \|\cdot\|_a) \to (\mathcal{P}_n, \|\cdot\|_a)$ is an isometric mapping, and $\|S_{k,a}\| \le 1$.

Indeed, the isometric nature of S is already mentioned in (2), thus the compression $S_{k,a}$ is a contraction. ∎

(4) The case where $D_n = 0$. Suppose $D_n = 0$, and let m ($1 \le m \le n$) be such that $D_k = 0$ for $m \le k \le n$ and $D_{m-1} > 0$. Then, $P_{m-1,a} = \mathrm{id}$, and S_{m-1} is a unitary mapping.

Indeed, for every k, $(\mathcal{P}_k, \|\cdot\|_a) \subset (\mathcal{P}_{k+1}, \|\cdot\|_a)$, and by the hypothesis

$$\dim(\mathcal{P}_{m-1}, \|\cdot\|_a) = \dim(\mathcal{P}_m, \|\cdot\|_a) = \cdots = \dim(\mathcal{P}_n, \|\cdot\|_a),$$

hence $(\mathcal{P}_{m-1}, \|\cdot\|_a) = (\mathcal{P}_n, \|\cdot\|_a)$. Thus $P_{m-1,a} = \mathrm{id}$ and $S_{m-1,a} = S: (\mathcal{P}_{m-1}, \|\cdot\|_a) \to (\mathcal{P}_n, \|\cdot\|_a)$. However, S is an isometry (thus unitary, since $m < \infty$). ∎

For the current approach to finite TMP, we consider the orthogonal polynomials in $(\mathcal{P}_n, \|\cdot\|_a)$ for the degenerate "limit case" where $D_n = 0$. By applying the standard orthogonalization procedure (see Appendix B), we obtain the following formulas.

(5) Orthogonal polynomials associated with a Toeplitz matrix. *With the preceding notation, suppose $D_n = 0$, and let m $(1 \leq m \leq n)$ be such that $D_k = 0$ for $m \leq k \leq n$ and $D_{m-1} > 0$. Then the polynomials p_k, $k = 0, \ldots, m$,*

$$p_k(z) = D_{k-1}^{-1} \det \begin{pmatrix} a_0 & a_{-1} & \cdots & a_{-k} \\ a_1 & a_0 & \cdots & a_{-k+1} \\ \cdots & \cdots & \cdots & \cdots \\ a_{k-1} & a_{k-2} & \cdots & a_{-1} \\ 1 & z & \cdots & z^k \end{pmatrix},$$

satisfy the following properties.

(i) $p_k(z) = z^k + \ldots$ $(0 \leq k \leq m)$.
(ii) $\langle p_k, p_j \rangle_a = 0$, $k \neq j$, $0 \leq j, k \leq m$.
(iii) *For $p \in \mathcal{P}_m$, $\|p\|_a = 0$ if and only if $p = \lambda p_m$ for some $\lambda \in \mathbb{C}$. In particular, $\langle p_m, f \rangle_a = 0$ $(\forall f \in \mathcal{P}_n)$. Moreover, $\|z^k p_m\|_a = 0$ for $0 \leq k \leq n - m$ (and hence $\langle z^k p_m, f \rangle_a = 0$, $\forall f \in \mathcal{P}_n$).*
(iv) *If $Z(p_m) = \{\zeta_1, \ldots, \zeta_m\}$ denotes the set of complex roots of p_m, then*

$$Z(p_m) = \sigma(S_{m-1,a})$$

(the spectrum of $S_{m-1,a}$). Moreover, each root ζ_j is simple and is located on $\mathbb{T}$, and the corresponding eigenvector is $p_{m,j} = p_m/(z - \zeta_j)$:

$$S_{m-1,a} p_{m,j} = \zeta_j p_{m,j}.$$

The characteristic polynomial of S_{m-1} is hence p_m:

$$\det(zI - S_{m-1}) = p_m(z).$$

Indeed, (i) is evident, and to verify (ii) it suffices to see that for $0 \leq j < k$, we have $\langle p_k, z^j \rangle_a = 0$. However, the products $\langle z^l, z^j \rangle_a = a_{j-l}$ $(l = 0, \ldots, k)$ form exactly the jth row of the determinant, and the result follows.

For (iii), note that $(z^j)_{0 \leq j < m}$ are independent in $(\mathcal{P}_{m-1}, \|\cdot\|_a)$, thus $\|p_k\|_a > 0$ for $0 \leq k < m$, and if $p \in \mathcal{P}_m$, $p = \sum_{k=0}^m c_k p_k$, then

$$\|p\|_a^2 = \sum_{k=0}^m |c_k|^2 \|p_k\|_a^2,$$

and hence $\|p\|_a^2 = 0 \Rightarrow c_k = 0$ $(0 \leq k < m)$, and $p = c_m p_m$. However, if we had $\|p_m\|_a > 0$, then we would have $\dim(\mathcal{P}_m, \|\cdot\|_a) = m+1$ (given the orthogonality of $(p_k)_{0 \leq k \leq m}$), which is not possible ($D_m = 0$), hence $\|p_m\|_a = 0$.

5.1 Positive Definiteness, Moment Problems, Orthogonal Polynomials 241

The fact that $\|z^k p_m\|_a = 0$ for $0 \leq k \leq n - m$ follows immediately from (2) above. There is another argument using the Toeplitz equations of §5.1.3(3):

$$\|z^k p_m\|_a^2 = (T_{a,n} S^k p_m, S^k p_m) = (S^{*k} T_{a,n} S^k p_m, p_m)$$
$$= (P_{n-k} T_{a,n} P_{n-k} p_m, p_m) = (T_{a,n} p_m, p_m) = \|p_m\|_a^2 = 0.$$

For (iv), let $\zeta \in \sigma(S_{m-1,a})$ and let $p \in (\mathcal{P}_{m-1}, \|\cdot\|_a)$, $p \neq 0$, be an eigenvector of $S_{m-1,a}$, i.e.

$$0 = S_{m-1,a} p - p\zeta = P_{m-1,a}((z - \zeta)p),$$

which (thanks to (4) and (iii)) is equivalent to $(z - \zeta)p = \lambda p_m$ where $\lambda \in \mathbb{C}$; since $\lambda \neq 0$, we have $\zeta \in Z(p_m)$, and hence $\sigma(S_{m-1,a}) \subset Z(p_m)$ (and it also follows that the eigenvalue ζ is geometrically simple: $\dim \mathrm{Ker}(S_{m-1,a} - \zeta_j I) = 1$); conversely, for every j, $p =: p_m/(z - \zeta_j)$ is in $\mathrm{Ker}(S_{m-1,a} - \zeta_j I)$, hence $Z(p_m) \subset \sigma(S_{m-1,a})$. We have already shown that the eigenvalues of $S_{m-1,a}$ are geometrically simple, but their algebraic multiplicities are also 1 since $S_{m-1,a}$ is unitary (there are no Jordan blocks); hence, $\sigma(S_{m-1,a})$ consists of m different unimodular numbers. ∎

5.1.5 The Truncated TMP: The Approach of Carathéodory (1911) and Szegő (1954)

The problem posed in §5.1.2 of the extension of a finite sequence $(a_k)_{1 \leq k \leq n}$ to a positive definite sequence $(a_k)_{k \in \mathbb{Z}}$ is equivalent (thanks to Theorem 5.1.1) to the existence of a measure $\mu \in \mathcal{M}(\mathbb{T})$, $\mu \geq 0$, such that $a_k = \widehat{\mu}(k)$ for $1 \leq k \leq n$. A brilliant idea, rich with links to the other fundamental subjects of analysis, was discovered and realized by Carathéodory [1911]: it involves using the Gauss–Lagrange quadrature formula (of interpolation) on the roots of the polynomials orthogonal with respect to the (future) μ. Another key observation is that these polynomials depend especially on the first few moments of μ, and not on the whole of μ (which is still unknown). This is why we will be able to use the techniques developed in §5.1.3–§5.1.4.

Theorem 5.1.3 (Carathéodory, 1911) *Let $c_1, c_2, \ldots, c_n$ be complex numbers, not all zero.*

(i) *There exists a unique positive measure μ of the form*

$$\mu = \sum_{1 \leq j \leq n} \rho_j \delta_{\zeta_j},$$

where $\rho_j \geq 0$ and $|\zeta_j| = 1$ for every j, such that

$$c_k = \widehat{\mu}(k) = \int_{\mathbb{T}} z^{-k}\, d\mu, \quad k = 1, 2, \ldots, n.$$

(ii) *The measure μ is constructed using the following algorithm:*

- $\rho_j = 0$ *for $m < j \leq n$ where $m = \mathrm{rank}(T_{a,n})$ and $a = (a_j)_{j=-n}^{n}$ is defined by $a_j = \overline{a}_{-j} = c_j$ for $1 \leq j \leq n$, and then*

$$a_0 = -\inf \sigma(T_{a',n}) \quad \text{where } a'_j = a_j \quad (0 < |j| \leq n), \quad a'_0 = 0,$$

- ζ_j, $1 \leq j \leq m$ *are the roots of the orthogonal polynomial $G =: p_m$ of §5.1.4(5),*

$$G(z) = D_{m-1}^{-1} \det \begin{pmatrix} a_0 & a_{-1} & \cdots & a_{-m} \\ a_1 & a_0 & \cdots & a_{-m+1} \\ \cdots & \cdots & \cdots & \cdots \\ a_{m-1} & a_{m-2} & \cdots & a_{-1} \\ 1 & z & \cdots & z^m \end{pmatrix},$$

- $\rho_j = \|G_j\|_a^2 > 0$, $G_j = \dfrac{G}{(z - \zeta_j) G'(\zeta_j)}$ *for $1 \leq j \leq m$.*

In particular, $\mu(\mathbb{T}) = \widehat{\mu}(0) = a_0 = -\inf \sigma(T_{a',n})$.

Proof (following Szegő, 1954) (i) We begin with the same definitions as in §5.1.2, by choosing a_0 in the manner mentioned in (ii): $a = (a_j)_{j=-n}^{n}$ where $a_j = \overline{a}_{-j} = c_j$ for $1 \leq j \leq n$, and

$$a_0 =: -\inf \sigma(T_{a',n}) \quad \text{where } a'_j = a_j \quad (0 < |j| \leq n), \quad a'_0 = 0.$$

Then, $T_{a,n} \gg 0$ and $D_n = 0$, and re-using the notation and conclusions of §5.1.4(5), we set $G = p_m$ and

$$G_j = \frac{G}{(z - \zeta_j) G'(\zeta_j)} = p_{m,j} \quad (1 \leq j \leq m),$$

so that the G_j are the eigenvectors of the unitary operator $S_{m-1,a}$: $S_{m-1,a} G_j = \zeta_j G_j$. They thus form an orthogonal basis of $(\mathcal{P}_{m-1}, \|\cdot\|_a)$, which implies, in particular,

$$1 = \sum_{j=1}^{m} \langle 1, G_j \rangle_a \frac{G_j}{\|G_j\|_a^2},$$

and then, for every polynomial $f \in \mathcal{P}_a$,

$$\langle f(S_{m-1,a}) 1, 1 \rangle_a = \sum_{j=1}^{m} f(\zeta_j) \frac{|\langle 1, G_j \rangle_a|^2}{\|G_j\|_a^2}.$$

5.1 Positive Definiteness, Moment Problems, Orthogonal Polynomials 243

Moreover, the Gauss–Lagrange interpolation formula tells us that for every polynomial $h \in \mathcal{P}_{m-1}$, $h = \sum_{j=1}^{m} h(\zeta_j) G_j$, therefore $h(\zeta_j) = \langle h, G_j \rangle_a / \|G_j\|_a^2$, and in particular (for $h = 1$), $1 = \langle 1, G_j \rangle_a / \|G_j\|_a^2$. For every polynomial $f \in \mathcal{P}_a$, we obtain

$$\langle f(S_{m-1,a})1, 1 \rangle_a = \sum_{j=1}^{m} f(\zeta_j) \|G_j\|_a^2,$$

and by setting $f = z^k$ ($0 \le k \le n$) and using $z^k(S_{m-1,a})1 = z^k$ (in the quotient space $(\mathcal{P}_{m-1}, \|\cdot\|_a) = (\mathcal{P}_n, \|\cdot\|_a)$),

$$a_k = (T_{a,n}1, z^k) = \langle 1, z^k \rangle_a = \langle 1, z^k(S_{m-1,a})1 \rangle_a = \sum_{j=1}^{m} \zeta_j^{-k} \|G_j\|_a^2 = \int_{\mathbb{T}} z^{-k} d\mu,$$

where $\mu = \sum_{j=1}^{m} \rho_j \delta_{\zeta_j}$, $\rho_j = \|G_j\|_a^2$. The uniqueness of μ is justified in (ii) below.

(ii) The algorithm was already justified in the course of the proof of (i). To show the uniqueness of μ, suppose that $\nu = \sum_{j=1}^{l} w_j \delta_{z_j}$ is such that

$\widehat{\nu}(k) = c_k$, $1 \le k \le n$, $w_j > 0$ $(1 \le j \le l \le n)$ and $|z_j| = 1$, $z_j \ne z_i$ $(j \ne i)$.

If $c = (c_j)_{|j| \le n}$, $c_j = \widehat{\nu}(j)$, then

$$(T_{c,n}p, p)_{\mathcal{P}_n} = (p, p)_{L^2(\nu)} = \int_{\mathbb{T}} |p|^2 \, d\nu \ge 0 \quad (\forall p \in \mathcal{P}_n),$$

and $(T_{c,n}q, q)_{\mathcal{P}_n} = 0$ where $q = \prod_{j=1}^{l}(z - z_j)$, which implies $T_{c,n} \gg 0$ and $\inf \sigma(T_{c,n}) = 0$. By the definition of the sequence a in (i), we have $a_0 = c_0$, and hence $a = c$. It follows that $l = \text{rank}(T_{c,n}) = \text{rank}(T_{a,n}) = m$, and then

$$\|q\|_a^2 = (T_{c,n}q, q) = \int_{\mathbb{T}} |q|^2 \, d\nu = 0.$$

By (iii) of §5.1.4(3), we have $q = \lambda p_m$ ($\lambda \in \mathbb{C}$, $\lambda \ne 0$), hence $\{z_1, \ldots, z_m\} = \{\zeta_1, \ldots, \zeta_m\}$.

And finally, for every k, $1 \le k \le m$, we have

$$w_k = \nu(\{z_k\}) = \int_{\mathbb{T}} |G_k|^2 \, d\nu = (T_{a,n}G_k, G_k) = \int_{\mathbb{T}} |G_k|^2 \, d\mu = \mu(\{z_k\}) = \rho_k. \quad \blacksquare$$

Remark After having identified $\{z_1, \ldots, z_m\} = \{\zeta_1, \ldots, \zeta_m\}$, the last line of the calculation could be replaced with a reference to Vandermonde's theorem, giving $w_k = \rho_k$, $\forall k$.

Corollary 5.1.4 (1) Another proof of Theorem 5.1.1.

We only need to consider the implication (ii) $\Rightarrow$ (i) since the converse is already evident. Let $a' = (a'_k)_{k=-n}^{n}$ where $a'_k = a_k$ ($k \ne 0$) and a_0 is replaced by $a'_0 =$

$a_0 - \inf \sigma(T_{a,n})$. By Theorem 5.1.3, there exists a positive measure μ_n such that $\widehat{\mu_n}(k) = a_k$ ($0 < |k| \le n$) and $\mu_n(\mathbb{T}) = \widehat{\mu_n}(0) = a_0 - \inf \sigma(T_{a,n}) \le a_0$. By the Helly selection theorem (weak compactness of the ball in $\mathcal{M}(\mathbb{T})$: see Appendix A), there exists a subsequence weak-∗-convergent to a limit $\lambda = \lim_k \mu_{n_k} \ge 0$, for which $\widehat{\lambda}(k) = a_k$ ($\forall k \in \mathbb{Z} \setminus \{0\}$) and $\lambda(\mathbb{T}) = \widehat{\lambda}(0) \le a_0$. The measure $\mu = \lambda + (a_0 - \widehat{\lambda}(0))m$ resolves the question (m is the normalized Lebesgue measure). ∎

Corollary 5.1.5 (2) A uniqueness theorem for the Fourier transform. *If μ and ν are two positive measures with finite support in $\mathbb{T}$, and*

$$n = \max(\#\mathrm{supp}(\mu), \#\mathrm{supp}(\nu)), \quad \text{if } \widehat{\mu}(k) = \widehat{\nu}(k) \text{ for } 1 \le k \le n,$$

then $\mu = \nu$.

In particular, if $|\zeta_j| = |z_l| = 1$ ($1 \le j \le m$, $1 \le l \le n$ and $m \le n$) and

$$\sum_{j=1}^{m} \zeta_j^k = \sum_{l=1}^{n} z_l^k \quad \text{for } 1 \le k \le n,$$

then $m = n$ and $\{z_1, \ldots, z_m\} = \{\zeta_1, \ldots, \zeta_m\}$.

Indeed, this is a consequence of the uniqueness part of Theorem 5.1.3. ∎

Gábor Szegő (1895–1985), a Hungarian–German–American mathematician, is known for his works in classical analysis, especially on orthogonal polynomials and Toeplitz matrices and operators. His famous collection of resolved problems (with George Pólya), *Aufgaben und Lehrsätze aus der Analysis*, vols I and II (1925), served for decades as an indispensable source of training for generations of analysts. He is the author of several reference monographs, including *Orthogonal Polynomials* [Szegő, 1959] and *Toeplitz Forms*, with Ulf Grenander [Grenander and Szegő, 1958]. It is impossible to overestimate the influence of many of Szegő's publications on the development of the theory of Toeplitz matrices/operators, particularly his founding articles, "Beiträge zur Theorie der Toeplitzsche Formen" [Szegő, 1920, 1921a], or his justification [Szegő, 1952] of Lars Onsager's conjecture (future Nobel Prize in Chemistry, 1968), crucial for the discovery of phase transition in the two-dimensional Lenz–Ising model.

After his first university years in Budapest, Szegő moved to Berlin and Göttingen in 1913–1914, where he took the courses of Schwarz, Hilbert, Landau, and Haar. The First World War erupted, and in 1915, rather unexpectedly, Szegő enlisted in the Austro–Hungarian cavalry, where he remained until 1919 (long after the end of the war). Passing through Vienna in 1918 he submitted his thesis (with its decisive result [Szegő, 1915] on the asymptotics of Toeplitz determinants which arose as a response to a question of his friend George Pólya). On his return from the army in 1919, but still in uniform, he married Anne Neményi, like him of Jewish origin and from an academic background (she submitted a thesis in chemistry the same year). Among the experiences of Szegő in Budapest at this time, he tutored a young child prodigy, János (Johanne, John) von Neumann; according to his wife, Szegő was delighted with his pupil and often returned with tears in his eyes after the sessions with the little János – so rapid and profoundly complete were the responses of this (future) genius. Life after the war was so hard in Hungary that Szegő accepted the support of Max von Neumann, a powerful banker and the father of János, who helped him move to Berlin, where, after his *Habilitation* thesis (1921), Szegő practiced as a *Privatdozent*. In 1926 Szegő obtained a position as professor in Königsberg (vacated by Konrad Knopp), but soon the climate changed with the Nazis' arrival to power. At first, the infamous "law of April 7, 1933" excluding Jews from any civil service positions did not touch Szegő, as he profited from an exemption accorded by the law to veterans of the First World War, but the hostility of the students and the daily threats made life impossible for the family (they already had two children). His friend George Pólya, in peaceful Switzerland, put him in contact with Jacob Tamarkin, a renowned American–Russian–Jewish mathematician, and professor at Brown University, who found him a position in St. Louis (1934). From 1938 to 1966 (when he retired) Szegő was professor and Dean of the Mathematics Department at Stanford University; in the words of Peter Lax of the Courant Institute in New York, "Szegő used his powers to turn the provincial mathematics department that Stanford had been ... into one of the leading departments of the country that Stanford is today." During the Second World War, Szegő again served in the army and was sent on a mission to Biarritz in 1945–1946 to teach at the American Army University.

Szegő was elected to several academies, including the National Academy of Sciences; his bust is installed in St. Louis and in Kunhegyes (his birth town in Hungary). His works are gathered in three volumes of *Collected Papers* (edited by Richard Askey).

5.2 Norm of a Toeplitz Matrix

As before, we regard a Toeplitz matrix of size $(n+1) \times (n+1)$,

$$T_{a,n} = T_{f,n} =: (a_{k-j})_{0 \le k, j \le n},$$

associated with a sequence $(a_k)_{k \in \mathbb{Z}} \in l^\infty(\mathbb{Z})$ or a distribution (on $\mathbb{T}$) $f = \sum_{k \in \mathbb{Z}} a_k z^k$, as a linear mapping on $\mathbb{C}^n$, $T_{f,n} \colon \mathbb{C}^n \to \mathbb{C}^n$, or on $\mathcal{P}_n = \mathrm{Vect}(z^k \colon 0 \le k \le n)$,

$$T_{f,n} \colon \mathcal{P}_n \to \mathcal{P}_n.$$

Clearly $T_{a,n}$ depends only on the $(a_k)_{|k| \le n}$ and hence, among the symbols f of such a matrix there is always a polynomial $f = \sum_{|k| \le n} a_k z^k$. This allows us to consider $T_{f,n}$ as a truncated Toeplitz operator $T_{f,n} = P_n T_f | \mathcal{P}_n$ where P_n is the orthogonal projection of H^2 onto $\mathcal{P}_n$. It is also clear that for every $f = \sum_{k \in \mathbb{Z}} a_k z^k$, we have

$$\sup_{n \ge 0} \|T_{f,n}\| = \lim_n \|T_{f,n}\| = \|f\|_{L^\infty(\mathbb{T})},$$

and in particular, the sup is finite if and only if $f \in L^\infty(\mathbb{T})$ (indeed, sup $< \infty$ easily implies $(a_k)_{k \in \mathbb{Z}} \in l^\infty(\mathbb{Z})$, and then for every $x \in \mathcal{P}_a$, $\lim_n \|T_{f,n} P_n x - T_f x\|_2 = \lim_n \|(P_+ - P_n) f x\|_2 = 0$, hence $\|f\|_{L^\infty(\mathbb{T})} = \|T_f\| \le \sup_{n \ge 0} \|T_{f,n}\|$; the converse is evident). ∎

The truncated Toeplitz (and Wiener–Hopf) operators have already appeared in Exercise 3.4.15 and in §3.5 and §4.4. In particular, we showed in Exercise 3.4.15(b) that by selecting a bounded symbol of the matrix $T_{\varphi,n}$ ($\varphi \in L^\infty(\mathbb{T})$), we have

$$\|T_{\varphi,n}\| \le d_n(\varphi) =: \mathrm{dist}_{L^\infty(\mathbb{T})}(\varphi, \bar{z}^n H_-^\infty + z^{n+1} H^\infty),$$

where, as before,

$$H^\infty = \{f \in L^\infty(\mathbb{T}) \colon \widehat{f}(k) = 0 \text{ for } k < 0\},$$
$$H_-^\infty = \{f \in L^\infty(\mathbb{T}) \colon \widehat{f}(k) = 0 \text{ for } k \ge 0\}.$$

The goal of this section is to establish a sort of converse to this inequality. More precisely, the following theorem holds.

Theorem 5.2.1 *For every $\varphi \in L^\infty(\mathbb{T})$,*
$$\frac{1}{3} d_n(\varphi) \leq \|T_{\varphi,n}\| \leq d_n(\varphi).$$
More precisely, there exists a polynomial $\psi = \sum_{|k| \leq 2n} \widehat{\psi}(k) z^k$ such that
$$\widehat{\psi}(k) = \widehat{\varphi}(k) \quad (|k| \leq n), \quad \text{and} \quad \|\psi\|_\infty \leq 3\|T_{\varphi,n}\|.$$

5.2.1 Comments and Special Cases

(a) The equality $\|T_{\varphi,n}\| = d_n(\varphi)$ does not hold in general. By analogy with the Hankel operators, where on one hand $H_f = 0$ if and only if $f \in H^\infty$, and on the other hand $\|H_\varphi\| = \text{dist}_{L^\infty(\mathbb{T})}(\varphi, H^\infty)$ (see Theorem 2.2.4), we could suspect that we always have $\|T_{\varphi,n}\| = d_n(\varphi)$. We show with an example that *this is not the case*. Let $n = 1$ and $\varphi(z) = \bar{z} + z$, and thus
$$T_{\varphi,1} = \begin{pmatrix} 0 & 1 \\ 1 & 0 \end{pmatrix}.$$

Then, for every $p, q \in \mathcal{P}$, we have
$$\|\varphi + \bar{z}^2 \bar{q} + z^2 p\|_\infty \geq \|\varphi + \bar{z}^2 \bar{q} + z^2 p\|_2 \geq 2^{1/2}$$
(by Parseval), and hence
$$d_1(\varphi) \geq 2^{1/2} > 1 = \|T_{\varphi,1}\|. \quad \blacksquare$$

(b) The Carathéodory–Fejér theorem (1911): the case of analytic φ. *For φ analytic, $\varphi \in H^\infty$, we have*
$$\|T_{\varphi,n}\| = \text{dist}_{L^\infty(\mathbb{T})}(\varphi, z^{n+1} H^\infty) = d_n(\varphi).$$

Indeed, by identifying $\mathcal{P}_n = H^2 \ominus z^{n+1} H^2$, we obtain
$$T_{\varphi,n} = T_\varphi^{z^{n+1}},$$
a truncated Toeplitz operator in the sense of Exercise 3.4.15, and hence, by Exercise 3.4.15(c),
$$\|T_{\varphi,n}\| = \text{dist}_{L^\infty(\mathbb{T})}(\varphi, z^{n+1} H^\infty),$$
thus $d_n(\varphi) \leq \text{dist}_{L^\infty(\mathbb{T})}(\varphi, z^{n+1} H^\infty) = \|T_{\varphi,n}\| \leq d_n(\varphi)$, and the result follows. $\blacksquare$

Note a curious phenomenon: while the Fourier spectrum of φ (which is in $[0, n]$) is quite far from that of $\bar{z}^n H_-^\infty$ (which is in $(-\infty, -n-1]$), this last component has no impact on the distance, i.e.

$$d_n(\varphi) = \mathrm{dist}_{L^\infty(\mathbb{T})}(\varphi, \bar{z}^n H_-^\infty + z^{n+1} H^\infty) = \mathrm{dist}_{L^\infty(\mathbb{T})}(\varphi, z^{n+1} H^\infty).$$

(c) The norm of $T_{\varphi,n}$ and $\|\varphi\|_\infty$ for the case of a polynomial φ. *In §5.2.1, it is not possible to replace a polynomial $\psi \in \mathcal{P}$ of $\deg(\psi) \leq 2n$ with a polynomial of degree n ($\deg(\psi) \leq n$) (even for analytic polynomials $\varphi \in \mathcal{P}_a$). Moreover,*

$$\sup\left\{\frac{\|\varphi\|_\infty}{\|T_{\varphi,n}\|} : \varphi \in \mathcal{P}_n\right\} = G_n,$$

where $G_n =: \|P_n : H^\infty \to \mathcal{P}_n\|$ is the nth "Landau constant." It is known that

$$G_n \sim \frac{1}{\pi}\log(n+1) \quad \text{(Landau, 1913),}$$

and

$$1 + \frac{1}{\pi}\log(n+1) \leq G_n < 1.0663 + \frac{1}{\pi}\log(n+1) \quad \text{(Brutman, 1982).}$$

See the Remarks below.

Indeed, given $\varphi \in \mathcal{P}_n$, we apply (b) and find a function $h \in H^\infty$ such that $\varphi = P_n h$ and $\|T_{\varphi,n}\| = \|h\|_\infty$, giving

$$\|\varphi\|_\infty \leq \|P_n : H^\infty \to \mathcal{P}_n\| \cdot \|T_{\varphi,n}\| = G_n \|T_{\varphi,n}\|.$$

Conversely, if $\epsilon > 0$ and $h \in H^\infty$ such that $\|P_n h\|_\infty \geq (G_n - \epsilon)\|h\|_\infty$, then for $\varphi = P_n h$ we have $\|\varphi\|_\infty \geq (G_n - \epsilon)\|h\|_\infty \geq (G_n - \epsilon)\|T_{h,n}\| = (G_n - \epsilon)\|T_{\varphi,n}\|$, and the result follows. ∎

Remark (i) A result of Landau [1913] gives the expression

$$G_n = \sum_{k=0}^n \left(\frac{1}{4^k} C_{2k}^k\right)^2$$

where $C_m^l = m!/(l!(m-l)!)$ are the binomial coefficients; see the comments in §5.8 for references and a comparison of the G_n with the (better-known) *Lebesgue constants L_n,*

$$L_n = \|P_n : L^\infty(\mathbb{T}) \to \mathcal{P}_n\| = \|\mathcal{D}_n\|_{L^1(\mathbb{T})} = \frac{4}{\pi^2}\log(n+1) + O(1) \quad \text{(Fejér, 1910)}$$

where

$$\mathcal{D}_n(e^{it}) = \sum_{k=0}^n e^{ikt} = e^{int/2} \frac{\sin((n+1)/2)t}{\sin(t/2)}$$

5.2 Norm of a Toeplitz Matrix

is the Dirichlet kernel. We provide here the elementary bounds

$$10^{-1}\log(n+1) \leq G_n \leq L_n \leq 2 + \frac{4}{\pi^2}\log(n+1) \quad (4/\pi^2 < 0.41 < 1/2).$$

Indeed, it is shown in Exercise 5.6.2(f) of [Nikolski, 2019] that for

$$f = \sum_{1 \leq k \leq n}(z^{n+k} - z^{n-k})/2k$$

($f \in \mathcal{P}_{2n}$), we have $\|P_n f\|_\infty \geq 10^{-1}\|f\|_\infty \log(n+1)$, which gives the inequality on the left.

For the inequality on the right, we rely on $G_n \leq L_n$ and

$$L_n = \|\mathcal{D}_n\|_{L^1(\mathbb{T})} = \frac{1}{\pi}\int_0^\pi \left|\frac{\sin((n+1)/2)t}{\sin(t/2)}\right|dt$$

$$= \frac{1}{\pi}\int_0^\pi \left|\sin\frac{n+1}{2}t\right|\left(\frac{2}{t} + \frac{1}{\sin(t/2)} - \frac{2}{t}\right)dt$$

$$\leq \frac{2}{\pi}\int_0^\pi t^{-1}\left|\sin\frac{n+1}{2}t\right|dt + \frac{1}{\pi}\int_0^\pi \left(\frac{1}{(t/2) - (1/6)(t/2)^3} - \frac{2}{t}\right)dt$$

$$\leq \frac{2}{\pi}\int_0^\pi t^{-1}\left|\sin\frac{n+1}{2}t\right|dt + 0.45$$

$$\leq 2.1 + \frac{2}{\pi}\sum_{1 \leq j \leq n/2}\int_{I(j)} t^{-1}\left|\sin\frac{n+1}{2}t\right|dt,$$

where

$$I(j) = \left(\frac{2\pi j}{n+1}, \frac{2\pi(j+1)}{n+1}\right), \quad j = 1,\ldots,\lfloor n/2 \rfloor.$$

Since

$$\int_{I(j)}\left|\sin\frac{n+1}{2}t\right|dt = \frac{4}{n+1},$$

we obtain

$$L_n \leq 2.1 + \frac{2}{\pi}\sum_{1 \leq j \leq n/2}\frac{4}{n+1}\left(\frac{2\pi j}{n+1}\right)^{-1}$$

$$\leq 2 + \frac{4}{\pi^2}\log(n+1). \qquad \blacksquare$$

(ii) We can state the same result by speaking of a "band extension" T_φ of a matrix $T_{\varphi,n}$ (see Figure 5.1 below):

$$\sup\left\{\frac{\|T_\varphi\|}{\|T_{\varphi,n}\|} : \varphi \in \mathcal{P}_n\right\} \geq \frac{1}{10}\log(n+1).$$

(d) The norm of a circulant. Let $T_{\varphi,n}$ be a circulant Toeplitz matrix, i.e. $\varphi = p + z^{-n-1}p$ where $p \in \mathcal{P}_n$, in other words

$$T_{\varphi,n} = \begin{pmatrix} a_0 & a_n & \cdots & a_2 & a_1 \\ a_1 & a_0 & \cdots & a_3 & a_2 \\ \cdots & \cdots & \cdots & \cdots & \cdots \\ a_{n-1} & a_{n-2} & \cdots & a_0 & a_n \\ a_n & a_{n-1} & \cdots & a_1 & a_0 \end{pmatrix}.$$

See §5.5 for information on circulants. In particular, it follows from §5.5.4 that

$$\|T_{\varphi,n}\| = \max_{0 \le k \le n} |p(e^{2\pi i k/(n+1)})| = \frac{1}{2} \max_{0 \le k \le n} |\varphi(e^{2\pi i k/(n+1)})|. \quad \blacksquare$$

(e) The "continuous" case. Theorem 5.2.1 is a discrete analog of Rochberg's Theorem 4.4.4; for comments see "Notes and Remarks," §5.8 below.

Our proof of Theorem 5.2.1 depends on the lemma below which, in fact, in itself represents a remarkable proposition on the following extremal matrix problem: let $\mathfrak{T}_n$ denote the set of strictly upper triangular $(n+1) \times (n+1)$ matrices; given a matrix A, find

$$\text{dist}(A, \mathfrak{T}_n) = \min_{B \in \mathfrak{T}_n} \|A + B\|.$$

The question is also equivalent to the problem of the best extension (completion) of a lower triangular matrix $A = (a_{ij})_{0 \le i,j \le n}$, $a_{ij} = 0$ for $j > i$, by a strictly upper triangular matrix $B = (b_{ij})_{0 \le i,j \le n}$, $b_{ij} = 0$ for $j \le i$, in such a manner that $\|A + B\|$ attains its minimum among all possible extensions. The proof of Lemma 5.2.2 will be given at the end of this section.

Lemma 5.2.2 (Arveson's distance formula, 1975) *Let $A = (a_{ij})_{0 \le i,j \le n}$ be a $(n+1) \times (n+1)$ matrix. Then*

$$\text{dist}(A, \mathfrak{T}_n) = \max_{0 \le k \le n} \left\| \begin{pmatrix} a_{k0} & \cdots & a_{kk} \\ \cdots & \cdots & \cdots \\ a_{n0} & \cdots & a_{nk} \end{pmatrix} \right\| = \max_{0 \le k \le n} \|(I - P_{k-1})AP_k\|,$$

where P_k is the orthogonal projection of $\mathbb{C}^{n+1}$ onto $\mathbb{C}^{k+1}$:

$$P_k x = P_k(x_0, \ldots, x_k, x_{k+1}, \ldots, x_n)$$
$$= (x_0, \ldots, x_k, 0, \ldots, 0), \quad x \in \mathbb{C}^{n+1} \quad (P_{-1} =: 0).$$

5.2 Norm of a Toeplitz Matrix

Figure 5.1 A band Toeplitz matrix. The gray is a Toeplitz portion $\widehat{\varphi}(k-j)$, $0 \le j$, $k \le n$; the white is zero.

5.2.2 Proof of Theorem 5.2.1

If necessary replacing φ with $\sum_{|k|\le n}\widehat{\varphi}(k)z^k$, we can suppose that $\varphi = \sum_{|k|\le n}\widehat{\varphi}(k)z^k$. Then, the Toeplitz operator

$$T_\varphi : l^2(\mathbb{Z}_+) \to l^2(\mathbb{Z}_+)$$

has a "banded matrix" of width $2n+1$: see Figure 5.1.

Let

$$P_{l,m}f = \sum_{k=l}^{m}\widehat{f}(k)z^k$$

be the orthogonal projection from $l^2(\mathbb{Z}_+)$ onto $\mathrm{Vect}(z^k : l \le k \le m)$, let $A =: T_{\varphi,n} = P_{0,n}T_\varphi P_{0,n}$ (the initial matrix), and let

$$\tilde{B} = P_{0,n}T_\varphi P_{n,2n}, \quad \tilde{C} = P_{n,2n}T_\varphi P_{0,n}$$

be lower and upper triangular matrices respectively. We apply Arveson's distance formula of Lemma 5.2.2 to the matrix $\tilde{B}$ to find a strictly upper triangular matrix $\Delta_1 \in \mathfrak{T}_n$ such that

$$\|B\| \le \max_{n\le k\le 2n} \|P_{k-n,n}T_\varphi P_{n,k}\| \quad \text{where } B = \tilde{B} + \Delta_1.$$

In the same manner, we can find a strictly lower triangular matrix Δ_2 such that

$$\|C\| \le \max_{0\le k\le n} \|P_{n,n+k}T_\varphi P_{k,n}\| \quad \text{where } C = \tilde{C} + \Delta_2.$$

Clearly, for $n \le k \le 2n$, we have

$$P_{k-n,n}T_\varphi P_{n,k} = P_{0,2n-k}T_\varphi P_{2n-k,n} = P_{0,2n-k}T_{\varphi,n}P_{2n-k,n},$$

and hence $\|B\| \le \|T_{\varphi,n}\|$. Similarly, $\|C\| \le \|T_{\varphi,n}\|$.

252   Toeplitz Matrices: Moments, Spectra, Asymptotics

We now consider the following infinite block tridiagonal matrix:

$$T = \begin{pmatrix} A & B & 0 & 0 & \cdots \\ C & A & B & 0 & \cdots \\ 0 & C & A & B & \cdots \\ 0 & 0 & C & A & \cdots \\ \cdots & \cdots & \cdots & \cdots & \cdots \end{pmatrix}.$$

We have $T = (t_{ij})_{i,j \geq 0}$ where $t_{ij} = \widehat{\varphi}(i - j)$ for every $|i - j| \leq n$. Applying the projection procedure of Exercise 2.3.2 to T, we obtain a Toeplitz matrix T_ψ such that $\widehat{\psi}(i - j) = \widehat{\varphi}(i - j)$ for $|i - j| \leq n$, $\widehat{\psi}(i - j) = 0$ for $|i - j| > 2n$ and $\|\psi\|_\infty = \|T_\psi\| \leq \|T\|$. But clearly $\|T\| \leq \|A\| + \|B\| + \|C\| \leq 3\|T_{\varphi,n}\|$. ∎

The proof of Lemma 5.2.2 presented here depends on another lemma, Stephen Parrott's "four-block lemma," already mentioned in §2.4.

Lemma 5.2.3 (four-block lemma) *Let H_i, K_i ($i = 1, 2$) be four Hilbert spaces and let $A: H_1 \to K_1$, $B: H_1 \to K_2$, and $C: H_2 \to K_2$ be three bounded operators. If $M_X: H_1 \oplus H_2 \to K_1 \oplus K_2$ is defined by the four-block matrix*

$$M_X = \begin{pmatrix} A & X \\ B & C \end{pmatrix},$$

where $X: H_2 \to K_1$, then

$$\min_X \|M_X\| = \max\{\|(A, B)^{\mathrm{col}}\|, \|(B, C)\|\}.$$

Proof Since $(A, B)^{\mathrm{col}}: H_1 \to K_1 \oplus K_2$ coincides with $M_X P_{H_1}$ and $(B, C) = P_{K_2} M_X$, clearly $\min_X \|M_X\| \geq \max\{\|(A, B)^{\mathrm{col}}\|, \|(B, C)\|\}$.

For the converse, suppose that $\max\{\|(A, B)^{\mathrm{col}}\|, \|(B, C)\|\} = 1$, which leads to $\|(A, B)^{\mathrm{col}}\| \leq 1$, or

$$I = \mathrm{id}_{H_1} \geq ((A, B)^{\mathrm{col}})^*(A, B)^{\mathrm{col}} = A^*A + B^*B.$$

Then, for every $x \in H_1$, $\|Ax\|^2 = (A^*Ax, x) \leq ((I - B^*B)x, x) = \|(I - B^*B)^{1/2}x\|^2$ (for the square root of a positive definite operator $I - B^*B \gg 0$, see Appendix D). Thus, the mapping $\alpha: (I - B^*B)^{1/2}x \mapsto Ax$ is well-defined on the range $\mathrm{Ran}(I - B^*B)^{1/2}$, and is linear and contracting; we complete its definition $\alpha: H_1 \to K_1$ by setting $\alpha = 0$ on $(\mathrm{Ran}(I - B^*B)^{1/2})^\perp = \mathrm{Ker}(I - B^*B)^{1/2}$, giving

$$A = \alpha(I - B^*B)^{1/2}.$$

5.2 Norm of a Toeplitz Matrix

In the same manner, $BB^* + CC^* \leq I_{K_2}$ implies $C^* = \beta(I - BB^*)^{1/2}$ for a contraction $\beta \colon K_2 \to H_2$, and hence $C = (I - BB^*)^{1/2}\beta^*$. Setting $X = -\alpha B^*\beta^* \colon H_2 \to K_1$ we obtain the following identity:

$$\begin{pmatrix} \alpha & 0 \\ 0 & I_{K_2} \end{pmatrix} \begin{pmatrix} -B^* & (I - B^*B)^{1/2} \\ (I - BB^*)^{1/2} & B \end{pmatrix} \begin{pmatrix} 0 & \beta^* \\ I_{H_1} & 0 \end{pmatrix}$$

$$= \begin{pmatrix} \alpha & 0 \\ 0 & I_{K_2} \end{pmatrix} \begin{pmatrix} (I - B^*B)^{1/2} & -B^*\beta^* \\ B & (I - BB^*)^{1/2}\beta^* \end{pmatrix} \begin{pmatrix} \alpha(I - B^*B)^{1/2} & -\alpha B^*\beta^* \\ B & C \end{pmatrix}$$

$$= \begin{pmatrix} A & X \\ B & C \end{pmatrix} = M_X,$$

where the two diagonal matrices are contracting and the Julia matrix U_{B^*},

$$U_{B^*} = \begin{pmatrix} -B^* & (I - B^*B)^{1/2} \\ (I - BB^*)^{1/2} & B \end{pmatrix}$$

is unitary (see Lemma 5.2.4 below). It follows that $\|M_X\| \leq 1$. ∎

Lemma 5.2.4 (Julia, 1944) *For every contracting operator between Hilbert spaces $A \colon H \to K$, the matrix U_A defines a unitary operator $U_A \colon H \oplus K \to K \oplus H$,*

$$U_A = \begin{pmatrix} -A & (I - AA^*)^{1/2} \\ (I - A^*A)^{1/2} & A^* \end{pmatrix}.$$

Proof By matrix multiplication,

$$U_A^* U_A = \begin{pmatrix} -A^* & (I - A^*A)^{1/2} \\ (I - AA^*)^{1/2} & A \end{pmatrix} \begin{pmatrix} -A & (I - AA^*)^{1/2} \\ (I - A^*A)^{1/2} & A^* \end{pmatrix}$$

$$= \begin{pmatrix} I & -A^*(I - AA^*)^{1/2} + (I - A^*A)^{1/2}A^* \\ -(I - AA^*)^{1/2}A + A(I - A^*A)^{1/2} & I \end{pmatrix}.$$

The off-diagonal elements are zero since $(I - AA^*)A = A(I - A^*A)$, and then $(I - AA^*)^n A = A(I - A^*A)^n$ (for every $n = 0, 1, \ldots$), hence $p((I - AA^*))A = Ap((I - A^*A))$ for every polynomial p. Choosing a sequence (p_n) uniformly convergent to $x \mapsto x^{1/2}$ on the interval $[0, 1]$, we obtain $p_n((I - AA^*))A = Ap_n((I - A^*A))$, and hence $(I - AA^*)^{1/2}A = A(I - A^*A)^{1/2}$. ∎

5.2.3 Proof of Lemma 5.2.2

Let D be the right-hand side of the identity; then $D \leq \operatorname{dist}(A, \mathfrak{T}_n)$ (since $(I - P_{k-1})(A + X)P_k = (I - P_{k-1})AP_k$ for every $X \in \mathfrak{T}_n$). We construct the required extension of the triangular matrix $A' = (a_{ij}, (j \leq i); a_{ij} = 0 \ (j > i))$, by iteratively applying Lemma 5.2.3. More precisely, to define x_{01}, we apply Lemma 5.2.3 to the $2 \times n$ matrix below on the left, to find x_{01} (a 1×1 matrix) such that $\|A_1\| \leq D$:

$$A_1 = \begin{pmatrix} a_{00} & \underline{x_{01}} \\ a_{10} & a_{11} \\ \cdots & \cdots \\ \cdots & \cdots \\ a_{n0} & a_{n1} \end{pmatrix}, \quad A_2 = \begin{pmatrix} a_{00} & x_{01} & \underline{x_{02}} \\ a_{10} & a_{11} & \underline{x_{12}} \\ a_{20} & a_{21} & a_{22} \\ \cdots & \cdots & \cdots \\ \cdots & \cdots & \cdots \\ a_{n0} & a_{n1} & a_{n2} \end{pmatrix}.$$

Then, to define x_{02}, x_{12} we apply it to $3 \times n$ the matrix on the right to find $(x_{02}, x_{12})^t$ (a 1×2 matrix), such that $\|A_2\| \leq D$, etc. Finally, we obtain $\|A_n\| \leq D$, $A_n = A' + X$ where $X \in \mathfrak{T}_n$. ∎

5.3 Inversion of a Toeplitz Matrix

The problem of inversion of a finite matrix Toeplitz $T_{\varphi,n}$ ($\varphi \in L^\infty(\mathbb{T})$) is quite different from that of an infinite matrix (an operator T_φ), and one of the reasons for this difference is immediately visible: the matrices $T_{\varphi,n}$ are not as tightly linked to the theory of functions on $\mathbb{T}$ as were the T_φ, and hence the language used in §3.3 loses its dominant role. It returns when we study the asymptotic inversion $n \to \infty$ (the limits $\lim_n T_{\varphi,n}^{-1}$, etc.): see §5.4, §5.6, §5.7. For the case of fixed n, the language of functions is replaced by that of vector spaces and matrices. Of course, for all matrices, the invertibility of $A\colon \mathbb{C}^n \to \mathbb{C}^n$, $A = (a_{ij})$, is equivalent to $\det(A) \neq 0$, or Ran$A =: A(\mathbb{C}^n) = \mathbb{C}^n$, or Ran$A^* = \mathbb{C}^n$; or the same criteria for a *transposed matrix*,

$$A^t = (a_{ij}^t), a_{ij}^t = a_{ji}.$$

In the case of a Toeplitz $A = T_{\varphi,n}$, this language remains theoretically quite inefficient (because of the problem of calculating $\det(T_{\varphi,n})$), but for numerical applications (of which there are many) it turns out to be very useful.

5.3.1 Two Matrix Inversion Theorems

The principal criterion of invertibility of a $T_{\varphi,n}$, found in the two theorems of this section, allude to two ideas of the theory of Toeplitz operators $T_\varphi\colon H^2 \to H^2$ already linked to the invertibility: the use of the equation $T_\varphi x = 1$ for the invertibility (see Exercise 3.4.11), and the utility of the Birkhoff–Wiener–Hopf factorization for an explicit form of the inverse $T_\varphi^{-1} = T_{1/\varphi_+} T_{1/\varphi_-}$ (see §3.3.1 and Theorem 3.3.6).

5.3 Inversion of a Toeplitz Matrix

We retain the previous notation for Toeplitz matrices,

$$T_{a,n} = T_{f,n} =: (a_{k-j})_{0 \le k, j \le n}, \quad \widehat{f}(k) = a_k,$$

and consider their actions on the spaces $\mathbb{C}^{n+1}$, $T_{a,n}: \mathbb{C}^{n+1} \to \mathbb{C}^{n+1}$, or $\mathcal{P}_n = \text{Vect}(z^k: 0 \le k \le n)$, $T_{f,n}: \mathcal{P}_n \to \mathcal{P}_n$. Let $(e_k)_{0 \le k \le n}$ denote the standard basis of the space $\mathbb{C}^{n+1}$, $e_k = (\delta_{jk})_{0 \le j \le n}$.

Theorem 5.3.1 (Gohberg and Sementsul, 1972) *Suppose there exist $x, y \in \mathcal{P}_n$ such that $T_{f,n}x = 1$, $T_{f,n}y = z^n$ and $\widehat{x}(0) = x_0 \ne 0$. Then the matrix $T_{f,n}: \mathcal{P}_n \to \mathcal{P}_n$ is invertible (a formula for $T_{f,n}^{-1}$ will be given in Theorem 5.3.2).*

Proof In this proof, we work in the space $\mathbb{C}^{n+1}$ with the notation $T_{a,n} = T_{f,n}$, and suppose $\mathbb{C}^n \subset \mathbb{C}^{n+1}$ by identifying a vector $x \in \mathbb{C}^n$ with $x = (x_j)_{j=0}^n = (x_0, \ldots, x_n, 0)$. To abbreviate, let

$$T = T_{a,n}^t$$

denote the transposed matrix; evidently, for every $x \in \mathbb{C}^{n+1}$ we have $T_{a,n}^* \overline{x} = \overline{(Tx)}$, where the bar is complex conjugation $\overline{x} = (\overline{x}_0, \ldots, \overline{x}_n)$.

Suppose that T is not invertible. Then, in the chain

$$\{0\} \subset T\mathbb{C}^1 \subset T\mathbb{C}^2 \subset \cdots \subset T\mathbb{C}^{n+1} \ne \mathbb{C}^{n+1},$$

for dimensional reasons, there must exist some l, $0 \le l \le n$ (at least one), such that $T\mathbb{C}^l = T\mathbb{C}^{l+1}$. Let $k = \max\{l\}$. We obtain a contradiction by showing that $k = n$ (the identity $T\mathbb{C}^n = T\mathbb{C}^{n+1}$ is impossible since $T_{a,n}y \perp \mathbb{C}^n$, hence $y \perp T_{a,n}^* \mathbb{C}^n = T\mathbb{C}^n$ ($*$ is the Hermitian conjugation), but $T_{a,n}y = e_n \ne 0$ is not orthogonal to $\mathbb{C}^{n+1}$).

We now show that $T\mathbb{C}^k = T\mathbb{C}^{k+1}$ with $k < n$ implies $T\mathbb{C}^{k+1} = T\mathbb{C}^{k+2}$. By hypothesis, there exists $\lambda_s \in \mathbb{C}$ such that

$$Te_k = \sum_{s=0}^{k-1} \lambda_s Te_s.$$

Let us show that

$$Te_{k+1} = \sum_{s=0}^{k-1} \lambda_s Te_{s+1},$$

which results in the stated identity. Compare the coordinates: for $1 \le i \le n$, we have

$$(Te_{k+1})_i = a_{k+1-i} = (Te_k)_{i-1} = \sum_{s=0}^{k-1} \lambda_s (Te_s)_{i-1} = \sum_{s=0}^{k-1} \lambda_s (Te_{s+1})_i.$$

For $i = 0$, use $T_{a,n}x = e_0$: for every $1 \le j \le n$, we have

$$0 = (e_0)_j = (T_{a,n}x)_j = \sum_{i=0}^{n} x_i(T_{a,n}e_i)_j = \sum_{i=0}^{n} x_i a_{j-i}.$$

In particular (since $k + 1 \ge 1$, $(T_{a,n}x)_{s+1} = 0$ for $0 \le s \le k - 1$, and $(T_{a,n}e_i)_j = (Te_j)_i$ for every i and j),

$$0 = (T_{a,n}x)_{k+1} = \sum_{i=0}^{n} x_i(T_{a,n}e_i)_{k+1} - \sum_{s=0}^{k-1} \lambda_s(T_{a,n}x)_{s+1}$$

$$= x_0(Te_{k+1})_i + \sum_{i=1}^{n} x_i(Te_{k+1})_i - \sum_{s=0}^{k-1} \lambda_s \sum_{i=0}^{n} x_i(Te_{s+1})_i$$

$$= \left(x_0(Te_{k+1})_0 - x_0 \sum_{s=0}^{k-1} \lambda_s(Te_{s+1})_0\right) + \sum_{i=1}^{n} x_i\left((Te_{k+1})_i - \sum_{s=0}^{k-1} \lambda_s(Te_{s+1})_i\right),$$

where the very last sum is also zero, since for $1 \le i \le n$ the equality $(Te_{k+1})_i = \sum_{s=0}^{k-1} \lambda_s(Te_{s+1})_i$ is already established. In conclusion, using $x_0 \ne 0$, we obtain $(Te_{k+1})_0 = \sum_{s=0}^{k-1} \lambda_s(Te_{s+1})_0$, and hence $Te_{k+1} = \sum_{s=0}^{k-1} \lambda_s Te_{s+1}$ and $T\mathbb{C}^{k+1} = T\mathbb{C}^{k+2}$, which implies the result. ∎

The following version of the Birkhoff–Wiener–Hopf factorization for Toeplitz matrices was found even before Theorem 5.3.1, but we state it without proof (one can be found in [Gohberg and Feldman, 1967], for example).

Theorem 5.3.2 (Baxter and Hirschman, 1964) *If $T_{f,n}: \mathcal{P}_n \to \mathcal{P}_n$ is invertible and $x, y \in \mathcal{P}_n$ such that $T_{f,n}x = 1$ ($x_0 \ne 0$), $T_{f,n}y = z^n$, then*

$$T_{f,n}^{-1} = \frac{1}{x_0}(T_{x,n}T_{z^{-n}y,n} - T_{zy,n}T_{z^{-n-1}x,n}).$$

Moreover, under the hypothesis $x(z) \ne 0$, $y(1/z) \ne 0$ ($|z| \le 1$), $T_{f,n} = T_{F,n}$, where

$$F(z) = F_1(z)F_2(z)$$

and $F_1(z) = 1/x(z)$, $F_2(z) = \widehat{y}(n)/z^{-n}y(z)$.

5.3.2 Comments

(a) In Theorem 5.3.1 the condition $x_0 \ne 0$ is *essential* but *not necessary* for the invertibility: the matrix $T^{(1)} = T_{2\cos(2\theta),2}$ ($T^{(1)} = T_{f,2}$, $f(e^{i\theta}) = e^{-2i\theta} + e^{2i\theta}$)

$$T^{(1)} = \begin{pmatrix} 0 & 0 & 1 \\ 0 & 0 & 0 \\ 1 & 0 & 0 \end{pmatrix},$$

is not invertible, but $T^{(1)}e_2 = e_0$, $T^{(1)}e_0 = e_2$, whereas the matrix $T^{(2)} = T_{f,2}$, where $f(e^{i\theta}) = e^{-i\theta} + e^{2i\theta} = 2\,e^{i\theta/2}\cos(3\theta/2)$,

$$T^{(2)} = \begin{pmatrix} 0 & 1 & 0 \\ 0 & 0 & 1 \\ 1 & 0 & 0 \end{pmatrix},$$

is invertible but $e_0 = T^{(2)}e_1$.

(b) The solutions $T_{f,n}x = 1$ ($x_0 \neq 0$), $T_{f,n}y = z^n$ are hence unique, they represent the first and last columns of the matrix $T_{f,n}^{-1}$, and this entire matrix is well-defined by these columns.

According to Cramer's rule, $x_0 = D_{n-1}/D_n$, where D_k are the Toeplitz determinants $D_k = \det(a_{l-j})_{0 \leq j, l \leq k}$ introduced in §5.1.4(1).

(c) Upper/lower bounds of the norm $\|T_{f,n}^{-1}\|$ as a function of the solutions $x, y \in \mathcal{P}_n$ such that $T_{f,n}x = 1$ ($x_0 \neq 0$), $T_{f,n}y = z^n$ that follow from Theorem 5.3.2:

$$\max(\|x\|_{L^2}, \|y\|_{L^2}) \leq \|T_{f,n}^{-1}\| \leq 2\|x\|_{L^\infty}\|y\|_{L^\infty},$$

where the norms $\|x\|_{L^\infty}$, $\|y\|_{L^\infty}$ can be replaced with the distances $d_n(x)$, $d_n(y)$ of Theorem 5.2.1.

(d) The triangular Toeplitz matrices $T_{f,n}$ corresponding to analytic symbols f (lower triangular), $\widehat{f}(k) = 0$, $k < 0$, or anti-analytic (upper triangular), $\widehat{f}(k) = 0$, $k > 0$. Clearly *such a $T_{f,n}$ is invertible if and only if $\widehat{f}(0) \neq 0$.*

It is also clear that, for f analytic, we have

$$T_{f,n} = f(S_n)$$

(see §5.1.3(3) for the truncated shift S_n). The mapping $f \mapsto f(S_n)$ ($f \in \mathcal{P}$), $\mathcal{P} \to L(\mathcal{P}_n)$, is a (polynomial) functional calculus of S_n: it is linear, multiplicative and sends $1(S_n) = \mathrm{id}$, $z(S_n) = S_n$. Moreover, given that $S_n^{n+1} = 0$, it is easy to extend it to the algebra $\mathrm{Hol}(\{0\})$ of functions holomorphic at the point zero, while keeping all properties of the calculus. Then, *$f(S_n)$ is invertible if and only if $f \in \mathrm{Hol}(\{0\})$ is invertible in $\mathrm{Hol}(\{0\})$* (i.e. $f(0) \neq 0$), and if this is the case, then

$$f(S_n)^{-1} = \frac{1}{f}(S_n).$$

All of the above remains correct for $T_{f,n}$ upper triangular (hence, f anti-analytic) by writing $T_{f,n} = F(S_n^*)$, $F(z) = f(1/z)$.

(e) Circulant Toeplitz matrices As in §5.2.1(d) (or in §5.5, with more details), let $T_{\varphi,n}$ be a circulant Toeplitz matrix, i.e. $\varphi = p + z^{-n-1}p$ where $p \in \mathcal{P}_n$. It follows from §5.5.4 that $T_{\varphi,n}$ is invertible if and only if $p(e^{2\pi ik/(n+1)}) \neq 0$ for every k, $0 \le k \le n$, and

$$\|T_{\varphi,n}^{-1}\| = \frac{1}{\min_k |p(e^{2\pi ik/(n+1)})|} = \frac{2}{\min_k |\varphi(e^{2\pi ik/(n+1)})|}.$$

The inverse is again a circulant and can be found with the formula $T_{\varphi,n}^{-1} = T_{\psi,n}$ with $\psi = q + z^{-n-1}q$, where q is the unique polynomial $q \in \mathcal{P}_n$ satisfying

$$q(e^{2\pi ik/(n+1)})p(e^{2\pi ik/(n+1)}) = 1 \quad \text{for every } k = 0, \ldots, n+1.$$

These assertions follow directly from the principal properties of circulants of §5.5.3–§5.5.4.

5.4 Inversion of Toeplitz Operators by the Finite Section Method

A numerical solution of a linear equation $Ax = y$ represents a difficult problem even if the existence of the solution is guaranteed. One of the effective methods is *inversion by finite-dimensional sections*: within a separable Hilbert space H, we select an increasing sequence of finite rank orthogonal projections $P_n \uparrow I$, rank $P_n = n+1$ ($n = 0, 1, \ldots$), and then show that the matrices $A_n = P_n A|(P_n H)$ are invertible and

$$\lim_n A_n^{-1} P_n x = A^{-1} x.$$

We realize this plan for certain Toeplitz operators $A = T_\varphi$ and the orthogonal projections P_n of the Hardy space H^2 onto the polynomials of $\mathcal{P}_n$ in such a way that $A_n = T_{\varphi,n}$. To simplify the notation, certain preparatory results are shown for a general Hilbert space H and with a fixed sequence $(P_n)_{n\ge 0}$.

5.4.1 The Finite Section Method

Definition Let H be a Hilbert space, $A \in L(H)$, and let $(P_n)_{n\ge 0}$ be an increasing sequence of orthogonal projections $P_n \uparrow I$ ($P_n H \subset P_{n+1} H$, $\forall n$; $\lim_n P_n x = x$, $\forall x \in H$), rank $P_n = n + 1$ ($n = 0, 1, \ldots$). A is said to be *invertible by the finite section method*, abbreviated

$$A \in \text{IFSM},$$

if the sections $A_n = P_n A|(P_n H)$ are invertible for sufficiently large n and if the sequence $(A_n^{-1} P_n x)_{n\ge 0}$ converges (in norm) for every $x \in H$.

5.4 Inversion of Toeplitz Operators by the Finite Section Method

Lemma 5.4.1

(i) $A \in \text{IFSM} \Rightarrow A \in L(H)^{-1}$, $\lim_n A_n^{-1} P_n x = A^{-1} x$ $(\forall x \in H)$ and

$$\|A^{-1}\| \leq l(A) =: \varlimsup_n \|A_n^{-1}\|.$$

(ii) If $A \in L(H)^{-1}$, then $A \in \text{IFSM}$ *if and only if*

$$l(A) = \varlimsup_n \|A_n^{-1}\| < \infty$$

(we suppose $\|A_n^{-1}\| = \infty$ if A_n is not invertible).

Proof (i) By the fundamental principles of operators (see Appendix D) the mapping $x \mapsto Bx =: \lim_n A_n^{-1} P_n x$ is in $L(H)$, and $\lim_n A_n P_n x = Ax$ (convergence in norm; verify first for $x \in \bigcup_n P_n H$). Passing to the limit in $A_n A_n^{-1} P_n x = P_n x$ and $A_n^{-1} A_n P_n x = P_n x$, we obtain $AB = \text{id}$ and $BA = \text{id}$, respectively, hence $B = A^{-1}$ and $\|A^{-1}\| \leq l(A)$.

(ii) The necessity of $l(A) = \varlimsup_n \|A_n^{-1}\| < \infty$ is already in (i). For the sufficiency, given $\varlimsup_n \|A_n^{-1} P_n\| < \infty$, it suffices to show the convergence of $(A_n^{-1} P_n x)_{n \geq 0}$ for $x \in \bigcup_n P_n H$ (a set dense in H). For n sufficiently large, we have

$$\begin{aligned}A_n^{-1} P_n x - A^{-1} x &= A_n^{-1} P_n x - P_n A^{-1} x + o(1) \\ &= A_n^{-1}(P_n x - A_n P_n A^{-1} x) + o(1) \\ &= A_n^{-1}(x - P_n A P_n A^{-1} x) + o(1) \\ &= A_n^{-1}(x - P_n A A^{-1} x + o(1)) + o(1) \\ &= A_n^{-1}(o(1)) + o(1) = o(1) \quad (\text{as } n \to \infty),\end{aligned}$$

where we write $x_n = y_n + o(1)$ if $\lim_n \|x_n - y_n\| = 0$. ∎

Lemma 5.4.2

(i) If $A \in L(H)^{-1}$ and $P_n A(I - P_n) = 0$ $(\forall n)$ *(the matrix of A is lower triangular)*, or $(I - P_n) A P_n = 0$ $(\forall n)$ *(matrix upper triangular)*, then $A \in \text{IFSM}$ and $l(A) = \|A^{-1}\|$.
(ii) If $A, B \in L(H)$, $A \in \text{IFSM}$ and $\|B\| < 1/l(A)$, then $A + B \in \text{IFSM}$ and $l(A + B) \leq l(A)(1 - \|B\| l(A))^{-1}$.
(iii) If $A, B \in L(H)$, $A \in \text{IFSM}$, $B \in \mathfrak{S}_\infty(H)$ and $A + B \in L(H)^{-1}$, then $A + B \in \text{IFSM}$ and $l(A + B) \leq l(A) \|(I + A^{-1} B)^{-1}\|$.
(iv) If $A, B \in L(H)$, $A, B \in \text{IFSM}$ and $\lim_n \|P_n A(I - P_n) B P_n\| = 0$, then $AB \in \text{IFSM}$ and $l(AB) \leq l(A) l(B)$.
(v) If $A, B \in L(H)^{-1}$, $AB \in \text{IFSM}$ and $[A, B] =: AB - BA \in \mathfrak{S}_\infty(H)$, then $BA \in \text{IFSM}$ and $l(BA) = l(AB) \|(BA)^{-1} AB\|$.

Proof (i) The adjoint operator A^* is upper triangular $(I - P_n)A^*P_n = 0$, thus $A^*P_nH \subset P_nH$, and moreover, since $\dim P_nH < \infty$ and A^* is invertible, we have $A^*P_nH = P_nH$ ($\forall n$). This implies that A^{-1} is lower triangular: since $(I - P_n)A^{*-1}P_n = 0$, we have $P_nA^{-1}(I - P_n) = 0$ ($\forall n$). Consequently, $P_nAP_n(P_nA^{-1}P_n) = P_nAA^{-1}P_n = P_n$, and similarly, $(P_nA^{-1}P_n)P_nAP_n = P_n$, hence $A_n^{-1} = (A^{-1})_n$, giving $l(A) \leq \|A^{-1}\|$. As the converse always holds, the result follows.

(ii) We have $\|B\|l(A) < 1$, thus $\|B_n\| \times \|A_n^{-1}\| < 1$ (for n sufficiently large), hence $(A + B)_n = A_n(I + A_n^{-1}B_n)$ and $\|(A + B)_n^{-1}\| \leq \|A_n^{-1}\|(1 - \|A_n^{-1}\| \times \|B_n\|)^{-1}$, giving $\overline{\lim}_n \|(A + B)_n^{-1}\| \leq l(A)(1 - \|B\|l(A))^{-1} < \infty$. Moreover, by Lemma 5.4.1, $\|A^{-1}\| \leq l(A)$ and thus $\|A^{-1}\| \times \|B\| < 1$, hence $A + B \in L(H)^{-1}$. The result follows from Lemma 5.4.1(ii).

(iii) Since $A, A + B \in L(H)^{-1}$, we also have $(I + A^{-1}B) \in L(H)^{-1}$. Thus, for every $x \in H$,

$$\|(A + B)_n P_n x\| = \|P_n A(I + A^{-1}B)P_n x\|$$
$$= \|P_n AP_n(I + A^{-1}B)P_n x + P_n A(I - P_n)(I + A^{-1}B)P_n x\|$$
$$\geq \|A_n^{-1}\|^{-1}\|(I + A^{-1}B)P_n x\| - \|P_n A(I - P_n)(I + A^{-1}B)P_n x\|$$
$$\geq \|A_n^{-1}\|^{-1}\|(I + A^{-1}B)^{-1}\|^{-1} \cdot \|P_n x\| - \epsilon_n \|P_n x\|,$$

where $\epsilon_n =: \|P_n A(I - P_n)A^{-1}B\| \to 0$ as $n \to \infty$ (since $B \in \mathfrak{S}_\infty(H)$). It follows that $(A + B)_n$ is invertible for n large enough, and

$$l(A + B) = \overline{\lim_n} \|(A + B)_n^{-1}\| \leq l(A)\|(I + A^{-1}B)^{-1}\|.$$

(iv) We have $P_n ABP_n = P_n AP_n BP_n + P_n A(I - P_n)BP_n$, hence

$$\|(AB)_n P_n x\| \geq \|P_n AP_n BP_n x\| - \|P_n A(I - P_n)BP_n x\|$$
$$\geq \|A_n^{-1}\| \times \|B_n^{-1}\| \times \|P_n x\| - \|P_n A(I - P_n)BP_n\| \times \|P_n x\|$$

(for every $x \in H$), which implies $l(AB) \leq l(A)l(B)$.

(v) This is a special case of (iii), since $BA = AB - [A, B]$ as well as $l(BA) \leq l(AB)\|(I - (AB)^{-1}[A, B])^{-1}\| = l(AB)\|(BA)^{-1}AB\|$. ∎

The following theorem contains most of the information known about the IFSM of Toeplitz operators on the space H^2; recall that on the Hardy space H^2, P_n is always an orthogonal projection onto $\mathcal{P}_n$ and the semi-commutator of T_φ, T_ψ is $[T_\varphi, T_\psi) = T_\varphi T_\psi - T_{\varphi\psi} = -H_{\bar\varphi}^* H_\psi$ (see Lemma 2.2.9). Given Lemma 5.4.1, the property $T_\varphi \in$ IFSM signifies that T_φ and $T_{\varphi,n}$ (after a certain rank) are invertible, and

$$\lim_n \|T_\varphi^{-1} x - T_{\varphi,n}^{-1} P_n x\|_2 = 0 \quad (\forall x \in H^2).$$

5.4 Inversion of Toeplitz Operators by the Finite Section Method

We prove each assertion of §5.4.2 just after its statement; for supplementary information and comments see §5.7 and §5.8.

5.4.2 Theorem (IFSM for Toeplitz Operators)

The theorem contains eight points; each is proved immediately after its statement.

(1) (Analytic/anti-analytic symbols) *If $\varphi^{\pm 1} \in H^\infty$ (or $\overline{\varphi}^{\pm 1} \in H^\infty$), then $T_\varphi \in$ IFSM and $l(T_\varphi) = \|1/\varphi\|_\infty$.*

Proof This is a special case of Lemma 5.4.2 (i), since $P_n T_\varphi (I - P_n) = 0$ ($\forall n$) for $\varphi \in H^\infty$ and $(I - P_n) T_\varphi P_n = 0$ ($\forall n$) for $\overline{\varphi} \in H^\infty$. ∎

(2) (Analytic/anti-analytic multipliers of IFSM) *Let $T_\varphi \in IFSM$ and $a^{\pm 1} \in H^\infty$. If $[T_a, T_\varphi] \in \mathfrak{S}_\infty(H^2)$ (for example, if $\overline{a} \in H^\infty + C(\mathbb{T})$, or $\varphi \in H^\infty + C(\mathbb{T})$), then*

$$T_{\varphi a} \in ISFM, \quad \text{and} \quad l(T_{\varphi a}) \leq \|1/a\|_\infty l(T_\varphi) \|T_a^{-1} T_\varphi^{-1} T_a T_\varphi\|.$$

For the anti-analytic version, assume that $T_\varphi \in ISFM$, $\overline{a}^{\pm 1} \in H^\infty$ and $[T_\varphi, T_a] \in \mathfrak{S}_\infty(H^2)$ (for example, $a \in H^\infty + C(\mathbb{T})$, or $\overline{\varphi} \in H^\infty + C(\mathbb{T})$), then

$$T_{\varphi a} \in ISFM \quad \text{and} \quad l(T_{\varphi a}) \leq \|1/a\|_\infty l(T_\varphi) \|T_\varphi^{-1} T_a^{-1} T_\varphi T_a\|.$$

Proof We have $T_\varphi, T_a \in$ IFSM (by (1)), and hence (by Lemma 5.4.2(iv), whose hypothesis for $A = T_a$ is satisfied by $P_n T_a (I - P_n) = 0$ ($\forall n$)) $l(T_a T_\varphi) \leq \|1/a\|_\infty l(T_\varphi)$. Moreover,

$$T_{\varphi a} = T_\varphi T_a = T_a T_\varphi - [T_a, T_\varphi),$$

and the result follows from Lemma 5.4.2(v).
The reasoning for $\overline{a}^{\pm 1} \in H^\infty$ is symmetric. ∎

(3) (Additive stability of IFSM) *Let $T_\varphi \in$ IFSM, $\psi \in L^\infty(\mathbb{T})$ such that $\|\psi\|_\infty < 1/l(T_\varphi)$. Then,*

$$T_{\varphi+\psi} \in \text{IFSM}, \quad \text{and} \quad l(T_{\varphi+\psi}) \leq l(T_\varphi)(1 - \|\psi\|_\infty l(T_\varphi))^{-1}.$$

In particular, $\|a - 1\|_\infty < 1 \Rightarrow T_a \in$ IFSM and $l(T_a) \leq (1 - \|a - 1\|_\infty)^{-1}$.

Proof This is a direct consequence of Lemma 5.4.2(ii). ∎

(4) ("Tight" multipliers of IFSM) *Let $T_\varphi \in$ IFSM, $a^{\pm 1} \in L^\infty(\mathbb{T})$ and $r = r_a > 0$ and $\lambda = \lambda_a \in \mathbb{C}\setminus\{0\}$ such that*

$$\operatorname{Ran}_{\operatorname{ess}}(a) \subset \overline{D}(\lambda, r|\lambda|),$$

where $\operatorname{Ran}_{\operatorname{ess}}(a)$ is the essential range of a: see Definitions 3.1.2. If $r < 1/l(T_{\varphi/\|\varphi\|_\infty})$, then

$$T_{\varphi a} \in \operatorname{IFSM}, \quad \text{and} \quad l(T_{\varphi a}) \leq \frac{l(T_\varphi)}{|\lambda_a|(1 - r_a l(T_{\varphi/\|\varphi\|_\infty}))}.$$

In particular:

(i) $l(T_a) \leq \dfrac{1}{|\lambda_a|(1 - r_a)}$.

(ii) *If a is real and positive, $a \geq 0$, then we can take*

$$r_a = \frac{\overline{M}(a) - \underline{M}(a)}{\overline{M}(a) + \underline{M}(a)} \quad \text{and} \quad \lambda_a = \frac{\overline{M}(a) + \underline{M}(a)}{2},$$

where $\overline{M}(a) = \operatorname{ess\,sup}(a)$, $\underline{M}(a) = \operatorname{ess\,inf}(a)$.

(iii) *If a is real and positive, $a \geq 0$, then $T_a \in$ IFSM and $l(T_a) \leq \|1/a\|_\infty$.*

Proof We have

$$\varphi a = \lambda \varphi + \varphi(a - \lambda)$$

and

$$\|\varphi(a - \lambda)\|_\infty \leq \|\varphi\|_\infty \|a - \lambda\|_\infty \leq r|\lambda| \cdot \|\varphi\|_\infty < |\lambda| \cdot \|\varphi\|_\infty / l(T_{\varphi/\|\varphi\|_\infty}) = 1/l(T_{\lambda\varphi}).$$

By (3), $T_{\varphi a} \in$ IFSM and

$$l(T_{\varphi a}) \leq l(T_{\lambda\varphi})(1 - \|\varphi(a - \lambda)\|_\infty l(T_{\lambda\varphi}))^{-1}$$

$$\leq \frac{l(T_\varphi)}{|\lambda|}(1 - r|\lambda| \cdot \|\varphi\|_\infty l(T_{\lambda\varphi}))^{-1} = \frac{l(T_\varphi)}{|\lambda|(1 - r l(T_{\varphi/\|\varphi\|_\infty}))}.$$

For a real and positive, it suffices to take $\lambda_a = (\overline{M}(a) + \underline{M}(a))/2$ and $r_a = (\overline{M}(a) - \underline{M}(a))/(\overline{M}(a) + \underline{M}(a)) < 1/l(T_{\varphi/\|\varphi\|_\infty})$. The case (iii) is established with $\varphi = 1$, $l(T_\varphi) = 1$. ∎

(5) (Sectorial symbols) *Let $\varphi^{\pm 1} \in L^\infty(\mathbb{T})$ be a sectorial function (see Examples 3.2.2) with sector value $2\alpha =: \operatorname{ess\,sup}_{t,t' \in \mathbb{T}} \arg(\varphi(t)/\varphi(t')) < \pi$. Then*

$$T_\varphi \in \operatorname{IFSM} \quad \text{and} \quad l(T_\varphi) \leq \frac{\|1/\varphi\|_\infty}{\cos(\alpha)}.$$

5.4 Inversion of Toeplitz Operators by the Finite Section Method

Proof Replacing φ with $e^{i\theta}\varphi$ if necessary, we have

$$\text{Ran}_{\text{ess}}(\varphi) \subset \{z \in \mathbb{C}: \|1/\varphi\|_\infty^{-1} \le |z| \le \|\varphi\|_\infty, |\arg(z)| \le \alpha\} \subset D(\lambda, r\lambda)$$

where $\lambda > 0$ is sufficiently large and $e^{\pm i\alpha}\|1/\varphi\|^{-1} \in \partial D(\lambda, r\lambda)$ (see the proof of Lemma 3.3.2 and its illustration). By (4(i)) we have $l(T_\varphi) \le (\lambda - r\lambda)^{-1}$, and the result is obtained by passing to the limit $\lambda \to \infty$ (as in Lemma 3.3.2). ∎

(6) (Factorization à la Wiener–Hopf; Baxter, 1963) *Suppose* $\varphi = \varphi_1 \psi \bar{\varphi}_2$ *where* $\varphi_{1,2}^{\pm 1} \in H^\infty$, $\|1 - \psi\|_\infty < 1$ *and* $\varphi_{1,2} \in \text{QC}(\mathbb{T})$. *Then,* $T_\varphi \in$ IFSM *and*

$$l(T_\varphi) \le \|\varphi_1^{-1}\|_\infty (1 - \|1 - \psi\|_\infty)^{-1} \|\varphi_2^{-1}\|_\infty \cdot \|T_{\varphi_1} T_\psi T_{\bar{\varphi}_2}(T_\varphi)^{-1}\| < \infty.$$

Proof We have $T_\varphi = T_{\varphi_1 \psi \bar{\varphi}_2} = T_{\varphi_1} T_{\psi \bar{\varphi}_2} + H_{\bar{\varphi}_1}^* H_{\psi \bar{\varphi}_2}$, where $H_{\bar{\varphi}_1} \in \mathfrak{S}_\infty(H^2)$ (see Theorem 2.2.8), and then $T_{\psi \bar{\varphi}_2} = T_\psi T_{\bar{\varphi}_2} + H_{\bar{\psi}}^* H_{\bar{\varphi}_2}$, where $H_{\bar{\varphi}_2} \in \mathfrak{S}_\infty(H^2)$. Hence,

$$T_\varphi = T_{\varphi_1} T_\psi T_{\bar{\varphi}_2} + K \quad \text{where } K \in \mathfrak{S}_\infty(H^2) \quad (K = T_{\varphi_1} H_{\bar{\psi}}^* H_{\bar{\varphi}_2} + H_{\bar{\varphi}_1}^* H_{\psi \bar{\varphi}_2}).$$

Since $T_\varphi = T_{\bar{\varphi}_2} T_\psi T_{\varphi_1}$ and $T_{\varphi_1}, T_\psi, T_{\bar{\varphi}_2}$ are invertible, it suffices to show that $l(T_{\varphi_1} T_\psi T_{\bar{\varphi}_2}) < \infty$ (by Lemmas 5.4.1(ii) and 5.4.2(ii)). Using $P_n T_{\varphi_1}(I - P_n) = 0$ and $P_n T_{1/\varphi_1}(I - P_n) = 0$ we observe that $(P_n T_{\varphi_1} P_n)(P_n T_{1/\varphi_1} P_n) = P_n$ and $(P_n T_{1/\varphi_1} P_n)(P_n T_{\varphi_1} P_n) = P_n$, hence $P_n T_{1/\varphi_1} \mathcal{P}_n = (P_n T_{\varphi_1} \mathcal{P}_n)^{-1}$. For every $x \in H^2$,

$$\|P_n T_{\varphi_1} T_\psi T_{\bar{\varphi}_2} P_n x\| = \|P_n T_{\varphi_1} P_n T_\psi T_{\bar{\varphi}_2} P_n x\|$$
$$= \|\varphi_1^{-1}\|_\infty^{-1} \|P_n T_\psi T_{\bar{\varphi}_2} P_n x\|$$

(for the same reason, $(I - P_n) T_{\bar{\varphi}_2} P_n = 0$)

$$= \|\varphi_1^{-1}\|_\infty^{-1} \|P_n T_\psi P_n T_{\bar{\varphi}_2} P_n x\|$$
$$\ge \|\varphi_1^{-1}\|_\infty^{-1} \|(P_n T_\psi P_n)^{-1}\|^{-1} \|P_n T_{\bar{\varphi}_2} P_n x\|$$
$$\ge c_n \|P_n x\|,$$

where $c_n = \|\varphi_1^{-1}\|_\infty^{-1} \|(P_n T_\psi P_n)^{-1}\|^{-1} \|\varphi_2^{-1}\|_\infty^{-1} \ge \|\varphi_1^{-1}\|_\infty^{-1} (1 - \|1 - \psi\|_\infty) \|\varphi_2^{-1}\|_\infty^{-1}$. Finally, by Lemma 5.4.2(ii),

$$l(T_\varphi) \le \|\varphi_1^{-1}\|_\infty (1 - \|1 - \psi\|_\infty)^{-1} \|\varphi_2^{-1}\|_\infty \|(1 + T_{1/\bar{\varphi}_2} T_\psi^{-1} T_{1/\varphi_1} K)^{-1}\|$$
$$\le \|\varphi_1^{-1}\|_\infty (1 - \|1 - \psi\|_\infty)^{-1} \|\varphi_2^{-1}\|_\infty \times \|T_{\varphi_1} T_\psi T_{\bar{\varphi}_2}(T_\varphi)^{-1}\| < \infty. \quad \blacksquare$$

(7) (Locally sectorial symbols; Gohberg and Feldman, 1967) *If* $T_\varphi \in L(H^2)^{-1}$ *and* φ *is locally sectorial (see Examples 3.2.2), then* $T_\varphi \in$ IFSM. *In particular,*

(i) *(Baxter, 1963) If* $\varphi \in W(\mathbb{T})$ *and* $T_\varphi \in L(H^2)^{-1}$, *then* $T_\varphi \in$ IFSM.

(ii) *(Gohberg and Feldman, 1967)* If $T_\varphi \in L(H^2)^{-1}$ and $\varphi \in C(\mathbb{T})$, or if $\varphi \in PC(\mathbb{T})$, then $T_\varphi \in$ IFSM.

Proof By Exercise 3.4.7(d) there exists $f \in C(\mathbb{T})$ and a sectorial function $g^{\pm 1} \in L^\infty(\mathbb{T})$ such that

$$\varphi = fg.$$

As in (5) above, let $\lambda \in \mathbb{C}\setminus\{0\}$ such that $\|1 - (g/\lambda)\|_\infty < 1$. Let $\epsilon > 0$ be sufficiently small that

$$2\epsilon + \|1 - (g/\lambda)\|_\infty < 1,$$

and then let $p \in \mathcal{P}$ be a polynomial such that $\|f - p\|_\infty < \epsilon\delta/2$ where $\delta = \min_\mathbb{T} |f| > 0$. Then,

$$|p| \geq |f| - |f - p| \geq \delta - \epsilon\delta/2 > \delta/2$$

and $|(f - p)/p| < \epsilon$, hence $f = p(1 + (f-p)/p)$, which implies wind$(p) =$ wind$(f) = 0$; the last point follows from

$$0 = \text{ind}(T_{fg}) = \text{ind}(T_f T_g + K) \quad (\text{where } K \in \mathfrak{S}_\infty)$$
$$= \text{ind}(T_f) + \text{ind}(T_g) = \text{ind}(T_f) = -\text{wind}(f).$$

Thus, $p = e^h$ where h is continuous and even $h \in C^\infty(\mathbb{T})$. The symbol φ can be written as $\varphi = \varphi_1 \psi \overline{\varphi}_2$, where

$$\varphi_1 = \lambda \cdot \exp\left(\sum_{k\geq 0} \widehat{h}(k)z^k\right), \quad \overline{\varphi}_2 = \exp\left(\sum_{k<0} \widehat{h}(k)z^k\right), \quad \psi = (g/\lambda)\left(1 + \frac{f-p}{p}\right).$$

Since

$$\|\psi - 1\|_\infty = \|1 - (g/\lambda) - (g/\lambda)\frac{f-p}{p}\|_\infty$$
$$\leq \|1 - (g/\lambda)\|_\infty + \|g/\lambda\|_\infty \epsilon < \|1 - (g/\lambda)\|_\infty + 2\epsilon < 1$$

and $\varphi_{1,2} \in C^\infty(\mathbb{T}) \cap H^\infty$, Item (6) gives $T_\varphi \in$ IFSM.

For (i) and (ii), clearly the symbols $\varphi \in PC(\mathbb{T})$ of the invertible operators T_φ are locally sectorial (see Exercise 3.4.5(a)). ∎

(8) *(Devinatz and Shinbrot, 1969)* If $\varphi \in H^\infty + C(\mathbb{T})$ *(or $\overline{\varphi} \in H^\infty + C(\mathbb{T})$) and* $T_\varphi \in L(H^2)^{-1}$, then $T_\varphi \in$ IFSM.

Proof We show that $\varphi = c\psi$ where $\psi^{\pm 1} \in H^\infty$ and $c \in C(\mathbb{T})$. Indeed, $\varphi = [\varphi]u$ where $[\varphi]$ is the outer part of φ and $u =: \varphi/[\varphi] \in QC(\mathbb{T})$ (see Exercise 3.4.3(e)). As $T_\varphi, T_{[\varphi]} \in L(H^2)^{-1}$, T_u is invertible. Then, by Exercise 3.4.3(g), $u = z^n e^{i(a+\tilde{b})}$ where $a, b \in C(\mathbb{T})$ and $n = -\text{ind}(T_u) = 0$, thus $u = e^{b+i\tilde{b}} e^{-b+ia}$. Setting $\psi = [\varphi] e^{b+i\tilde{b}}$ and $c = e^{-b+ia}$, we obtain the proposed representation.

Since $T_\varphi = T_c T_\psi$ and $T_\psi \in L(H^2)^{-1}$, we have $T_c \in L(H^2)^{-1}$, and using (7(ii)) and then (2) for T_c, we reach the conclusion. ■

5.4.3 Comment: A Counter-Example of Treil (1987)

There exists a function $\varphi \in L^\infty(\mathbb{T}) \cap C(\mathbb{T} \setminus \{1\})$ such that $T_\varphi \in L(H^2)^{-1}$, but $l(T_\varphi) = \infty$ (hence, $T_\varphi \notin$ IFSM). Later, Grudsky (1996) gave an explicit example: $\varphi(e^{it}) = e^{ia(\operatorname{ctg}(t/2))}$, where $a \in C(\mathbb{R})$ is the 2π-periodic function defined by $a(x) = 2|x|, |x| \leq \pi$.

5.5 Theory of Circulants

The circulants are Toeplitz matrices with an additional structure of "cyclicity" of the columns (and the rows). They are useful for many calculations with general Toeplitz matrices, and in particular, for the spectral asymptotics of $T_{\varphi,n}$ as $n \to \infty$: see §5.6 below.

5.5.1 Cyclic Shift

The mapping $C_n: \mathbb{C}^{n+1} \to \mathbb{C}^{n+1}$ defined by

$$C_n x = C_n(x_0, \ldots, x_n)^{\operatorname{col}} =: (x_n, x_0, \ldots, x_{n-1})^{\operatorname{col}}, \quad x \in \mathbb{C}^{n+1},$$

is called a *cyclic shift* (of order $n + 1$). The matrix of C_n is

$$C_n = \begin{pmatrix} 0 & 0 & \cdots & 0 & 1 \\ 1 & 0 & \cdots & 0 & 0 \\ 0 & 1 & \cdots & 0 & 0 \\ \vdots & \vdots & \ddots & \vdots & \vdots \\ 0 & 0 & \cdots & 1 & 0 \end{pmatrix}.$$

The following properties of C_n are almost evident.

(1) C_n is a Toeplitz matrix, $C_n = T_{z+z^{-n},n}$, it is *unitary*, $(C_n x, C_n y) = (x, y)$ for every $x, y \in \mathbb{C}^{n+1}$ (where $(\cdot, \cdot)$ denotes the standard scalar product in $\mathbb{C}^{n+1}$), and hence, $C_n^* C_n = C_n C_n^* = I$.

(2) $C_n^{n+1} = I$, thus the *eigenvalues* of C_n satisfy the equation $z^{n+1} = 1$, and by setting

$$\epsilon_k = e^{2\pi k i/(n+1)}, \quad k = 0, \ldots, n, \quad \text{and} \quad x_k = (1, \epsilon_k, \epsilon_k^2, \ldots, \epsilon_k^n)^{\operatorname{col}},$$

we obtain

$$C_n x_k = \epsilon_k^{-1} x_k, \quad (k = 0, \ldots, n).$$

The eigenvectors are orthogonal $(x_k, x_j) = 0$ $(k \neq j)$ and the sequence $(x_k/\sqrt{n+1})_{0 \leq k \leq n}$ is an orthonormal basis of $\mathbb{C}^{n+1}$.

(3) *The diagonalizing matrix* of C_n is the *discrete Fourier transform* $\mathcal{F}_d = \mathcal{F}_{d,n}$ defined by $\mathcal{F}_d e_k = x_k/\sqrt{n+1}$, $k = 0, \ldots, n$,

$$\mathcal{F}_d = \frac{1}{\sqrt{n+1}} \begin{pmatrix} 1 & 1 & \cdots & 1 & 1 \\ 1 & \varepsilon_1 & \cdots & \varepsilon_{n-1} & \varepsilon_n \\ 1 & \varepsilon_1^2 & \cdots & \varepsilon_{n-1}^2 & \varepsilon_n^2 \\ \vdots & \vdots & \ddots & \vdots & \vdots \\ 1 & \varepsilon_1^n & \cdots & \varepsilon_{n-1}^n & \varepsilon_n^n \end{pmatrix},$$

hence

$$\mathcal{F}_d^{-1} C_n \mathcal{F}_d e_k = \epsilon_k^{-1} e_k,$$

where $(e_k)_0^n$ is a standard basis of $\mathbb{C}^{n+1}$, and

$$\mathcal{F}_d^{-1} C_n \mathcal{F}_d = \mathrm{diag}(1, \epsilon_1^{-1}, \ldots, \epsilon_n^{-1}).$$

It is easy to see that $\mathcal{F}_d$ is unitary $(\mathcal{F}_d^* \mathcal{F}_d = \mathcal{F}_d \mathcal{F}_d^* = I)$, and that $\mathcal{F}_d^2 e_0 = e_0$ and $(\mathcal{F}_d^2 e_t, e_s) = \delta_{t+s, n+1}$ for $s + t > 0$, hence $\mathcal{F}_d^2$ is a symplectic matrix of order n on $e_0^\perp$; in particular, $\mathcal{F}_d^4 = I$.

(4) *The functional calculus of* C_n, $f \mapsto f(C_n)$, is easily defined with the aid of the diagonalization (2,3), by setting for $f \in C(\mathbb{T})$

$$f(C_n) x_k = f(\epsilon_k^{-1}) x_k \quad (0 \leq k \leq n),$$

or $f(C_n) = \mathcal{F}_d f(D) \mathcal{F}_d^{-1}$ where $D = \mathrm{diag}(1, \epsilon_1^{-1}, \ldots, \epsilon_n^{-1})$. Clearly, for the trigonometric polynomials $f \in \mathcal{P}$, $f = \sum_{|j| \leq N} \widehat{f}(j) z^j$, we have

$$f(C_n) = \sum_{|j| \leq N} \widehat{f}(j) C_n^j.$$

Given the periodicity $C_n^{n+1} = I$, we can give a "standard" form to every $f(C_n)$ ($f \in C(\mathbb{T})$). If the Fourier series of f converges absolutely, $f \in W(\mathbb{T})$ (the *Wiener algebra*)

$$W(\mathbb{T}) =: \left\{ f = \sum_{k \in \mathbb{Z}} \widehat{f}(k) z^k : \sum_{k \in \mathbb{Z}} |\widehat{f}(k)| < \infty \right\},$$

we have $f(C_n) = \sum_{j=0}^n c_j C_n^j$ where $c_j = \sum_{k \in \mathbb{Z}} \widehat{f}(j + k(n+1))$ ($j = 0, \ldots, n$), and for the general case $f \in C(\mathbb{T})$ the same formula can be written as follows:

$$f(C_n) = \sum_{j=0}^n c_j C_n^j \quad \text{where } c_j = \frac{1}{n+1} \sum_{l=0}^n \epsilon_1^{-jl} f(\epsilon_1^l) \quad (j = 0, \ldots, n).$$

5.5.2 Definition of Circulants

Let $c = (c_k)_{0 \leq k \leq n}^{\text{col}} \in \mathbb{C}^{n+1}$. A matrix (a linear mapping) $T_n: \mathbb{C}^{n+1} \to \mathbb{C}^{n+1}$ of the form

$$T_n e_k = C_n^k c, \quad k = 0, \ldots, n,$$

is said to be *circulant*; we have

$$T_n = \begin{pmatrix} c_0 & c_n & \cdots & c_2 & c_1 \\ c_1 & c_0 & \cdots & c_3 & c_2 \\ \vdots & \ddots & \ddots & \ddots & \vdots \\ c_{n-1} & c_{n-2} & \cdots & c_0 & c_1 \\ c_n & c_{n-1} & \cdots & c_1 & c_0 \end{pmatrix}.$$

The *polynomial* $p(z) = \sum_{k=0}^{n} c_k z^k$ ($p \in \mathcal{P}_n$), or the vector c itself, is called the *representing polynomial (vector)* of T_n; we write

$$T_n = T_n(c) \quad \text{or} \quad T_n = T_n(p).$$

Let $\text{Circ}(\mathbb{C}^{n+1})$ denote the *set of circulants*.

5.5.3 Basic Properties

(i) *Every $T_n(c)$ is a Toeplitz matrix, i.e. $T_n(c) = T_n(p) = T_{\varphi,n}$ where $\varphi = p + z^{-n-1} p$, where p is the representing polynomial of $T_n(c)$.*

(ii) $I, C_n \in \text{Circ}(\mathbb{C}^{n+1})$ *and for every $T_n(p) \in \text{Circ}(\mathbb{C}^{n+1})$ we have*

$$T_n(p) = p(C_n) = \sum_{k=0}^{n} c_k C_n^k \quad (p \in \mathcal{P}_n).$$

Indeed, for every k, $1 \leq k \leq n$, we have

$$C_n^k = \begin{pmatrix} 0 & \cdots & 1 & \cdots & \cdots & 0 \\ 0 & \cdots & 0 & 1 & \cdots & 0 \\ \cdots & \cdots & \cdots & \cdots & \ddots & \cdots \\ \cdots & \cdots & \cdots & \cdots & \cdots & 1 \\ 1 & 0 & \cdots & \cdots & 0 & 0 \\ \cdots & \ddots & \cdots & \cdots & \cdots & \cdots \\ 0 & \cdots & 1 & \cdots & 0 & 0 \end{pmatrix}. \quad \blacksquare$$

(iii) *For every $f \in C(\mathbb{T})$, the matrix $f(C_n)$ is circulant; more precisely, if $f \in W(\mathbb{T})$, then*

$$f(C_n) = T_n(c) \quad \text{where } c_k = \sum_{j \equiv k \,(\mathrm{mod}\, n+1)} \widehat{f}(j), \quad 0 \leq k \leq n,$$

and

$$c_j = \frac{1}{n+1} \sum_{l=0}^{n} \epsilon_1^{-jl} f(\epsilon_1^l)$$

in the general case. The eigenvalues of $f(C_n)$ are $f(\epsilon_1^{-k})$ ($0 \leq k \leq n$, in the order of the eigenvectors x_k (common for all the circulants); a circulant with given eigenvalues is unique.

(iv) Example: circulants with a band structure. Let $f \in \mathcal{P}$ be a trigonometric polynomial $f = \sum_{|k| \leq l} a_k z^k$ and $2l < n$. Then $f(C_n)$ is a *Toeplitz matrix with a band structure* (a band-structured Toeplitz matrix $T_{f,n}$, with a modification by "small triangles" in the south-west and north-east corners of the matrix):

$$f(C_n) = \begin{pmatrix} a_0 & a_{-1} & \cdots & \cdots & a_{-l} & 0 & \cdots & \cdots & 0 & a_l & \cdots & a_2 & a_1 \\ a_1 & a_0 & a_{-1} & \cdots & \cdots & a_{-l} & 0 & \cdots & \cdots & 0 & \cdots & \cdots & a_2 \\ \cdots & a_1 & a_0 & a_{-1} & \cdots & \cdots & \cdots & \cdots & \cdots & \cdots & \cdots & \cdots & \cdots \\ \cdots & \cdots & \cdots & \cdots & \cdots & \cdots & \cdots & \cdots & \cdots & \cdots & \cdots & \cdots & a_l \\ a_l & \cdots & \cdots & \cdots & \cdots & \cdots & \cdots & \cdots & \cdots & \cdots & \cdots & \cdots & \cdots \\ 0 & a_l & \cdots & \cdots & \cdots & \cdots & \cdots & \cdots & \cdots & \cdots & \cdots & \cdots & \cdots \\ \cdots & \cdots & \cdots & \cdots & \cdots & \cdots & \cdots & \cdots & \cdots & \cdots & \cdots & \cdots & \cdots \\ \cdots & \cdots & \cdots & \cdots & \cdots & \cdots & \cdots & \cdots & \cdots & \cdots & \cdots & \cdots & \cdots \\ 0 & \cdots & \cdots & \cdots & \cdots & \cdots & \cdots & \cdots & \cdots & \cdots & \cdots & \cdots & a_{-l} \\ a_{-l} & 0 & \cdots & \cdots & \cdots & \cdots & \cdots & \cdots & \cdots & \cdots & \cdots & \cdots & a_{-l+1} \\ \cdots & \cdots & \cdots & \cdots & \cdots & \cdots & \cdots & \cdots & \cdots & \cdots & \cdots & \cdots & \cdots \\ a_{-2} & \cdots & \cdots & \cdots & \cdots & \cdots & \cdots & \cdots & \cdots & \cdots & \cdots & a_0 & a_{-1} \\ a_{-1} & a_{-2} & \cdots & a_{-l} & 0 & \cdots & \cdots & 0 & a_l & \cdots & a_1 & a_0 \end{pmatrix}.$$

The eigenvalues of the matrix are $f(\epsilon_n^k) = f(e^{-2\pi i k/(n+1)})$, $0 \leq k \leq n$.

(v) All matrix operations leave the set of circulants invariant: the sum or the product of two circulants is a circulant; the transpose, the complex conjugate or the adjoint of a circulant is a circulant. In particular,

$$T_n(p)T_n(q) = T_n(pq),$$

hence $T_n(p)$ and $T_n(q)$, as well as $T_n(p)$ and $T_n(p)^* = T_n(z^{n+1}\overline{p})$ commute; *every circulant is normal.*

5.5.4 Spectrum and Diagonalization of Circulants

(1) *The discrete Fourier transform $\mathcal{F}_d$ diagonalizes every circulant $T_n(p)$; more precisely,*
$$\mathcal{F}_d^{-1} T_n(p) \mathcal{F}_d = \operatorname{diag}(p(1), p(\bar{\epsilon}_1), \ldots, p(\bar{\epsilon}_n)).$$
Indeed, this is immediate by (ii) and §5.5.1(3). ∎

Consequently,
$$\sigma(T_n(p)) = \{p(1), p(\bar{\epsilon}_1), \ldots, p(\bar{\epsilon}_n)\}, \quad \det(T_n(p)) = \prod_k p(\epsilon_k),$$
$$\|T_n(p)\| = \max_{0 \le k \le n} |p(\bar{\epsilon}_k)|, \qquad \|T_n(p)^{-1}\| = 1/\min_{0 \le k \le n} |p(\bar{\epsilon}_k)|.$$

(2) *A circulant whose spectrum is prescribed in advance.* Such a matrix is given in §5.3.2(iii), namely, to have $\sigma(T_n(p)) = \{\lambda_0, \ldots, \lambda_n\}$, we set
$$p(z) = \sum_{k=0}^{n} z^k \frac{1}{n+1} \sum_{l=0}^{n} \epsilon_1^{-kl} \lambda_l.$$

We conclude with an application of circulant matrices; for other applications, in particular, to the asymptotic distribution of the spectra of $T_{\varphi,n}$, $n \to \infty$, see §5.6.4.

5.5.5 An Inequality of Wirtinger (1904)

Let $z_0, \ldots, z_n$ be complex numbers, $z = (z_k) \in \mathbb{C}^{n+1}$ such that $\sum_{k=0}^{n} z_k = 0$. Then,
$$\sum_{k=0}^{n} |z_k - z_{k+1}|^2 \ge 4 \sin^2 \frac{\pi}{n+1} \sum_{k=0}^{n} |z_k|^2,$$
where $z_{n+1} = z_0$; equality is attained if and only if $z_k = \alpha \epsilon_1^k + \beta \epsilon_1^{n-k}$, where $\alpha, \beta \in \mathbb{C}$ and $\epsilon_1 = e^{2\pi i/(n+1)}$ is defined in §5.5.1(2).

Indeed, with the notation of §5.5.1(2), we have $(z, x_0) = 0$ and hence
$$z = \sum_{k=1}^{n} (z, x_k) \frac{x_k}{n+1}, \quad \|z\|^2 = \sum_{k=1}^{n} \frac{|(z, x_k)|^2}{n+1},$$
and then
$$\sum_{k=0}^{n} |z_k - z_{k+1}|^2 = \|(I - C_n)z\|^2 = \sum_{k=1}^{n} |1 - \bar{\epsilon}_1^k|^2 \frac{|(z, x_k)|^2}{n+1}$$
$$\ge |1 - \bar{\epsilon}_1|^2 \sum_{k=1}^{n} \frac{|(z, x_k)|^2}{n+1} = |1 - \bar{\epsilon}_1|^2 \|z\|^2,$$
where
$$|1 - \bar{\epsilon}_1|^2 = |1 - e^{-2\pi i/(n+1)}|^2 = 4 \sin^2 \frac{\pi}{n+1}.$$

Since $|1-\bar{\epsilon}_1|^2 = |1-\bar{\epsilon}_1^n|^2 < |1-\bar{\epsilon}_1^k|^2$ for every k, $1 < k < n$, the case of equality follows also. ∎

Remark We can then deduce the *Wirtinger integral inequality*

$$\int_\mathbb{T} |f|^2\, dm \le \int_\mathbb{T} |f'|^2\, dm,$$

where $f' \in L^2(\mathbb{T})$, $\int_\mathbb{T} f\, dm = 0$ and

$$f'(e^{it}) = \frac{d}{dt} f(e^{it}),$$

if we rewrite §5.5.5 as

$$\frac{1}{n+1} \sum_{k=0}^{n} |z_k|^2 \le \left(\frac{\pi/(n+1)}{\sin(\pi/n+1)}\right)^2 \frac{1}{n+1} \sum_{k=0}^{n} \left|\frac{z_k - z_{k+1}}{2\pi/(n+1)}\right|^2$$

and pass to the limit $n \to \infty$ (supposing $f' \in C(\mathbb{T})$). (Of course, the justification is even simpler with the Fourier series, $f' = \sum_{k\ne 0} k\widehat{f}(k)\, e^{ikt}$.)

5.6 Toeplitz Determinants and Asymptotics of Spectra

This section is devoted to three remarkable results of Gábor Szegő – two asymptotic formulas for the Toeplitz determinants $D_n = \det(T_{\varphi,n})$ and a formula of the limit distribution of the spectra $\sigma(T_{\varphi,n})$. As we have seen in Chapter 1, all three have numerous important applications (see also §5.8 below), and all three depend heavily on a theorem on Hardy spaces, the *Szegő–Verblunsky formula*: given a Borel measure $\mu \in \mathcal{M}(\mathbb{T})$, $\mu \ge 0$, we have

$$d^2(\mu) =: \operatorname{dist}^2_{L^2(\mu)}(1, H_0^2(\mu)) = \inf_{p \in \mathcal{P}_a} \int_\mathbb{T} |1-p|^2\, d\mu = \exp\left(\int_\mathbb{T} \log\left|\frac{d\mu}{dm}\right| dm\right);$$

see Appendix F, as well as [Nikolski, 2019] Chapter 2, particularly Theorem 2.7.1. The right-hand side of the formula is a sort of "geometric mean" of μ (the same as for its Radon–Nikodym derivative $d\mu/dm$). We denote this mean as

$$G(\mu) = \exp\left(\int_\mathbb{T} \log\left|\frac{d\mu}{dm}\right| dm\right).$$

5.6.1 The First Szegő Asymptotic Formula (1915)

Let $\mu \in M(\mathbb{T})$, $\mu \geq 0$ and $T_{\mu,n} = (\hat{\mu}(k-j))_{0 \leq k,j \leq n}$ its Toeplitz matrices, $D_n = D_n(\mu) =: \det(T_{\mu,n})$, and

$$d_n = d_n(\mu) = D_n/D_{n-1},$$

where we set $d_n = 0$ if $D_{n-1} = 0$.

(i) The sequence (d_n) is decreasing, $d_{n+1} \leq d_n$, and

$$\lim_n d_n(\mu) = G(\mu), \quad \lim_n D_n^{1/(n+1)} = G(\mu).$$

(ii) If $\mu = (1/|p|^2)m$ where $p \in \mathcal{P}_a$ and $p(z) \neq 0$ for $z \in \mathbb{T}$, then for $n \geq r =: \deg(p)$,

$$d_n(\mu) = G(\mu), \quad D_n = D_{r-1} \cdot (G(\mu))^{n-r+1}.$$

Proof (i) Since $T_{\mu,n} = ((z^j, z^k)_{L^2(\mu)})_{0 \leq k,j \leq n}$ is a Gram matrix of the sequence $(z^k)_0^n$ in the space $L^2(\mathbb{T}, \mu)$, we have (see Appendix B)

$$d_n(\mu) = D_n/D_{n-1} = \text{dist}^2_{L^2(\mu)}(z^n, \mathcal{P}_{n-1}) = \text{dist}^2_{L^2(\mu)}(1, z\mathcal{P}_{n-1})$$

(the last equality is evident in the form of an integral, as well as in the form of Toeplitz matrices). Hence, $d_n(\mu) \downarrow d^2(\mu)$, which reduces (i) to the Szegő–Verblunsky formula.

(ii) Without loss of generality, we suppose $p(z) \neq 0$ for $|z| \leq 1$ (if $p(z_1) = 0$, $|z_1| < 1$ we replace the polynomial p with

$$q = p\frac{1 - \bar{z}_1 z}{z - z_1}$$

having the same modulus on $\mathbb{T}$, $|q| = |p|$). Then $(1/p)H_0^2 = H_0^2$ and

$$d^2(\mu) = \inf_{h \in H_0^2} \int_{\mathbb{T}} |1 - h|^2 \frac{1}{|p|^2} dm = \inf_{h \in H_0^2} \int_{\mathbb{T}} \left|\frac{1}{p} - \frac{h}{p}\right|^2 dm = \frac{1}{|p(0)|^2}.$$

Moreover,

$$d^2(\mu) \leq d_n(\mu) = \text{dist}^2_{L^2(\mu)}(1, z\mathcal{P}_{n-1}) \leq \int_{\mathbb{T}} \left|\frac{p}{p(0)}\right|^2 \frac{1}{|p|^2} dm = |p(0)|^{-2},$$

hence $G(\mu) = d^2(\mu) = d_n(\mu)$. ∎

5.6.2 Equidistribution of Sequences, after Weyl (1910)

Definitions 5.6.1 Let $a < b$ ($a, b \in \mathbb{R}$), and let w be a measurable real function on a probability space (Ω, μ), $\mu\Omega = 1$, such that

$$a \leq w \leq b \quad \mu\text{-a.e. on } \Omega.$$

Sequences $(t_{n,j})_{j=0}^n \subset [a,b]$ $(n = 0, 1, \ldots)$ are said to be *(asymptotically) distributed like w with respect to a family of functions* $F \subset C[a,b]$ if for every $f \in F$ we have

$$\lim_n \frac{1}{n+1} \sum_{j=0}^n f(t_{n,j}) = \int_a^b f(t) \, d\lambda_w(t) = \int_\Omega (f \circ w) \, d\mu,$$

where

$$\lambda_w(t) = \mu\{x \in \Omega : w(x) < t\}, \quad a \le t \le b.$$

The sequences $(t_{n,j})_{j=0}^n$ and $(u_{n,j})_{j=0}^n$ $(n = 0, 1, \ldots)$ *are (asymptotically) equidistributed* on the interval $[a,b]$ *with respect to F* if $t_{n,j}, u_{n,j} \in [a,b]$ for every n, j, and for every $f \in F$, we have

$$\lim_n \frac{1}{n+1} \sum_{j=0}^n (f(t_{n,j}) - f(u_{n,j})) = 0;$$

the sequences are *distributed like w, or equidistributed on* $[a,b]$ if they are (equi-) distributed like w with respect to the whole of $C[a,b]$.

Remark Clearly a sequence $(t_{n,j})_{j=0}^n \subset [a,b]$, $n = 0, 1, \ldots$, is asymptotically distributed like w if and only if the measures

$$\nu_n = \frac{1}{n+1} \sum_{j=0}^n \delta_{t_{n,j}}$$

converge $\sigma(\mathcal{M}[a,b], C[a,b])$ *weak-* to* λ_w.

Hermann Weyl (1885–1955) was a German–American mathematician and physicist, whose contribution transformed or, frankly, gave birth to several domains of mathematics, such as spectral analysis, geometric function theory, Riemann surfaces, complex manifolds, representation theory, the theories of gauge fields and relativity, the foundations of mathematics, and certain others. He was perhaps the last "universalist" in the model of Hilbert and Poincaré. He left 15 fundamental monographs, including three of the key illuminating works of the twentieth century: *Die*

Idee der Riemannschen Flächen (1913), *Raum, Zeit, Materien* (1918), and *Gruppentheorie und Quantenmechanik* (1928).

His doctoral thesis (1908, Göttingen, under Hilbert) was inspired by Hilbert's famous course on the theory of integral operators (which he took in the company of Erhard Schmidt, Ernst Hellinger, and Otto Toeplitz), and was then extended in his *Habilitation* thesis (1910) to a spectral theory of the Sturm–Liouville operators. It was here that the famous *Weyl criterion (of continuous spectrum)* first appeared (see Appendix E.3, and then the discussion in Appendix E.10, as well as Lemma 3.1.1 and Theorem 3.1.3), along with the "limit-point, limit-circle" classification of vibrating strings, and certain other tools of classical spectral theory. Then came the idea and the formula of the asymptotic distribution of eigenvalues (for the Laplacian (1911), and later for certain other operators). For Riemann surfaces, Weyl introduced (1913) the coverings, the differential manifolds, the duality of differentials and the 1-cocycles, and also the separation axiom T_2 (usually attributed to Hausdorff). In the theory of relativity and matter he defined the notion of the *(Weyl) gauge metric* and constructed the first (tentative) unified theory of electromagnetic and gravitational fields (without quantum effects), introduced the *Weyl tensor* and other tools. Then, in the theory of representations (created in parallel with Élie Cartan), Weyl introduced the irreducible matrix representations, group algebras, approximate identities, the Peter–Weyl theorem (!), etc., not to mention his stroke of genius in analytic number theory – the *Weyl sums*, the uniform distributions of sequences "mod 1" (the *Weyl criterion*)! Weyl was co-founder of *Intuitionism* (with Luitzen Brouwer). At least 40 mathematical terms bear Weyl's name.

Weyl's life coincided with a turbulent period of history including two world wars, the Nazi regime in Germany, etc. Nonetheless, he had the good luck to avoid the worst circumstances: he spent the years 1904–1913 in Göttingen (as a doctoral student, and then as *Privatdozent*), and 1913–1930 in Zürich as professor. (All of Weyl's biographers point out a certain light and libertine character of the intellectual elite of the epoch. In particular, thanks to the "sexual revolution" coming from the United States, it became fashionable to show off one's extramarital affairs. Among other consequences, Weyl maintained for years a "legal" love affair with the wife of his best friend Erwin Schrödinger, whereas she took care of Erwin's daughter by another woman.) During the three years 1930–1933, Weyl inherited the prestigious chair of Hilbert; even so, he was pushed out by the Nazi law of April 7, 1933 – or more precisely its modification on June 30, 1933, forbidding public service to anyone whose spouse was "non-Aryan" (Weyl's wife, Helene Joseph, was of Jewish origin). In 1933 Weyl found himself among the very first recruits of the new Institute for Advanced Study (with Veblen, Einstein, von Neumann, and Gödel). This last period as a professor at Princeton (1933–1955, the last three years after his retirement shared between Princeton and Zürich) was the most productive of his career (among others, he wrote six books, which all became classics, such as *Symmetry* (1952) (his "swan song" as he called it himself) or *The Classical Groups* (1939)). Weyl died suddenly (in Zürich) while walking home from the post office where he had posted *thank you* letters for a celebration of his 70th birthday.

Weyl was a foreign member of the Royal Society (1936), a recipient of the Lobachevsky Prize (1927), the plenary speaker of the 1928 ICM, President of the Deutschen Mathematiker-Vereinigung, and *Doctor Honoris Causa* of several universities. There exists an extensive literature with biobibliographies, interpretations, and analyses of the heritage of Hermann Weyl. We limit ourselves to the following three references: a fascinating overview by Chevalley and Weil [1957], some selected works edited by Arnold and Parshin [Weyl, 1984], and the proceedings of the colloquium for Weyl's centenary [Chandrasekharan, 1986].

We conclude with a few famous quotations from Weyl:

My work always tried to unite the truth with the beautiful, but when I had to choose one or the other, I usually chose the beautiful.

Abstract algebra ... leads to such a level of generality that we finish by losing all mathematical content.   (Speech in honor of Felix Klein, 1929)

In these days the angel of topology and the devil of abstract algebra fight for the soul of each individual mathematical domain.

("Invariants," *Duke Math. J.* **5**:3 (1939), pages 489–502)

The Gods have imposed upon my writing the yoke of a foreign language that was not sung at my cradle.

(*The Classical Groups* (1939); Weyl was particularly gifted at languages)

Lemma 5.6.2   (i)   *If* $(\Omega, \mu) = ([0, 1], dx)$ *and* $w \in C[0, 1]$, $a \leq w \leq b$, *then a sequence* $(t_{n,j})_{j=0}^{n} \subset [a, b]$ *is distributed like* w *with respect to* $F \subset C[a, b]$ *if and only if* $(t_{n,j})_{j=0}^{n}$ *and* $(u_{n,j} = w(j/(n+1)))_{j=0}^{n}$ *are equidistributed with respect to* F.

5.6 Toeplitz Determinants and Asymptotics of Spectra

(ii) If $(t_{n,j})_{j=0}^n$ and $(u_{n,j})_{j=0}^n$ ($\subset [a,b]$ equipped with the normalized measure $dx/(b-a)$) are equidistributed (respectively, distributed like w) with respect to $F \subset C[a,b]$, then so are they with respect to $\mathrm{span}_{C[a,b]}(F)$.

(iii) For every $a < b$ ($a, b \in \mathbb{R}$) and every $0 < \epsilon < 1/\max(|a|,|b|)$,

$$\mathrm{span}_{C[a,b]}(L_u: -\epsilon < u < \epsilon) = C_0[a,b] =: \{f \in C[a,b]: f(0) = 0\},$$

where $L_u(t) = \log(1 + ut)$ (if $0 \notin [a,b]$, then $C_0[a,b] = C[a,b]$).

(iv) If $(t_{n,j})_{j=0}^n$ and $(u_{n,j})_{j=0}^n$ ($\subset [a,b]$, equipped with $dx/(b-a)$) are equidistributed (respectively, distributed like w) with respect to L_u ($\forall u, -\epsilon < u < \epsilon < 1/\max(|a|,|b|)$), then they are equidistributed (respectively, distributed like w) on $[a,b]$.

Proof (i) Indeed,

$$\int_\Omega (f \circ w)\, dx = \lim_n \sum_{j=0}^n \frac{1}{n+1} f(w(j/n+1)), \quad \forall f \in C[a,b].$$

(ii) Clearly the equidistribution (or the distribution like w) extends from F to $\mathrm{Vect}(F)$. Suppose $f \in \mathrm{clos}_{C[a,b]}(\mathrm{Vect}(F))$, and let $\epsilon > 0$. Then there exists $g \in \mathrm{Vect}(F)$ such that $\|f - g\|_\infty < \epsilon$, hence

$$\overline{\lim_n} \left| \sum_{j=0}^n \frac{1}{n+1} f(t_{n,j}) - \int_a^b f(t)\, d\lambda_w(t) \right|$$

$$\leq \overline{\lim_n} \left| \sum_{j=0}^n \frac{1}{n+1} (f(t_{n,j}) - g(t_{n,j})) \right| + \int_a^b |f - g|\, d\lambda_w < 2\epsilon,$$

and since $\epsilon > 0$ is arbitrary, we obtain the result. The same holds for equidistribution.

(iii) Suppose $v \in \mathcal{M}[a,b]$ and $\int_a^b L_u\, dv = 0$, $\forall u \in (-\epsilon, \epsilon)$. The function

$$L(z) =: \int_a^b \log(1 + zt)\, dv(t)$$

is holomorphic in a neighborhood of 0 and is zero on $(-\epsilon, \epsilon)$, hence $L = 0$. By integrating the normally convergent series $\log(1 + zt) = \sum_{k\geq 1} (-1)^{k+1} t^k z^k / k$, we obtain

$$0 = \widehat{L}(k) = \frac{(-1)^{k+1}}{k} \int_a^b t^k\, dv(t) \quad \text{for } k = 1, 2, \ldots.$$

The Hahn–Banach theorem implies $t^k \in \mathrm{span}_{C[a,b]}(L_u: -\epsilon < u < \epsilon)$ ($\forall k \geq 1$); this concerns functions $t \mapsto t^k$ on $[a,b]$), giving $\mathrm{span}_{C[a,b]}(L_u: -\epsilon < u < \epsilon) \supset C_0[a,b]$. The result follows, as the reverse inclusion is evident.

(iv) This is a consequence of (ii) and (iii), as the equalities of Definition 5.6.2(1) are evident for constant functions f. ∎

5.6.3 Asymptotic Distribution of Spectra

Here we prove and comment on one of the principal results of this chapter, as follows.

Theorem 5.6.3 (Szegő, 1920) *Let $\varphi \in L^\infty(\mathbb{T}) = (L^\infty(0, 2\pi), dx/(2\pi))$ be a real function, $a = \operatorname{ess\,inf}(\varphi)$, $b = \operatorname{ess\,sup}(\varphi)$, and $\lambda_{n,k}$ ($k = 0, \ldots, n$) the eigenvalues of the self-adjoint matrix $T_{\varphi,n}$ ($n = 0, 1, \ldots$),*

$$\sigma(T_{\varphi,n}) = \{\lambda_{n,k} : k = 0, \ldots, n\}.$$

Then, the sequences $(\lambda_{n,k})_{k=0}^n$ are asymptotically distributed on $[a, b]$ like φ, that is, for every $f \in C[a, b]$,

$$\lim_n \sum_{j=0}^n \frac{1}{n+1} f(\lambda_{n,j}) = \frac{1}{2\pi} \int_0^{2\pi} f(\varphi(e^{ix})) \, dx,$$

and if $\varphi \in C(\mathbb{T})$, the sequences $(\lambda_{n,k})_{k=0}^n$ and $(u_{n,j} = \varphi(e^{2\pi i j/(n+1)}))_{j=0}^n$ are equidistributed.

Proof For $L_u(t) = \log(1 + ut)$ and $u \in \mathbb{R}$ sufficiently small, we have

$$\lim_n \sum_{j=0}^n \frac{1}{n+1} L_u(\lambda_{n,j}) = \lim_n \log \left(\prod_j (1 + u\lambda_{n,j}) \right)^{1/(n+1)}$$

$$= \lim_n \log(\det(T_{1+u\varphi,n}))^{1/(n+1)}$$

$$= \frac{1}{2\pi} \int_0^{2\pi} L_u(\varphi) \, dx$$

by §5.6.1(i). The lemma of §5.6.2(2) completes the proof. ∎

Remarks

(1) Theorem 5.6.3 can be extended to Riemann-integrable functions f by using the standard approximation $g \leq f \leq h$ where $g, h \in C([a, b])$ and $\int_a^b (h - g) \, d\lambda_\varphi < \epsilon$. In place of a general treatment of this question, we present the special case of Corollary 5.6.4 below.

(2) We thus observe that the spectra $\sigma(T_{\varphi,n})$ converge weakly in mean with respect to $C([a, b])$, but in fact to $\sigma(M_\varphi)$, $M_\varphi: L^2(\mathbb{T}) \to L^2(\mathbb{T})$, where $\sigma(M_\varphi) = \operatorname{Ran}_{\text{ess}}(\varphi)$, and not to $\sigma(T_\varphi)$, $T_\varphi: H^2(\mathbb{T}) \to H^2(\mathbb{T})$. An explanation of this fact (partial and not completely convincing) is that the matrix of

5.6 Toeplitz Determinants and Asymptotics of Spectra

$T_{\varphi,2n}$ coincides with $(\widehat{\varphi}(i-j))_{-n \le i,j \le n}$, which is also the matrix of the compression $P_{-n,n} M_\varphi P_{-n,n}$ which converges to M_φ as $n \to \infty$. Nonetheless, we can say that a certain "pseudo-spectrum" of $T_{\varphi,n}$ converges to $\sigma(T_\varphi)$, see an example and the explanations of Remark (1) of Corollary 5.6.4 below.

Corollary 5.6.4 (asymptotic density of spectra (Toeplitz, 1911), (Szegő, 1920)) *Under the hypotheses and notation of Theorem 5.6.3, for every interval $[\alpha,\beta) \subset [a,b]$ where α is a point of continuity of λ_φ (see §5.6.2 for the definition),*

$$\lim_n \frac{1}{n+1} \mathrm{card}\{j: \lambda_{n,j} \in [\alpha,\beta)\} = m\{x \in [0,2\pi]: \alpha \le \varphi(e^{ix}) < \beta\},$$

where m is the normalized Lebesgue measure.

Indeed, to justify the stated formula, it suffices to extend the asymptotic relation of Theorem 5.6.3 to the indicator functions $f = \chi_{[\alpha,\beta)}$. Let $g_k, h_k \in C[a,b]$ be such that

$$g_k(a) = g_k(\alpha) = 0, \quad g_k\left(\alpha + \frac{1}{k}\right) = 1, \quad g_k\left(\beta - \frac{1}{k}\right) = 1,$$

$g_k(\beta) = g_k(b) = 0$ and g_k is linear between these points, and (symmetrically)

$$h_k(a) = h_k\left(\alpha - \frac{1}{k}\right) = 0, \quad h_k(\alpha) = 1, \quad h_k(\beta) = 1, \quad g_k\left(\beta + \frac{1}{k}\right) = h_k(b) = 0$$

(and h_k is linear between these points). Then $g_k \le f \le h_k$ and, given $\epsilon > 0$, we can find k such that

$$0 \le \int_a^b (h_k - f)\, d\lambda_\varphi < \epsilon, \quad 0 \le \int_a^b (f - g_k)\, d\lambda_\varphi < \epsilon.$$

Then, by Theorem 5.6.3, by taking n sufficiently large, for

$$Q_n =: \frac{1}{n+1} \mathrm{card}\{j: \lambda_{n,j} \in [\alpha,\beta)\}$$

we will have

$$\int_a^b f\, d\lambda_\varphi - \epsilon < \sum_{j=0}^n \frac{1}{n+1} g_k(\lambda_{n,j}) \le Q_n \le \sum_{j=0}^n \frac{1}{n+1} h_k(\lambda_{n,j}) < \int_a^b f\, d\lambda_\varphi + \epsilon,$$

and the result follows. ■

Remarks

(1) In particular, by the definition of $a = \operatorname{ess\,inf}(\varphi)$, $b = \operatorname{ess\,sup}(\varphi)$, we have, for every $\epsilon > 0$,

$$\lim_n \frac{1}{n+1} \operatorname{card}\{j: a \le \lambda_{n,j} < a + \epsilon\} > 0,$$

$$\lim_n \frac{1}{n+1} \operatorname{card}\{j: b - \epsilon \le \lambda_{n,j} \le b\} > 0.$$

It is interesting to regard the behavior of the spectra $\sigma(T_{\varphi,n})$ for φ having gaps in $\operatorname{Ran}_{\text{ess}}(\varphi)$. For example, for φ taking on only two values, say $a < b$ (such as $\varphi(e^{ix}) = a\chi_{(-\pi,0)} + b\chi_{(0,\pi)}$), we obtain a zero distribution density of $\lambda_{n,j}$ on every interval $[\alpha, \beta] \subset (a, b)$, whereas by Theorem 3.3.8 (or Lemma 3.3.7) the spectrum of the limit operator $\sigma(T_\varphi)$ is the entire segment $[a, b]$. By contrast, for $\varphi \in C(\mathbb{T})$ (real), the spectrum is $\sigma(T_\varphi) = [a, b]$, where $a = \inf(\varphi)$, $b = \sup(\varphi)$, and the asymptotic density of the spectra $\sigma(T_{\varphi,n})$ is everywhere positive.

Nonetheless, if φ is real, the spectra $\sigma(T_{\varphi,n})$ are *asymptotically everywhere dense* on $[a, b]$, $a = \operatorname{ess\,inf}(\varphi)$, $b = \operatorname{ess\,sup}(\varphi)$. To justify this density, note that for every $\lambda \in [a, b] \setminus \sigma(T_{\varphi,n})$, we have

$$\|R_\lambda(T_{\varphi,n})\| = \frac{1}{\operatorname{dist}(\lambda, \sigma(T_{\varphi,n}))}$$

(since $T_{\varphi,n}$ is self-adjoint), and if we suppose that $\underline{\lim}_n \|R_\lambda(T_{\varphi,n})\| =: M < \infty$, we would have

$$\|P_n(\varphi - \lambda)f\|_2 = \|(T_{\varphi,n} - \lambda I_n)f\|_2 \ge (1/M)\|f\|_2$$

for every polynomial $f \in \mathcal{P}_a$ and for an infinite sequence of values n ($n \to \infty$), which implies

$$\|(T_\varphi - \lambda I)f\|_2 = \|P_+(\varphi - \lambda)f\|_2 \ge (1/M)\|f\|_2$$

(and hence, contradicts Theorem 3.3.8). Hence $\lim_n \operatorname{dist}(\lambda, \sigma(T_{\varphi,n})) = 0$. ∎

The above reasoning shows that in the case of two values, notwithstanding the said density 0 of the $\sigma(T_{\varphi,n})$ on (a, b) and Remark (2) after Theorem 5.6.3, the ϵ-*pseudo-spectra* $\sigma_\epsilon(T_{\varphi,n}) :=: \{\lambda: \|R_\lambda(T_{\varphi,n})\| \ge 1/\epsilon\}$ converge to the ϵ-*pseudo-spectrum* $\sigma_\epsilon(T_\varphi)$ (and not to that of M_φ). We do not linger here on the precise definition of "pseudo-spectra," nor on their general properties, and refer the reader to [Böttcher and Silbermann, 1999], [Böttcher, 1994], and [Trefethen and Embree, 2005].

It must be noted (for the same reason as before) that the same asymptotic density $\sigma(A) \subset \lim_n \sigma(P_n A P_n)$, where $P_n \uparrow \operatorname{id}$ is a monotonic sequence of orthogonal projections, holds for every *self-adjoint (or normal)* operator A on a Hilbert space. Of course, this is far from being correct for an arbitrary operator

5.6 Toeplitz Determinants and Asymptotics of Spectra

A, $A \in L(H)$ (a standard example: $A = S$ the shift operator on the Hardy space $H = H^2$ and P_n the orthogonal projection onto $\mathcal{P}_n$ with $\sigma(S) = \overline{\mathbb{D}}$ and $\sigma(P_n S P_n) = \{0\}$).

(2) The case $\varphi \in L^1(\mathbb{T})$. We can directly justify a weaker form of the *concentration of the eigenvalues* $\lambda_{n,j}$ *towards the extremities of the spectral interval* (a, b) of Remark (1), including the cases $a = -\infty$ and/or $b = \infty$, without using the theorem of asymptotic distribution of Corollary 5.6.4. More precisely, *let* $\varphi \in L^1(\mathbb{T})$ *be a real function*, $a = \operatorname{ess\,inf}(\varphi)$, $b = \operatorname{ess\,sup}(\varphi)$, *and*

$$\min(\sigma(T_{\varphi,n})) = \lambda_{n,0} \leq \lambda_{n,1} \leq \cdots \leq \lambda_{n,n} = \max(\sigma(T_{\varphi,n})).$$

Then, $\lim_n \lambda_{n,0} = a$ *and* $\lim_n \lambda_{n;n} = b$ *(including the cases $a = -\infty$ and/or $b = \infty$)*.

Indeed, since

$$\lambda_{n,0} = \min(\sigma(T_{\varphi,n})) = \min\{(\varphi f, f)_{L^2(\mathbb{T})} : f \in \mathcal{P}_n, \|f\|_2 = 1\},$$
$$\lambda_{n,n} = \max(\sigma(T_{\varphi,n})) = \max\{(\varphi f, f)_{L^2(\mathbb{T})} : f \in \mathcal{P}_n, \|f\|_2 = 1\},$$

we have

$$\lambda_{n,0} \geq \lambda_{n+1,0} \quad \text{and} \quad \lim_n \lambda_{n,0} = \inf\left\{\int_{\mathbb{T}} \varphi |f|^2 \, dm : f \in \mathcal{P}, \, \|f\|_2 = 1\right\} = a,$$

and similarly for $\lim_n \lambda_{n,n} = b$. ∎

Example 5.6.5 ($\varphi(e^{ix}) = 2\cos(x)$) Let $\varphi = z + z^{-1}$. Then

$$T_{\varphi,n} = \begin{pmatrix} 0 & 1 & 0 & \cdots & 0 & 0 \\ 1 & 0 & 1 & \cdots & 0 & 0 \\ 0 & 1 & 0 & \cdots & 0 & 0 \\ \vdots & \ddots & \ddots & \ddots & \ddots & \vdots \\ \vdots & & \ddots & \ddots & 0 & 1 \\ 0 & 0 & 0 & \cdots & 1 & 0 \end{pmatrix}.$$

The characteristic equation $\Delta_n(\lambda) = 0$, where $\Delta_n(\lambda) = \det(\lambda I - T_{\varphi,n})$, can be made explicit by expanding the determinant $\Delta_n(\lambda)$ by the first row

$$\Delta_n(\lambda) = \lambda \Delta_{n-1}(\lambda) - \Delta_{n-2}(\lambda) \quad (n = 2, 3, \ldots),$$

with the initial values $\Delta_0(\lambda) = \lambda$, $\Delta_1(\lambda) = \lambda^2 - 1$. In fact, the preceding recurrence formula also remains correct for $n = 1$ if we set $\Delta_{-1} =: 1$. To find the polynomial $\Delta_n(\lambda)$ and its roots, we first consider the general solution of the recurrence chain

$$\Delta_n(\lambda) = A y_1^n + B y_2^n, \quad n = 0, 1, \ldots,$$

where $A, B \in \mathbb{C}$ and $y_{1,2}$ are the roots of the equation $y^2 - \lambda y + 1 = 0$. Knowing that λ is a real number in the interval $[-2, 2]$ ($2 = \max(\varphi) = -\min(\varphi)$), we obtain $y_{1,2} = (\lambda \pm i\sqrt{4 - \lambda^2})/2$, hence unimodular numbers, i.e.

$$y_{1,2} = e^{\pm ix} \quad \text{where } x \in [0, \pi).$$

With this notation, we have $\lambda = 2\cos(x)$ and, by using the initial values $\Delta_0 = \lambda$, $\Delta_{-1} = 1$, we find $A + B = 2\cos x$, $1 = (A + B)\cos x + i(B - A)\sin x$, giving

$$A = 2\cos x - B, \quad B = \frac{1 - 2e^{-ix}\cos x}{2i\sin x},$$

$$\Delta_n(2\cos x) = 2(\cos nx + i\sin nx)\cos x - 2iB\sin nx$$

$$= \frac{\sin 2x \cos nx - \sin nx + 2\cos^2 x \sin nx}{\sin x} = \frac{\sin(n+2)x}{\sin x}.$$

The solutions of $\Delta_n = 0$ are $x_{n,k} = k\pi/(n+2)$, $k = 1, \ldots, n+1$, so that (in the sense of increasing eigenvalues)

$$\lambda_{n,j} = 2\cos\frac{(n+2-j)\pi}{n+2}, \quad j = 1, \ldots, n+1.$$

The asymptotic density of the spectra follows the rule of Corollary 5.6.4:

$$\operatorname{card}\left\{k \in [1, n+1] : \alpha \leq 2\cos\frac{\pi k}{n+2} \leq \beta\right\} \approx \frac{n+2}{\pi}\left(\arccos\frac{\alpha}{2} - \arccos\frac{\beta}{2}\right),$$

$$m\{x \in (-\pi, \pi) : \alpha \leq 2\cos x \leq \beta\} = \frac{2}{2\pi}\left(\arccos\frac{\alpha}{2} - \arccos\frac{\beta}{2}\right). \quad \blacksquare$$

5.6.4 Asymptotic Distribution Meets the Circulants

Here we give another proof of the principal Theorem 5.6.3, using the technique of circulants of §5.5. This approach was already sketched out in the pioneering work of Toeplitz (1911), then developed by Szegő (1958). The new proof is no simpler than the first, but it contains other ideas hidden in the relations of Theorem 5.6.3. In particular, it also works for complex symbols φ and not just for positive real symbols. To avoid repeating exactly the same computations, we now present the statement for the complex case, but the proof below is presented for symbols $\varphi \geq 0$ (the modifications for the complex case, noted in the text, are minimal and absolutely transparent).

Theorem 5.6.6 (asymptotic distribution of spectra: the case of complex φ) *Let $\varphi \in L^\infty(\mathbb{T})$ and let $\lambda_{n,k}$ ($k = 0, \ldots, n$) be the eigenvalues of a matrix $T_{\varphi,n}$ ($n = 0, 1, \ldots$) repeated according to their algebraic multiplicity,*

$$\sigma(T_{\varphi,n}) = \{\lambda_{n,k} : k = 0, \ldots, n\}.$$

(1) $\sigma(T_{\varphi,n}) \subset \text{conv}(\text{Ran}_{\text{ess}}(\varphi)) =: \Sigma(\varphi)$.
(2) *The sequences $(\lambda_{n,k})_{k=0}^{n}$ are asymptotically distributed in $\Sigma(\varphi)$ with respect to the holomorphic test functions $f \in \text{Hol}(\Sigma(\varphi))$ like φ, i.e. for every $f \in \text{Hol}(\Sigma(\varphi))$,*

$$\lim_{n} \frac{1}{n+1} \sum_{j=0}^{n} f(\lambda_{n,j}) = \int_{0}^{2\pi} f \circ \varphi \, dm.$$

Proof (of Theorems 5.6.6 and 5.6.3) (1) It suffices to show that $\sigma(T_{\varphi,n}) \subset D(\lambda, R)$ where $D(\lambda, R)$ is any disk containing $\text{Ran}_{\text{ess}}(\varphi)$. However, $\text{Ran}_{\text{ess}}(\varphi) \subset D(\lambda, R)$ means that $\|\varphi - \lambda\|_{\infty} < R$, which implies $\|T_{\varphi-\lambda,n}\| < R$, and hence $\sigma(T_{\varphi,n}) \subset D(\lambda, R)$.

(2) We present the proof of this part under the hypotheses of Theorem 5.6.3 and note the modifications to make for the complex case. Let φ_l be a Fejér sum of φ:

$$\varphi_l = \Phi_l * \varphi = \sum_{|k| \leq l} \left(1 - \frac{|k|}{l}\right) \widehat{\varphi}(k) z^k$$

(where Φ_l is the Fejér kernel of order l). Then, φ_l is a real trigonometric polynomial with $\deg(\varphi_l) \leq l$ and $a \leq \varphi_l(\zeta) \leq b$ ($\zeta \in \mathbb{T}$; a and b are defined in Theorem 5.6.3 (in the complex case $\varphi_l(\zeta) \in \Sigma(\varphi)$). By regarding a circulant $\varphi_l(C_n)$ for $n \geq l$, we have, by §5.5.4, $\sigma(\varphi_l(C_n)) = \{\lambda_{n,k} = \varphi_l(\epsilon_n^k): 0 \leq k \leq n\}$ where $\epsilon_n = e^{-2\pi i/(n+1)}$, and hence the mean

$$\frac{1}{n+1} \sum_{k} \lambda_{n,k} = \frac{1}{n+1} \text{Trace}(\varphi_l(C_n))$$

is a Riemann sum of the integral $\int_{\mathbb{T}} \varphi_l \, dm$, giving the relation

$$\lim_{n} \frac{1}{n+1} \text{Trace}(\varphi_l(C_n)) = \int_{\mathbb{T}} \varphi_l \, dm.$$

Since for every $s = 1, 2, \ldots$, we have $(\varphi_l(C_n))^s = \varphi_l^s(C_n))$ and $\lambda_{n,k}(\varphi_l(C_n)^s) = (\lambda_{n,k}(\varphi_l(C_n)))^s$, we equally obtain

$$\lim_{n} \frac{1}{n+1} \text{Trace}((\varphi_l(C_n))^s) = \int_{\mathbb{T}} \varphi_l^s \, dm.$$

Now, by using the approximation properties of the φ_l, we replace φ_l with φ, and $\varphi_l(C_n)$ with $T_{\varphi,n}$, and then $f(x) = x^s$ ($a \leq x \leq b$) with arbitrary $f \in C[a, b]$.

By realizing this plan for the right-hand side of the last asymptotics, we first remark that $\varphi_l^{s+1} - \varphi^{s+1} = \varphi_l^s(\varphi_l - \varphi) + \varphi(\varphi_l^s - \varphi^s)$, and hence, by induction,

$\lim_l \|\varphi_l^s - \varphi^s\|_{L^2} = 0$ (in fact, we have the convergence in $L^p(\mathbb{T})$ for every p, $1 \leq p < \infty$), and then

$$\lim_l \int_{\mathbb{T}} \varphi_l^s \, dm = \int_{\mathbb{T}} \varphi^s \, dm \quad (\forall s = 0, 1, \ldots).$$

For the left-hand side, recall that for every matrix $A = (a_{i,j})$ of order $(n+1) \times (n+1)$, we have

$$|\mathrm{Trace}(A)| = \left|\sum_{k=0}^n a_{k,k}\right| \leq \sqrt{n+1}\left(\sum_{k,j} |a_{k,j}|^2\right)^{1/2} = \sqrt{n+1}\|A\|_{\mathfrak{S}_2},$$

where $\|A\|_{\mathfrak{S}_2}$ is the Hilbert–Schmidt (or Frobenius) norm of A: see Appendix D. Hence, to replace

$$\lim_n \frac{1}{n+1} \mathrm{Trace}((\varphi_l(C_n))^s)$$

with

$$\lim_n \frac{1}{n+1} \mathrm{Trace}(T_{\varphi,n}^s),$$

it suffices to show that, for every $s \geq 1$ and certain $l = l(n) \to \infty$,

$$\|(\varphi_l(C_n))^s - T_{\varphi,n}^s\|_{\mathfrak{S}_2} = o(\sqrt{n+1}) \quad (\text{as } n \to \infty).$$

Using

$$(\varphi_l(C_n))^{s+1} - T_{\varphi,n}^{s+1} = \varphi_l(C_n)^s(\varphi_l(C_n) - T_{\varphi,n}) + (\varphi_l(C_n))^s - T_{\varphi,n}^s)T_{\varphi,n}$$

and $\sup_{n,l} \|\varphi_l(C_n)^s\| < \infty$ (since $\|\varphi_l^s\|_\infty \leq \|\varphi\|_\infty^s$), $\sup_n \|T_{\varphi,n}^s\| < \infty$, and then by the inequality $\|AB\|_{\mathfrak{S}_2} \leq \|A\|_{\mathfrak{S}_2} \|B\|$ (for every pair of operators or matrices A, B: see Appendix D), we can reduce our claim to the case $s = 1$. In this case,

$$\|\varphi_l(C_n) - T_{\varphi,n}\|_{\mathfrak{S}_2} \leq \|\varphi_l(C_n) - T_{\varphi_l,n}\|_{\mathfrak{S}_2} + \|T_{\varphi_l,n} - T_{\varphi,n}\|_{\mathfrak{S}_2}.$$

For the first term, we have the matrix of Example 5.5.3(iv) (for $f = \varphi_l$ and $l = l(n) \leq n/2$), giving

$$\|\varphi_l(C_n) - T_{\varphi_l,n}\|_{\mathfrak{S}_2}^2 = \sum_{|k|\leq l} |k|\left(1 - \frac{|k|}{l}\right)^2 |\widehat{\varphi}(k)|^2 \leq \sum_{|k|\leq l} |k| \cdot |\widehat{\varphi}(k)|^2 = o(n),$$

when $n \to \infty$ and $l = l(n) \leq n$ (we have, of course, $\sum_k |\widehat{\varphi}(k)|^2 < \infty$). For the second term, by summing along the diagonals, we obtain

$$\|T_{\varphi_l,n} - T_{\varphi,n}\|_{\mathfrak{S}_2}^2 = \sum_{0\leq k,j\leq n, |k-j|\leq l} \frac{|k-j|^2}{l^2} |\widehat{\varphi}(k-j)|^2 + \sum_{0\leq k,j\leq n, |k-j|>l} |\widehat{\varphi}(k-j)|^2$$

$$\leq (n+1)\sum_{|s|\leq l} \frac{|s|^2}{l^2} |\widehat{\varphi}(s)|^2 + (n+1)\sum_{|s|>l} |\widehat{\varphi}(s)|^2 = o(n)$$

as $n > l = l(n) \to \infty$ (for the first sum, as for the first term of the preceding estimate, it is the dominated convergence theorem that implies that it is $o(1)$).
We have thus proved that for every $s = 0, 1, \ldots$, we have

$$\lim_n \frac{1}{n+1} \operatorname{Trace}(T_{\varphi,n}^s) = \int_\mathbb{T} \varphi^s \, dm.$$

As the polynomials are dense in $C[a,b]$ (in $\operatorname{Hol}(\Sigma(\varphi))$ in the complex case), Lemma 5.6.2(2) completes the proof. ∎

For the more restrictive hypotheses on φ, we can reverse a portion of the reasoning of Lemma 5.6.2(2), and deduce from §5.6.4 Szegő's first asymptotic formula.

Corollary 5.6.7 *If φ is a sectorial function and $1/\varphi \in L^\infty(\mathbb{T})$ (T_φ is invertible), then*

$$\lim_n (D_n)^{1/n+1} = \exp\left(\int_\mathbb{T} \log(\varphi) \, dm \right).$$

Indeed, there exists a holomorphic branch of the logarithm $f = \log z$ on $\Sigma(\varphi)$; applying the theorem to f, we obtain the formula. ∎

Remark In fact, it is known that the formula is valid for certain other classes of symbols; for example, this is the case for $\varphi \in H^\infty + C(\mathbb{T})$ (and T_φ is invertible) (Widom, 1976): see §5.8 for some references. More precisely, under the latter conditions, the formula of Corollary 5.6.7 is known in an even stronger form

$$\lim_n \frac{D_{n+1}}{D_n} = \exp\left(\int_\mathbb{T} \log(\varphi) \, dm \right),$$

where $\log(\varphi)$ is uniquely defined by the factorization $\varphi = [\varphi] e^{i(a+\tilde{b})}$ (with real $a, b \in C(\mathbb{T})$) from Exercises 3.4.3(e) and 3.4.3(g).

However, the greatest difference between Theorems 5.6.3 and 5.6.6 is in the supply of test functions: the weak convergence of the spectra

$$\lim_n \frac{1}{n+1} \sum_{\lambda \in \sigma(T_{\varphi,n})} f(\lambda) = \int_\mathbb{T} f \circ \varphi \, dm$$

with respect to $f \in C([a,b])$ in Theorem 5.6.3 is a much stronger assertion than the similar convergence in Theorem 5.6.6 but with respect to $f \in \operatorname{Hol}(\Sigma(\varphi))$. Indeed, the first algebra contains decompositions of the unity, and hence possesses conclusions of the type "asymptotic density of the spectra" (see Corollary 5.6.4), whereas a weak convergence with respect to the holomorphic functions does not mean very much. Consider the following example.

Example. Let $\varphi(z) = z$; then, $\sigma(T_{\varphi,n}) = \{0\}$ (counted $n+1$ times according to the multiplicity), and $\sigma(T_\varphi) = \overline{\mathbb{D}}$, $\sigma(M_\varphi) = \mathbb{T}$, $\Sigma(\varphi) = \operatorname{conv}(\operatorname{Ran}_{\mathrm{ess}}(\varphi)) = \overline{\mathbb{D}}$, and for every $f \in \text{Hol}(\overline{\mathbb{D}})$ and every $n \in \mathbb{Z}_+$ we have the identity

$$\frac{1}{n+1} \sum_{\lambda \in \sigma(T_{\varphi,n})} f(\lambda) = f(0) = \int_{\mathbb{T}} f \circ z \, dm = \widehat{f}(0).$$

However, it is difficult to interpret this as a sort of approximation of $\sigma(T_\varphi)$ or $\sigma(M_\varphi)$ by $\sigma(T_{\varphi,n})$.

5.6.5 The Second Term of the Szegő Asymptotics

According to §5.6.1, for every measure $\mu \in \mathcal{M}(\mathbb{T})$, $\mu \geq 0$, there exists the limit

$$\lim_n D_n^{1/(n+1)} = G(\mu)$$

where $D_n = D_n(\mu) =: \det(T_{\mu,n})$ and $T_{\mu,n} = (\widehat{\mu}(k-j))_{0 \leq k,j \leq n}$. In 1950, Lars Onsager (an American–Norwegian theoretical physicist, Nobel Prize 1968), while seeking his resolution of the two-dimensional Lenz–Ising model (see Chapter 1) proposed the second term of the Szegő asymptotics

$$\frac{1}{n+1} \log D_n = \log G(\mu) + o(1)$$

in the form

$$\frac{1}{n+1} \log D_n = \log G(\mu) + \frac{E(\mu)}{n+1} + o(1/n),$$

i.e. $D_n \sim G(\mu)^{n+1} E(\mu)$ when $n \to \infty$, where $E(\mu)$ is the Dirichlet (energy) integral of the outer function $[d\mu/dm]$, see §5.6.5.4(2) below, as well as Appendix F. Very soon after, Szegő found the proof of Onsager's conjecture for the measures $\mu = \varphi m$ with positive density φ "smooth" enough (1952).

In this handbook, we treat the question of the second term of the asymptotics following the approach of Widom (1976) in the framework of the Krein algebra (see §5.6.5.1), in particular, without the hypothesis of positivity. For the case of positive symbols φ the question had been resolved much earlier by Golinskii and Ibragimov (1971) under the widest conditions possible: see §5.8 for a history of the problem. Szegő's direct calculatory method is sketched in Exercise 5.7.4 below.

Remark also that in the positive case $\varphi \geq 0$ the sole existence of the limit $\lim_n (D_n/(G(\mu)^{n+1}))$ follows immediately from §5.6.1, according to which $D_n/D_{n-1} \geq G(\mu)$, meaning that the sequence $(D_n/G(\mu)^{n+1})_n$ is increasing:

$$\frac{D_n}{G(\mu)^{n+1}} \geq \frac{D_{n-1}}{G(\mu)^n}, \quad n \geq 0.$$

The proof of the principal result (Theorem 5.6.8) depends on the definition and on certain properties of the class $\mathfrak{S}_1$ of trace class operators on a Hilbert space,

5.6 Toeplitz Determinants and Asymptotics of Spectra

as well as the properties of the determinant $A \mapsto \det A$, where $A - I \in \mathfrak{S}_1$, and of Trace on $\mathfrak{S}_1$; see Appendix D for necessary details.

The bulk of this section is divided into four parts (1–4) introducing new techniques needed for the proof of the "Strong Szegő Theorem" (Theorem 5.6.8 below).

5.6.5.1 The Krein Algebra (1966)

The Dirichlet space $\mathcal{D}(\mathbb{T})$ is defined by

$$\mathcal{D}(\mathbb{T}) = \left\{ f \in L^1(\mathbb{T}) \colon \|f\|_{\mathcal{D}}^2 =: \sum_{k \in \mathbb{Z}} |k| \cdot |\widehat{f}(k)|^2 < \infty \right\},$$

its analytical portion by $\mathcal{D}_a(\mathbb{T}) = \{ f \in \mathcal{D}(\mathbb{T}) \colon \widehat{f}(k) = 0, \forall k \leq 0 \}$, and the *Krein space* $K(\mathbb{T})$ by

$$K(\mathbb{T}) = L^\infty(\mathbb{T}) \cap \mathcal{D}(\mathbb{T}),$$

equipped with the norm $\|f\|_{K(\mathbb{T})} = \|f\|_\infty + \|f\|_{\mathcal{D}}$. If $f \in \mathcal{D}_a(\mathbb{T})$, then obviously

$$\|f\|_{\mathcal{D}}^2 = \frac{1}{\pi} \int_{\mathbb{D}} |f'(x+iy)|^2 \, dx \, dy.$$

(1) Formula of Douglas (1931) *For every $f \in \mathcal{D}(\mathbb{T})$,*

$$\|f\|_{\mathcal{D}}^2 = \int_{\mathbb{T}} \int_{\mathbb{T}} \left| \frac{f(z_1) - f(z_2)}{z_1 - z_2} \right|^2 dm(z_1) \, dm(z_2).$$

Indeed,

$$\frac{f(z_1) - f(z_2)}{z_1 - z_2} = \sum_{k \in \mathbb{Z}} \widehat{f}(k) \frac{z_1^k - z_2^k}{z_1 - z_2},$$

where, for $k \geq 1$,

$$\frac{z_1^k - z_2^k}{z_1 - z_2} = z_1^{k-1} + z_1^{k-2} z_2 + \cdots + z_2^{k-1}$$

and, for $k \leq -1$,

$$\frac{z_1^k - z_2^k}{z_1 - z_2} = -(z_1^k z_2^{-1} + z_1^{k+1} z_2^{-2} + \cdots + z_1^{-1} z_2^k).$$

Since $(z_1^\alpha z_2^\beta)_{\alpha,\beta}$ is an orthonormal basis in $L^2(\mathbb{T} \times \mathbb{T})$, the result follows from Parseval's identity. ∎

(2) (Krein, 1966) *$K(\mathbb{T})$ is a Banach algebra (with the standard pointwise multiplication): it is called the "Krein algebra."*

Clearly $K(\mathbb{T})$ is a Banach space; moreover, $\|1\|_{K(\mathbb{T})} = 1$ and the inequality $\|fg\|_{K(\mathbb{T})} \leq \|f\|_{K(\mathbb{T})}\|g\|_{K(\mathbb{T})}$ easily follows from

$$|fg(z_1) - fg(z_2)| \leq |f(z_1)| \cdot |g(z_1) - g(z_2)| + |g(z_2)| \cdot |f(z_1) - f(z_2)|. \quad \blacksquare$$

(3) Lipschitz invariance. *The spaces $\mathcal{D}(\mathbb{T})$ and $K(\mathbb{T})$ are stable with respect to Lipschitz composition, i.e. if $\varphi \in K(\mathbb{T})$ and F satisfies the Lipschitz condition on a domain including $\mathrm{Ran}_{\mathrm{ess}}\varphi = \varphi(\mathbb{T})$, then $F \circ \varphi \in K(\mathbb{T})$. In particular, $\mathrm{ess\,inf}|\varphi| > 0$, $\varphi \in K(\mathbb{T}) \Rightarrow (1/\varphi) \in K(\mathbb{T})$, and if there exists a Lipschitz branch of $\log$ on $\mathrm{Ran}_{\mathrm{ess}}\varphi = \varphi(\mathbb{T})$, then $\varphi \in K(\mathbb{T}) \Rightarrow \log(\varphi) \in K(\mathbb{T})$.*

Indeed, if $|F(w_1) - F(w_2)| \leq c|w_1 - w_2|$ on $\mathrm{Ran}_{\mathrm{ess}}\varphi = \varphi(\mathbb{T})$, then

$$|F(\varphi(z_1)) - F(\varphi(z_2))| \leq c|\varphi(z_1) - \varphi(z_2)|,$$

and the property follows from (1). $\blacksquare$

(4) $K(\mathbb{T})$ and the Hilbert–Schmidt Hankel operators. $\varphi \in K(\mathbb{T}) \Rightarrow H_\varphi, H_{\overline{\varphi}} \in \mathfrak{S}_2(H^2)$ *and* $\|H_\varphi\|_{\mathfrak{S}_2}^2 = \sum_{k<0} |k| \cdot |\widehat{\varphi}(k)|^2$, $\|H_{\overline{\varphi}}\|_{\mathfrak{S}_2}^2 = \sum_{k>0} |k| \cdot |\widehat{\varphi}(k)|^2$. *In particular, $K(\mathbb{T}) \subset QC(\mathbb{T})$.*

The principal portion of the property is Exercise 2.3.7. The rest follows from $\mathfrak{S}_2(H^2) \subset \mathfrak{S}_\infty(H^2)$ and from Theorem 2.2.8 applied to φ and $\overline{\varphi}$. $\blacksquare$

5.6.5.2 Widom's Form of the Strong Szegő Theorem

The following theorem is the principal result of this section (and perhaps of the whole of Chapter 5). For extensive comments on the subject, and in particular on Szegő's initial form of the theorem, see §5.8 below.

Theorem 5.6.8 (Szegő, 1954; Widom, 1976) *Suppose $\varphi \in K(\mathbb{T})$ and T_φ invertible. Then $I - T_\varphi T_{1/\varphi} \in \mathfrak{S}_1(H^2)$, and moreover*

(1) *(Widom, 1976)*

$$\lim_n \frac{D_n(\varphi)}{G(\varphi)^{n+1}} = \det(T_\varphi T_{1/\varphi}),$$

where $G(\varphi) = \exp(\int_\mathbb{T} \log(\varphi)\, dm)$ and $\log(\varphi)$ is chosen as in the Remark to Corollary 5.6.7.

(2) *(Widom, 1976)*

$$\det(T_\varphi T_{1/\varphi}) = E(\varphi) =: \exp\left(\sum_{k \in \mathbb{N}} k\, \widehat{\log(\varphi)}(k) \cdot \widehat{\log(\varphi)}(-k)\right)$$

and if φ is real then, in addition,

$$\det(T_\varphi T_{1/\varphi}) = \exp(\|P_+(\log(\varphi))\|_{\mathcal{D}(\mathbb{T})}^2) = \exp\left(\frac{1}{4\pi} \int_\mathbb{D} \left|\frac{[\varphi]'}{[\varphi]}\right|^2 dx\, dy\right),$$

where $[\varphi]$ is an outer function generated by φ (see Appendix F).

5.6.5.3 Some Key Indications for the Strategy of the Proof
Suppose $\varphi = \varphi_1 \varphi_2$ where $\overline{\varphi}_1^{\pm 1}, \varphi_2^{\pm 1} \in H^\infty$.

(a) T_φ is invertible and

$$\det(P_n T_\varphi^{-1} P_n) = G(\varphi)^{-n-1}.$$

Indeed, $T_\varphi = T_{\varphi_1} T_{\varphi_2}$ and $T_\varphi^{-1} = T_{1/\varphi_2} T_{1/\varphi_1}$, and since $1/\varphi_2 \in H^\infty$, $P_n T_\varphi^{-1} P_n = P_n T_{1/\varphi_2} P_n T_{1/\varphi_1} P_n$ where the matrices $P_n T_{1/\varphi_2} P_n$, $P_n T_{1/\varphi_1} P_n$ are triangular, thus

$$\det(P_n T_\varphi^{-1} P_n) = \det(P_n T_{1/\varphi_2} P_n) \det(P_n T_{1/\varphi_1} P_n)$$
$$= \frac{1}{\overline{\varphi}_2(0)^{n+1}} \times \frac{1}{\overline{\varphi}_1(0)^{n+1}}$$
$$= \exp\left(-(n+1) \int_\mathbb{T} \log(\varphi_2)\, dm\right) \times \exp\left(-(n+1) \int_\mathbb{T} \log(\varphi_1)\, dm\right)$$
$$= G(\varphi)^{-n-1}. \qquad\blacksquare$$

(b) We have

$$P_n T_\varphi P_n = (P_n T_{1/\varphi}^{-1} P_n) \cdot (P_n T_{1/\varphi_1} P_n)(P_n T_{1/\varphi_2} P_n)(P_n T_{\varphi_1} T_{\varphi_2} P_n),$$

and by (a) it only remains to treat the determinants of the last three factors (which will be done using $\log(\overline{\varphi}_1), \log(\varphi_2) \in H^\infty$, see §5.6.5.4).

(c) We have

$$((P_n T_\varphi P_n)^{-1})_{0,0} = \frac{D_{n-1}(\varphi)}{D_n(\varphi)}$$

and $(P_n T_\varphi^{-1} P_n)_{0,0} = G(\varphi)^{-1}$.

Indeed, the first identity is Cramer's rule for the calculation of the inverse matrix $(a_{ij})^{-1}$. The second follows from $T_\varphi^{-1} = T_{1/\varphi_2} T_{1/\varphi_1}$ where the operator T_{1/φ_2} has a lower triangular matrix and T_{1/φ_1} has an upper triangular matrix, hence

$$(T_\varphi^{-1})_{0,0} = (T_{1/\varphi_2})_{0,0}(T_{1/\varphi_1})_{0,0} \quad \text{and}$$
$$(T_{1/\varphi_j})_{0,0} = 1/\varphi_j(0) = \exp(-\log(\varphi_j(0))) = G(\varphi_j)^{-1};$$

note that $G(\varphi_1)G(\varphi_2) = G(\varphi)$. $\qquad\blacksquare$

This last calculation also suggests the possibility of the first Szegő asymptotics formula for the symbols having a factorization.

5.6.5.4 Proof of Theorem 5.6.8

The fact that $I - T_\varphi T_{1/\varphi} \in \mathfrak{S}_1(H^2)$ is trivial, as $I - T_\varphi T_{1/\varphi} = H_{\overline\varphi}^* H_{1/\varphi}$, and since $\overline\varphi, 1/\varphi \in K(\mathbb{T}) \Rightarrow H_{\overline\varphi}, H_{1/\varphi} \in \mathfrak{S}_2(H^2)$ (§5.6.5.1(4)), we have $H_{\overline\varphi}^* H_{1/\varphi} \in \mathfrak{S}_1(H^2)$ (Appendix D).

(1) A brilliant observation by Harold Widom is that a finite Toeplitz matrix has a hidden "almost block-diagonal structure," where on the principal diagonal there are "large" operators of type $T_{\varphi,m}$ whereas the rest consists of "small" operators of Hankel type. More precisely, for $m, n \in \mathbb{Z}_+$ and $c_k = \widehat\varphi(k)$, we write

$$T_{\varphi,n+1+m} = \begin{pmatrix} A & B \\ C & D \end{pmatrix} = \left(\begin{array}{ccc|ccc} c_0 & \cdots & c_{-n} & c_{-n-1} & \cdots & c_{-n-1-m} \\ \vdots & \ddots & \vdots & \vdots & \cdots & \vdots \\ c_n & \cdots & c_0 & c_{-1} & \cdots & c_{-m-1} \\ \hline c_{n+1} & \cdots & c_1 & c_0 & \cdots & c_{-m} \\ \vdots & \cdots & \vdots & \vdots & \ddots & \vdots \\ c_{n+1+m} & \cdots & c_{m+1} & c_m & \cdots & c_0 \end{array}\right),$$

where

$$A = P_n M_\varphi P_n, \qquad B = P_n M_\varphi (P_{n+1+m} - P_n),$$
$$C = (P_{n+1+m} - P_n) M_\varphi P_n, \qquad D = (P_{n+1+m} - P_n) M_\varphi (P_{n+1+m} - P_n)$$

with the usual notation P_n for an orthogonal projection on $\mathcal{P}_n$. To simplify the calculation we naturally identify

$$\mathcal{P}_{n+m+1} = \mathcal{P}_n \oplus \mathcal{P}_m, \quad x = (x_0, \ldots, x_{n+m+1}) = (x_0, \ldots, x_n) \oplus (x_{n+1}, \ldots, x_{n+m+1}),$$

and to reveal the Hankel nature of C and D, we use a permutation of the coordinates $J_n(x_0, \ldots, x_n) = (x_n, \ldots, x_0)$ (a symplectic matrix of order n) and the notation

$$P'_n = J_n P_n.$$

Clearly $A = T_{\varphi,n}$ and $D = T_{\varphi,m}$. For B, we obtain (with the identification mentioned) $\mathcal{P}_{n+m+1} \ominus \mathcal{P}_n = \mathcal{P}_m$

$$B = P_n M_\varphi z^{n+1} P_m = J_n P_n J_n P_n M_{\varphi z^{n+1}} P_m = P'_n \Gamma_\varphi P_m$$

since (using matrix notation)

$$P_n M_{\varphi z^{n+1}} P_m = (c_{i-j-n-1})_{0 \le j \le m, 0 \le i \le n},$$
$$J_n P_n M_{\varphi z^{n+1}} P_m = (c_{-i-j-1})_{0 \le j \le m, 0 \le i \le n} = P_n \Gamma_\varphi P_m.$$

5.6 Toeplitz Determinants and Asymptotics of Spectra

Similarly, for C,
$$C = P_m \Gamma_{\tilde{\varphi}} P'_n,$$
where $\tilde{\varphi}(z) = \varphi(1/z)$, $z \in \mathbb{T}$. By §5.4.2(8), the matrices A and D are invertible for n and m sufficiently large, hence
$$T_{\varphi,n+1+m} = \begin{pmatrix} A & B \\ C & D \end{pmatrix} = \begin{pmatrix} A & 0 \\ 0 & D \end{pmatrix} \begin{pmatrix} I_n & A^{-1}B \\ D^{-1}C & I_m \end{pmatrix},$$
and thus (with the notation $D_n(\varphi) = \det(T_{\varphi,n})$)
$$D_{n+m+1}(\varphi) = D_n(\varphi) d_m(\varphi) \det(I_m - D^{-1}CA^{-1}B) =: D_n(\varphi) d_m(\varphi) d_{m,n},$$
where $d_{m,n} = \det(I_m - D^{-1}CA^{-1}B)$. We have
$$D^{-1}CA^{-1}B = T^{-1}_{\varphi,m} P_m \Gamma_{\tilde{\varphi}} P'_n T^{-1}_{\varphi,n} P'_n \Gamma_\varphi P_m$$
where (as is easy to see) $P'_n T^{-1}_{\varphi,n} P'_n = T^{-1}_{\tilde{\varphi},n}$, therefore
$$D^{-1}CA^{-1}B = T^{-1}_{\varphi,m} P_m \Gamma_{\tilde{\varphi}} T^{-1}_{\tilde{\varphi},n} P_n \Gamma_\varphi P_m.$$

Now note that
$$\lim_{m,n \to \infty} \|D^{-1}CA^{-1}B - P_m T^{-1}_\varphi \Gamma_{\tilde{\varphi}} T^{-1}_{\tilde{\varphi}} \Gamma_\varphi P_m\|_{\mathfrak{S}_1} = 0,$$
which follows from Appendix D.7 and the fact that $\Gamma_{\tilde{\varphi}}, \Gamma_\varphi \in \mathfrak{S}_2(H^2)$, and then (by §5.4.2(8)) $\lim_m T^{-1}_{\varphi,m} P_m f = T^{-1}_\varphi f$ ($\forall f \in H^2$) and $\lim_n T^{-1}_{\tilde{\varphi},n} P_n f = T^{-1}_{\tilde{\varphi}} f$ ($\forall f \in H^2$). By the continuity of the determinant (Appendix D.7), we have
$$\lim_{m,n \to \infty} d_{m,n} = \det(I - T^{-1}_\varphi \Gamma_{\tilde{\varphi}} T^{-1}_{\tilde{\varphi}} \Gamma_\varphi).$$
We apply Exercise 2.3.12(a) (a final *coup de théâtre*!):
$$\Gamma_\varphi T_{1/\varphi} + T_{\tilde{\varphi}} \Gamma_{1/\varphi} = \Gamma_1 = 0,$$
so that $T^{-1}_{\tilde{\varphi}} \Gamma_\varphi = -\Gamma_{1/\varphi} T^{-1}_{1/\varphi}$, and thus
$$\lim_{m,n \to \infty} d_{m,n} = \det(I + T^{-1}_\varphi \Gamma_{\tilde{\varphi}} \Gamma_{1/\varphi} T^{-1}_{1/\varphi}).$$
Again, by Exercise 2.3.12(a), $1 - T_\varphi T_{1/\varphi} = \Gamma_{\tilde{\varphi}} \Gamma_{1/\varphi}$, hence
$$I + T^{-1}_\varphi \Gamma_{\tilde{\varphi}} \Gamma_{1/\varphi} T^{-1}_{1/\varphi} = I + T^{-1}_\varphi (I - T_\varphi T_{1/\varphi}) T^{-1}_{1/\varphi} = T^{-1}_\varphi T^{-1}_{1/\varphi},$$
which shows that
$$\lim_{m,n \to \infty} d_{m,n} = \det(T^{-1}_\varphi T^{-1}_{1/\varphi}) = \det(T^{-1}_{1/\varphi} T^{-1}_\varphi) = 1/\det(T_\varphi T_{1/\varphi}).$$
Then, to deduce the theorem, we use Corollary 5.6.7 (and its Remark): the limit
$$\lim_m \frac{D_{m+1}(\varphi)}{D_m(\varphi)} = G(\varphi) = \exp\left(\int_\mathbb{T} \log(\varphi) \, dm \right)$$

exists, and hence so do the limits ($\forall n \geq 0$):

$$\lim_m \frac{D_{m+n+1}(\varphi)}{D_m(\varphi)} = G(\varphi)^{n+1}.$$

Writing

$$\frac{D_n(\varphi)}{G(\varphi)^{n+1}} = \frac{D_{m+n+1}(\varphi)}{D_m(\varphi)G(\varphi)^{n+1}d_{m,n}},$$

letting $n \to \infty$, and choosing a suitable $m = m(n) \gg n$, we obtain

$$\lim_n \frac{D_n(\varphi)}{G(\varphi)^{n+1}} = \lim_n d_{m(n),n}^{-1} = \det(T_\varphi T_{1/\varphi}),$$

which concludes the proof of (1).

(2) Suppose first that $\varphi \in W(\mathbb{T}) \cap K(\mathbb{T})$ (and T_φ invertible). Then $\mathrm{wind}(\varphi) = 0$ and a continuous branch of $L =: \log(\varphi)$ is in $W(\mathbb{T}) \cap K(\mathbb{T})$. Moreover, $\varphi = \varphi_-\varphi_+$ where

$$\varphi_\pm = \exp(L_\pm) \in (H^\infty)^{-1}, \quad L_\pm =: P_\pm(\log(\varphi)).$$

As for every $h \in H^\infty$ we have $T_h^n = T_{h^n}$ for every $n = 0, 1, \ldots$, and then

$$\exp(T_h) = T_{\exp(h)},$$

we obtain

$$T_\varphi T_{1/\varphi} = T_{\varphi_-} T_{\varphi_+} T_{1/\varphi_-} T_{1/\varphi_+} = e^A e^B e^{-A} e^{-B},$$

where $A = T_{L_-}$, $B = T_{L_+}$. We can apply Lemma 5.6.9 below, since $H_{\overline{L_+}}^*, H_{L_-} \in \mathfrak{S}_2$ (see §5.6.5.1(4)) and $[A, B] = AB - BA = H_{\overline{L_+}}^* H_{L_-} \in \mathfrak{S}_1$ (for the definitions and properties of $\mathfrak{S}_1$ and the Trace see Appendix D):

$$\det(T_\varphi T_{1/\varphi}) = \det(e^A e^B e^{-A} e^{-B}) = \exp(\mathrm{Trace}([A, B]))$$
$$= \exp\left(\mathrm{Trace}(H_{\overline{L_+}}^* H_{L_-})\right) = \exp\left(\langle H_{L_-}, H_{\overline{L_+}} \rangle_{\mathfrak{S}_2}\right)$$
$$= \exp\left(\sum_{k \geq 1} k\widehat{L_-}(-k)\widehat{L_+}(k)\right),$$

which is the desired formula.

The general case of $\varphi \in K(\mathbb{T})$ (and T_φ invertible) follows from an approximation by the Poisson means $\varphi_r = P_r * \varphi$, $(\varphi_\pm)_r = (\varphi_r)_\pm$ and a passage to the limit in

$$\det(T_{\varphi_r} T_{1/\varphi_r}) = \exp\left(\sum_{k \geq 1} k\widehat{L_-}(-k)r^k \widehat{L_+}(k)r^k\right)$$

(on the left, we use $\lim_{r \to 1} \|\varphi_r - \varphi\|_{\mathcal{D}(\mathbb{T})} = 0$, $\lim_{r \to 1} \|(1/\varphi_r) - (1/\varphi)\|_{\mathcal{D}(\mathbb{T})} = 0$, which comes from $1/\varphi \in K(\mathbb{T}) \subset H^\infty + C(\mathbb{T})$ and implies $\lim_{r \to 1} \|H_{\varphi_r} - H_\varphi\|_{\mathfrak{S}_2} = 0$ and $\lim_{r \to 1} \|H_{1/\varphi_r} - H_{1/\varphi}\|_{\mathfrak{S}_2} = 0$).

Finally, if φ is a real function, we have $\widehat{L_-}(-k)\widehat{L_+}(k) = |\widehat{L}(k)|^2$, $L = \log(\varphi)$, and on the other hand, since φ is either positive or negative on $\mathbb{T}$, we have $L = \ln |\varphi|$ (up to a choice of logarithm), then

$$\log[\varphi] = \int_{\mathbb{T}} \frac{\zeta + z}{\zeta - z} \ln |\varphi(\zeta)| \, dm(\zeta)$$

$$= \int_{\mathbb{T}} \left(1 + 2 \sum_{k \geq 0} z^{k+1} \bar{\zeta}^{k+1}\right) \ln |\varphi(\zeta)| \, dm(\zeta)$$

$$= \widehat{L}(0) + 2 \sum_{k \geq 0} \widehat{L}(k+1) z^{k+1}.$$

Finally, the second formula for $\det(T_\varphi T_{1/\varphi})$ follows from the initial Remarks of §5.6.5.1. ∎

5.6.6 A Formula for Determinant and Trace

Lemma 5.6.9 (Helton and Howe, 1973) *Let $A, B \in L(H)$ where H is a Hilbert space, such that $[A, B] \in \mathfrak{S}_1(H)$. Then $(I - e^A e^B e^{-A} e^{-B}) \in \mathfrak{S}_1(H)$ and*

$$\det(e^A e^B e^{-A} e^{-B}) = \exp(\mathrm{Trace}[A, B]).$$

The proof of the above formula is sketched in Exercise 5.7.7 below.

5.6.7 Some Formulas for Trace$[T_\varphi, T_\psi]$ (Following Helton and Howe, and Berger and Shaw, 1973)

The following formulas, and especially the elementary ways to prove them, elucidate the links between the traces of Toeplitz commutators and certain objects of a different nature, such as the surfaces of images, indices, or differential forms. Recall that for $A \in \mathfrak{S}_1(H^2)$ the trace of A can be expressed as Trace $A = \sum_{k \geq 0}(Az^k, z^k)$, that for $\varphi, \psi \in K(\mathbb{T})$, $[T_\varphi, T_\psi] \in \mathfrak{S}_1(H)$ and that $T_{z^n} = S^n$ for $n \geq 0$ (the translation (shift) operator) and $T_{z^n} = S^{*|n|}$ for $n < 0$ (the inverse translation (backward shift) operator).

(1) *For every $n, k \in \mathbb{Z}$ and $j \in \mathbb{Z}_+$, $(T_{z^n} T_{z^k} z^j, z^j) = (P_+ z^{j+k}, P_+ z^{j-n})$, and hence*

$(T_{z^n} T_{z^k} z^j, z^j) = 0$ if $n \neq -k$ ($\forall j \geq 0$), or $n = -k$ and $j + k < 0$,

$(T_{z^n} T_{z^k} z^j, z^j) = 1$ if $n = -k$ and $j + k \geq 0$. ∎

(2) Trace$([T_{z^n}, T_{z^k}]) = 0$ if $n \neq -k$, and Trace$([T_{z^n}, T_{z^k}]) = -n$ if $n = -k$. ∎

(3) *If* $\varphi, \psi \in K(\mathbb{T})$,

$$\mathrm{Trace}([T_\varphi, T_\psi]) = \sum_{n \in \mathbb{Z}} n\widehat{\varphi}(-n)\widehat{\psi}(n).$$

In the case where $\varphi \in K(\mathbb{T}) \cap H^\infty$,

$$\mathrm{Trace}([T_\varphi^*, T_\varphi]) = \frac{1}{\pi} \int_{\mathbb{D}} |\varphi'|^2 \, dx \, dy.$$

Indeed, first suppose that $\varphi, \psi \in \mathcal{P}$. Then,

$$[T_\varphi, T_\psi] = \sum_{n,k \in \mathbb{Z}} n\widehat{\varphi}(n)\widehat{\psi}(k)[T_{z^n}, T_{z^k}]$$

(a finite sum), and (2) implies the formula. We then remark that the two sides of the formula are continuous for the norm $\|\varphi\|^2_{\mathcal{D}(\mathbb{T})} = \sum |n| \cdot |\widehat{\varphi}(n)|^2$ and that the polynomials $\mathcal{P}$ are dense in $\mathcal{D}(\mathbb{T})$.

The analytic case also follows since

$$\sum_{n \geq 1} n|\widehat{\varphi}(n)|^2 = \frac{1}{\pi} \int_{\mathbb{D}} |\varphi'|^2 \, dx \, dy. \qquad \blacksquare$$

(4) *If* $\varphi \in K(\mathbb{T})$, $\psi' \in L^2(\mathbb{T})$,

$$\mathrm{Trace}([T_\varphi, T_\psi]) = \int_{\mathbb{T}} z\varphi\psi' \, dm(z),$$

where $\psi'(z) = d\psi/dz$ *(so that* $z(z^n)' = nz^n$, $\forall n \in \mathbb{Z}$*)*. $\qquad \blacksquare$

(5) *Let* φ *and* ψ *be two real functions,* $\varphi', \psi' \in L^2(\mathbb{T})$ *and let* Φ, Ψ *be smooth extensions of* φ *and* ψ *in the disk* $\overline{\mathbb{D}}$ *(for example, the Poisson extensions,* $\Phi = \varphi * P_z$, $\Psi = \psi * P_z$*) and let* $F = \Phi + i\Psi : \mathbb{D} \to \mathbb{C} = \mathbb{R}^2$ *be the corresponding mapping. Finally, let* $\deg_F(w)$ *(*$w \in F(\mathbb{D})$*) denote the degree of* F *with respect to* w,

$$\deg_F(w) = \mathrm{card}\{z : F(z) = w, J_F(z) > 0\} - \mathrm{card}\{z : F(z) = w, J_F(z) < 0\},$$

where $J_F(z)$ *is the Jacobian of* $(x,y) \mapsto F(x,y)$. *Then*

$$\mathrm{Trace}([T_\varphi, T_\psi]) = \frac{1}{2\pi i} \int_{\mathbb{D}} \left(\frac{\partial \Phi}{\partial x} \cdot \frac{\partial \Psi}{\partial y} - \frac{\partial \Psi}{\partial x} \cdot \frac{\partial \Phi}{\partial y} \right) dx \, dy$$

$$= \frac{1}{\pi} \int_{F(\mathbb{D})} \deg_F(w) \, dx \, dy,$$

and

$$\mathrm{Trace}([T_F^*, T_F]) = \frac{1}{2\pi i} \int_{\sigma(T_F) \setminus F(\mathbb{T})} \mathrm{ind}(T_F - wI) \, dw \, d\overline{w}.$$

Indeed, the first two expressions for Trace($[T_\varphi, T_\psi]$) follow from (4) and Stokes' formula (see, for example, the classic text by Spivak [1971]); for the third we refer directly to [Helton and Howe, 1973]. ∎

5.6.8 Conclusion

As we have already mentioned, Theorem 5.6.8, which gives the asymptotics

$$D_n(\varphi) = G(\varphi)^{n+1} E(\varphi)(1 + o(1)), \quad n \to \infty$$

for T_φ invertible and $\varphi \in K(\mathbb{T})$, is not the most general form known: the same asymptotics is proved in [Johansson, 1988] under the natural condition (necessary just for writing the formula) that $\varphi = \exp(f) \in L^1(\mathbb{T})$ where $f \in \mathcal{D}(\mathbb{T})$ (the Dirichlet space of §5.6.5.1). (Under the hypotheses of Theorem 5.6.8 this condition is automatic, since the invertibility of T_φ implies $\inf |\varphi| > 0$ and wind(φ) = 0, hence we obtain, by §5.6.5.1(3), that $\varphi = \exp(f)$, $f \in K(\mathbb{T})$.) For positive symbols $\varphi \geq 0$, the asymptotics of the $D_n(\varphi)$,

$$\lim_n \frac{D_n(\varphi)}{G(\varphi)^{n+1}} = E(\varphi),$$

is established in [Golinskii and Ibragimov, 1971] under the sole condition $\varphi \in L^1(\mathbb{T})$, $\log(\varphi) \in L^1(\mathbb{T})$, independent of whether the expression

$$E(\varphi) = \exp\left(\sum_{k \geq 1} k\, |(\log(\varphi))\hat{}\,(k)|^2 \right)$$

is finite or not. For several more details on the subject, see the research monograph by Simon [2005b].

A curious circumstance is noted in [Simon, 2005b], page 346: Theorem 5.6.8 gives

$$\det(T_\varphi T_{1/\varphi}) = E(\varphi),$$

and hence

$$\lim_n \det(T_\varphi T_{1/\varphi})_n = E(\varphi),$$

whereas a double application of the same §5.6.1 and the fact that $G(\varphi)G(1/\varphi) = 1$, $E(\varphi) = E(1/\varphi)$ implies

$$\lim_n \det(T_{\varphi,n} T_{1/\varphi,n}) = \lim_n \det(T_{\varphi,n}) \det(T_{1/\varphi,n}) = E(\varphi)^2.$$

It would be very interesting to know the asymptotics of $\det(T_\varphi T_\psi)_n$ (as $n \to \infty$) for general $\varphi, \psi \in K(\mathbb{T})$ (having wind(φ) = wind(ψ) = 0; otherwise, the anomalies already begin with $\varphi, \psi = \bar{z}, z$), and even, perhaps, of the $\det(T)_n$ for $T \in \mathrm{alg}\mathcal{T}_{K(\mathbb{T})}$ (as function of Sym(T)?).

5.7 Exercises

5.7.0 Basic Exercises: Volumes, Distances, and Approximations

(a) Volumes and Gram determinants. An n-dimensional Hilbert space (or following another terminology, a Hermitian space) H can be equipped with a translation-invariant *measure (volume)* Vol $=$ Vol$_n$: it suffices to choose an orthonormal basis (ONB) $(b_j)_{j=1}^n$ and an associated unitary isomorphism $U: \mathbb{C}^n \to H$, $U(\lambda_j) = \sum_{j=1}^n \lambda_j b_j$, $(\lambda_j)_{j=1}^n \in \mathbb{C}^n$, and to set

$$\text{Vol}(A) = \mu_{2n}(U^{-1}A)$$

for every Borel set A in H where μ_{2n} is the standard Lebesgue measure in $(\mathbb{R}^2)^n = \mathbb{C}^n$ (since U is continuous, $U^{-1}A$ is a Borel set). Recall also that to every linear mapping $V: H \to H$ is associated a *determinant*

$$\det(V) =: \det((Vb_i, b_j)_{i,j=1}^n)$$

(which does not depend on the choice of the ONB $(b_j)_{j=1}^n$).

(1) *Show that* Vol *does not depend on the choice of the ONB* (b_j).

SOLUTION: Indeed, if $(b'_j)_{j=1}^n$ is another ONB and $U': \mathbb{C}^n \to H$ its associated isomorphism, then for every Borel set A,

$$\text{Vol}'(A) = \mu_{2n}(U'^{-1}A) = \mu_{2n}(U'^{-1}UU^{-1}A) = \mu_{2n}(U^{-1}A) = \text{Vol}(A)$$

(the second-to-last equality holds by the change of variables theorem in $\mathbb{R}^{2n}$ – the mapping $U'^{-1}U$ is unitary and, hence, $|\det(U'^{-1}U)| = 1$; see (2) below for more details on the change of variables).

(2) *Let* $(b_j)_{j=1}^n$ *be an ONB,* $\mathcal{X} = (x_j)_{j=1}^n$ *a sequence of vectors of H,* $V: H \to H$ *a linear mapping defined by* $Vb_j = x_j$, $1 \le j \le n$, *and $G(\mathcal{X})$ the Gram matrix of $\mathcal{X}$,*

$$G(\mathcal{X}) = G(x_1, \ldots, x_n) =: ((x_i, x_j))_{i,j=1}^n$$

(see also Appendix B.4). Also, denote

$$V(\mathbb{D}^n) = \sum_{j=1}^n x_j \mathbb{D} = \left\{ \sum_{j=1}^n x_j \lambda_j : \lambda_j \in \mathbb{D} \right\}.$$

Show that

$$\text{Vol}(V(\mathbb{D}^n)) = |\det(V)|^2 \text{Vol}(\mathbb{D}^n) = |\det(V)|^2 \pi^n,$$

where $\mathbb{D}^n = \sum_{j=1}^n b_j \mathbb{D}$, *and*

$$|\det(V)|^2 = \det(V^*V) = \det(G(\mathcal{X})).$$

SOLUTION: Clearly $V^*V = G(X)$, which implies the second statement. For the first, use the change of variables formula in $(\mathbb{R}^2)^n = \mathbb{C}^n$: for every differentiable mapping $\varphi\colon \mathbb{R}^{2n} \to \mathbb{R}^{2n}$ and every Borel subset $A \subset \mathbb{R}^{2n}$,

$$\int_{\varphi(A)} f\, d\mu_{2n} = \int_A (f \circ \varphi) |\det(J(\varphi))|\, d\mu_{2n},$$

where $J(\varphi)$ is the Jacobian matrix of φ (thus, the matrix of the differential $d\varphi$). As is well known (and easy to verify), if φ is complex holomorphic as a mapping $(\mathbb{R}^2)^n = \mathbb{C}^n \to (\mathbb{R}^2)^n = \mathbb{C}^n$, $\varphi = \varphi(\lambda_1, \ldots, \lambda_n)$, $\lambda_j \in \mathbb{C}$, then $|\det(J(\varphi))| = |\det(\varphi')|^2$ where $\varphi' = (\partial \varphi_i / \partial \lambda_j)$ is the complex differential of φ. By the definition of the volume Vol, the change of variables formula can be transferred to H, hence for our mapping V (holomorphic since linear and satisfying $V' = V$), for every Borel subset $A \subset H$, we have

$$\int_{V(A)} f\, d\mathrm{Vol}_n = \int_A (f \circ V) |\det(V)|^2\, d\mathrm{Vol}_n.$$

We obtain the result with $A = \mathbb{D}^n$ and $f = 1$.

(3) *Show the Hadamard inequality (1893): if $(b_j)_{j=1}^n$ is an ONB of H and $V\colon H \to H$ a linear mapping, then*

$$|\det(V)| \leq \prod_{j=1}^n \|Vb_j\|.$$

Hint Without loss of generality, $\det(V) \neq 0$. Then V can be represented in the form $V = UT$, U unitary, T triangular.

SOLUTION: Using the Hint,

$$|\det(V)| = |\det(U)| \cdot |\det(T)| = |\det(T)|$$

$$= \prod_1^n |(Tb_j, b_j)| \leq \prod_1^n \|Tb_j\| = \prod_1^n \|Vb_j\|.$$

NOTE *The factorization of the Hint is a simple and classical fact: if $(e_j)_1^n$ is the orthogonalization of (Vb_j) (see Appendix B.3) and $Ub_j = e_j$, $1 \leq j \leq n$ a unitary operator, then $(U^*Vb_j, b_k) = (Vb_j, e_k) = 0$ for $k > j$, and so $V = UT$, $T = U^*V$ is the needed factorization.*

We can also mention that the left-hand side of Hadamard's inequality does not depend on (b_j); the equality holds for the ONB of eigenvectors of the Hermitian square V^*V: if $V^*Vb_j = s_j b_j$ ($s_j \geq 0$), then

$$|\det(V)|^2 = \det(V^*V) = \prod_1^n s_j = \prod_1^n \|Vb_j\|^2.$$

Beckenbach and Bellman [1961] page 89 mentioned that for Hadamard's inequality, "there are perhaps a hundred proofs available in published and unpublished form."

(b) Distances, determinants and approximation. *Here, we present a few applications of the techniques of (a); we use the notation of (a(1)) and (a(2)).*

(1) *Let* $\dim H \geq n+1$, $X_n = (x_1, \ldots, x_n)$ *a sequence of n vectors of H*, $H_n = \text{Vect}(x_1, \ldots, x_n)$, *and then* $x \in H$ *and* $X_{n+1} = (X, x) = (x_1, \ldots, x_n, x)$. *Show that*

$$\text{dist}(x, H_n)^2 \det(G(X_n)) = \det(G(X_{n+1})).$$

SOLUTION: Since the conclusion of (a(2)) above does not depend on the choice of the basis (b_j), we can suppose that $\text{Vect}(b_j) = H_n$, and hence for a general element $y = \sum_{j=1}^{n} x_j \lambda_j + \lambda_{n+1} x$ $(\lambda_j \in \mathbb{D})$ of $V(\mathbb{D}^{n+1})$, we have $(y, b_{n+1}) = \lambda_{n+1}(x, b_{n+1}) =: \lambda$ and

$$y = \sum_{j=1}^{n} x_j \lambda_j + \lambda_{n+1} P_n x + \lambda b_{n+1} = y' + \lambda_{n+1} P_n x + \lambda b_{n+1},$$

where y' is an arbitrary element of $V(\mathbb{D}^n)$, $V b_j = x_j$, $|\lambda| < |(x, b_{n+1})| = \|(I - P_n)x\|$ and P_n denotes the orthogonal projection on H_n. By Fubini's theorem,

$$\text{Vol}_{n+1}(V(\mathbb{D}^{n+1})) = \int_D \text{Vol}_n(V(\mathbb{D}^n) + \lambda_{n+1} P_n x) \, d\mu_2(\lambda)$$
$$= \int_D \text{Vol}_n(V(\mathbb{D}^n)) \, d\mu_2(\lambda),$$

where $D = \{\lambda \in \mathbb{C} : |\lambda| < \|(I - P_n)x\|\}$, and hence

$$\text{Vol}_{n+1}(V(\mathbb{D}^{n+1})) = \text{Vol}_n(V(\mathbb{D}^n)) \pi \|(I - P_n)x\|^2.$$

Taking into account $\|(I - P_n)x\|^2 = \text{dist}(x, H_n)^2$, and again applying (a(2)), we obtain

$$\det(G(X_{n+1})) \pi^{n+1} = \pi^n \det(G(X_n)) \pi \, \text{dist}(x, H_n)^2$$

and the result follows.

(2) *The best approximation of power functions. Given a complex number* $\alpha \in \mathbb{C}$, *the power function* x^α *is defined by* $x^\alpha(t) = t^\alpha = e^{\alpha \log(t)}$, $0 < t < 1$. *Noting that* $x^\alpha \in L^2(0, 1) \Leftrightarrow \text{Re}(\alpha) > -1/2$, *show that for every family of powers* $\alpha_0, \alpha_1, \ldots, \alpha_n$ ($\text{Re}(\alpha_j) > -1/2$, $\alpha_j \neq \alpha_k$ *for* $j \neq k$),

$$\text{dist}(x^{\alpha_0}, \text{Vect}(x^{\alpha_j} : 1 \le j \le n)) = \frac{1}{(2\,\text{Re}(\alpha_0) + 1)^{1/2}} \prod_{j=1}^{n} \left| \frac{\alpha_0 - \alpha_j}{\alpha_0 + \overline{\alpha}_j + 1} \right|,$$

where $\text{dist} = \text{dist}_{L^2(0,1)}$.

Hint Use (1) and the value of the Cauchy determinant of Exercise 5.7.5(a) below.

SOLUTION: By (1),

$$\text{dist}(x^{\alpha_0}, \text{Vect}(X_n))^2 = \det(G(X_{n+1})) / \det(G(X_n)),$$

where $X_n = (x^{\alpha_j} : 1 \le j \le n)$, $X_{n+1} = (x^{\alpha_j} : 0 \le j \le n)$. For the Gram determinants, we use the scalar products

$$(x^\alpha, x^\beta)_{L^2(0,1)} = \int_0^1 t^\alpha t^{\overline{\beta}}\, dt = \frac{1}{\alpha + \overline{\beta} + 1},$$

giving

$$\det(G(X_n)) = \det\left(\frac{1}{\alpha_j + \overline{\alpha}_k + 1}\right)_1^n, \quad \det(G(X_{n+1})) = \det\left(\frac{1}{\alpha_j + \overline{\alpha}_k + 1}\right)_0^n,$$

where we recognize the Cauchy determinants $C_n(a, b)$ of Exercise 5.7.5(a) with $a_j = \alpha_j$, $b_k = \overline{\alpha}_k + 1$. By the calculation of Exercise 5.7.5(a), we have

$$C_n(a, b) = \frac{\prod_{1 \le j < k \le n}(a_k - a_j)(b_k - b_j)}{\prod_{j,k}(a_j + b_k)},$$

therefore

$$\det(G(X_n)) = \frac{\prod_{1 \le j < k \le n}|\alpha_k - \alpha_j|^2}{\prod_{j,k}(\alpha_j + \overline{\alpha}_k + 1)}$$

and

$$\det(G(X_{n+1})) = \frac{\prod_1^n |\alpha_0 - \alpha_j|^2 \cdot \prod_{1 \le j < k \le n}|\alpha_k - \alpha_j|^2}{(\alpha_0 + \overline{\alpha}_0 + 1)\prod_1^n |\alpha_0 + \overline{\alpha}_j + 1|^2 \prod_1^n(\alpha_j + \overline{\alpha}_k + 1)},$$

and the result follows.

(3) *Theorem of Müntz (1914) and Szász (1916).* Let α_j ($j = 1, 2, \ldots$), $\text{Re}(\alpha_j) > -1/2$, $\alpha_j \ne \alpha_k$ for $j \ne k$, and $X = (x^{\alpha_j} : j \ge 1)$. Show that

$$\text{span}_{L^2(0,1)}(X) = L^2(0,1) \Leftrightarrow \sum_{j \ge 1} \frac{\text{Re}(\alpha_j + 1/2)}{1 + |\alpha_j + 1/2|^2} = \infty.$$

SOLUTION: By Weierstrass's theorem, the polynomials $\mathcal{P}_a$ are dense in $C[0,1]$ for the norm $\|f\|_\infty = \max_{0 \le t \le 1}|f(t)|$, and thus for the norm $\|f\|_2$, and $C[0,1]$ is dense in $L^2(0,1)$ (for example, every $f \in L^2(0,1)$ is the sum of its Fourier series $f = \sum_{k \in \mathbb{Z}} c_k e^{2\pi i k x}$). Hence $\mathcal{P}_a$ is dense in $L^2(0,1)$, and thus $\text{span}_{L^2(0,1)}(X) = L^2(0,1)$ if and only if $x^a \in \text{span}_{L^2(0,1)}(\mathcal{X})$ for every $a > 0$. By (2) above, the last inclusion is equivalent to

$$\lim_n \prod_{j=1}^n \left|\frac{a-\alpha_j}{a+\overline{\alpha}_j+1}\right| = 0;$$

this, in turn, is equivalent to the divergence (towards 0) of the product

$$\prod_{j\geq 1} \left|\frac{b-b_j}{b+\overline{b}_j}\right|^2,$$

where $b_j = \alpha_j + 1/2$ ($\text{Re}(b_j) > 0$), $b = a + 1/2$. Since

$$\left|\frac{b-b_j}{b+\overline{b}_j}\right|^2 < 1,$$

this product diverges if and only if

$$\sum_j \left(1 - \left|\frac{b-b_j}{b+\overline{b}_j}\right|^2\right) = \infty.$$

However,

$$1 - \left|\frac{b-b_j}{b+\overline{b}_j}\right|^2 = \frac{4b\,\text{Re}(\overline{b}_j)}{|b+\overline{b}_j|^2} = \frac{4b\,\text{Re}(b_j)}{(b+\text{Re}\,b_j)^2 + (\text{Im}\,b_j)^2},$$

and the nature of the series having this general term is the same as that of term

$$\frac{Re(b_j)}{1 + (Re\,b_j)^2 + (Im\,b_j)^2}.$$

NOTE *The function*

$$\epsilon \frac{z-b_j}{z+\overline{b}_j}, \quad |\epsilon| = 1$$

is none other than the elementary Blaschke factor (see Appendix F) for the half-plane $\text{Re}(z) > 0$, *and the convergence*

$$\sum_{j \geq 1} \frac{Re(b_j)}{1 + (\text{Re}\,b_j)^2 + (\text{Im}\,b_j)^2} < \infty$$

is called the Blaschke condition (for $\text{Re}(z) > 0$). *Müntz proved the theorem for real positive* α_j, $\alpha_j > 0$, *satisfying* $\alpha_j \to \infty$, *in which case the condition of completeness is equivalent to* $\sum_{j \geq 1} (1/\alpha_j) = \infty$. *It is easy to see that the same condition (but for* $\text{Re}(\alpha_j) > 0$ $(j > 1)$, $\alpha_1 = 0$) *is responsible of the completeness of the powers* (x^{α_j}) *in the space* $C[0,1]$ *(a direct generalization of Weierstrass's theorem corresponding to the case* $\alpha_j = j$, $j = 0, 1, 2, \ldots$;

under the general form of Szász, the theorem had been conjectured previously by S. Bernstein).

5.7.1 Positive Definite Sequences and Holomorphic Functions

(a) Positive harmonic functions (Herglotz, 1911). Let u be a harmonic function in $\mathbb{D}$, $u \geq 0$. Show that there exists a unique Borel measure $\mu \in \mathcal{M}(\mathbb{T})$, $\mu \geq 0$, such that $u = P * \mu$, that is,

$$u(z) = \int_{\mathbb{T}} \frac{1 - |z|^2}{|\zeta - z|^2} \, d\mu(\zeta), \quad z \in \mathbb{D}.$$

SOLUTION: Let $0 < r < 1$, $u_r(z) = u(rz)$ and $\mu_r = u_r m$; since $\text{Var}(\mu_r) = \mu_r(\mathbb{T}) = u_r(0) = u(0) < \infty$ and by the weak compactness of the ball in the space $\mathcal{M}(\mathbb{T})$ (Appendix A), there exists a sequence $(\mu_{r_n})_{n \geq 1}$ weakly convergent to a measure $\mu \in \mathcal{M}(\mathbb{T})$. Clearly $\mu \geq 0$ (as $f \in C(\mathbb{T})$, $f \geq 0$ implies $\int f \, d\mu = \lim_n \int f u_{r_n} \, dm \geq 0$), and since the Poisson kernel is continuous, we have

$$u(z) = \lim_n u(r_n z) = \lim_n \int_{\mathbb{T}} \frac{1 - |z|^2}{|\zeta - z|^2} \, d\mu_{r_n}(\zeta) = \int_{\mathbb{T}} \frac{1 - |z|^2}{|\zeta - z|^2} \, d\mu(\zeta)$$

for every $z \in \mathbb{D}$. The uniqueness of μ follows from the Fourier representation of $P * \mu$,

$$(P * \mu)(r\zeta) = \sum_{n \in \mathbb{Z}} r^{|n|} \widehat{\mu}(n) \zeta^n,$$

and hence if $P * \mu = P * \nu$, then $\widehat{\mu}(n) = \widehat{\nu}(n)$ for every $n \in \mathbb{Z}$, thus $\mu = \nu$.

(b) Another characterization of positive definite sequences (Carathéodory, 1911; Toeplitz, 1911). Let $(a_k)_{k \in \mathbb{Z}} \in l^\infty(\mathbb{Z})$ and for every k, $a_{-k} = \overline{a_k}$. Show that the following assertions are equivalent.

(1) There exists a measure $\mu \in \mathcal{M}(\mathbb{T})$, $\mu \geq 0$, such that $a_k = \widehat{\mu}(k)$ ($\forall k \in \mathbb{Z}$).
(2) $\text{Re}(F(z)) \geq 0$ for every $z \in \mathbb{D}$ where $F(z) = \sum_{k \geq 0} b_k z^k$ and $b_k = a_k$ ($k \neq 0$), $b_0 = a_0/2$.
(3) The Toeplitz matrices $T_{a,n}$ are positive definite: $T_{a,n} \gg 0$ ($\forall n \in \mathbb{Z}_+$), i.e.

$$(T_{a,n} x, x) = \sum_{0 \leq j,k \leq n} a_{j-k} x_k \overline{x}_j \geq 0 \quad (\forall x = (x_j)_0^n \in \mathbb{C}^{n+1}, \, \forall n \in \mathbb{Z}_+).$$

Hint Proceed by (1) $\Rightarrow$ (3) $\Rightarrow$ (2) $\Rightarrow$ (1). For (2) $\Rightarrow$ (1) use (a).

SOLUTION: (1) $\Rightarrow$ (3) since, for every $x \in \mathbb{C}^{n+1}$,

$$0 \leq \int_{\mathbb{T}} \left| \sum_{0 \leq k \leq n} x_k z^k \right|^2 d\mu = \sum_{0 \leq j,k \leq n} x_k \overline{x}_j \int_{\mathbb{T}} z^{k-j} \, d\mu = \sum_{0 \leq j,k \leq n} a_{j-k} x_k \overline{x}_j = (T_{a,n} x, x).$$

(3) ⇒ (2) since, for every $z \in \mathbb{D}$, we have

$$\frac{2\operatorname{Re} F(z)}{1 - |z|^2} = \sum_{j \geq 0} |z|^{2j}(F(z) + \overline{F(z)})$$

$$= \sum_{j \geq 0} z^j \bar{z}^j \left(\sum_{k \geq 0} b_k z^k + \sum_{k \geq 0} b_{-k} \bar{z}^k \right)$$

$$= \sum_{j \geq 0, k \geq 0} b_k z^{k+j} \bar{z}^j + \sum_{j \geq 0, k \geq 0} b_{-k} z^j \bar{z}^{k+j}$$

$$= \sum_{j \geq 0} \sum_{l \geq j} b_{l-j} z^l \bar{z}^j + \sum_{l \geq 0, m \geq l} b_{l-m} z^l \bar{z}^m$$

$$= \sum_{j \geq 0, k \geq 0} a_{k-j} z^k \bar{z}^j$$

$$= \lim_n \sum_{0 \leq j, k \leq n} a_{k-j} z^k \bar{z}^j$$

$$\geq 0$$

(as the series $\sum_{j \geq 0, k \geq 0} |a_{k-j} z^k \bar{z}^j|$ converges).

(2) ⇒ (1) since, by the theorem (a) applied to $u = 2\operatorname{Re} F$, there exists a measure $\mu \geq 0$ such that, for every $z = r\zeta \in \mathbb{D}$ ($|\zeta| = 1$),

$$2\operatorname{Re} F(z) = \sum_{k \geq 0} b_k r^k \zeta^k + \sum_{k \geq 0} b_{-k} r^k \bar{\zeta}^k$$

$$= \int_{\mathbb{T}} \frac{1 - |zv|^2}{|t - z|^2} d\mu(t)$$

$$= \int_{\mathbb{T}} \left(\sum_{k \in \mathbb{Z}} r^{|k|} \zeta^k \bar{t}^k \right) d\mu(t)$$

$$= \sum_{k \in \mathbb{Z}} \widehat{\mu}(k) r^{|k|} \zeta^k,$$

therefore $a_k = \widehat{\mu}(k)$ ($\forall k \in \mathbb{Z}$).

(c) Theorem of Schur (1917). Let $\varphi = \sum_{k \geq 0} a_k z^k \in \operatorname{Hol}(\mathbb{D})$ and $\lambda > 0$. Show that the following assertions are equivalent.

(i) $|\varphi(z)| \leq \lambda$ ($\forall z \in \mathbb{D}$).

(ii) For every $n \geq 0$, the matrices A_n are positive definite,

$$A_n = \begin{pmatrix} \lambda I_n & T_{\bar{\varphi},n} \\ T_{\varphi,n} & \lambda I_n \end{pmatrix} \gg 0.$$

(iii) For every $n \geq 0$, $\det(A_n) \geq 0$.

(iv) For every $n \geq 0$ and every ϵ, $|\epsilon| = 1$, $\det(T_{2\lambda - \epsilon\varphi - \bar{\epsilon}\bar{\varphi},n}) \geq 0$.

SOLUTION: According to the comments at the beginning of §5.2, it suffices to show that (ii)–(iv) are equivalent to $\|T_{\varphi,n}\| \leq \lambda$; however, $A_n \gg 0$ signifies that, for every

$x, y \in \mathcal{P}_n$, we have $\lambda(x, x) + (T^*_{\varphi,n} y, x) + (T_{\varphi,n} x, y) + \lambda(y, y) \geq 0$, which is equivalent to $\|T_{\varphi,n}\| \leq \lambda$ (hence (i) ⇔ (ii)). It is the same for (iv): by the Sylvester criterion, the inequalities of (iv) characterize the positivity of the matrices $T_{2\lambda-\epsilon\varphi-\overline{\epsilon\varphi},n}$ ($\forall n \geq 0$), which reduces to $|\text{Re}(\epsilon\varphi)| \leq \lambda$ for every ϵ, $|\epsilon| = 1$, hence the equivalence (iv) ⇔ (i). Finally, to verify (iii) ⇒ (i) (the converse (ii) ⇒ (iii) is evident), it suffices to show that (iii) implies $B_n =: \lambda^2 I_n - T^*_{\varphi,n} T_{\varphi,n} \gg 0$ (for every $n \geq 0$); however, for $0 \leq k \leq n$, we have $0 \leq \det(A_k) = \det(\lambda^2 I_k - T^*_{\varphi,k} T_{\varphi,k}) = \det(B_k)$ (a well-known identity for determinants: if P, Q, R, S are $k \times k$ matrices and $PR = RP$, then $\det((P, R)^{\text{col}}, (Q, S)^{\text{col}}) = \det(PS - RQ)$; see e.g. [Gantmacher, 1966], Chapter II §5.3), and then $B_k = P_k B_n P_k$, which implies $B_n \gg 0$ by the Sylvester criterion.

5.7.2 Semi-Commutators of Finite Toeplitz Matrices

Let $\varphi, \psi \in L^\infty(\mathbb{T})$.

(i) *Show that*

$$T_{\varphi\psi,n} - T_{\varphi,n} T_{\psi,n} = P_n H^*_{\overline{\varphi}} H_\psi P_n + P_n(z^{n+1} H_\varphi H^*_{\overline{\psi}} \overline{z}^{n+1}) P_n$$

(Widom, 1976).

(ii) *By modifying the definition of a finite Hankel matrix to $\tilde{H}_{\varphi,n} = (I-P_n)M_\varphi P_n$, we would have the identity*

$$T_{\varphi\psi,n} - T_{\varphi,n} T_{\psi,n} = \tilde{H}^*_{\overline{\varphi},n} \tilde{H}_{\psi,n}.$$

Hint Calculation similar to Lemma 2.2.9.

5.7.3 Inversion of Wiener–Hopf Operators by the Finite Section Method

Definition Let $s > 0$ and $P_s f = \chi_{(0,s)} f$ an orthogonal projection of $L^2(\mathbb{R}_+)$ onto a subspace $L^2(0, s)$. An operator $A \colon L^2(\mathbb{R}_+) \to L^2(\mathbb{R}_+)$ is said to be *invertible by the finite section method*, abbreviated

$$A \in \text{IFSM},$$

if the finite sections $A_s = P_s A | L^2(0, s)$ (which we identify with $P_s A P_s$) are invertible after a certain rank and $\lim_{s \to \infty} A_s^{-1} P_s f$ exists (for the norm) for every $f \in L^2(\mathbb{R}_+)$. For the integral Wiener–Hopf operators (see §4.2), there is a theory of invertibility by truncations completely analogous to that of §5.4 (for the Toeplitz operators). We cite here only the beginning and end of the theory.

(a) *Let $A \in L(L^2(\mathbb{R}_+))$.*

(i) *Show that $A \in$ IFSM $\Rightarrow A$ invertible and $\lim_{s \to \infty} A_s^{-1} P_s f = A^{-1} f$ ($\forall f \in L^2(\mathbb{R}_+)$) and*
$$\|A^{-1}\| \le l(A) =: \overline{\lim_s} \|A_s^{-1}\|.$$

(ii) *If $A \in L(L^2(\mathbb{R}_+))^{-1}$, then $A \in$ IFSM if and only if $l(A) = \overline{\lim}_s \|A_s^{-1}\| < \infty$ (we suppose $\|A_s^{-1}\| = \infty$ if A_s is not invertible).*

(b) *State and prove analogs of the assertions of Theorem §5.4.2. In particular, prove the following theorem.*

Theorem (Devinatz and Shinbrot, 1969) *Let k be a pseudo-measure on $\mathbb{R}$ and $\Phi =: 2\pi \mathcal{F}^{-1} k \in H^\infty(\mathbb{C}_+) + C(\overline{\mathbb{R}})$ (or $\overline{\Phi} \in H^\infty(\mathbb{C}_+) + C(\overline{\mathbb{R}})$) and $W_k \in L(L^2(\mathbb{R}_+))^{-1}$; then $W_k \in$ IFSM.*

Hint Follow the techniques of §5.4.2 and §4.2.

5.7.4 When the Second Szegő Asymptotics Stabilize (Szegő, 1952)

Let $\varphi = 1/|p|^2 \in L^1(\mathbb{T})$, $p \in \mathcal{P}_a$, $m = \deg p$ and let $[\varphi]$ be an outer function of modulus $|\varphi|$. Show that, for every $n \ge m$,
$$\frac{D_n(\varphi)}{G(\varphi)^{n+1}} = \exp\left(\frac{1}{4\pi} \int_\mathbb{D} \left|\frac{[\varphi]'}{[\varphi]}\right|^2 dx\,dy\right),$$
where $G(\varphi) = \exp\left(\int_\mathbb{T} \log(\varphi)\,dm\right)$.

Hint Follow the calculations of §5.6.1(ii).

SOLUTION: By §5.6.1(ii), for every $n \ge m$ we have
$$\frac{D_n(\varphi)}{G(\varphi)^{n+1}} = \frac{D_{m-1}(\varphi)}{G(\varphi)^m},$$
where $D_{m-1}(\varphi) = \det(T_{\varphi, m-1})$; without changing $|p|$ on $\mathbb{T}$ we can suppose that all the roots of p are in $|z| > 1$ and $p(z) = a(z_0 - z)(z_1 - z)\dots(z_{m-1} - z)$ so that $p(0) = az_0 \dots z_{m-1} > 0$. We first show that $T_{\varphi, m-1} = V^* A V$ where the mapping V is suggested by the Lagrange interpolation formula in the following manner: for $x = \sum_{0 \le k < m} x_k z^k \in \mathcal{P}_{m-1}$ we have (when supposing the z_j pairwise distinct)
$$\frac{x(z)}{p(z)} = \sum_{0 \le k < m} \frac{x(z_k)}{p'(z_k)} \times \frac{1}{z - z_k} = \sum_{0 \le k < m} \frac{y_k}{z - z_k},$$
where
$$y = (y_k)_0^{m-1} = Vx,$$

$$Vx = \left(\sum_{0\le k<m} x_k z_j^k/p'(z_j)\right)_{0\le j<m} = (x(z_j)/p'(z_j))_{0\le j<m}.$$

Therefore

$$(T_{\varphi,m-1}x, x)_{\mathcal{P}_{m-1}} = \int_{\mathbb{T}} \left|\frac{x(z)}{p(z)}\right|^2 dm = \sum_{0\le k,j<m} y_k \bar{y}_j \left(\frac{1}{z-z_k}, \frac{1}{z-z_j}\right) = (Ay, y),$$

where

$$A = \left(\frac{1}{z-z_k}, \frac{1}{z-z_j}\right)_{0\le k,j<m}$$

and $((1-\bar{\lambda}z)^{-1}$ is the reproducing kernel of H^2: see Appendix F)

$$\left(\frac{1}{z-z_k}, \frac{1}{z-z_j}\right) = -\bar{z}_j^{-1} \frac{1}{\bar{z}_j^{-1} - z_k} = \frac{1}{\bar{z}_j z_k - 1}.$$

Thus $T_{\varphi,m-1} = V^*AV$, and it suffices to calculate $\det V$ and $\det A$; the matrix of V is $(z_j^k/p'(z_j))_{0\le k,j<m}$ and by calculating $p'(z_j)$ we find

$$|\det V| = \left|a^{-m}\prod_{k<j}(z_k-z_j)^{-1}\right|.$$

For $\det(A)$, we have (by reducing the calculation to that of the det of a classic Cauchy matrix $\det((a_j - b_k)^{-1})$: see Exercise 5.7.5 below)

$$\det A = \frac{1}{z_0 z_1 \ldots z_{m-1}} \det\left(\frac{1}{\bar{z}_j - z_k^{-1}}\right)$$

$$= \frac{1}{z_0 z_1 \ldots z_{m-1}} \prod_{j<k}((\bar{z}_k - \bar{z}_j)(z_j^{-1} - z_k^{-1})) \times \prod_{j,k}(\bar{z}_k - z_j^{-1})^{-1}$$

$$= \prod_{j<k}|z_k-z_j|^2 \times \prod_{j,k}(\bar{z}_k z_j - 1)^{-1}$$

$$= \prod_{j<k}|z_k-z_j|^2 \times \prod_k \frac{1}{|z_k|^2-1} \times \prod_{j<k}|\bar{z}_k z_j - 1|^{-2},$$

and finally

$$D_{m-1}(\varphi) = \det(T_{\varphi,m-1}) = |\det V|^2 \det A = \left(|a|^{2m}\prod_{k,j}(\bar{z}_k z_j - 1)\right)^{-1}.$$

Since $G(\varphi) = [\varphi](0) = 1/|p(0)|^2 = (az_0 z_1 \ldots z_{m-1})^{-2}$, we obtain

$$\frac{D_{m-1}(\varphi)}{G(\varphi)^m} = |z_0 z_1 \ldots z_{m-1}|^{2m} \prod_{k,j}(\bar{z}_k z_j - 1)^{-1}.$$

We compare the last quantity with $\|(\log[\varphi])'\|_{\mathbb{D}}^2$: since $1/p$ is an outer function, we have $[\varphi] = 1/p^2$, and thus $(\log[\varphi])' = -2p'/p = 2\sum_k (z_k - z)^{-1}$, implying

$$\frac{1}{4\pi} \int_{\mathbb{D}} \left|\frac{[\varphi]'}{[\varphi]}\right|^2 dx\,dy = \sum_{k,j} \frac{1}{\pi} \int_{\mathbb{D}} \frac{dx\,dy}{(z - z_k)(\bar{z} - \bar{z}_j)}$$

$$= \sum_{k,j} \frac{1}{\pi} \int_{\mathbb{D}} \frac{r\,dr\,d\theta}{(r e^{i\theta} - z_k)(r e^{-i\theta} - \bar{z}_j)}$$

$$= \sum_{k,j} \frac{1}{\pi} \int_0^1 \frac{2\pi r\,dr}{z_k \bar{z}_j - r^2}$$

$$= \sum_{k,j} [-\log(z_k \bar{z}_j - r^2)]_0^1$$

$$= \sum_{k,j} \log \frac{z_k \bar{z}_j}{z_k \bar{z}_j - 1}$$

$$= \log \left(\prod_{k,j} \frac{z_k \bar{z}_j}{z_k \bar{z}_j - 1} \right)$$

$$= \log \left(|z_0 z_1 \ldots z_{m-1}|^{2m} \prod_{k,j} (\bar{z}_k z_j - 1)^{-1} \right),$$

which proves the result in the case of pairwise distinct roots z_k. As the two parts of the final equation are obviously continuous in $z_0, \ldots, z_{m-1}$, the general result also follows.

5.7.5 Cauchy Determinants (1841)

(a) *Let $a_j, b_k \in \mathbb{C}$ ($1 \le j, k \le n$), and define the Cauchy determinants as*

$$C_n(a,b) = \det \left(\frac{1}{a_j + b_k} \right)_{1 \le j,k \le n},$$

(where we suppose $a_j + b_k \ne 0$ ($\forall j, k$)), and

$$D_n(a,b) = \det \left(\frac{1}{a_j - b_k} \right)_{1 \le j,k \le n}$$

(where we suppose $a_j - b_k \ne 0$ ($\forall j, k$)). Show that

$$C_n(a,b) = \frac{\prod_{1 \le j < k \le n}(a_k - a_j)(b_k - b_j)}{\prod_{j,k}(a_j + b_k)},$$

$$D_n(a,b) = \frac{\prod_{1 \le j < k \le n}(a_k - a_j)(b_k - b_j)}{\prod_{j,k}(a_j - b_k)}.$$

Hint Induction on n by transforming the first column into $(1, 0, \ldots, 0)^{\text{col}}$.

SOLUTION: Subtract the first column from the columns $2 \leq k \leq n$,

$$c_{jk} \mapsto c'_{jk} = \frac{1}{a_j+b_k} - \frac{1}{a_j+b_1} = \frac{b_1-b_k}{a_j+b_1} \times \frac{1}{a_j+b_k}, \quad c'_{j1} = c_{j1},$$

and then remove the factor $1/(a_j+b_1)$ from each row and the factor b_1-b_k from the columns $2 \leq k \leq n$:

$$\det(c_{jk}) = \det(c'_{jk}) = \left(\prod_1^n \frac{1}{a_j+b_1}\right) \times \left(\prod_2^n (b_1-b_k)\right) \det(c^{(2)}_{jk}),$$

where

$$(c^{(2)}_{j1})_j = (1,1,\ldots,1)^{\text{col}} \quad \text{and} \quad (c^{(2)}_{jk}: 1 \leq j \leq n, 2 \leq k \leq n) = \left(\frac{1}{a_j+b_k}\right).$$

Next, subtract the row $c^{(2)}_{1k}$ from the rows $2 \leq j \leq n$ (which, again, does not change the determinant),

$$c^{(2)}_{jk} \mapsto c^{(3)}_{jk} = \frac{a_1-a_j}{a_1+b_k} \times \frac{1}{a_j+b_k}, \quad 2 \leq k \leq n,$$

$(c^{(3)}_{j1})_j = (1, 0, \ldots 0)^{\text{col}}$. Then, again, remove the factors $a_1 - a_j$ from each column $2 \leq k \leq n$ and $1/(a_1+b_k)$ from each row $2 \leq j \leq n$, giving (by expanding the det by the first column)

$$\det(c_{jk}) = \left(\prod_1^n \frac{1}{a_j+b_1}\right) \times \left(\prod_2^n (b_1-b_k)\right) \det(c^{(2)}_{jk})$$

$$= \left(\prod_1^n \frac{1}{a_j+b_1}\right) \times \left(\prod_2^n \frac{1}{a_1+b_k}\right)$$

$$\times \left(\prod_2^n (b_1-b_k)(a_1-a_k)\right) \times \det\left(\frac{1}{a_j+b_k}\right),$$

where the last determinant is C_{n-1} on the indices $2 \leq j,k \leq n$. The induction concludes the calculation.

(b) Hilbert determinants (1894). *From (a), deduce*

$$H_n := \det\left(\frac{1}{j+k-1}\right)_{1 \leq j,k \leq n} = \frac{c_n^4}{c_{2n}} \quad \text{where } c_n = \prod_{k=1}^{n-1} k!.$$

SOLUTION: Setting $a_j = j$, $b_k = k-1$ in (a), we obtain

$$H_n = \frac{\prod_{1 \leq j < k \leq n}(k-j)^2}{\prod_{j,k}(j+k-1)} = \frac{c_n^2}{c_{2n}/c_n^2} = \frac{c_n^4}{c_{2n}}.$$

5.7.6 The Second Term of the Asymptotic Distribution of Spectra (Libkind (1972), Widom (1976))

If $\varphi \in K(\mathbb{T})$ (see §5.6.5.1) is an absolutely continuous function (a primitive of its derivative) and $f \in \mathrm{Hol}(\Sigma(\varphi))$ ($\Sigma(\varphi) = \mathrm{conv}(\mathrm{Ran}_{\mathrm{ess}}(\varphi))$), then

$$\frac{1}{n+1} \sum_{0 \le j \le n} f(\lambda_{n,j}) = \int_0^{2\pi} f \circ \varphi \, dm + \frac{E(f, \varphi)}{n+1} + o(1/n) \quad (n \to \infty),$$

where

$$E(f, \varphi) = \frac{1}{2} \sum_{k \in \mathbb{Z}} \int_{\mathbb{T}} \int_{\mathbb{T}} \zeta_1^{-k} \zeta_2^k R(\zeta_1, \zeta_2) \, dm(\zeta_1) \, dm(\zeta_2),$$

$$R(\zeta_1, \zeta_2) = \frac{f(\varphi(\zeta_1)) - f(\varphi(\zeta_2))}{\varphi(\zeta_1) - \varphi(\zeta_2)} (\varphi'(\zeta_1) - \varphi'(\zeta_2)),$$

and $\varphi'(\zeta) = \zeta(d\varphi/d\zeta)$.

Hint See [Libkind, 1972], [Widom, 1976], and [Simon, 2005b] for details (which are not far from the special cases already treated in Theorem 5.6.8 with $f(z) = \log(1 + uz)$).

5.7.7 The Helton and Howe Formula of Lemma 5.6.9

If $A, B \in L(H)$ and $[A, B] \in \mathfrak{S}_1(H)$ (H is a Hilbert space), then $(I - e^A e^B e^{-A} e^{-B}) \in \mathfrak{S}_1(H)$ and

$$\det(e^A e^B e^{-A} e^{-B}) = \exp(\mathrm{Trace}[A, B]).$$

STEPS OF THE PROOF For details, see [Simon, 2005b], page 342.

(1) By induction,

$$[A^n, B] = A^{n-1}[A, B] + [A^{n-1}, B]A \in \mathfrak{S}_1(H)$$

and

$$\|[A^n, B]\|_{\mathfrak{S}_1} \le n\|A\|^{n-1}\|[A, B]\|_{\mathfrak{S}_1},$$

and again by induction

$$\|[A^n, B^m]\|_{\mathfrak{S}_1} \le nm\|A\|^{n-1}\|B\|^{m-1}\|[A, B]\|_{\mathfrak{S}_1},$$

hence (by developing in a power series)

$$\|[e^A, e^B]\|_{\mathfrak{S}_1} \le \|[A, B]\|_{\mathfrak{S}_1} e^{\|A\| + \|B\|}.$$

(2) Thus $e^A e^B e^{-A} e^{-B} - I = [e^A, e^B] e^{-A} e^{-B} \in \mathfrak{S}_1(H)$.

(3) For $s \in \mathbb{C}$, we set $F(s) = e^{sA} e^{sB} e^{-sA} e^{-sB}$ (a holomorphic function with values in $I + \mathfrak{S}_1$), and noting that $X(s) =: I - F(s)$ is small in $\|\cdot\|_{\mathfrak{S}_1}$ for $|s| < \epsilon$, we obtain that $\log(F(s)) =: -\sum_{k \geq 1} X(s)^k/k$ converges for the norm $\|\cdot\|_{\mathfrak{S}_1}$. Hence by Lidskii's theorem (see e.g. [Gohberg, Goldberg, and Kaashoek, 1990]), $\log(\det(F(s))) = \text{Trace}(\log(F(s)))$ for $|s| < \epsilon$.

(4) Thus $(d/ds)\log(\det(F(s))) = \text{Trace}(F'(s)F(s)^{-1})$.

(5) (The decisive step) By an intensive computation using $\text{Trace}(CD) = \text{Trace}(DC)$, we obtain $\text{Trace}(F'(s)F(s)^{-1}) = 2s \, \text{Trace}[A, B]$, which, with step (4), gives the desired formula $\det(F(s)) = \exp(s^2 \, \text{Trace}[A, B])$. ∎

5.7.8 The Formula of Borodin and Okounkov (2000) (and Geronimo and Case (1979))

(a) Let $\varphi \in K(\mathbb{T})$ (see §5.6.5.1), $\varphi(\zeta) \neq 0$ for $\zeta \in \mathbb{T}$. Show that for every $n = 0, 1, \ldots$,

$$\frac{\det(T_{\varphi,n})}{G(\varphi)^{n+1}} = \frac{\det(I - (I - P_n)\Gamma_{\bar{b}}\Gamma_c(I - P_n))}{\det(I - \Gamma_{\bar{b}}\Gamma_c)},$$

where $b = \overline{\varphi}_1 \varphi_2^{-1}$, $c = \overline{\varphi}_1^{-1} \varphi_2$ (where $\varphi = \overline{\varphi}_1 \varphi_2$ is a Birkhoff–Wiener–Hopf factorization of φ; $\varphi_j^{\pm 1} \in K(\mathbb{T}) \cap H^\infty$) and P_n (as always) an orthogonal projection onto $\mathcal{P}_n$.

SOLUTION: (Basor and Widom (2000, see Notes and Remarks), Böttcher [2001]). The C. Jacobi formula for the minors of the inverse A^{-1} of an $m \times m$ matrix states that, for a principal minor of size $n < m$, we have

$$\det(P_n A^{-1} P_n) = \frac{\det((I - P_n)A(I - P_n))}{\det A}$$

(see [Gantmacher, 1966] §1.4). Given an operator $K \in \mathfrak{S}_1(H^2)$ such that $I - K$ is invertible, we apply the formula to $A = P_m(I - K)P_m$ where $m > n$, and obtain

$$\det(P_n(I - P_m K P_m)^{-1} P_n) = \frac{\det((I - P_n) - (I - P_n)P_m K P_m (I - P_n))}{\det(I - P_m K P_m)}.$$

With $\lim_{m \to \infty} \|P_m K P_m - K\|_{\mathfrak{S}_1} = 0$ and the continuity of det for the norm $\|\cdot\|_{\mathfrak{S}_1}$ (see Appendix D), we have

$$\det(P_n(I - K)^{-1} P_n) = \frac{\det(I - (I - P_n)K(I - P_n))}{\det(I - K)},$$

where we set $K = \Gamma_{\bar{b}}\Gamma_c$ (which is in $\mathfrak{S}_1(H^2)$ since $\Gamma_{\bar{b}}, \Gamma_c \in \mathfrak{S}_2(H^2)$). By Lemma 2.2.9(1),

$$I - K = T_b T_c = T_{\overline{\varphi}_1 \varphi_2^{-1}} T_{\overline{\varphi}_1^{-1} \varphi_2} = T_{\overline{\varphi}_1} T_\varphi^{-1} T_{\varphi_2},$$

hence

$$P_n(I - K)^{-1}P_n = P_n(T_{\varphi_2^{-1}}T_\varphi T_{\overline{\varphi}_1^{-1}})P_n = P_n T_{\varphi_2^{-1}} P_n T_\varphi P_n T_{\overline{\varphi}_1^{-1}} P_n,$$

and then

$$\det(P_n(I - K)^{-1}P_n) = \det(P_n T_{\varphi_2^{-1}} P_n)\det(T_{\varphi,n})\det(P_n T_{\overline{\varphi}_1^{-1}} P_n)$$

$$= G(\varphi)^{-n-1}\det(T_{\varphi,n})$$

(see §5.6.5.3(a)), which completes the proof.

(b) Derivation of the strong Szegő theorem. *Under the hypotheses of (a), show that*

$$\frac{\det(T_{\varphi,n})}{G(\varphi)^{n+1}} = (1 + o(1))\det(T_\varphi T_{1/\varphi});$$

see Theorem 5.6.8 for other expressions for the right-hand side.

SOLUTION: (Böttcher and Widom, 2006) Since $\Gamma_{\tilde b}\Gamma_c \in \mathfrak{S}_1$ and $\lim_n \|(I - P_n)x\|_2 = 0$ ($\forall x \in H^2$), we obtain

$$\frac{\det(I - (I - P_n)\Gamma_{\tilde b}\Gamma_c(I - P_n))}{\det(I - \Gamma_{\tilde b}\Gamma_c)} = \frac{1 + o(1)}{\det(I - \Gamma_{\tilde b}\Gamma_c)}$$

$$= \frac{1 + o(1)}{\det(T_b T_c)}$$

$$= (1 + o(1))\det(T_c^{-1} T_b^{-1}),$$

and then (by a calculation similar to that of (a)),

$$\det(T_c^{-1} T_b^{-1}) = \det((T_{\overline{\varphi}_1^{-1}} T_{\varphi_2})^{-1}(T_{\overline{\varphi}_1} T_{\varphi_2^{-1}})^{-1})$$

$$= \det(T_{\varphi_2^{-1}} T_\varphi T_{\overline{\varphi}_1^{-1}})$$

$$= \det(T_\varphi T_{\overline{\varphi}_1^{-1}} T_{\varphi_2^{-1}})$$

$$= \det(T_\varphi T_{1/\varphi}).$$

5.8 Notes and Remarks

The contents of this chapter provide a fairly realistic panorama of the theory of "Toeplitz forms" in its historical development. As discussed in Chapter 1, the theory was initiated by applications to problems of trigonometric interpolation (Carathéodory, Toeplitz) and to holomorphic functions (Schur, Herglotz), and then extended as an illustration of Hilbert's spectral theory (Toeplitz). The true formation of the theory and the rise of its role in analysis began with Szegő's discoveries of the tight links between Toeplitz matrices and extremal problems of approximation theory, orthogonal polynomials, and complex analysis. The Wiener–Hopf integral equations and their principal techniques (such as the Birkhoff–Wiener–Hopf factorization) which appeared soon after, were for a long time developed independently, only to be reunited with Toeplitz theory in the 1950s. Before commenting on the contents of this chapter point by point, we note that the introductory character of this handbook did not allow the development of each theme in proportion to its actual value, but we have tried to minimize the damage with a few references and indications to ulterior results. The most regrettable omissions are, perhaps, the absence of numerical methods and algorithms for the inversion of Toeplitz matrices, local principles for the inversion by the finite section method, generalizations to abstract *Toeplitz-like* operators and quadratic forms, profound links recently discovered with the asymptotics of orthogonal polynomials, as well as Toeplitz operators on $\mathbb{T}^n$, $\mathbb{R}^n$ and on other groups and manifolds (often treated with the techniques of C^*-algebras).

We now proceed with specific remarks and comments.

Theorem 5.1.1 and §5.1.1. Moment problem. This theorem is a great classic of harmonic analysis, opening the door between interpolation and the properties of quadratic forms, then to positive extensions of functionals, and then to the representations of groups. Theorem 5.1.1 was found by Carathéodory [1907, 1911], but the language of Toeplitz forms was suggested by the fundamental work of Toeplitz [1911b] and published independently by these two authors (*loc. cit.*) in the same volume of *Rendiconti del Circ. Mat. Palermo* (1911). The proof in §5.1.1 came much later [Stone, 1932] with the ideas of spectral theory, also developed (independently) by Akhiezer: see [Akhiezer and Krein, 1938]. In general, the interpolation theorems remain a cornerstone of Toeplitz theory (operators and forms), as well as of spectral theory; see for example Theorem 5.1.3, a true gem of the interaction between the disciplines mentioned.

See also the comments below concerning Exercise 5.7.1.

§5.1.2, §5.1.5 and Theorem 5.1.3. Positive definite functions. The theme of positive definite functions, triggered by Theorems 5.1.1 and 5.1.3 in [Carathéodory, 1907, 1911], very rapidly developed into one of the principal subjects of harmonic analysis, first on the groups $\mathbb{T}$, $\mathbb{Z}$, and $\mathbb{R}$, next in several variables (with a famous analog of Theorem 5.1.1 in $\mathbb{R}^n$ by Bochner in 1933), in 1933 then on locally compact groups (for certain aspects of all these generalizations see [Weil, 1940], a seminal work in classical harmonic analysis, as well as [Rudin, 1962], [Berg, Christensen, and Ressel, 1984], and [Gelfand and Vilenkin, 1961]), and finally in a general theory of positive definite kernels

[Aronszajn, 1950]. In particular, the problem of extension to a positive definite function attracted much attention. The Carathéodory Theorem 5.1.3, the very first of its kind, stands out by virtue of its perfection and beauty (an explicit construction, the uniqueness, the optimality regarding the dimension of the spaces involved). The magnificent proof of Theorem 5.1.3 following Szegő [1954] (also presented in [Grenander and Szegő, 1958]) is a foundational result, rich with links to numerous subjects of analysis – once more to the spectral theory, and then to Gauss–Lagrange interpolation, to the distribution of the zeros of orthogonal polynomials (and their additional properties), etc. Incidentally, this last subject has played an eminent role in potential theory (see [Stahl and Totik, 1992]), which is not so very far from the theory of "generalized Toeplitz forms"; see the final comments of these Notes, below.

§5.1.4 and Theorem 5.1.3. Szegő's proof. The "operator" approach to the moment problem (in particular, the trigonometric moment problem) was developed especially by Akhiezer and Krein [1938], but parts (4) and (5) (key to Szegő's proof of Theorem 5.1.3) follow the ideas of [Szegő, 1954]. We repeat that this proof strikes us as a rare masterpiece because of its beauty and the richness of its allusions to other major themes of classical analysis.

For the (vast) theory of orthogonal polynomials and their applications we refer to [Szegő, 1959] and [Simon, 2005b], where the reader can find a rich source of information on the subject (the latter text contains more than 1000 reference titles) which has become central to today's analysis. It is curious to note that so many germs of future grand theories are easily recognizable in a proof of the very particular proposition which is Theorem 5.1.3.

§5.2. Norm of a Toeplitz matrix. As already mentioned in the text (§5.2.1(e)), a continuous analog of the main theorem of §5.2 was proved long before the result of this section (estimates for the norm of a truncated Wiener–Hopf operator $W_k^a = P_a W_k | L^2(0, a)$, Rochberg's Theorem 4.4.4 [1987]). By way of a comment, we can remark that even if it is likely that the arguments of [Rochberg, 1987] could be adapted to the discrete case, clearly the constants of the resulting estimates of such an adaptation would be difficult to control (the reasoning of [Rochberg, 1987] passes several times through upper bounds of $\|P_+\|$ for different weighted norms). For whatever reason, the estimates for the norm of $T_{\varphi,n} = P_n T_\varphi | P_n$ lay in waiting for more than ten years, and finally appeared in [Bakonyi and Timotin, 2001] (with a slightly different statement: see below) and independently in [Nikolski, 2002c]. The reasoning in §5.2.2 follows an unpublished note [Nikolski, 2002c] written at the request of Alexander Volberg (Michigan State) for his work on singular integrals, then included in the article by Nikol'skaya and Farforovskaya [2003]. Our reasoning gives a

"band extension" T_ψ of width $2n$ with the norm $\|T_\psi\|$ controllable by $3\|T_{\varphi,n}\|$; a comparison with the comment §5.2.1(c) suggests, perhaps, the existence of an explicit extension of the same size (with the Fejér kernel, for example).

We can also mention a result in [Böttcher and Grudsky, 2005], Section 5.3, stating
$$\|\sigma_n(\varphi)\|_\infty \leq \|T_{\varphi,n}\| \leq \|P_{-n,n}\varphi\|_\infty,$$
where $P_{-n,n}\varphi$ and
$$\sigma_n(\varphi) = \sum_{|k|\leq n}\left(1 - \frac{|k|}{n+1}\right)\hat{\varphi}(k)z^k$$
are, respectively, a partial sum (see §5.2.2 for the notation) and a Fejér–Cèsaro mean of the Fourier series of φ. Of course, the inequality on the right is obvious, whereas the one on the left – in a slightly weakened form $\frac{1}{3}\|\sigma_n(\varphi)\|_\infty \leq \|T_{\varphi,n}\|$ – easily follows from Theorem 5.2.1. On the other hand (and this is also mentioned in [Böttcher and Grudsky, 2005]), $\sup_{\varphi,n}\|P_{-n,n}\varphi\|_\infty/\|T_{\varphi,n}\| = \infty$ (which follows from §5.2.1(c)) and $\sup_{\varphi,n}\|T_{\varphi,n}\|/\|\sigma_n(\varphi)\|_\infty = \infty$ (just consider $\varphi = z^n$).

The result of Bakonyi and Timotin [2001] compares, for an analytic polynomial $p \in \mathcal{P}_n$, the distance
$$d'_n(p) =: \text{dist}_{L^\infty(\mathbb{T})}(p, H^\infty_- + z^{n+1}H^\infty)$$
with the norms of the Arveson submatrices (see §5.2.2),
$$\delta(p) = \max_{0\leq k\leq n}\|P_{n-k,n}T_pP_k\|.$$
Clearly $\delta(p) \leq d'_n(p)$, and the principal result of [Bakonyi and Timotin, 2001] states that
$$d'_n(p) \leq 2\frac{n+1}{n+2}\delta(p).$$
Comparing this inequality with that of Theorem 5.2.1, we remark that for a polynomial $\varphi \in \mathcal{P}$, $\deg\varphi \leq n$, certainly, we have $d_n(\varphi) = d'_{2n}(z^n\varphi)$, but $\|T_{\varphi,n}\| = \|P_{n,2n}(T_{z^n\varphi})P_n\| \leq \delta(z^n\varphi)$ (and even $\|T_{\varphi,n}\| < \delta(z^n\varphi)$ in the generic case). Nonetheless, we can deduce a result of type Theorem 5.2.1 by noting that for $p = z^n\varphi$ we have
$$\delta(z^n\varphi) = \max_{0\leq k\leq 2n}\|P_{2n-k,2n}T_{z^n\varphi}P_k\| \leq 2\|P_{n,2n}T_{z^n\varphi}P_n\| = 2\|T_\varphi\|$$
(to verify, let $0 \leq k \leq n$, then
$$\|P_{2n-k,2n}T_{z^n\varphi}P_k\| = \|P_{2n-k,n-1}T_{z^n\varphi}P_k + P_{n,2n}T_{z^n\varphi}P_k\|$$
$$\leq 2\|P_{n,2n}T_{z^n\varphi}P_n\| = 2\|T_\varphi\|$$

given the Toeplitz character of the matrices in question), and hence [Bakonyi and Timotin, 2001] implies

$$d_n(\varphi) = d'_{2n}(z^n\varphi) \le 2\frac{2n+1}{2n+2}\delta(z^n\varphi) \le 4\frac{2n+1}{2n+2}\|T_\varphi\| \le 4\|T_\varphi\|.$$

See also [Bakonyi and Woerdeman, 2011] for an exhaustive treatment of the matrix completion problem, as well as [Bessonov, 2014] for a different expression of the norm $\|T_{\varphi,n}\|$ which uses an analog of the space BMO on the cyclic group $\{e^{2\pi ik/n+1} : 0 \le k \le n\}$.

§5.2.1(b) and Exercise 5.7.1(c). *Theorems of Carathéodory–Fejér and Schur.* The first proposition is proved in [Carathéodory and Fejér, 1911], and the second in [Schur, 1917, 1918]. These articles served as models for the resolution of many other extremal problems in complex and harmonic analysis: see [Garnett, 1981], [Foias and Frazho, 1990], and [Foias, Frazho, Gohberg, and Kaashoek, 1998]. For a general form, see Exercise 3.4.15(c) and the comments in §3.5.

§5.2.1(c). *The Landau and Lebesgue constants.* Lebesgue [1906] showed that, for every $f \in C(\mathbb{T})$,

$$\left\|\sum_{|k|\le n}\widehat{f}(k)z^k\right\|_\infty \le L_{2n+1}\|f\|_\infty$$

and the constant L_{2n+1} (defined in Remark (i) of §5.2.1(c)) is sharp, and Fejér found in 1910 that $L_n = (4/\pi^2)\log(n+1) + O(1)$ as $n \to \infty$ (see [Zygmund, 1959]). The question of the exact expressions for the constants L_n and G_n (defined in §5.2.1(c)) attracted the attention of Fejér, Gronwall, Landau, Szegő, Hardy, and many others; a very good overview of the subject (with precise references) and of new approaches/results can be found in [Alzer, 2002] (see also [Lorch, 1954]). We add the following to the information given in §5.2.1(c):

$$0,98943\cdots + \frac{4}{\pi^2}\log(n+1) \le L_n \le 1 + \frac{4}{\pi^2}\log(n+1) \quad (n \ge 0)$$

(due to Galkin, 1971: see [Alzer, 2002] for a reference),

$$L_n = \frac{16}{\pi^2}\sum_{k\ge 1}\left(\frac{1}{4k^2-1}\sum_{m=1}^{(n+1)k}\frac{1}{2m-1}\right) \quad (\text{Szegő, 1921})$$

(see [Pólya and Szegő, 1925], vol. 2, VI§4),

$$\frac{4}{\pi} + \frac{A}{\log(n+1)} < \frac{L_n}{G_n} < \frac{4}{\pi} + \frac{B}{\log(n+1)} \quad (n \ge 1),$$

where $A = -1,15671\ldots, B = -0,17650\ldots$ (the constants A and B given in the form of series are known to be sharp [Alzer, 2002]).

5.8 Notes and Remarks

The short Remark §5.2.1(c(ii)) is linked to a rich subject of research around the "method of band extension" (see [Rodman, Spitkovsky, and Woerdeman, 2002]).

Lemma 5.2.2 and §5.2.3. Arveson's formula. Arveson's distance formula Lemma 5.2.2 [Arveson, 1975] plays a major role in the theory of operator algebras (the C^*-algebras, as well as the "triangular algebras": see [Davidson, 1988]), in problems of completion, and in extremal problems. The proof in the text follows a presentation of Davidson [1988].

Lemmas 5.2.3 and 5.2.4. Completion of matrices and Julia's lemma. The four-block lemma 5.2.3 was published in [Parrott, 1978], but the principal idea and important special cases were already found in [Adamyan, Arov, and Krein, 1968]; see also a comment in §2.4 above. Note that in [Arsene and Gheondea, 1982] there is a complete description of the completions M_X satisfying $\|M_X\| \leq 1$ (where X must run over an operator space ball $X = K + GYD$, $\|Y\| \leq 1$ with left and right "half-radii" $G \geq 0$, $D \geq 0$ depending on A, B, C). For more information see [Foias and Frazho, 1990]. Julia's Lemma 5.2.4 [Julia, 1944] (which already appeared in §2.4 above), *de facto*, was the first step towards the construction of unitary dilations by Szőkefalvi-Nagy and Foias (see [Szőkefalvi-Nagy and Foias, 1967] and [Nikolski, 2002b]), techniques that changed the landscape of twentieth-century operator theory.

Theorem 5.3.1 and §5.3.2. Inversion of a Toeplitz matrix. The results of Theorems 5.3.1 and 5.3.2 are somewhat isolated in the context of this chapter, but they play an important role in numerical matrix analysis; they contain the most efficient algorithm for the inversion of a Toeplitz matrix. Initially, they were proved in [Baxter and Hirschman, 1964] (including the principal observation that the inverse matrix $T_{f,n}^{-1}$ is completely defined by its first and last columns x and y, where $T_{f,n}x = 1$, $T_{f,n}y = z^n$) but under a more restrictive hypothesis with respect to Theorem 5.3.1, namely: by supposing $x(z) \neq 0$ and $y(z) \neq 0$ for $|z| < 1$. Under the current hypotheses $x(0) \neq 0$, $y(0) \neq 0$, Theorems 5.3.1 and 5.3.2 appeared in [Gohberg and Sementsul, 1972]; see also the more accessible sources by Iokhvidov [1974] and Heinig and Rost [1984]. Given their interest for applications, these results generated a substantial literature with many variations in the hypotheses and the forms of writing the inverses; see the monographs by Gohberg and Feldman [1967], Iokhvidov [1974], Heinig and Rost [1984], Prössdorf and Silbermann [1991], and the references therein. In particular, the role of the essential conditions (but not necessary, see §5.3.2(a)) $x(0) \neq 0$, $y(0) \neq 0$ have been well studied.

However, we would like to emphasize that here we have no intention of entering into a discussion of the numerous algorithms for the inversion of Toeplitz matrices. This subject has its own long history beginning with works by Levinson [1947], Durbin [1960], and Trench [1964] (who has results quite close to those of Gohberg and Sementsul [1972]); for a systematic treatment of these algorithms and for references see [Golub and Van Loan, 1996].

Note also that practically all the publications in question, from the very first [Baxter and Hirschman, 1964], contain *continuous analogs of Theorem 5.3.1 and §5.3.2*. To give an example, here is a statement from [Gohberg and Feldman, 1967].

Let $k \in L^1(-a, a)$ such that the equations

$$x - W_k^a x = P_a k, \quad y - W_{\tilde{k}}^a y = P_a \tilde{k} \quad (\tilde{k}(t) = k(-t))$$

admit solutions $x, y \in L^1(0, a)$ satisfying $1 + \widehat{x}(\lambda) \neq 0$, $1 + \widehat{y}(\lambda) \neq 0$ (Im$(\lambda) \leq 0$). Then, the operator

$$I - W_k^a : L^1(0, a) \to L^1(0, a)$$

is invertible and its inverse is

$$(I - W_k^a)^{-1} f = f + \int_0^a K(t, s) f(s)\, ds \quad (t \in (0, a)),$$

where

$$K(t, s) = x(t-s) + y(s-t) + \int_0^{\min(t,s)} x(t-r) y(s-r)\, dr - \int_a^{a+\min(t,s)} y(r-t) x(r-s)\, dr.$$

See also Exercise 5.7.3.

§5.3.2(d), triangular Toeplitz matrices. This special case, because of its holomorphic nature, is linked to the Bézout equations. More precisely (and this is well known in the model theory of operators: see e.g. [Nikolski, 1986, 2002b]), the inverse of the matrix $T_{f,n} = f(S_n)$ (see the notation of §5.3.2(d)) is $T_{f,n}^{-1} = g(S_n)$, where

$$fg + z^{n+1} h = 1 \quad (h, g \in H^\infty),$$

and

$$\|T_{f,n}^{-1}\| = \min\{\|g\|_\infty : fg + z^{n+1} h = 1\}.$$

Recall that the best known estimate for solutions of the general Bézout equation $\sum_{k=1}^m F_k G_k = 1$ is (according to Uchiyama and Wolff) of the order of

$$|G(z)| \leq a\delta^{-2} \log \frac{e}{\delta},$$

where a is a numerical constant ($a = 30$ is sufficient) and $0 < \delta \le |F(z)| =:$ $(\sum_k |F_k(z)|^2)^{1/2} \le 1$: see [Nikolski, 1986], Appendix 3. For the sharpness of this last upper bound, there exist $F = (F_1, F_2)$, $\delta \le |F| \le 1$, such that every solution G satisfies $\|G\|_\infty \ge b\delta^{-2} \log\log(c/\delta)$ where $b > 0$, $c > 0$ are numerical constants [Treil, 2002]. It would be quite interesting to make explicit the calculation in the case of $T_{f,n}^{-1} = g(S_n)$ where $F = (f, z^{n+1})$. Note also that $\|f(S_n)\| \le 1$ obviously implies $\|f(S_n)^{-1}\| \le 1/|f(0)|^n$.

5.4. Inversion by the finite section method (the term proposed in Baxter's founding article [Baxter, 1963] and now universally accepted) is one of the central themes in the theory of Toeplitz matrices and operators and their applications. In [Baxter, 1963] this was treated for T_φ on the space $l^1(\mathbb{Z}_+)$ (hence with $\varphi \in \mathcal{F}l^1(\mathbb{Z}) = W(\mathbb{T})$), with the aid of a factorization of Birkhoff–Wiener–Hopf type (a simplified realization of the technique is presented in §5.4.2(6)). For §5.4.2(7), see [Gohberg and Feldman, 1967], and for §5.4.2(8), see the fundamental work of Devinatz and Shinbrot [1969] where an abstract operator version of the Wiener–Hopf techniques is developed. This last perspective strongly influenced the ulterior evolution of the theory: see for example Chapter 7 of [Böttcher and Silbermann, 1990] and a discussion on pages 336–338 therein. Regarding the result of §5.4.2(7(ii)), note that Böttcher [1994] later showed that in the case when $\varphi \in PC(\mathbb{T})$, we have $l(T_\varphi) = \|T_\varphi^{-1}\|$ and $\overline{\lim}_n$ of the definition of $l(T_\varphi)$ is, in fact, $\lim_n$; see [Böttcher and Silbermann, 1999] Section 3.13(c).

Our presentation of inversion by the finite section method (IFSM, according to the abbreviation of §5.4) follows the classical scheme mentioned, but also advances the idea of *multipliers of IFSM* ($f \in$ IFSM $\Rightarrow \varphi f \in$ IFSM) which simplifies and clarifies several arguments. With regard to these multipliers of IFSM, it is curious that according to Theorem 3.3.6(5), a symbol φ of an invertible Toeplitz operator T_φ is always of the form $\varphi = \psi a$, where $\psi^{\pm 1} \in H^\infty$ and a is sectorial, but the inclusion $T_\varphi \in$ IFSM holds uniquely under a hypothesis of a range $\text{Ran}_{\text{ess}}(a)$ "sufficiently tight" (§5.4.2(4,5)), whereas in general the property is not correct: see the example of §5.4.3 (moreover, it is probable that even a positive real function $a^{\pm 1} \in L^\infty$ is not necessarily a multiplier of IFSM). The example of [Treil, 1987] mentioned in §5.4.3 is a fundamental element of the theory; it is quite surprising that it was discovered so late (besides the original article, proofs can also be found in [Böttcher and Silbermann, 1990] and [Böttcher and Grudsky, 1996]). For the explicit example see [Böttcher and Grudsky, 1996].

The explicit estimations of the principal quantity of *IFSM* $l(T_\varphi)$ probably seem to be a novelty in the presentation. For the detailed general theory of

IFSM (including for abstract operators on a Hilbert space, or even on a Banach space) we refer to [Gohberg and Feldman, 1967], [Böttcher and Silbermann, 1990], [Prössdorf and Silbermann, 1991], [Böttcher and Silbermann, 1999], and [Simon, 2005b]. Perhaps the greatest omission in our treatment of IFSM is the absence of the *local principles of IFSM* discovered by Kozak [1973] (inspired by the local theory of Simonenko: see §3.2) and then developed into a very efficient tool for the subject: see [Gohberg and Feldman, 1967], [Böttcher and Silbermann, 1990], [Prössdorf and Silbermann, 1991], and [Böttcher and Silbermann, 1999]. Note also that the *continuous version of IFSM* for the Wiener–Hopf operators W_k indicated in Exercise 5.7.3, namely

$$\lim_{a\to\infty}(W_k^a)^{-1}P_a f = W_k^{-1} f \quad (\forall f \in L^2(\mathbb{R}_+)),$$

is a special case of the inversion of Toeplitz operators by the finite section method with respect to an increasing family of more general model subspaces $K_{\theta_a} = H^2 \ominus \theta_a H^2$ where $\{\theta_a\}$ is a family of inner functions such that $0 < a < b$ $\Rightarrow K_{\theta_a} \subset K_{\theta_b}$ (or equivalently, θ_a *divides* θ_b) and $\mathrm{span}(K_{\theta_a} : a > 0) = H^2$; here IFSM means (by definition) that $(P_a T_\varphi | K_{\theta_a})^{-1} P_a$ exists and converges to T_φ^{-1} as $a \to \infty$. The classical case of §5.4.2 corresponds to a family $\{z^n : n = 1, 2, \ldots\}$, and the Wiener–Hopf case of Exercise 5.7.3 corresponds to a family $\theta_a = \exp(-a((1+z)/(1-z))), a > 0$. According to [Treil, 1987], each invertible operator T_φ is IFSM with respect to a certain family of subspaces $\{K_{\theta_a}\}$, but no complete theory for such an inversion is known apart from the two classical cases mentioned above. See also the comments on Exercise 5.7.3 below.

As we know by Chapter 1, the theme of the asymptotic behavior of the truncated matrices $T_{\varphi,n}$ is not limited to the convergence of the inverses $T_{\varphi,n}^{-1}$, but also manifests itself in the questions of determinants $\det(T_{\varphi,n})$, of the distribution of the spectra $\sigma(T_{\varphi,n})$, etc. Among these "*et cetera*" there lies the problem of the asymptotic behavior of the matricial elements,

$$(T_{\varphi,n}^{-1})_{j,k} \quad (0 \le j, k \le n).$$

The problem arises in numerous applications: in statistical physics (such as the Lenz–Ising model: see Chapter 1), random walks on $\mathbb{Z}$, differential equations, etc. The distinctive feature of all these applications is that their corresponding Toeplitz matrices have "singular" symbols in the sense that either $\varphi \notin L^\infty$ or $1/\varphi \notin L^\infty$. We repeat the key references already mentioned in Chapter 1: [Courant, Friedrichs, and Lewy, 1928], [Spitzer and Stone, 1960], [Kesten, 1961], [Fisher and Hartwig, 1968], [Widom, 1973], [Bleher, 1979], as well as the more recent [Böttcher and Widom, 2007], [Rambour and Rinkel, 2007], [Böttcher and Silbermann, 1999], and [Böttcher, 1995] (the last three sources contain important lists of references), and – just to give a taste of the results in the case of singular symbols and without entering into the details – we cite the classical result of Spitzer and Stone [1960] for the symbol $\varphi = |1 - z|^2 f$:

$$(T_{\varphi,n}^{-1})_{j,k} = \frac{G(x,y)}{f(1)} \cdot n + o(n) \quad \text{where } \frac{j}{n} \to x, \frac{k}{n} \to y \quad (n \to \infty),$$

and where $x, y \in]0, 1[$, f is a sufficiently smooth positive function, $f(z) > 0$, and $G(x, y)$ is the Green's function corresponding to the differential operator $\partial^2/\partial x^2$ on the interval $(0, 1)$ associated with the boundary conditions $F(0) = F(1) = 0$. See also the comments on §5.6.5–§5.6.7 below.

§5.5. *The theory of circulants*, as part of that of Toeplitz matrices/ operators, appeared already in the founding article by Toeplitz [1911a]. Initially, the circulants were known for their role in the theory of regular polygons, isoperimetric inequalities, and other subjects where the roots of unity play a role. The "theory" §5.5.1–§5.5.4 in itself is completely elementary, but Toeplitz [1911a] had already shown its efficacy in the treatment of the asymptotics of the spectra $\sigma(T_{\varphi,n})$ – a subject much less evident. For example, by an approximation by circulants, he deduced that for a positive real function φ the number of positive eigenvalues is proportional to $m\{z \in \mathbb{T}: \varphi(z) \geq 0\}$:

$$\lim_n \frac{1}{n+1} \sum_{0 \leq k \leq n} \chi_{[0,a]}(\lambda_{n,k}(\varphi)) = \int_{\mathbb{T}} \chi_{[0,a]} \circ \varphi \, dm = m\{z \in \mathbb{T}: \varphi(z) \geq 0\},$$

where $a = \|\varphi\|_\infty$, a true precursor (1911) to theorems of distribution of spectra that came a few years later (1915). On the utility of circulants for the question of distribution of spectra, see also the comments for §5.6.4 and Corollary 5.6.7 (below). The Fourier matrix §5.5.1(3) is used for the fast Fourier transform (FFT).

The principal sources on the circulants and their place in the theory of Toeplitz matrices/operators are [Grenander and Szegő, 1958], [Davis, 1994], [Gray, 2005], and [Kra and Simanca, 2012].

§5.5.5. *The Wirtinger integral inequality* (1904) is part of the theory of isoperimetric inequalities; see [Hardy, Littlewood, and Pólya, 1934], in particular §7.7, n° 258, where it is also attributed to Hurwitz [1901]. We also quote the generalizations

$$\int_0^1 |f|^2 \, dx \leq C^2 \int_0^1 |f^{(n)}|^2 \, dx,$$

for f satisfying the limit conditions $\int_0^1 f \, dx = 0$, $f^{(k)}(0) = f^{(k)}(1)$ ($0 \leq k < n - 1$), where the best constant is $C = 1/(2\pi)^n$, and a study of the dependence of this constant with respect to different choice of limit conditions in [Böttcher and Widom, 2007]. For several other generalizations see also Stechkin's Appendix V in the Russian edition (1948) of [Hardy, Littlewood, and Pólya, 1934]. In several variables, this fact is known as the *Poincaré inequality*,

$$\|u\|_{L^p(\Omega)} \le C\|\nabla u\|_{L^p(\Omega)},$$

where $\Omega \subset \mathbb{R}^n$ is a bounded open connected subset and u a smooth function with support in Ω satisfying $\int_\Omega u\,dx = 0$. The question of the exact constant C (called the *Poincaré constant*) is delicate, but for $p = 2$ and for domains Ω with smooth boundary we have $C = \lambda_{\min}^{-1/2}$, where $\lambda_{\min} > 0$ is the minimal eigenvalue of the Laplacian $-\Delta$ on the Sobolev space $W_0^{1,2}(\Omega)$. See [Acosta and Durán, 2004] for convex Ω, and for further information.

5.6. Asymptotics of determinants and spectra. In reality, this is the major theme of the theory of Toeplitz matrices/operators, the real motor of its development during a century of existence. It is already present in the very first article on the spectral theory of "Toeplitz forms" [Toeplitz, 1911a]; then Gábor Szegő's important articles [Szegő, 1915, 1920, 1921a] introduced nearly all of the important ideas of the future theory. Indeed, these articles, whose influence on future developments is difficult to overestimate, contained in a clear and explicit form the following elements (at least): (a) the links between the best approximation $\text{dist}(1, z\mathcal{P}_n)$ and the asymptotics of the D_n; (b) the links between orthogonal polynomials and Toeplitz forms, and in particular, the identification of the zeros of these polynomials with the spectrum of the nth truncation of the form; (c) the introduction of the techniques of equidistribution (proposed earlier by Bohl (in 1909), Sierpinski (in 1910), and (especially) Weyl (1910)) in the problem of the asymptotics of the spectra $\sigma(T_{\varphi,n})$.

§5.6.1. The first Szegő formula and its proof are taken from [Szegő, 1915, 1920]. The latter article, among other things, cites an interesting formula of Pólya and Landsberg ([Pólya, 1914]; [Landsberg, 1910], page 231) for the determinant $D_n(\mu)$:

$$D_n(\mu) = \frac{1}{(n+1)!} \int_{\mathbb{T}^{n+1}} \prod_{0 \le i < j \le n} |\zeta_i - \zeta_j|^2 \, d\mu(\zeta_0)\, d\mu(\zeta_1) \ldots d\mu(\zeta_n).$$

See also the comments on Theorem 5.6.3 below.

§5.6.2. Equidistribution of sequences and the associated techniques are very useful in spectral analysis and dynamics. The principal definitions were invented (independently) by Bohl (in 1909), Sierpinski (in 1910), and Weyl [1910b], but the theory is usually associated with Weyl, given the spectacular success he obtained when applying it to the distribution of spectra in perturbation theory, to harmonic analysis, and to number theory. See [Kuipers and

Niederreiter, 1974] for the current state of the theory and its applications; for applications to Toeplitz matrices, see [Grenander and Szegő, 1958] and [Gray, 2005].

Theorem 5.6.3 and Corollary 5.6.4. Asymptotic distribution of the $\sigma(T_{\varphi,n})$. These fundamental classical theorems and their proofs are from [Szegő, 1920]. As we have already noted, the formula of the density of spectra in Corollary 5.6.4 was proved in [Toeplitz, 1911a] by an asymptotic comparison with circulants (see §5.6.4 and Theorem 5.6.6 for this method).

A very interesting link of Theorem 5.6.3 and §5.6.1 with the "sharp Littlewood conjecture" (SLiC) is found in [Klemes, 2001]. The SLiC (still open today) affirms that the Dirichlet kernel $\mathcal{D}_n(e^{it}) = \sum_{k=0}^{n} e^{ikt}$ is extremal for the norm $\|\cdot\|_{L^p(\mathbb{T})}$ among all the trigonometric polynomials with unimodular coefficients: if

$$F = F_\sigma = \sum_{k \in \sigma} \epsilon_k z^k, \quad |\epsilon_k| = 1 \quad \text{and} \quad \text{card}(\sigma) = n+1, \quad \sigma \subset \mathbb{Z},$$

then

$$\|\mathcal{D}_n\|_{L^p(\mathbb{T})} \leq \|F\|_{L^p(\mathbb{T})} \quad \text{for } 0 \leq p \leq 2,$$

and

$$\|\mathcal{D}_n\|_{L^p(\mathbb{T})} \geq \|F\|_{L^p(\mathbb{T})} \quad \text{for } 2 \leq p \leq \infty;$$

by definition,

$$\|F\|_0 = G(|F|) = \exp\left(\int_\mathbb{T} \log|F|\, dm\right).$$

The classical Littlewood conjecture demanded to show that $c\|\mathcal{D}_n\|_{L^1(\mathbb{T})} \leq \|F\|_{L^1(\mathbb{T})}$ with a constant $c > 0$ independent of n and F; this conjecture was confirmed by Konyagin (in 1981) and McGehee, Pigno, and Smith (in 1981); see for example [DeVore and Lorentz, 1993]. Klemes observed that the SLiC is a consequence of a simple application of Theorem 5.6.3 (with $f(\lambda) = \lambda^p$, $\lambda > 0, 0 < \lambda < \infty$) and a new "Toeplitz SLiC" (TSLiC):

$$\|T_{|\mathcal{D}_n|^2, N}\|_{\mathfrak{S}_p} \leq \|T_{|F|^2, N}\|_{\mathfrak{S}_p} \quad \text{for } 0 \leq p \leq 1 \ (\forall n, N \geq 1),$$

and

$$\|T_{|\mathcal{D}_n|^2, N}\|_{\mathfrak{S}_p} \geq \|T_{|F|^2, N}\|_{\mathfrak{S}_p} \quad \text{for } 1 \leq p \leq \infty \ (\forall n, N \geq 1);$$

here $\|A\|_{\mathfrak{S}_p} = \left(\sum_{k=0}^n \lambda_k(A)^p\right)^{1/p}$ is the Schatten–von Neumann norm of a positive definite matrix A ($\lambda_k(A)$ are the eigenvalues). Then, he proved the TSLiC for $0 \leq p \leq 2$ and for F_σ of the following particular form:

$$\sigma = [0, n], \quad \epsilon_k = \pm 1.$$

The principal step of the proof is an upper bound on determinants: for every $\lambda \geq 0$,

$$\det(\lambda I_N + T_{|\mathcal{D}_n|^2, N}) \leq \det(\lambda I_N + T_{|F|^2, N}).$$

We briefly sketch how we can then deduce the TSLiC and the SLiC: applying

$$\lambda^p = a_p \int_0^\infty t^{-1-p} \log(1 + t\lambda)\, dt, \quad 0 < p < 1, \quad a_p^{-1} = \int_0^\infty t^{-1-p} \log(1 + t)\, dt,$$

we obtain $\|T_{|\mathcal{D}_n|^2, N}\|_{\mathfrak{S}_p} \leq \|T_{|F|^2, N}\|_{\mathfrak{S}_p}$ for $0 \leq p \leq 1$, and applying

$$\lambda^p = b_p \int_0^\infty t^{-1-p}\{t\lambda - \log(1 + t\lambda)\}\, dt \quad (1 < p < 2),$$

where

$$b_p^{-1} = \int_0^\infty t^{-1-p}\{t\lambda - \log(1 + t)\}\, dt,$$

we have $\|T_{|\mathcal{D}_n|^2, N}\|_{\mathfrak{S}_p} \geq \|T_{|F|^2, N}\|_{\mathfrak{S}_p}$ for $1 \leq p \leq 2$. Next, by §5.6.1,

$$\int_{\mathbb{T}} \log(\lambda + |\mathcal{D}_n|^2)\, dm \leq \int_{\mathbb{T}} \log(\lambda + |F|^2)\, dm \quad (\lambda \geq 0),$$

and then, by Theorem 5.6.3, $\|\mathcal{D}_n\|_{L^p(\mathbb{T})} \leq \|F\|_{L^p(\mathbb{T})}$ for $0 \leq p \leq 2$, $\|\mathcal{D}_n\|_{L^p(\mathbb{T})} \geq \|F\|_{L^p(\mathbb{T})}$ for $2 \leq p \leq 4$ (hence, the SLiC for $0 \leq p \leq 4$). ∎

We conclude by noting that [Klemes, 2001] provides even more interesting information on the matrices $T_{|\mathcal{D}_n|^2, N}$ and on the TSLiC.

By a tradition dating back to Toeplitz (1911), the authors sometimes examine the question of the distribution of the *spectra of quadratic forms* $f \mapsto (T_\varphi f, f)$ $(f \in \mathcal{P}_n)$ in the space $\mathcal{P}_n$ equipped with a norm defined by another Toeplitz form $(T_\psi f, f) =: (f, f)_\psi$ (positive definite). The "operator" point of view permits us to resolve the question "for free." Indeed, this concerns an operator $A_n: \mathcal{P}_n \to \mathcal{P}_n$ with bilinear form $(A_n f, g)_\psi = (T_\varphi f, g)$, giving $A_n = T_{\psi,n}^{-1/2} T_{\varphi,n} T_{\psi,n}^{-1/2}$, and hence

$$\sigma(A_n) = \{\lambda \in \mathbb{C}: \det(\lambda I_n - A_n) = 0\} = \{\lambda \in \mathbb{C}: \det(\lambda T_{\psi,n} - T_{\varphi,n}) = 0\},$$

then

$$\det(A_n) = \det(T_{\varphi,n})/\det(T_{\psi,n}), \quad \lim_n \det(A_n)^{1/n+1} = G(\varphi)/G(\psi) = G(\varphi/\psi),$$

and

$$\lim_n \frac{1}{n+1} \sum_{j=0}^n f(\lambda_{n,j}(A_n)) = \int_0^{2\pi} f \circ (\varphi/\psi)\, dm.$$

5.8 Notes and Remarks

See [Szegő, 1920] and [Grenander and Szegő, 1958], where the question is resolved by repeating (with the necessary adaptations) the calculations already effected in the case of the Euclidean norm.

To conclude with the asymptotic distribution of spectra, note that this part of Toeplitz theory has recently found important new applications to asymptotic behavior of the two-dimensional Lenz–Ising models. This led to an explosion of research and results on the $\sigma(T_{\varphi,n})$ and $\det(T_{\varphi,n})$, in particular on the asymptotics of the individual eigenvalues $\lambda_{n,k}$ (and not only the extremal values $\lambda_{n,0}$, $\lambda_{n,n}$ as before), e.g. for $k/n \to x$. Without trying to cite and/or analyze the results, note that it was shown that for φ with two values ± 1 (as mentioned in the Remarks on Theorem 5.6.3 and Corollary 5.6.4), in every interval $(-1+\epsilon, 1-\epsilon)$, $\epsilon > 0$, there are $\approx \log(n+1)$ eigenvalues $\lambda_{n,k}$ distributed uniformly (with an increment $\approx 2/\log(n+1)$); for the rest, see for example [Deift, Its, and Krasovsky, 2013] (and its 12-page list of references!) as well as [Böttcher, Bogoya, Grudsky, and Maximenko, 2017] for one of the best introductory texts on the subject.

Example 5.6.5. This very nice calculation is from [Grenander and Szegő, 1958].

§5.6.4 and Corollary 5.6.7. Circulants and the distribution of spectra. The idea of using circulants for the study of the distribution of spectra comes from [Toeplitz, 1911a] where (among others) the special case of Corollary 5.6.4 was settled, by showing that for φ positive and regular and $-\infty < c \leq \varphi \leq C < \infty$,

$$\lim_n \frac{1}{n+1} \text{card}\{k: \lambda_{n,k}(T_{\varphi,n}) \geq 0\} = m\{z: \varphi(z) \geq 0\}.$$

A remarkable point of this approach is that it works well for the complex-valued symbols. Our presentation follows [Grenander and Szegő, 1958] up to the modifications necessary for the passage from positive symbols to complex symbols (in particular, in point (1) of Theorem 5.6.6).

Historically, the first results on the asymptotics of the spectra $\sigma(T_{\varphi,n})$ in the complex case are due to [Widom, 1976] (with different techniques).

§5.6.5–§5.6.7. The "strong Szegő theorem" and its evolution under the influence of theoretical physics. This is one of the main themes of Toeplitz analysis of the last few decades, which greatly stimulated the development of the subject by attracting dozens of mathematicians. A masterly panorama of the corresponding research has been drawn by Böttcher [1995]. This history is very instructive as an example of the direct interaction between theoretical physics and classical analysis, as the problem is rooted in the statistical theory of gases, more precisely in the "two-dimensional Lenz–Ising model" and the behavior of spontaneous magnetization below the critical temperature (the

"Curie point," or the phase transition). For references, see [Böttcher, 1995] and the Wikipedia entry for "Ising model"; note that a formula for such behavior was conjectured by Onsager in 1948, then completed in 1950 by the indication of a possible usage of Toeplitz determinants. In fact, this was G. Szegő, who justified in 1952 [Szegő, 1952] Onsager's formula for the measures $\mu = \varphi m$ with a rather smooth density $\varphi > 0$:

$$\frac{1}{n+1} \sum_{k=0}^{n} \log(\lambda_{n,k}) = \int_{\mathbb{T}} \log(\varphi) \, dm + \frac{E(\varphi)}{n+1} + o(1/n),$$

with a proof via the phenomenal computation presented in Exercise 5.7.4 in the case where $1/\varphi$ is a positive polynomial, followed by a polynomial approximation of $(\log(\varphi))'$. Later on, the formula was justified under the least restrictive conditions possible (just the existence of the quantities involved) [Golinskii and Ibragimov, 1971], always in the case of a positive measure μ. In the case where φ is a smooth function, the remainder $o(1/n)$ of the asymptotics can have a qualified order of decrease, for example for $\beta > 1/2$ ($\beta \notin \mathbb{N}$) and $\varphi \in C^\beta$ (i.e. the derivative $f^{([\beta])}$ satisfies the condition Lip($\beta - [\beta]$)) the remainder is $O(1/n^{2\beta})$ as $n \to \infty$ [Böttcher and Widom, 2006].

In the same way that Szegő's first asymptotics formula §5.6.1 led to a law of distribution of spectra, namely Theorem 5.6.3, Theorem 5.6.8 implies

$$\frac{1}{n+1} \sum_{k=0}^{n} f(\lambda_{n,k}) = \int_{\mathbb{T}} f \circ \varphi \, dm + \frac{E(f, \varphi)}{n+1} + o(1/n),$$

where $E(f, \varphi)$ is a constant to be defined. For $f(z) = z^k$ ($k = 0, 1, \ldots$), it was found by Kac [1954] and for every f holomorphic and $\varphi' \in W(\mathbb{T})$ by Libkind [1972], see Exercise 5.7.6 for this formula. The particular values of $E(f, \varphi)$ remained somewhat mysterious until Widom [1976] found a proof and much more illuminating expressions (in particular, for the first time, the formulas were justified for complex-valued symbols φ and under much less restrictive hypotheses); principally, these are the arguments given in Theorem 5.6.8 and §5.6.5.4 above. Among others, Widom [1976] provided a new derivation of Libkind's formula, Exercise 5.7.6 [Libkind, 1972], which is explained as a consequence of Theorem 5.6.8 and Cauchy's formula, hence passing by

$$\frac{1}{n+1} \sum_{0 \leq j \leq n} f(\lambda_{n,j}) = \int_0^{2\pi} f \circ \varphi \, dm - \frac{1}{2\pi i(n+1)} \int_{\partial \Omega} f(\lambda) L(\lambda) \, d\lambda + o(1/n)$$

when $n \to \infty$ and where

$$L(\lambda) = \frac{d}{d\lambda} \log \det(T_{\varphi - \lambda} T_{1/\varphi - \lambda}),$$

Ω is a bounded open set containing $\sigma(T_\varphi)$ and $\sigma(T_{\tilde\varphi})$, and f is holomorphic.

5.8 Notes and Remarks

The research that followed and/or accompanied the results described previously concentrated, principally, on the three following themes:

(i) new approaches to qualified asymptotics of spectra (Szegő-type strong theorem),
(ii) new classes of symbols φ, especially singular symbols (where $\inf |\varphi| = 0$ and/or $\sup |\varphi| = \infty$),
(iii) asymptotics of matrix elements of the inverses $T_{\varphi,n}^{-1}$.

For theme (i), this is especially about the systematic usage of general operator tools for an analysis of finite matrices (as is already the case in §5.6.5.4 and Exercise 5.7.8). A somewhat dramatic history of what is called the "Borodin–Okounkov formula" (Exercise 5.7.8) is described in detail in [Böttcher, 1995]. The question of linking the Toeplitz determinants with the general theory of Fredholm determinants $\det(I - K)$, $K \in \mathfrak{S}_1$ was brought up in 1999 by Its and Deift and resolved by Borodin and Okounkov [2000]. Then, several other proofs were found. (That of Exercise 5.7.8, taken from [Böttcher, 2001], is perhaps the most elementary. It is based on an earlier argument by Basor and Widom: see [Böttcher, 2001] for references.) Finally, it became clear that the formula and its links with the Szegő asymptotics were already known in 1979 (in a form nonetheless somewhat implicit) [Geronimo and Case, 1979].

Theme (ii), as with all of the story of the second term of the Szegő asymptotics, emerged at the demand of statistical physics where (always on the Lenz–Ising model: see Chapter 1) Toeplitz matrices with singular symbols appeared for the case of the critical temperature (of phase transition) or greater; in fact, the symbols of "Fisher–Hartwig type"

$$\varphi(\zeta) = \prod_{j=1}^{M} (\omega_{j,\beta_j}(\zeta)|\zeta - \zeta_j|^{2\alpha_j})$$

where $|\zeta_j| = 1$ (pairwise distinct), $\text{Re}(\alpha_j) > -1/2$ and each ω_{j,β_j} has a single singularity – a jump at point ζ_j having one-sided limits equal to $e^{\pm i\pi\beta_j}$. Fisher and Hartwig [1968] conjectured that for a symbol of this type, we have

$$\det(T_{\varphi,n}) = (1 + o(1)) \times G(\varphi)^{n+1}(n+1)^{\sum_j(\alpha_j^2 - \beta_j^2)} \exp(\tilde{E}(\varphi)),$$

where $\tilde{E}(\varphi)$ is a constant depending on φ, or equivalently

$$\frac{1}{n+1} \sum_{k=0}^{n} \log(\lambda_{n,k}) = \int_{\mathbb{T}} \log(\varphi)\, dm + q \frac{\log(n+1)}{n+1} + \frac{\tilde{E}(\varphi)}{n+1} + o(1/n),$$

where $q = \sum_j(\alpha_j^2 - \beta_j^2)$ (an exponent whose nature remains somewhat mysterious). The Fisher–Hartwig conjecture was successively confirmed by Widom [1973] ($\beta_j = 0$, $\forall j$), Basor (1978 (see Notes and Remarks)), $\text{Re}(\beta_j) = 0$, $\forall j$;

1979, $\alpha_j = 0$ and $|\mathrm{Re}(\beta_j)| < 1/2$, $\forall j$ and Böttcher and Silbermann (1985), $|\mathrm{Re}(\alpha_j)| < 1/2$ and $|\mathrm{Re}(\beta_j)| < 1/2$, $\forall j$; this last result is, in a sense, the best possible). We refer to the survey articles by Böttcher [1995] and Deift, Its, and Krasovsky [2013] for details and numerous references.

Theme (iii) mentioned above also arose from problems concerning the two-dimensional Lenz–Ising model since other matrix elements $(T_{\varphi,n}^{-1})_{jk}$, $0 \le j, k \le n$ of the matrices $T_{\varphi,n}$ also have physical interpretations (and not just $(T_{\varphi,n}^{-1})_{0,0} = D_{n-1}(\varphi)/D_n(\varphi)$). Mathematically, the problem (probably) appeared in [Fisher and Hartwig, 1968], however, at the "physics level" it was also treated by Wu (in 1966); see [Deift, Its, and Krasovsky, 2013] for the precise references. Of course, if the asymptotics of the $(T_{\varphi,n}^{-1})_{jk}$, $0 \le j, k \le n$ is uniform in j, k, this also implies the asymptotics of the traces and the distribution of the spectra. We note a recent publication by Rambour and Seghier [2012] as well as the numerous sources mentioned therein and in the comments on §5.4 above (where a typical result is also cited).

§5.6.5.1. The Krein–Dirichlet algebra: §5.6.5.1(1) is a formula from antiquity, at least according to Douglas [1931] §2.1. The Krein algebra $K(\mathbb{T})$ and the theorem in §5.6.5.1(2) are introduced in [Krein, 1966].

Theorem 5.6.8 and §5.6.5.4 are borrowed from [Widom, 1976] and have already been discussed above. Once again we emphasize the principal idea behind the formal calculation in the proof of Theorem 5.6.8(1): the decomposition into four blocks

$$T_{\varphi,n+1+m} = \begin{pmatrix} A & B \\ C & D \end{pmatrix}$$

reveals a structure with a surprising interpretation: the north-west and south-east parts A and D are Toeplitz (hence "large" operators, the principal diagonal portion of $T_{\varphi,n+1+m}$) and the north-east and south-west parts B and C are Hankel (hence "small" operators, making a negligible contribution to $T_{\varphi,n+1+m}$). This unexpected point of view and its technical consequences led to a proof (presented in the text) remarkable in its clarity and transparency. The idea of using the Helton–Howe formula of Lemma 5.6.9 is also due to Widom [1976]. For a simplified treatment of the second-term Szegő asymptotics, see also [Böttcher and Silbermann, 1999], pages 125–126.

Lemma 5.6.9, §5.6.7, and Exercise 5.7.7. These results are published in the famous *Proceedings of the Conference on Operator Theory (Halifax, 1973)* (vol. 345 of Lecture Notes in Mathematics, Springer (1973)), which played an important role in the development of operator theory during the final decades of the twentieth century (containing, among other things, the "BDF (Fredholm)

5.8 Notes and Remarks

theory" of Brown, Douglas, and Fillmore [1973] for essentially normal operators T, which are defined by $[T, T^*] \in \mathfrak{S}_\infty$). In particular, Lemma 5.6.9 and Exercise 5.7.7 appeared in [Helton and Howe, 1973] and §5.6.7 in [Helton and Howe, 1973] and [Berger and Shaw, 1973].

§5.7. Exercises (a few supplements to chapter 5).

Exercise 5.7.0. Basic exercises. The volumes, determinants, distances and Jacobians of changes of variables are tightly linked according to their origins. For their usage in approximation theory (and especially for the "best approximations"), see [Akhiezer, 1965], [DeVore and Lorentz, 1993], where, among others, the references to the original publications of Müntz and Szász can be found. The Hadamard inequality appeared in [Hadamard, 1893].

Exercise 5.7.1. Part (b) is an alternative proof of Theorem 5.1.1 taken from [Carathéodory, 1911] and [Toeplitz, 1911b] and based on part (a) by Herglotz [1911]. Schur's theorem of Exercise 5.7.1(c) [Schur, 1917, 1918] was long underestimated by the community of analysts (see for example the comments in [Privalov, 1941] Chapter II, §11.1), but finally became (along with the theorems of Carathéodory–Fejér and Nevanlinna–Pick) a basis and the model to follow for a vast theory (well appreciated by electrical and electronic engineers) of extremal problems resolved by the method of matricial inequalities: see [Rosenblum and Rovnyak, 1985], [Foias and Frazho, 1990], [Gohberg, Goldberg, and Kaashoek, 1990], and [Foias, Frazho, Gohberg, and Kaashoek, 1998]. We remark in passing that the principal point of the article [Schur, 1917, 1918] is not in Exercise 5.7.1(c) but in what is now known as "Schur's algorithm" and the "Schur parameters" of holomorphic functions; see [Simon, 2005b] for the current state of the method.

Exercise 5.7.2(i) is a detail from [Widom, 1976].

Exercise 5.7.3. Inversion of Wiener–Hopf operators by the finite section method. In principle, this is quite a vast theory, parallel to (but independent from) that of IFSM for Toeplitz; see [Gohberg and Feldman, 1967] and [Devinatz and Shinbrot, 1969] (where the theorem of Exercise 5.7.3 is proved), and then [Böttcher and Silbermann, 1990], [Prössdorf and Silbermann, 1991], and [Böttcher, Karlovich, and Spitkovsky, 2002].

Exercise 5.7.4. The case $1/\varphi \in \mathcal{P}$ where the asymptotics become an equality. This is part of Szegő's original approach [Szegő, 1952] to the proof of Onsager's conjecture on the second term asymptotics for Szegő determinants. After this incredible calculation for the case where $1/\varphi \geq 0$ is a trigonometric polynomial, Szegő passed to an approximation of functions satisfying

$(\log(\varphi))' \in \text{Lip}(\alpha)$, $0 < \alpha < 1$, to justify for this case the same asymptotics for $D_n(\varphi)$.

Exercise 5.7.5(a). The Cauchy determinants were introduced and calculated in the famous mathematics course by Cauchy [1841] (there are four volumes in total, edited between 1840 and 1847) as tools to resolve the problem of rational interpolation $F(-b_k) = c_k$ where $F = \sum_{j=1}^{n} (x_j/(z - a_j))$; see also [Pólya and Szegő, 1925] vol. 2. Given the importance of Cauchy and "Cauchy-like" matrices $C = ((r_j s_k/(a_j + b_k)))_{jk}$, there exists a vast literature on algorithmic computations of $\det C$ and the inverses C^{-1}, as well as for even more general matrices with "displacement structure" $XC - CY = r^{\text{col}} \cdot s$.

Exercise 5.7.5(b). The Hilbert determinants H_n were calculated in [Hilbert, 1894] (to minimize $\int_a^b p(x)^2 \, dx$ among the polynomials $p \neq 0$ with integer coefficients); it follows, in particular, that $1/H_n$ is an integer. Using Stirling's formula, we can show that $H_n = a_n n^{-1/4} (2\pi)^n 4^{-n^2}$, where $\lim_n a_n = e^{1/4} 2^{1/12} A^{-3} \approx 0,6450\ldots$ (A is called the Glaisher–Kinkelin constant); see [Choi, 1983].

Exercise 5.7.6. Libkind's formula. The references for this formula for the second term of the Szegő asymptotics are given in the text. Note only that between the striking result of Szegő [1952] (the case $f(z) = \log(1 + uz)$) of the formula in Exercise 5.7.6) and Libkind's formula lies an influential article of Kac [1954] that treated the case $f(z) = z^n$, $n = 0, 1, \ldots$, reinterpreted the result in probabilistic terms, and provided the following *analog of Theorem 5.6.8 for the Wiener–Hopf operators* $\tilde{W}_k^{2a} f = (k * f)|(-a, a)$, $f \in L^2(-a, a)$, $a > 0$:

$$\lim_{a \to \infty} D_a(\delta_0 - \lambda k) G^{-a} = \exp\left(\int_0^\infty x \left| \int_\mathbb{R} \log(1 - \lambda F(t)) e^{itx} \frac{dt}{2\pi} \right|^2 dx \right),$$

where $D_a(\delta_0 - \lambda k) =: \det(I - \lambda \tilde{W}_k^{2a})$ (the "Fredholm determinant": see Appendix E),

$$G = G(\delta_0 - \lambda k) =: \exp\left(\frac{1}{\pi} \int_0^\infty \log(1 - \lambda F) \, dt \right),$$

$F = \widehat{k}$, $\lambda \in \mathbb{R}$ is sufficiently small, and $k \geq 0$ is an even function such that

$$\int_\mathbb{R} k \, dx = 1, \quad \int_\mathbb{R} |x| k(x) \, dx < \infty.$$

Exercise 5.7.7. The Helton–Howe formula. The original proof [Helton and Howe, 1973] is different; that of the text is taken from [Simon, 2005b], where, in turn, it is a modification of the proof by Ehrhardt [2003]. The formula in Exercise 5.7.7 was used in several other non-commutative situations, for example, in the Pincus–Xia theory of "principal functions" of "almost normal" operators [Xia, 1983].

Exercise 5.7.8(a). The Borodin–Okounkov formula has already been discussed above (in the general overview of §5.6.5–§5.6.7), with its references.

Exercise 5.7.8(b) comes from [Böttcher and Widom, 2006] and [Böttcher, 2001].

We conclude these comments on Chapter 5 by repeating that it is merely a quick introduction to the vast theory of Toeplitz matrices and the associated "structured" matrices (band matrices, Hankel matrices, block Toeplitz and Hankel matrices, etc.). Here is a list (no doubt also incomplete) of research subjects (or immense theories of great importance for applications) that are not found in the pages of this handbook.

(1) *Complexity of algorithms* for the calculation of the eigenvalues $\lambda_{n,k}(T_\varphi)$, the elements $(T_{\varphi,n}^{-1})_{ij}$, the inverses $T_{\varphi,n}^{-1}$, etc. See [Heinig and Rost, 1984], [Bini, 1995], [Kailath and Sayed, 1999] [Gray, 2005], and [Chan and Jin, 2007].

(2) *"Generalized Toeplitz"* matrices of all kinds, such as

$$(\varphi z^k, z^l)_{L^2(\mu)} \quad (k, l \in \mathbb{Z}_+),$$

where μ is a positive Borel measure on $\mathbb{C}$, or even more generally,

$$(\varphi f_k, f_l)_{L^2(X,\mu)}$$

where (f_k) is a sequence of functions in $L^2(X, \mu)$ $((X, \mu)$ a general measure space), orthogonal or not. The corresponding theories were initiated by Szegő [Grenander and Szegő, 1958] and extensively developed (including results on the asymptotic distribution of spectra) for μ supported by $\mathbb{R}$ (Hankel matrices) or by a rectifiable curve, or when the (f_k) are systems of classical polynomials (Hermite, Bessel, Haar).

(3) *Upper/lower bounds for the "condition numbers"* $CN(T_{\varphi,n}) =: \|T_{\varphi,n}\| \cdot \|T_{\varphi,n}^{-1}\|$ of the $T_{\varphi,n}$. The theory of *pseudo-spectra*: for $\epsilon > 0$, the ϵ-*pseudo-spectrum* $\sigma_\epsilon(A)$ is the set

$$\{\lambda \in \mathbb{C} : \|\lambda x - Ax\| < \epsilon \|x\| \text{ for a certain } x\}.$$

See [Böttcher and Silbermann, 1999], [Böttcher and Grudsky, 2005], and [Trefethen and Embree, 2005].

(4) *The theorems of distribution of spectra* in their natural generality: for $T_{\varphi,n}$ with matrix-valued φ (results of this type exist, even for the strong Szegő asymptotics, beginning with [Widom, 1976]); for $T \in \text{alg}(\mathcal{T}_X)$ where

$X = C(\mathbb{T})$, $PC(\mathbb{T})$, $H^\infty + C(\mathbb{T})$, etc. (in particular, in this handbook we have not touched on the asymptotic distributions of the *singular numbers* $s_k(T_{\varphi,n})$ of the $T_{\varphi,n}$, which correspond to $\lambda_k(T^*_{\varphi,n}T_{\varphi,n}) =: s_k(T_{\varphi,n})^2$); for continuous analogs of the $T_{\varphi,n}$ (the truncated Wiener–Hopf operators W^a_k). For these subjects see [Böttcher and Silbermann, 1990], [Böttcher and Silbermann, 1999], and [Böttcher, Karlovich, and Spitkovsky, 2002].

Appendix A
Key Notions of Banach Spaces

We refrain from recalling the basic definitions of linear algebra and functional analysis, and refer the reader, for example, to [Rudin, 1991]; the more specialized topics presented below can also be found in the same source. By default, every vector space is over the complex field $\mathbb{C}$, and every normed space is separable (if the contrary does not follow from the context, as, for example, for the spaces l^∞, L^∞, etc.).

In what follows, the letter X denotes a Banach space (a complete normed space). If $A \subset X$, Vect(A) denotes the linear hull of A, and $\text{span}_X(A)$ denotes the closure of Vect(A).

A.1 Duality

For a normed space X, X^* denotes its *dual space*, i.e. the space of bounded linear functionals $L \colon X \to \mathbb{C}$ equipped with the norm

$$\|L\| = \sup\{|L(x)| \colon x \in X, \|x\| \leq 1\}.$$

X^* is always a Banach space. For reasons of symmetry we also use the notation

$$\langle x, L \rangle = L(x) \quad (x \in X, L \in X^*).$$

- *The Hahn–Banach Theorem (1932).*
  (i) Let $E \subset X$ be a vector subspace and $L_0 \in E^*$. Then there exists $L \in X^*$ such that $L|E = L_0$ and $\|L\| = \|L_0\|$.
  (i) Let $E \subset X$ be a vector subspace and $x \in X$. Then, for every $L \in X^*$ such that $L|E = 0$, we have $|\langle x, L \rangle| \leq \text{dist}_X(x, E) \cdot \|L\|$. If $x \notin \overline{E}$ then there exists $L \in X^*$ such that $L|E = 0$ and $1 = |\langle x, L \rangle| = \text{dist}_X(x, E) \cdot \|L\|$.

- *Corollary (norm in duality)*. For every $x \in X$,

$$\|x\| = \sup\{|\langle x, L\rangle| : L \in X^*, \|L\| \leq 1\}.$$

- *Corollary (approximation by linear combinations)*. Let $A \subset X$, $x \in X$. The following assertions are equivalent:
  (i) $x \in \mathrm{span}_X(A)$.
  (ii) $\forall L \in X^*,\ L|A = 0 \Rightarrow L(x) = 0$.

- *Corollary (norm of a restriction)*. Let $L \in X^*$, $E \subset X$ be a vector subspace and $E^\perp = \{L \in X^* : L|E = 0\}$ (the annihilator of E). Then

$$\|L|E\| = \mathrm{dist}_{X^*}(L, E^\perp).$$

- *Weak topologies*. A *base of the "weak topology"* $\sigma(X, X^*)$ is defined by

$$\{x \in X : |\langle x - x_0, L_j\rangle| < \epsilon,\ j = 1, \ldots, n\} \text{ where } n \in \mathbb{N},\ \epsilon > 0,\ L_j \in X^*,\ x_0 \in X.$$

A *base of the "weak-star (weak-*) topology"* $\sigma(X^*, X)$ is defined by

$$\{L \in X^* : |\langle x_j, L - L_0\rangle| < \epsilon,\ j = 1, \ldots, n\} \text{ where } n \in \mathbb{N},\ \epsilon > 0,\ x_j \in X,\ L_0 \in X^*.$$

- *Weak-* convergence*. Let X be a Banach space and let $A \subset X$ be such that $\mathrm{span}_X(A) = X$. A countable sequence (L_k) converges $\sigma(X^*, X)$ to 0 if and only if $\sup_k \|L_k\| < \infty$ and $\lim_k L_k(x) = 0\ \forall x \in A$.

- *Weak-* compactness*. The ball $\overline{B}_{X^*}(0, 1) = \{L \in X^* : \|L\| \leq 1\}$ is $\sigma(X^*, X)$ compact.

- *Reflexivity*. For every $x \in X$, the formula $j(x)L = \langle x, L\rangle$, $L \in X^*$, defines a functional $j(x) \in (X^*)^*$ such that $\|j(x)\| = \|x\|$. The space X is said to be *reflexive* if $j(X) = (X^*)^*$; X is reflexive if and only if the ball $\overline{B}_X(0, 1) = \{x \in X : \|x\| \leq 1\}$ is $\sigma(X, X^*)$ compact.

A.2 Sum of Subspaces

Let $E_j \subset X$, $j = 1, 2$, be closed subspaces such that $E_1 \cap E_2 = \{0\}$. The sum $E_1 + E_2 = \{x_1 + x_2 : x_j \in E_j\}$ is closed if and only if there exists $c > 0$ such that

$$\frac{1}{c}(\|x_1\| + \|x_2\|) \leq \|x_1 + x_2\| \leq c(\|x_1\| + \|x_2\|) \quad (\forall x_j \in E_j),$$

and if and only if the linear mapping $P: E_1 + E_2 \to E_1$, $x_1 + x_2 \mapsto x_1$ is bounded (the *projection onto E_1 parallel to E_2*).

- *Complemented subspaces.* A closed subspace $E \subset X$ is said to be *complementable* if there exists a subspace $F \subset X$ such that $E \cap F = \{0\}$ and $E + F = X$ (F is then said to be a *complement* of E in X).

A.3 Subspaces of Finite Dimension or Codimension

A vector subspace $E \subset X$ is said to be of *finite codimension* if the quotient space X/E is finite-dimensional; we denote $\dim(X/E) = \mathrm{codim} E$.

- *A vector subspace $E \subset X$ is of finite codimension $n = \mathrm{codim} E < \infty$* if and only if there exist linear functionals L_j on X, $1 \le j \le n$, such that

$$E = \bigcap_j \mathrm{Ker} L_j,$$

and if and only if there exists a subspace $F \subset X$, $\dim F = n$, such that $E + F = X$ (e.g. $F = \mathrm{Vect}(x_j \in X: L_j(x_j) = 1, 1 \le j \le n)$); E is closed if and only if there exists $L_j \in X^*$ *satisfying* $E = \bigcap_j \mathrm{Ker} L_j$ (for $n = 1$, this is the Hahn–Banach theorem above).

- A vector subspace $E \subset X$ of finite dimension, $\dim E < \infty$, is always closed.

- If $E \subset X$ is a closed subspace and $F \subset X$ is finite-dimensional, then $E + F$ is closed.

- A closed subspace $E \subset X$ of finite dimension or codimension is always complimentable.

A.4 Schauder Bases

A pair of sequences $(x_j)_{j \ge 1} \subset X$, $(f_j)_{j \ge 1} \subset X^*$ is called a *Schauder basis of X* if $\langle x_j, f_k \rangle = \delta_{jk}$ and for every $x \in X$, $x = \sum_{j \ge 1} \langle x, f_j \rangle x_j$, where the series converges in norm: $\lim_n \|x - \sum_{j=1}^n \langle x, f_j \rangle x_j\| = 0$.

Many classical spaces (always separable) possess a Schauder basis, such as $X = L^p(\Omega, \mu)$ and the Hardy spaces H^p for $1 \le p < \infty$, but there are subspaces of l^p, $p \ne 2$, $X \subset l^p$ that do not have any (discovered by Enflo in 1973). If the series above converges for every permutation of terms

($\lim_\sigma \|x - \sum_{j \in \sigma} \langle x, f_j \rangle x_j\| = 0$, where σ runs over the filter of finite subsets of $\mathbb{N}$ ordered by inclusion), we speak of an *unconditional basis*. Every normalized unconditional basis ($\|x_j\| = 1$) in a Hilbert space is a Riesz basis (i.e. it is the range of an orthonormal basis by an isomorphism of the space). See [Pietsch, 2007] for more information.

A.5 Notes and Remarks

All the above information is standard. See also [Dunford and Schwartz, 1958] or [Lindenstrauss and Tzafriri, 1977, 1979] for more details.

Appendix B
Key Notions of Hilbert Spaces

All the vector spaces below are over the complex field $\mathbb{C}$. For information common to all Banach spaces, see Appendix A.

B.1 Semi-Inner Products and Hilbert Spaces

Let H be a vector space. A complex function $(\cdot,\cdot) = (\cdot,\cdot)_H$ on $H \times H$ is called a *semi-inner product* if it satisfies the following properties:

(i) for every $y \in H$, the mapping $x \mapsto (x, y)$ is a linear functional on H,
(ii) $(x, y) = \overline{(y, x)}$ for every $x, y \in H$,
(iii) $(x, x) \geq 0$ for every $x \in H$.

If, in addition, it is not degenerate, i.e.

(iv) $(x, x) = 0 \Leftrightarrow x = 0$,

it is called an *inner product*.

- *Cauchy–Schwarz inequality.* For every semi-inner product,

$$|(x, y)|^2 \leq (x, x) \cdot (y, y).$$

- Given a semi-inner product $(\cdot,\cdot)$, the function $x \mapsto \|x\| = (x, x)^{1/2}$ is a seminorm on H (a norm if $(\cdot,\cdot)$ is an inner product). A semi-inner product defines an inner product on the *quotient space* $H/L = \{X = x + L : x \in H\}$ where $L = \mathrm{Ker}(\cdot,\cdot) = \{x \in H : (x, x) = 0\}$,

$$(X, Y) = (x, y) \quad \text{where } x \in X,\ y \in Y \quad (\text{and } X, Y \in H/L).$$

- A vector space H equipped with an inner product $(\cdot,\cdot) = (\cdot,\cdot)_H$ and with the associated norm is called a *pre-Hilbert space* (or *Euclidean*, or *Hermitian*); if it is complete as a normed space, it is said to be a *Hilbert space*.

- *Example.*

$$H = \mathcal{L}^2(\Omega,\mu) = \left\{f\colon \Omega \to \mathbb{C}\colon \int_\Omega |f|^2\, d\mu < \infty\right\},$$

with the semi-inner product

$$(f,g) = \int_\Omega f\overline{g}\, d\mu \quad (f,g \in L^2(\Omega,\mu));$$

$H/L = L^2(\Omega,\mu)$ is a Hilbert space where

$$L = \operatorname{Ker}(\cdot,\cdot) = \{f\colon f = 0\ \mu\text{-a.e.}\}.$$

In particular,

$$l^2(J) = \left\{(x_j)_{j\in J}\colon x_j \in \mathbb{C},\ \sum_j |x_j|^2 < \infty\right\}$$

where $(x,y) = \sum_{j\in J} x_j \overline{y}_j$ is a Hilbert space.

In what follows, H is a Hilbert space.

B.2 Orthogonal Decompositions

Let $x, y \in H$. x is said to be *orthogonal to* y (written $x \perp y$) if $(x, y) = 0$. Subspaces $E, F \subset H$ are orthogonal ($E \perp F$) if $x \perp y$ for every $x \in E, y \in F$.

- *"Pythagorean" Theorem (580–495 BCE).* If $x_j \in H$ and $x_j \perp x_k$ ($j \neq k$), then

$$\left\|\sum_1^n x_j\right\|^2 = \sum_1^n \|x_j\|^2.$$

- *Corollary.* The vector sum of closed orthogonal subspaces is closed: if $E, F \subset H$ are closed and $E \perp F$, then $E + F$ is closed (which is not necessarily the case for general E, F).

- *Orthogonal complement* of a vector subspace $E \subset H$. This is $E^\perp = \{y \in H\colon x \perp y,\ \forall x \in E\}$.

  If E is closed, then $E = (E^\perp)^\perp$ and $H = E + E^\perp$ (this is often written $H = E \oplus E^\perp$ to highlight the orthogonality), hence every $x \in H$ can be uniquely written in the form $x = x' + x''$ where $x' \in E$, $x'' \in E^\perp$.

  The mapping $P_E\colon x \mapsto x'$ is called the *orthogonal projection onto* E. Clearly P_E is linear, $P_E^2 = P_E$, and for every $x \in H$, $\|P_E x\| \leq \|x\|$.

- A *projection* $P: H \to H$ is a linear mapping satisfying $P^2 = P$; P is an orthogonal projection onto a closed subspace if and only if $\|P\| \leq 1$, and if and only if $P = P^*$ (where P^* is the adjoint operator: see Appendix D).

- *Corollary.* Let $A \subset H$. Then, $\mathrm{span}_H(A) = H \Leftrightarrow (x \perp A \Rightarrow x = 0)$.

- *Convergence of an orthogonal series.* Let $x_j \in H$ ($j = 1, 2, \ldots$) and $x_j \perp x_k$ ($j \neq k$). The series $\sum_{j \geq 1} x_j$ converges in H if and only if $\sum_j \|x_j\|^2 < \infty$; if this is case, then $\|x\|^2 = \sum_j \|x_j\|^2$ where $x = \sum_j x_j$. A convergent orthogonal series $\sum_j x_j$ *converges unconditionally*: for any $\epsilon > 0$ there exists a finite subset $\sigma_\epsilon \subset \mathbb{N}$ such that, for every finite $\sigma \supset \sigma_\epsilon$, $\|x - \sum_{j \in \sigma} x_j\| < \epsilon$.

- *Orthogonal decomposition.* Let $H_j \subset H$ ($j = 1, 2, \ldots$) be subspaces with $H_j \perp H_k$ ($j \neq k$). Then, the closed linear hull of the family (H_j) is

$$\mathrm{span}_H(H_j: j = 1, 2, \ldots) = \Big\{ x = \sum_{j \geq 1} x_j : x_j \in H_j \, (\forall j) \text{ and } \sum_j \|x_j\|^2 < \infty \Big\}.$$

This last set is denoted $\sum_{j \geq 1} \oplus H_j$.

- *Orthogonal decomposition (part II).* We have $\sum_{j \geq 1} \oplus H_j = H$ if and only if

$$(x \perp H_j, \forall j) \Rightarrow x = 0$$

and when this is the case, then for every $x \in H$,

$$x = \sum_j P_{H_j} x, \quad P_{H_j} x \in H_j, \quad \|x\|^2 = \sum_j \|P_{H_j} x\|^2 \quad \text{(Parseval's equality)}.$$

B.3 Orthogonal Bases

A special case of the preceding decompositions is when each H_j is generated by a single vector $e_j \neq 0$; hence (e_j) is an orthogonal sequence $((e_j, e_k) = 0$, $j \neq k)$, *complete in* H if $(x \perp e_j, \forall j) \Rightarrow x = 0$. In this last case, for every $x \in H$ there exists a unique convergent series of the form $\sum_j a_j e_j$ whose sum is x; in fact,

$$a_j e_j = P_{H_j} x = \frac{(x, e_j)}{\|e_j\|^2} e_j,$$

and thus

$$\forall x \in H: \quad x = \sum_j \frac{(x, e_j)}{\|e_j\|^2} e_j, \quad \|x\|^2 = \sum_j \frac{|(x, e_j)|^2}{\|e_j\|^2}.$$

Such an orthogonal and complete sequence is called an *orthogonal basis of H*; if, in addition, $\forall j$, $\|e_j\| = 1$, we speak of *an orthonormal basis*.

- *The existence of an orthonormal basis (Gram–Schmidt orthogonalization theorem).* Let $(x_j)_{j\geq 1} \subset H$ be a free sequence, i.e. $\forall k$, $x_k \notin \text{Vect}(x_j : j \neq k)$. Then there exists one, and only one, *orthonormal sequence* $(e_j)_{j\geq 1}$ satisfying the following properties:
  (i) for every $n = 1, 2, \ldots$, $\text{Vect}(e_j : 1 \leq j \leq n) = \text{Vect}(x_j : 1 \leq j \leq n) =: L_n$,
  (ii) for every j, $(x_j, e_j) > 0$.

  The explicit formula is
  $$e_n = \frac{x_n - P_{L_{n-1}} x_n}{\|x_n - P_{L_{n-1}} x_n\|}, \quad P_{L_{n-1}} x = \sum_{j=1}^{n-1} (x, e_j) e_j \quad (\forall x \in H).$$

- *Corollary.* In every separable Hilbert space there exists an orthonormal basis.

- *Example.* Let μ be a finite Borel measure on $\mathbb{T}$, with $\text{supp}(\mu)$ an infinite set. Then there exists a unique orthonormal basis $(\varphi_k)_{k\geq 1}$ of trigonometric polynomials φ_k such that $\deg \varphi_k = [k/2]$, $k = 1, 2, \ldots$ (We apply the theorem to the sequence $(x_k)_{k\geq 1} = (1, e^{ix}, e^{-ix}, e^{2ix}, e^{-2ix}, \ldots)$ for which we know that $\text{span}_{L^2(\mu)}(x_k)_{k\geq 1} = L^2(\mu)$.) The φ_k are called the *orthogonal polynomials with respect to μ*.

- *Corollary.* All separable Hilbert spaces of the same dimension are unitarily isomorphic: if $\dim H_1 = \dim H_2$, and the H_j are separable, there exists a unitary mapping $U: H_1 \to H_2$, i.e. linear, bijective, and isometric.

B.4 Gram Matrices

Given a finite family of vectors $\mathcal{X} = (x_j)_{0 \leq j \leq n} \subset H$, the matrix
$$G(\mathcal{X}) = ((x_i, x_j))_{0 \leq j, i \leq n}$$
is called the *Gram matrix* of $\mathcal{X}$.

- A Gram matrix is *positive definite*, $G(\mathcal{X}) \gg 0$, i.e. for every $a = (a_j)_0^n \in \mathbb{C}^{n+1}$,
$$(Ga, a) = \sum_{i,j}(x_i, x_j) a_i \bar{a}_j \geq 0$$
(since this is $= \|\sum_i a_i x_i\|^2$). In particular, $\det(G(\mathcal{X})) \geq 0$.

Conversely, if $A\colon \mathbb{C}^{n+1} \to \mathbb{C}^{n+1}$ and $A \gg 0$, there exists $\mathcal{X} = (x_j)_{0 \le j \le n} \subset H$ such that $G(\mathcal{X}) = A$ (it suffices to take the columns of the square root B of A ($B^2 = A$, $B \gg 0$), $x_j = Be_j$, where (e_j) is the standard basis of $\mathbb{C}^{n+1}$).

- Given two families of vectors $\mathcal{X} \subset H$, $\mathcal{Y} \subset K$, then there exists a unitary mapping $U\colon (\mathcal{X}) \longrightarrow (\mathcal{Y})$ (see Appendix D) such that $U x_j = y_j$ ($\forall j$) if and only if $G(\mathcal{X}) = G(\mathcal{Y})$.

- Let $\mathcal{X} = (x_j)_{0 \le j \le n} \subset H$ be a free family (i.e. linearly independent), $E = \mathrm{Vect}(\mathcal{X})$ and $x \in H$. Then

$$\mathrm{dist}(x, E)^2 = \frac{\det(G(x_0, \ldots, x_n, x))}{\det(G(\mathcal{X}))}.$$

In particular, a family $\mathcal{X}$ is not free and $\dim E < n + 1$ if and only if $\det(G(\mathcal{X})) = 0$.

B.5 The Riesz Representation Theorem

Every bounded continuous linear functional L on a Hilbert space H is of the form $L(x) = (x, y)$ ($\forall x \in H$); such a $y \in H$ is unique and $\|L\| = \|y\|$.

B.6 Reproducing Kernel Hilbert Spaces (RKHS)

Let H be a Hilbert space of functions on a set Ω. H is said to be a *reproducing kernel Hilbert space (RKHS)* if, for every $x \in \Omega$, there exists $k_x \in H$ such that $f(x) = (f, k_x)$ ($\forall f \in H$). If such a k_x exists, then it is unique, and the family $\{k_x\}$ (or the mapping $x \mapsto k_x$, $x \in \Omega$) is called the *reproducing kernel* of H.

- A space of functions on Ω is an RKHS if and only if, for every $x \in \Omega$, the linear functional $f \mapsto f(x)$ ($f \in H$) is continuous on H.

- *Example(s)*.

  (a) The Sobolev space

  $$W_1^2(0, 1) = \left\{ f\colon \int_0^1 |f'|^2 \, dx < \infty,\ f(0) = 0 \right\}$$

  (the completion of differentiable functions satisfying the said condition); $k_x(y) = y$ for $0 \le y \le x$ and $k_x(y) = x$ for $y \le x \le 1$.

(b) The Hardy space

$$H^2 = \left\{ f(z) = \sum_{k \geq 0} \widehat{f}(k) z^k \ (z \in \mathbb{D}) : \ \sum_{k \geq 0} |\widehat{f}(k)|^2 < \infty \right\},$$

$$k_z(\zeta) = (1/(1 - \bar{z}\zeta)) \ (z \in \mathbb{D}, \zeta \in \mathbb{D}).$$

- A reproducing kernel $k_x(y)$ is *a positive definite function on* $\Omega \times \Omega$: for every finite family $(x_i) \subset \Omega$ the matrix $(k_{x_i}(x_j))_{ij}$ is positive definite (it is a Gram matrix $G(\mathcal{X})$, $\mathcal{X} = (k_{x_i})_i$).

The converse is also true: a function on $\Omega \times \Omega$ positive definite in the above sense is a reproducing kernel of an RKHS.

B.7 Notes and Remarks

For generalities and historical information, see [Riesz and Szőkefalvi-Nagy, 1955], [Dieudonné, 1981], or [Pietsch, 2007]. For more information and applications of Gram matrices, see [Akhiezer, 1965]. For an overview of RKHS, see [Nikolski, 2002a].

Appendix C
An Overview of Banach Algebras

Biography Israel Gelfand.

All the topics below can be found in [Rudin, 1991].

A *Banach algebra* A is a Banach space equipped with a multiplication $A \times A \to A$, associative and distributive with respect to vector operations, and such that $\|xy\| \le \|x\| \cdot \|y\|$ ($\forall x, y \in A$); the algebra is *unital* if it contains a unit element e satisfying $\|e\| = 1$.

In what follows, A always denotes a Banach algebra (unital unless dictated by the context).

An element $x \in A$ is said to be invertible if there exist $y \in A$, $z \in A$ such that $yx = e$, $xz = e$; in this case, $y = z$ and it is denoted x^{-1} (the *inverse* of x). The set of invertible elements of A is denoted

$$A^{-1} = \{x \in A : \exists x^{-1}\}.$$

C.1 Spectrum and the Holomorphic Functional Calculus

The set A^{-1} is open and the mapping $x \mapsto x^{-1}$ is continuous on this set. Given $x \in A$, the set

$$\sigma(x) = \sigma_A(x) =: \{\lambda \in \mathbb{C} : \lambda e - x \notin A^{-1}\},$$

called the *spectrum of* x, is compact and non-empty. Moreover, the limit

$$\lim_n \|x^n\|^{1/n} =: r(x) \le \|x\|,$$

called the *spectral radius of* x, exists, and if $|\lambda| > r(x)$, then $\lambda \notin \sigma(x)$ and

$$(\lambda e - x)^{-1} = \sum_{k \ge 0} \frac{x^n}{\lambda^{n+1}}$$

(absolute convergence). In particular, $\|x\| < 1 \Rightarrow (e - x)^{-1} = \sum_{k \ge 0} x^k$.

339

Always, $\sigma(x) \cap \{\lambda: |\lambda| = r(x)\} \neq \emptyset$.

- *Resolvent and functional calculus.* The function $\lambda \mapsto (\lambda e - x)^{-1} =: R_\lambda(x)$, called the *resolvent of x*, is holomorphic on $\mathbb{C} \setminus \sigma(x)$, and for a function $f \in \mathrm{Hol}(\sigma(x))$ (the algebra of functions holomorphic on an open neighborhood V_f of $\sigma(x)$) the mapping

$$f \mapsto f(x) =: \frac{1}{2\pi i} \int_\gamma R_\lambda(x) f(\lambda) \, d\lambda$$

(where $\gamma \subset V_x \setminus \sigma(x)$ is a curve separating $\sigma(x)$ and ∞) is a homomorphism $\mathrm{Hol}(\sigma(x)) \to A$ satisfying $1(x) = e$, $z(x) = x$ (functional calculus for x); z stands for the identity map, $z(\zeta) = \zeta$, $\forall \zeta \in \mathbb{C}$.

- *Spectral mapping theorem:* $f \in \mathrm{Hol}(\sigma(x)) \Rightarrow \sigma(f(x)) = f(\sigma(x))$. Moreover, the equality also holds for $f(x) = \sum_{k\geq 0} a_k x^k$ if $\sum_{k\geq 0} \|a_k x^k\| < \infty$.

- We always have $\|R_\lambda(x)\| \geq 1/\mathrm{dist}(\lambda, \sigma(x))$, or equivalently,

$$|\lambda - \mu| < \|R_\lambda(x)\|^{-1} \Rightarrow \mu \in \mathbb{C} \setminus \sigma(x).$$

C.2 C^*-algebras

A *Banach algebra A is said to be a C^*-algebra* if it is equipped with an involution $j(x) =: x^*$, $j: A \to A$ satisfying $j^2 = \mathrm{id}$, $(xy)^* = y^* x^*$, $(\lambda x + \mu y)^* = \overline{\lambda} x^* + \overline{\mu} y^*$ and $\|xx^*\| = \|x\|^2$.

- *Example.* $A = L(H)$, the algebra of bounded operators on a Hilbert space H, is a C^*-algebra where the involution T^* is the adjoint operator: $(x, T^* y) = (Tx, y)$ ($\forall x, y \in H$). A sub-algebra $A \subset L(H)$, closed and stable for the adjoint operation (i.e. $T \in A \Rightarrow T^* \in A$), is again a C^*-algebra.

  According to a 1943 theorem of Gelfand and Naimark (see [Gelfand, Raikov, and Shilov, 1960] and [Murphy, 1990]), every C^*-algebra is isometrically isomorphic to a sub-C^*-algebra of $L(H)$.

- Let A be a C^*-algebra. If $x \in A$, $x^* = x$ (a *Hermitian element*), and $\sigma(x) \subset \mathbb{R}_+ = [0, \infty)$, then there exists one, and only one, Hermitian $y \in A$ ($\sigma(y) \subset \mathbb{R}_+$) such that $y^2 = x$. An element $p \in A$ is said to be an (ortho-) *projection* if $p^* = p$ and $p^2 = p$; every C^*-algebra is topologically generated by the projections it contains: $A = \mathrm{span}_A(p: p \in A$ is a projection$)$.

C.3 Commutative Banach Algebras: Theory of Israel Gelfand

Here, A is a unital and *commutative* Banach algebra: $xy = yx$ ($\forall x, y \in A$).

- Every homomorphism $L: A \to \mathbb{C}$ (called a *complex homomorphism*) is continuous, and if $L \neq 0$, then $L(e) = 1$ and $\|L\| = 1$.

- A subspace $I \subset A$ is a maximal ideal of A (i.e. a proper ideal, $I \neq A$, contained in no other proper ideal of A) if and only if $I = \mathrm{Ker}L$, where L is a non-null complex homomorphism.

- The set $\mathfrak{M}(A) =: \{L \in A^* : L \text{ homomorphism } A \to \mathbb{C}, L \neq 0\}$ is compact for the weak topology $\sigma(A^*, A)$ (Appendix A), and is called the *maximal ideal space of A*.

- *Gelfand transform.* Given $x \in A$, the Gelfand transform $\widehat{x}$ is the complex function on $\mathfrak{M}(A)$ defined by $\widehat{x}(L) = \langle x, L \rangle$, $L \in \mathfrak{M}(A)$.
  We have $\widehat{x} \in C(\mathfrak{M}(A))$ and $x \in A^{-1} \Leftrightarrow \widehat{x}(L) \neq 0, \forall L \in \mathfrak{M}(A)$, hence $x \mapsto \widehat{x}$ is a homomorphism $A \to C(\mathfrak{M}(A))$, and

$$\sigma(x) = \widehat{x}(\mathfrak{M}(A)) \quad \text{and} \quad \|\widehat{x}\|_\infty = r(x).$$

The second of these equalities is *Gelfand's formula*, a name also given to a more general formula, valid in every Banach algebra:

$$\lim_{n \to \infty} \|x^n\|^{1/n} = \max_{\lambda \in \sigma(x)} |\lambda|.$$

If the Gelfand transform $x \mapsto \widehat{x}$ is injective, A is said to be *semi-simple* (without radical; in general, the *radical* $R(A)$ is defined by $R(A) = \{x \in A : r(x) = 0\}$).

- If A is a *commutative C^*-algebra*, the Gelfand transform $x \mapsto \widehat{x}$ is an isometric homomorphism of A onto $C(\mathfrak{M}(A))$ (due to Gelfand and Naimark, 1943: see [Gelfand, Raikov, and Shilov, 1960] and [Murphy, 1990]).

- *Functional calculus and the Gelfand transform.* If $f \in \mathrm{Hol}(\sigma(x))$, then $\widehat{f(x)} = f(\widehat{x})$.

- *Local functional calculus.* Suppose that A is semi-simple, $x \in A$ and f is a holomorphic function locally injective on an open neighborhood of $\sigma(x)$; if there exists $b \in C(\mathfrak{M}(A))$ satisfying $f \circ b = \widehat{x}$, then there is also an $a \in A$ such that $f \circ a = x$. In particular, if there is a continuous logarithm $e^b = \widehat{x}$

($b \in C(\mathfrak{M}(A))$), then there exists $a \in A$ satisfying $e^a = x$. (due to Shilov, 1953: see [Gelfand, Raikov, and Shilov, 1960] for a proof and references).

Israel Gelfand (in Russian Израиль Моисеевич Гельфанд, 1913–2009) was a Russian (Soviet period) and American mathematician and biomathematician, creator of *Gelfand theory* (of commutative Banach algebras) and founder of a dozen directions of research. He led the important "Gelfand seminar," which functioned for almost 50 years and from which emanated several dozen extraordinary mathematicians. He was the author of more than 800 articles and 30 monographs. A short list of the domains where Gelfand's impact was decisive or where he was quite simply the founder includes: Banach algebras ("normed rings," maximal ideals), C^*-algebras (algebras with an involution and the GNS construction, with Naimark and Raikov), group representations (with Naimark and Graev), integral geometry (a generalization of the Radon transform), the solution of the inverse Sturm–Liouville problem (with Levitan), the approach to inverse problems for the Lax and Korteweg–de Vries equations (with Dikyi), the topological classification of elliptic operators (leading directly to the theory of Atiyah and Singer), and cohomology of infinite-dimensional Lie groups (with Fuchs), discriminants, dilogarithms, and hypergeometric functions. From 1958 onwards, Gelfand worked extensively in biology and medicine; he published more than 100 articles and monographs on the neurophysiology of movement, on cellular migration, and on several experimental questions. Several dozen mathematical objects bear Gelfand's name, such as the *Gelfand transform* (on the maximal ideal space), the *Gelfand–Mazur, Gelfand–Naimark* and *Gelfand–Levitan theorems*, the *Gelfand–Naimark–Segal construction, Gelfand triplets*, the *Gelfand–Fuchs cohomology*, etc. Among the numerous monographs written by Gelfand (with co-authors), a particularly important role was played by *Commutative Normed Rings* (1960) and the six volumes of *Generalized Functions* (1958–1964; a thorough multifaceted development of Schwartz's theory of distributions, including "equipped Hilbert spaces" or "Gelfand triplets"). Gelfand was rewarded by being elected to a dozen academies, including the US National Academy, the Royal Society, the Académie des Sciences (Paris), the Accademia dei Lincei, etc., and being awarded a *Doctor Honoris Causa* degree by several universities (including the Sorbonne), as well as the Wolf Prize, the Kyoto Prize, the AMS Leroy Steel Prize, etc. He was plenary speaker at the International Congress of Mathematicians on three occasions (1954, 1962, 1970). Gelfand's studies and early career suffered from the typical oppressions of the Soviet era. In 1928 he was excluded from a provincial technical school (in the region of Vinnitsa, now in Ukraine) with the right to continue at neither high school nor university, given that he was the son of "an element of the parasite class" (his father was a miller).

After moving to Moscow, where he subsisted on casual work, Gelfand continued to teach himself, and in 1931 he succeeded in establishing contact with the Kolmogorov seminar at the University of Moscow. His thesis "Functions and operators in abstract spaces" (1935, under the direction of Kolmogorov) was fundamental for the development of infinitesimal analysis in infinite dimensions. Professor at the University of Moscow from 1941, and then member of several of the most prestigious academies in the world, Gelfand was long denied access to the Soviet Academy because of the anti-Semitic position of the Academy's Department of Mathematics, directed by Pontryagin and Vinogradov (this even provoked protests by other Academy departments, such as Physics). Gelfand was one of the rare university figures of the USSR who had the courage to sign protest letters against the politics of the Kremlin (in 1968, against the Soviet invasion of Czechoslovakia, and then letters against the persecution of dissidents to the regime, such as mathematician Alexander Esenin-Volpin). Gelfand's activities included several more initiatives that were important both socially and professionally: the "Gelfand correspondence school" (destined for junior and senior high school students throughout the USSR) functioned effectively for decades, as did the famous "Gelfand seminar" at the University of Moscow – a true Research Institute in its own right!

The Gelfand seminar, a "seminar across all mathematics," was held at the University of Moscow for almost 50 years, starting from the mid-1940s. The participants included almost all the elite of Russian/Soviet analysis of the time; it is impossible to list here all the names: the seminar was attended by 100–120 people every Monday, including 30–40 world-renowned professors, heads of their own research schools. By tradition, each group had its own seating: the first rows on the left were reserved for high school students [sic!] (more often than not, the winners of the Olympiads); those in the center were reserved for Gelfand's own young students, and those just beyond for his students of previous generations; rows 6–7 onward were reserved for other researchers grouped according to their interests, and finally the doctoral students, visitors, etc. Formally, the seminar was supposed to start at 6.00 pm., but Gelfand would never arrive before 7.00 pm. at the earliest, and then the seminar went on without interruption until 10.30 and sometimes until midnight. In reality, this was a weekly congress at the highest level of mathematics, but also a one-man show for a single actor, Gelfand himself. The goal was always to go to the heart of the result or the theory being explained, right to the roots, right to the truth (of which the only bearer and herald was Gelfand: the lecturer did not have the right to speak of motivations, evaluations, or any other "philosophy" of the subject). This style was perhaps borrowed from another great Moscow seminar, that of Lev Landau (Nobel Prize in Physics, 1962), even more authoritarian and legendary (the rivalry between the mathematicians and the physicists was very strong at the time). The following behavior was frequent during the seminar: improvisations could at any moment dramatically divert the course of the presentation (and leave the lecturer speechless in front of the blackboard for several dozen minutes); a "test attendant," chosen randomly by Gelfand from any category present in the room – from a high school student to a renowned senior professor – to go to the blackboard and explain in front of everyone what he had understood of the current presentation; Gelfand had the privilege to suddenly "not understand a thing" while everyone else found it all perfectly clear, but also to stop a question from the audience that did not please him, with a brusque "don't ask stupid questions." But the clarifications and the profoundness of the comments were so exceptional that the participants travelled dozens, and even hundreds, of kilometers to attend this spectacle with such an extraordinary actor. Isadore Singer (who shares with Atiyah the attribution of the index theorem), in his speech for Gelfand's 80th birthday, said: "We have a giant in our midst. I look to other domains to find comparable examples: Balanchine for the dance, or Thomas Mann for literature, or Stravinsky, or even better – Mozart for music; but for me, the best comparison is with artists such as Cézanne and Matisse."

In 1967, Gelfand also founded and then edited the very high-level journal *Functional Analysis and its Applications* (in Russian, translated by Springer), especially strong in the 1960s and 1970s (and directed with more or less the same enthusiasm as the seminar). There are numerous testimonials and recognitions concerning Gelfand, with an analysis of his activities in mathematics, biology, and medicine, but we will mention only four (on which the current note is based, and where there are references to dozens of other sources): P. Duren and S. Zdravkovska (eds.), *Golden Years of Moscow Mathematics* (AMS, 1993); S. Gelfand and S. Gindikin, *I. M. Gelfand Seminar* (vol. 16 of Advances in Soviet Mathematics, part 1, AMS, 1993); and two particularly rich websites, http://israelmgelfand.com/ and https://ru.wikipedia.org/wiki/Гельфанд,_Израиль_Моисеевич (the Russian Wikipedia page devoted to Gelfand). From these sources, one can find, among other very serious information, numerous anecdotes and jokes linked with Gelfand. We cite a few here.

In 1973, Gelfand was invited to receive his degree of *Doctor Honoris Causa* at Oxford University, but the Soviets refused him an "exit visa"; Sir Michael Atiyah managed to contact Queen Elizabeth, who in turn called the Soviet ambassador – and a visa was delivered! At the beginning of the 1960s, during a festive reunion around a (well-lubricated) meal, Anatoli Vitushkin – a great expert in "hard analysis" – addressed Gelfand with the following remark: "Me, I am a wolf – when I see a mathematical prey, I pursue it with all my force up to exhaustion, and I capture it; and you, Israel Moiseevich, you are a lynx, extremely well educated and wise – you climb to the top of a tree where you have a large (intellectual) horizon; you wait until the prey wanders under the tree, and you jump down on it!" Vladimir I. Arnold, during the celebration of Gelfand's 70th birthday, compared Gelfand and Kolmogorov by saying: "If mathematics could be compared to a mountainous landscape, then Kolmogorov would be an exploring mountaineer tracing the trails to new summits, whereas Gelfand would be the builder of motorways in this terrain." And to finish – the word *Gelfand* in Yiddish means "Elephant." Gelfand married twice and had three children. Newspapers around the world marked his death by publishing detailed obituaries (the title in the *New York Times*: "Israel Gelfand, Math Giant, Dies at 96").

C.4 Examples

(a) $A = C(X)$, X a compact topological space; here, $X = \mathfrak{M}(A)$, $\widehat{x} = x$ ($\forall x \in A$).

(b) $A = \text{AP}(\mathbb{R})$, the Bohr algebra of almost periodic functions. By definition,

$$\text{AP}(\mathbb{R}) = \text{span}_{L^\infty(\mathbb{R})}(e^{ist} : s \in \mathbb{R})$$

(closure of the trigonometric polynomials with real frequency for the uniform norm $\|\cdot\|_\infty$). Every $t \in \mathbb{R}$ defines a complex homomorphism $f \mapsto f(t)$ of $\text{AP}(\mathbb{R})$, and hence $\mathbb{R} \subset \mathfrak{M}(\text{AP}(\mathbb{R}))$. It is well known that $\mathbb{R}$ is dense in $\mathfrak{M}(\text{AP}(\mathbb{R}))$: see [Katznelson, 1976]. Moreover, the invertible functions f of $\text{AP}(\mathbb{R})$ "admit a logarithm" up to an exponential factor: $f(t) = e^{iwt} e^{a(t)}$ where $a \in \text{AP}(\mathbb{R})$ and $w = w_f$ is the "(perpetual) mean motion"

$$w_f = \lim_{T\to\infty} (\arg(f(T)) - \arg(f(-T)))$$

(Bohr's theorem: see [Katznelson, 1976] and [Nikolski, 1986] for proofs).

(c) $A = \mathcal{M}(G)$, the algebra of finite complex measures on a locally compact and Abelian group G equipped with the norm $\|\mu\| = \text{Var}(\mu)$, the total variation of μ,

$$\text{Var}(\mu) = \sup\left\{\sum |\mu(\sigma_j)| : \{\sigma_j\} \text{ is a finite partition of } G\right\},$$

and with convolution as multiplication

$$\mu * \nu(\sigma) = \int_G \mu(\sigma - t)\, d\nu(t), \quad \sigma \subset G.$$

Every $s \in \widehat{G}$ (the dual group of bounded characters of G) defines a complex homomorphism by the *Fourier transform*

$$\mathcal{F}\mu(s) = \int_G \langle s, -t\rangle\, d\mu(t),$$

which means $\widehat{G} \subset \mathfrak{M}(\mathcal{M}(G))$.

If G is a discrete group (such as $G = \mathbb{Z}^n$, $n = 1, 2, \ldots$), then

$$\widehat{G} = \mathfrak{M}(\mathcal{M}(G)).$$

In particular (by adding to the preceding property another, of symmetry), we obtain that if $\mu \in \mathcal{M}(G)$ such that $\mathcal{F}\mu(s) \neq 0$ ($s \in \widehat{G}$) and if there exists a continuous logarithm a of $\mathcal{F}\mu$ (i.e. $\mathcal{F}\mu = e^a$; for $G = \mathbb{Z}$, it is equivalent to say that the Cauchy index of $\mathcal{F}\mu$ on $\widehat{G} = \mathbb{T}$ is zero), then there exists a logarithm $\nu \in \mathcal{M}(G)$ of μ: $\mu = e^{*\nu}$, where

$$e^{*\nu} = \sum_{n \geq 0} \frac{\nu^{*n}}{n!}, \quad \nu^{*n} = \nu * \nu * \cdots * \nu \quad (n \text{ times}).$$

If G is non-discrete (such as $G = \mathbb{R}^n$, important for the Wiener–Hopf operators), then $\mathfrak{M}(\mathcal{M}(G))$ is much larger: $\widehat{G}$ is not dense in $\mathfrak{M}(\mathcal{M}(G))$, and there even appears what is known as the "Wiener–Pitt phenomenon," i.e. there exist measures $\mu \in \mathcal{M}(G)$ such that $\inf\{|\mathcal{F}\mu(s)|: s \in \widehat{G}\} > 0$ but that are not invertible; see [Rudin, 1962] and [Gelfand, Raikov, and Shilov, 1960] for references and proofs. Nonetheless, if in the Lebesgue decomposition with respect to the Haar measure $\mu = wm + \mu_{sc} + \mu_d$ (m is the Haar measure on G, μ_{sc} the singular continuous portion and μ_d the discrete portion of μ) we have $\|\mu_{sc}\| < \|wm + \mu_d\|$ (in particular, if $\mu_{sc} = 0$), then $\inf |\mathcal{F}\mu| > 0 \Rightarrow \mu \in \mathcal{M}(G)^{-1}$: see [Gelfand, Raikov, and Shilov, 1960]. Joseph Taylor (see [Taylor, 1973] for details and references) found a profound analog to Bohr's theorem, cited in Example (b) above: a measure $\mu \in \mathcal{M}(\mathbb{R})$ is invertible if and only if $\mu = \delta_c * \rho^{*n} * e^{*\nu}$ where $\nu \in \mathcal{M}(\mathbb{R})$, $n \in \mathbb{Z}$ and $c \in \mathbb{R}$ (a sort of mean motion), and $\rho = \delta_0 + 2 e^{-t}\chi_{(0,\infty)} dt$ (see also the comments on §4.2.5–§4.2.6 in §4.6).

(d) $A = L^1(0, 1) + \mathbb{C}\delta_0$ with the restriction to $(0, 1)$ of the convolution $\mu * \nu$ as multiplication operation on $\mu = a\delta_0 + f$, $\nu = b\delta_0 + g$. The radical is $R(A) = L^1(0, 1)$ and $\mathfrak{M}(A) = \{L_0\}$, the only point defined by $L_0(a\delta_0 + f) = a$ ($a\delta_0 + f \in A$).

C.5 Notes and Remarks

In addition to the sources mentioned in the text, see [Weil, 1940], [Loomis, 1953], [Dunford and Schwartz, 1958], and [Katznelson, 1976].

Appendix D
Linear Operators

For the definitions and basic notions, see Appendices A–C and [Rudin, 1991]. In what follows, X and Y are Banach spaces, always over the complex field $\mathbb{C}$.

D.1 Bounded Linear Operators

The *space of bounded linear operators* $X \to Y$ is denoted $L(X, Y)$, abbreviated as $L(X)$ when $Y = X$. $L(X, Y)$ is a Banach space equipped with the standard norm

$$\|T\| = \sup\{\|Tx\|_Y : \|x\|_X \leq 1\}.$$

It is a unital Banach algebra for $Y = X$. All the concepts and results on Banach algebras (Appendix C) are applicable to $L(X)$.

- *One-sided invertibility.* $T \in L(X, Y)$ is *left-invertible* if there exists $S_l \in L(Y, X)$ such that $S_l T = I_X \ (= \mathrm{id}_X)$, and *right-invertible* if there exists $S_d \in L(Y, X)$ such that $T S_d = I_Y$. If the two exist, then $S_l = S_d = T^{-1}$ and T is *invertible* (an isomorphism).

D.2 Adjoints, Polar Representation, and Duality

The *adjoint* T^* of an operator $T \in L(X, Y)$ is defined as the $T^* \in L(Y^*, X^*)$ satisfying

$$\langle Tx, \varphi \rangle_Y = \langle x, T^* \varphi \rangle_X$$

for every $x \in X, \varphi \in Y^*$. For $E \subset X$, we define $E^\perp = \{y \in X^* : \langle x, y \rangle = 0, \forall x \in E\}$, and for $F \subset X^*$, $F_\perp = \{x \in X : \langle x, y \rangle = 0, \forall y \in F\}$. We have $(E^\perp)_\perp = \mathrm{span}_X(E)$, $(F_\perp)^\perp = \mathrm{span}_{\sigma(X^*, X)}(F)$.

- For every $T \in L(X, Y)$, the *range of X* is denoted $\text{Ran}(T) = TX$. We have
$$(\text{Ran}(T))^\perp = \text{Ker}(T^*) = \{\varphi \in Y^* : T^*\varphi = 0\}, \quad (\text{Ker}(T))^\perp = \text{clos}_\sigma(\text{Ran}(T^*))$$
and
$$(\text{Ran}(T^*))_\perp = \text{Ker}(T), \quad (\text{Ker}(T^*))_\perp = \text{clos}(\text{Ran}(X)),$$
where $\sigma = \sigma(X^*, X)$ is the weak-* topology (Appendix A).

- If $T \in L(X)$, then $\sigma(T^*) = \sigma(T)^* =: \{\bar{\lambda} : \lambda \in \sigma(T)\}$, $\|T\| = \|T^*\|$ (for a conjugate linear duality X, X^*; it is simply $\sigma(T) = \sigma(T^*)$ for a bilinear duality).

D.3 Fundamentals of Linear Operators (Banach and Schauder)

(a) Closed graph theorem. If $T: X \to Y$ is linear and its *graph* $G(T) =: \{(x, Tx) : x \in X\} \subset X \times Y$ is closed, then $T \in L(X, Y)$.

(b) Open mapping theorem. If $T \in L(X, Y)$ and $TX = Y$, then the range $T(V)$ of any open $V \subset X$ is open.

(c) Banach isomorphism theorem. If $T \in L(X, Y)$ is bijective, $T^{-1} \in L(Y, X)$.

(d) Banach surjectivity theorem. Let $T \in L(X, Y)$; then, $TX = Y \Leftrightarrow \exists \epsilon > 0$ such that, for every $\varphi \in Y^*$, $\|T^*\varphi\| \geq \epsilon\|\varphi\|$. Moreover, the sharpest (largest) ϵ in this estimate is the same as in $TB_X(0, 1/\epsilon) \supset B_Y(0, 1)$ where $B_X(x, r) = \{x' \in X : \|x - x'\| < r\}$, the ball of center x and radius $r > 0$. The dual statement, $\|Tx\| \geq \epsilon\|x\|(\forall x \in X) \Leftrightarrow T^*Y^* = X^*$ is also true.

- Given $T \in L(X, Y)$, $\text{Ran}(T)$ is closed $\Leftrightarrow \text{Ran}(T^*)$ is closed.

- Given $T \in L(X, Y)$ the following assertions are equivalent:
  (i) $T \in L(X, Y)^{-1}$,
  (ii) $\text{Ker}(T) = \{0\}$ and $\|T^*\varphi\| \geq \epsilon\|\varphi\|$ ($\epsilon > 0$; $\forall \varphi \in Y^*$),
  (iii) $\text{Ker}(T^*) = \{0\}$ and $\|Tx\| \geq \epsilon\|x\|$ ($\epsilon > 0$; $\forall x \in X$).

- T is left-invertible if and only if $\|Tx\| \geq \epsilon\|x\|$ ($\forall x \in X$) and $\text{Ran}T$ is complementable in Y.

Hint If R is a left inverse, TR is a bounded projection on $\text{Ran}T$.

(e) Banach–Steinhaus theorem. If $T_n \in L(X, Y)$ and, for every $x \in X$, $\sup_n \|T_n x\| < \infty$, then $\sup_n \|T_n\| < \infty$.

D.4 Integral Operators, Schur Test

Let (Ω_j, μ_j), $j = 1, 2$, be two measure spaces and let k be a measurable function on $\Omega_2 \times \Omega_1$ such that, for certain $\varphi > 0$ and $\psi > 0$,

$$\int_{\Omega_1} |k(x,y)|\varphi(y)\, d\mu_1(y) \leq a\psi(x), \quad \int_{\Omega_2} |k(x,y)|\psi(x)\, d\mu_2(x) \leq b\varphi(y).$$

Then the operator

$$J_k f(x) = \int_{\Omega_1} k(x,y) f(y)\, d\mu_1(y), \quad x \in \Omega_2,$$

is bounded $L^2(\Omega_1, \mu_1) \to L^2(\Omega_2, \mu_2)$ and $\|J_k\| \leq \sqrt{ab}$.

Indeed,

$$\left(\int_{\Omega_1} |k(x,y)| \cdot |f(y)|\, d\mu_1(y) \right)^2$$

$$= \left(\int_{\Omega_1} |k(x,y)\varphi(y)|^{1/2} \cdot |k(x,y)/\varphi(y)|^{1/2} |f(y)|\, d\mu_1(y) \right)^2$$

$$\leq a\psi(x) \int_{\Omega_1} |k(x,y)/\varphi(y)| \cdot |f(y)|^2 \, d\mu_1(y),$$

so that

$$\|J_k f\|_2^2 \leq a \int_{\Omega_2} \int_{\Omega_1} |f(y)|^2 |k(x,y)/\varphi(y)|\psi(x)\, d\mu_1(y)\, d\mu_2(x)$$

$$= a \int_{\Omega_1} |f(y)|^2/\varphi(y) \int_{\Omega_2} |k(x,y)|\psi(x)\, d\mu_2(x)\, d\mu_1(y)$$

$$\leq ab \int_{\Omega_1} |f(y)|^2 \, d\mu_1(y). \qquad \blacksquare$$

- *Example.*

$$Tf(x) = \frac{1}{x} \int_0^x f(y)\, dy \text{ is bounded } L^2(\mathbb{R}_+) \to L^2(\mathbb{R}_+) \text{ and } \|T\| = 2.$$

Indeed, $T = J_k$ where $k(x,y) = 1/x$ for $0 < y < x$ and $k(x,y) = 0$ for $y > x > 0$. Taking $\varphi = \psi = x^{-1/2}$, we obtain the condition of the Schur test above with $a = b = 2$, thus $\|T\| \leq 2$. The direct calculation for $f_\alpha(x) = x^\alpha \chi_{(0,1)}(x)$ ($\alpha > -1/2$), shows that $PTf_\alpha = (1/(1+\alpha))f_\alpha$, where $Ph = \chi_{(0,1)}h$, hence $\|T\| \geq \|PT\| \geq 1/(1+\alpha)$ for every $\alpha > -1/2$, thus $\|T\| \geq 2$. $\blacksquare$

D.5 Compact Operators

An operator $T \in L(X, Y)$ is said to be compact if $\mathrm{clos}_Y(TB_X(0,1))$ is compact. The set of compact operators of X to Y is denoted $\mathfrak{S}_\infty(X, Y)$, or $\mathfrak{S}_\infty(X)$ if $X = Y$, or simply $\mathfrak{S}_\infty$ if the spaces are clearly known.

- $\mathfrak{S}_\infty(X, Y)$ is a "closed bilateral ideal": a closed subspace of $L(X, Y)$ such that $ABC \in \mathfrak{S}_\infty(W, Z)$ for every $B \in \mathfrak{S}_\infty(X, Y)$, $C \in L(W, X)$, $A \in L(Y, Z)$.

- $T \in \mathfrak{S}_\infty(X, Y) \Leftrightarrow T^* \in \mathfrak{S}_\infty(Y^*, X^*)$.

- $T \in \mathfrak{S}_\infty(X, Y) \Leftrightarrow (\varphi_n \in Y^*$ and $\sigma^*\text{-}\lim_n \varphi_n = 0 \Rightarrow \lim_n \|T^*\varphi_n\|_{X^*} = 0)$; in particular, if X, Y are reflexive then $T \in L(X, Y)$ is compact if and only if it transforms weakly convergent sequences to sequences convergent in norm.

- (Riesz and Schauder.) Let $T \in \mathfrak{S}_\infty(X)$, $\dim X = \infty$. Then,
   (i) $0 \in \sigma(T)$,
   (ii) $\sigma(T)$ is at most countable with a single possible accumulation point $\lambda = 0$,
   (iii) $\lambda \in \sigma(T) \setminus \{0\} \Leftrightarrow \mathrm{Ker}(T - \lambda I) \neq \{0\}$; moreover $\dim \mathrm{Ker}(T - \lambda I) = \dim \mathrm{Ker}(T^* - \overline{\lambda} I) < \infty$ and $\mathrm{Ran}(T - \lambda I)$ is closed (hence, $T - \lambda I$ is Fredholm and $\mathrm{ind}(T - \lambda I) = 0$: see Appendix E); also, $\dim(\bigcup_{n \geq 1} \mathrm{Ker}(T - \lambda I)^n) < \infty$.

- *Finite rank operators.* The rank of a $T \in L(X, Y)$ is $\mathrm{rank}(T) =: \dim(\mathrm{Ran}(T))$; the set of finite rank operators $\mathfrak{F}(X, Y) = \{T \in L(X, Y) : \mathrm{rank}(T) < \infty\}$ is a "bilateral ideal" of $L(X, Y)$,

$$\mathfrak{F}(X, Y) \subset \mathfrak{S}_\infty(X, Y),$$

and if Y admits a Schauder basis (see Appendix A.4 above), then

$$\mathrm{clos}_{L(X,Y)} \mathfrak{F}(X, Y) = \mathfrak{S}_\infty(X, Y).$$

In fact, this last equality, if it holds for every Banach space X, defines an *approximation property* (AP) for Y (according to Grothendieck). It is also equivalent to saying that for every finite set $K \subset Y$ and every $\epsilon > 0$ there exists $T \in \mathfrak{F}(X)$ satisfying $\|x - Tx\| < \epsilon$, $\forall x \in K$. The existence of a Schauder basis implies (AP) (in particular, for a Hilbert space and all the $L^p(\Omega, \mu)$), but the converse is false (found by Szarek in 1987); the example of Enflo mentioned in Appendix A.4 gives, in reality, a subspace of l^p, $p \neq 2$ without (AP). We refer to [Pietsch, 2007] for more information on the property (AP).

- *Example.* An operator $T \in L(L^2(\Omega_1), L^2(\Omega_2))$ is in $\mathfrak{F}(L^2(\Omega_1), L^2(\Omega_2))$ if and only if $T = J_k$ (of Appendix D.4) where $k(x, y) = \sum_{j=1}^{n} u_j(x) v_j(y)$ (a "degenerate kernel") with $v_j \in L^2(\Omega_1)$, $u_j \in L^2(\Omega_1)$. Also, if $k \in L^2(\Omega_2 \times \Omega_1)$, then $J_k \in \mathfrak{S}_\infty(L^2(\Omega_1), L^2(\Omega_2))$, since by Cauchy's inequality, $\|J_k\| \le \|k\|_{L^2(\Omega_2 \times \Omega_1)}$, and to conclude, it suffices to approximate $k \in L^2(\Omega_2 \times \Omega_1)$ by degenerate kernels k_n (leading to $\lim_n \|J_k - J_{k_n}\| = 0$), which is possible since

$$\mathrm{span}_{L^2}(\chi_{E_2 \times E_1} = \chi_{E_2}(x) \chi_{E_1}(y) : \mu_j(E_j) < \infty) = L^2(\Omega_2 \times \Omega_1).$$

On the other hand, the operator T of the *example in Appendix D.4 is not compact*, as $PT|L^2(0, 1)$ is not compact: it has an uncountable family of eigenvectors f_α, $\alpha > -1/2!$). ∎

D.6 The Calkin Algebra $\mathcal{K} = L(X) / \mathfrak{S}_\infty(X)$

The quotient mapping $\Pi \colon L(X) \to \mathcal{K}$, $\Pi(T) = T + \mathfrak{S}_\infty(X)$ is a homomorphism defining the "essential properties" of an operator (properties up to a compact operator), such as:

- *essential norm* $\|T\|_{\mathrm{ess}} = \|\Pi(T)\| = \mathrm{dist}_{L(X)}(T, \mathfrak{S}_\infty(X))$,

- *essential (Fredholm) spectrum* $\sigma_{\mathrm{ess}}(T) =: \sigma(\Pi(T))$,

- *an essential inverse (regularizer)* $ST - I \in \mathfrak{S}_\infty(X)$, $TS - I \in \mathfrak{S}_\infty(X)$,

- *essentially normal operator* in a Hilbert space $X = H$: $[T, T^*] \in \mathfrak{S}_\infty(H)$ (see [Brown, Douglas, and Fillmore, 1973] for the theory of these operators).

If X is a Hilbert space, then $\mathcal{K}$ is a C^*-algebra.

See Appendix E below for Fredholm theory (spectral theory in the algebra $\mathcal{K}$).

D.7 Operators on a Hilbert Space

Let H be a Hilbert space.

- $L(H)$ is a C^*-algebra (see Appendix C.2).

- An operator $T \in L(H)$ is said to be *self-adjoint*, *unitary*, or *normal* if it satisfies, respectively, $T = T^*$, $TT^* = T^*T = I$, $TT^* = T^*T$. The *unitary*

D.7 Operators on a Hilbert Space

operators $T: H \to K$ between different spaces are defined by the same equations $TT^* = I_K$, $T^*T = I_H$.

- T is self-adjoint, unitary, or normal if and only if, for every $x, y \in H$, it satisfies, respectively, $(Tx, x) \in \mathbb{R}$, $(Tx, Ty) = (x, y)$ (and T is "onto"), or $\|Tx\| = \|T^*x\|$.

- $T \in L(H)$ is said to be *positive definite*, denoted $T \gg 0$, if, for every $x \in H$, $(Tx, x) \geq 0$.

- Every $T \gg 0$ admits a unique positive square root (and roots of any degree) $T^{1/2}$: a unique $A \gg 0$ such that $A^2 = T$.

- *Polar representation (factorization).* Every $T \in L(H)$ admits a unique representation

$$T = V|T| \quad \text{where } |T| = (T^*T)^{1/2} \gg 0 \quad (\textit{module of } T)$$

and V is a "partial isometry" such that $V: H = (\text{Ker}T)^\perp \oplus (\text{Ker}T) \to H = (\text{Ker}T^*)^\perp \oplus (\text{Ker}T^*)$ where $V|(\text{Ker}T)^\perp$ is unitary $(\text{Ker}T)^\perp \to (\text{Ker}T^*)^\perp$ and $V|(\text{Ker}T) = 0$.

- *Unitarily equivalent operators.* $T \in L(H)$ and $S \in L(K)$ are unitarily equivalent if there exists a unitary operator $U \in L(H, K)$ such that $UT = SU$.

- *Example of a normal operator.* Let E be a Hilbert space, μ a positive Borel measure on $\mathbb{C}$ with compact support $\text{supp}(\mu)$, and let $H = L^2(\mathbb{C} \to E, \mu)$ be the L^2 space of functions with values in E equipped with the norm

$$\|f\|^2 = \int \|f(z)\|_E^2 \, d\mu(z) < \infty.$$

The multiplication operator $M_z: H \to H$, $M_z f(\zeta) = z(\zeta) f(\zeta)$, $z(\zeta) = \zeta$ is normal; it is self-adjoint if and only if $\text{supp}(\mu) \subset \mathbb{R}$, and positive definite if and only if $\text{supp}(\mu) \subset \mathbb{R}_+$.

- *Spectral theorem.* An operator $T \in L(H)$ is normal if and only if it is unitarily equivalent to a restriction $M_z|F$ of the operator of the preceding example on a subspace $F \subset L^2(\mathbb{C} \to E, \mu)$ reducing M_z: $M_z F \subset F$, $M_z F^\perp \subset F^\perp$; such a space is always of the form

$$F = \{f \in L^2(\mathbb{C} \to E, \mu): f(z) \in E(z) \; \mu\text{-a.e. } z\},$$

354   Linear Operators

where $E(z) \subset E$ is a family of closed subspaces; μ can always be chosen in such a manner that $\mathrm{supp}(\mu) = \sigma(T)$.

If T is an operator with *a simple spectrum*, i.e. if there exists a cyclic vector $x \in H$, $\mathrm{span}_H(T^k T^{*l} x : k \geq 0, l \geq 0) = H$, then $E = E(z) = \mathbb{C}$ μ-a.e. z, and hence $F = L^2(\mu)$, and T *is unitarily equivalent to the multiplication operator*

$$M_z : L^2(\mu) \to L^2(\mu).$$

- *Case of a compact normal operator.* If T is normal and $T \in \mathfrak{S}_\infty(H)$, then in the above example, $\mu = \sum_{k\geq 0} \delta_{\lambda_k}$ where (λ_k) is a sequence (finite or not) tending to 0 if it is infinite, and $\lambda_0 = 0$, which gives $H = \sum_{k\geq 0} \oplus H_k$ where $H_k = \mathrm{Ker}(T - \lambda_k I)$, hence $T|H_k = \lambda_k I_{H_k}$, $\dim H_k < \infty$ (if $k \geq 1$).

D.8 Schatten–von Neumann Classes $\mathfrak{S}_p(H)$

Let H be a Hilbert space and $1 \leq p < \infty$; the *Schatten–von Neumann p-class* is defined by

$$\mathfrak{S}_p(H) = \left\{ T \in \mathfrak{S}_\infty(H) : \|T\|_{\mathfrak{S}_p}^p = \sum_{k \geq 0} s_k(T)^p < \infty \right\},$$

where $s_k(T)^2 = \lambda_k(T^*T)$ (the eigenvalues of a positive operator T^*T in decreasing order, $\lambda_0(T^*T) \geq \lambda_1(T^*T) \geq \cdots \geq \lambda_k(T^*T) \geq \ldots$, each repeated according to its multiplicity, i.e. $\dim \mathrm{Ker}(T^*T - \lambda I)$); we have $s_k(T) = \lambda_k(|T|)$. Note that $\|T\| = s_0(T) = \max_{k \geq 0} s_k(T)$.

An operator $T \in \mathfrak{S}_2(H)$ is called a *Hilbert–Schmidt* operator, and $T \in \mathfrak{S}_1(H)$ is a *finite trace operator*.

- $\mathfrak{S}_p(H)$ is a vector space, $\|\cdot\|_{\mathfrak{S}_p}$ is a norm (the *Schatten–von Neumann* norm, or, if $p = 2$, the Hilbert–Schmidt or Frobenius norm), and $(\mathfrak{S}_p(H), \|\cdot\|_{\mathfrak{S}_p})$ is a Banach space such that $A, C \in L(H)$, $B \in \mathfrak{S}_p(H) \Rightarrow ABC \in \mathfrak{S}_p(H)$ and $\|ABC\|_{\mathfrak{S}_p} \leq \|A\| \cdot \|B\|_{\mathfrak{S}_p} \|C\|$; $\mathfrak{F}(H)$ is a dense subset of $\mathfrak{S}_p(H)$.

- *Compact operators.*

  (i) $T \in \mathfrak{S}_\infty(H)$ if and only if there exist two orthonormal sequences, finite or not, $(x_k)_{k\geq 0}$, $(y_k)_{k\geq 0}$ and a monotone sequence of positive numbers $s_0 \geq s_1 \geq \ldots$, such that $\lim_l s_k = 0$,

  $$T = \sum_{k \geq 0} s_k(\cdot, x_k) y_k \quad \text{(Riesz decomposition)},$$

D.8 Schatten–von Neumann Classes $\mathfrak{S}_p(H)$

convergent for the operator norm. In fact, $T^*Tx_k = s_k^2 x_k$, $y_k = Tx_k/s_k$ and $TT^*y_k = s_k y_k$ (orthonormal sequences of eigenvectors corresponding to the non-zero eigenvalues), and hence $s_k = s_k(T)$ ($k \geq 0$).

Indeed, it suffices to define (x_k) as stated, set $y_k = Tx_k/s_k$, and then easily verify the other properties with a direct calculation. ∎

(ii) *Corollary.* Let $T \in \mathfrak{S}_\infty(H)$ and $k \in \mathbb{Z}_+$, then $s_k(T) = \inf\{\|T - A\| : A \in L(H), \mathrm{rank}(A) \leq k\}$ (which implies all the properties of the preceding bullet point).

(iii) *Corollary.* If $T \in \mathfrak{S}_p(H)$, $1 \leq p \leq \infty$ and $A_n \in L(H)$, $\lim_n \|A_n x\| = 0$ ($\forall x \in H$), then $\lim_n \|A_n T\|_{\mathfrak{S}_p} = 0$.

Indeed, by using the decomposition (i), we can write

$$T = \sum_{0 \leq k < N} s_k(\cdot, x_k) y_k + \sum_{k \geq N} s_k(\cdot, x_k) y_k = S_N + T_N$$

where $\lim_N \|T_N\|_{\mathfrak{S}_p} = 0$, and then $\|A_n T_N\|_{\mathfrak{S}_p} \leq (\sup_n \|A_n\|) \|T_N\|_{\mathfrak{S}_p}$ ($\sup_n \|A_n\| < \infty$ by Banach–Steinhaus) and, for every fixed N,

$$\|A_n T_N\|_{\mathfrak{S}_p} \leq \sum_{0 \leq k < N} s_k \|x_k\| \cdot \|A_n y_k\|,$$

hence $\lim_n \|A_n T_N\|_{\mathfrak{S}_p} = 0$. Finally, we obtain $\lim_n \|A_n T\|_{\mathfrak{S}_p} = 0$. ∎

- *Remark.* The hypotheses of (iii) do not lead to $\lim_n \|TA_n\|_{\mathfrak{S}_p} = 0$, even for an operator T such that $\mathrm{rank}(T) = 1$ (a standard example: $A_n = S^{*n}$, where S^* is the backward shift, or inverse translation, in $l^2(\mathbb{Z}_+)$, $A_n(x_0, x_1, \ldots) = (x_n, x_{n+1}, \ldots)$, $T = (\cdot, x)y$ and $\|TA_n\|_{\mathfrak{S}_p} = \|A_n^* x\| \cdot \|y\| = \|x\| \cdot \|y\| \not\to 0$).

- *Hilbert–Schmidt operators.*

  (i) Let (e_k), (e'_j) be two orthonormal bases of H, and $T \in L(H)$. Then, $\sum_k \|Te_k\|^2 = \sum_j \|T^* e'_j\|^2$, and the sums are finite if and only if $T \in \mathfrak{S}_2(H)$ and

  $$\|T\|^2_{\mathfrak{S}_2} = \sum_k \|Te_k\|^2 = \|T^*\|^2_{\mathfrak{S}_2}.$$

  Indeed, by Parseval's equality (Appendix B),

  $$\sum_k \|Te_k\|^2 = \sum_j \sum_k |(Te_k, e'_j)|^2 = \sum_k \sum_j |(Te_k, e'_j)|^2 = \sum_j \|T^* e'_j\|^2,$$

  and the result follows. ∎

(ii) *Example.* If $k \in L^2(\Omega \times \Omega, \mu \times \mu)$, the operator J_k of Appendix D.4 is in $\mathfrak{S}_2(L^2(\Omega, \mu))$: given an orthonormal basis (e_j) of $L^2(\Omega, \mu)$, we have

$$\|J_k\|_{\mathfrak{S}_2}^2 = \sum_j \|J_k e_j\|^2 = \sum_j \sum_i |(J_k e_j, e_i)_{L^2(\Omega, \mu)}|^2$$

(the family $\bar{e}_j e_i = \bar{e}_j(y) e_i(x)$ is an orthonormal basis of $L^2(\Omega \times \Omega, \mu \times \mu)$), thus

$$\|J_k\|_{\mathfrak{S}_2}^2 = \sum_{j,i} |(k, \bar{e}_j e_i)_{L^2(\Omega \times \Omega, \mu \times \mu)}|^2 = \|k\|_{L^2(\Omega \times \Omega, \mu \times \mu)}^2 < \infty. \quad \blacksquare$$

- *Trace class operators.*

  (i) (i) $T \in \mathfrak{S}_1(H)$ if and only if there exist two sequences (finite or not) $(x_k)_{k \geq 0}$, $(y_k)_{k \geq 0}$ in H such that

  $$T = \sum_{k \geq 0} (\cdot, x_k) y_k, \quad \sum_{k \geq 0} \|x_k\| \cdot \|y_k\| < \infty.$$

  We have $\|T\|_{\mathfrak{S}_1} = \min \sum_{k \geq 0} \|x_k\| \cdot \|y_k\|$ (over all the representations of T). Indeed,

  $$\|T\|_{\mathfrak{S}_1} \leq \sum_{k \geq 0} \|(\cdot, x_k) y_k\|_{\mathfrak{S}_1} = \sum_{k \geq 0} \|x_k\| \cdot \|y_k\| < \infty,$$

  hence $T \in \mathfrak{S}_1(H)$. For a $T \in \mathfrak{S}_1(H)$, the existence of such a representation follows from the Riesz decomposition. $\quad \blacksquare$

  (ii) $A, B \in \mathfrak{S}_2(H) \Rightarrow AB \in \mathfrak{S}_1(H)$, and $\|AB\|_{\mathfrak{S}_1} \leq \|A\|_{\mathfrak{S}_2} \|B\|_{\mathfrak{S}_2}$. Indeed, if (e_k) is an orthonormal basis of H and $x \in H$, then

  $$ABx = A\left(\sum_k (Bx, e_k) e_k\right) = \sum_k (x, B^* e_k) A e_k \quad (\forall x \in H),$$

  and $\sum_k \|B^* e_k\| \cdot \|A e_k\| \leq \|A\|_{\mathfrak{S}_2} \|B\|_{\mathfrak{S}_2}$ by Cauchy's inequality. $\quad \blacksquare$

- *Trace of an operator* $T \in \mathfrak{S}_1(H)$. Given an orthonormal basis (e_k) of H, the series

  $$\text{Trace}(T) = \sum_{k \geq 0} (T e_k, e_k)$$

  converges absolutely and is called the *trace of* T. The value is independent of the choice of (e_k); $|\text{Trace}(T)| \leq \|T\|_{\mathfrak{S}_1}$.

Indeed, if $T = \sum_{k\geq 0}(\cdot, x_k)y_k$, $\sum_{k\geq 0} \|x_k\| \cdot \|y_k\| < \infty$, then, by Parseval,

$$\sum_{j\geq 0}(Te_j, e_j) = \sum_{j\geq 0}\sum_{k\geq 0}(e_j, x_k)(y_k, e_j) = \sum_{k\geq 0}(y_k, x_k).$$

The absolute convergence and the inequality are evident. ∎

- *Analytic trace and geometric trace (Lidskii, 1959).* Let $T \in \mathfrak{S}_1(H)$ and let $\lambda_k(T)$ be the eigenvalues of T counted with their multiplicity (a value $\lambda \neq 0$ is repeated m times, where $m = \dim(\bigcup_{l\geq 1} \text{Ker}(T - \lambda I)^l)$. Then,

$$\text{Trace}(T) = \sum_k \lambda_k(T),$$

where the series on the right converges absolutely.

Remark. In the case where T possesses an absolutely convergent spectral decomposition

$$T = \sum_k \lambda_k(\cdot, y_k)x_k, \quad (x_j, y_k) = \delta_{jk}, \quad \sum_k |\lambda_k| \cdot \|y_k\| \cdot \|x_k\| < \infty$$

(for example, T is of finite rank and simple spectrum), the equality is clear, by Parseval:

$$\sum_{j\geq 0}(Te_j, e_j) = \sum_k \lambda_k(x_k, y_k) = \sum_k \lambda_k(T).$$

The proof in the general case has nothing to do with this naive argument: see [Davidson, 1988], [Gohberg, Goldberg, and Kaashoek, 1990], and [Simon, 2005a].

- *Corollary.* $A \in \mathfrak{S}_1(H)$, $B \in L(H) \Rightarrow \text{Trace}(AB) = \text{Trace}(BA)$.
  Indeed, it is easy to see that $\sigma(AB) \setminus \{0\} = \sigma(BA) \setminus \{0\}$. ∎

- *The Weyl inequalities (1949).* Let $A \in \mathfrak{S}_\infty(H)$ and $\lambda_k(A)$ be the eigenvalues of A counted with their multiplicity. Then, for every $p > 0$ and $n \geq 0$,

$$\sum_{k=0}^n |\lambda_k(A)|^p \leq \sum_{k=0}^n s_k(A)^p,$$

and in particular, $\sum_{k\geq 0} |\lambda_k(A)| \leq \|A\|_{\mathfrak{S}_1}$.

- *Determinant.* Let $A \in \mathfrak{S}_1(H)$ and $\lambda_k(A)$ be the eigenvalues of A counted with their multiplicity. Define

$$\det(I + A) = \prod_k (1 + \lambda_k(A)).$$

- If rank$(A) < \infty$, then det$(I+A)$ coincides with the determinant of the matrix of the restriction $(I + A)|E$ on a subspace dim $E < \infty$, $E \supset$ Ran(A) (with respect to any basis of E).

- If $A \in \mathfrak{S}_1(H)$, then $|\det(I + A)| \leq \prod_k(1 + |\lambda_k(A)|) \leq \exp(\|A\|_{\mathfrak{S}_1})$ and

    $$|1 - \det(I + A)| \leq \|A\|_{\mathfrak{S}_1} \exp(\|A\|_{\mathfrak{S}_1}).$$

    Moreover, $A \mapsto \det(I + A)$ is a continuous function on $\mathfrak{S}_1(H)$.

- *Corollary.*
    (i) If $A \in \mathfrak{S}_1(H)$ and $(e_k)_{k \geq 0}$ is an orthonormal basis of H, then

    $$\det(I + A) = \lim_n \det(\delta_{ij} + (Ae_i, e_j))_{0 \leq j, i \leq n}.$$

    Indeed, $\lim_n \|A - P_n A P_n\|_{\mathfrak{S}_1} = 0$ where P_n is an orthogonal projection onto Vect$(e_j: 0 \leq j \leq n)$. ∎
    (ii) If $A, B \in \mathfrak{S}_1(H)$, then $(I + A)(I + B) = I + C$, $C \in \mathfrak{S}_1(H)$, and $\det(I + A)(I + B) = \det(I + A)\det(I + B)$, $\det(I + AB) = \det(I + BA)$ and

    $$\det(I + zA) = \exp((z + o(1))\text{Trace}(A)) \quad (\text{as } z \to 0).$$

    Hint $\log \det(I + zA) = \sum_k \log(1 + z\lambda_k(A))$ and $\sum_k |\lambda_k(A)| < \infty$.

D.9 Notes and Remarks

The contents of this appendix can be found in numerous texts; for the extensions, in addition to the sources already mentioned, as well as for the attribution of the results and the original references, see [Dunford and Schwartz, 1958], [Reed and Simon, 1972, 1978], [Lindenstrauss and Tzafriri, 1977, 1979], and [Simon, 2005a].

The example of the operator T presented in Appendix D.4 was the subject of the article by Brown, Halmos, and Shields [1965], who showed that $I - T$ is unitarily equivalent to the unitary left translation (shift) operator on $l^2(\mathbb{Z})$, and the compression $(I - PT)|L^2(0, 1)$ – to the operator S^* of inverse translation on $l^2(\mathbb{Z}_+)$ (backward shift) (hence, it is a Toeplitz operator!), from which many other properties of T can be deduced.

Appendix E
Fredholm Operators and the Noether Index

Biographies Erik Ivar Fredholm, Frederick Atkinson, Fritz Noether, Felix Hausdorff.

One of the "golden rules" of linear algebra determines the relation between the *conditions of resolvability* and the number of *free parameters* for a solution of the linear equation,

$$Tx = y,$$

where $T \in L(\mathbb{C}^m, \mathbb{C}^n)$. More precisely,

(1) there exist d^* linear functionals f_j, $1 \le j \le d^*$, on $\mathbb{C}^n$, where $d^* = \dim \ker T^*$, such that the conditions $\langle y, f_j \rangle = 0$ ($1 \le j \le d^*$) are necessary and sufficient for the resolvability of the equation;
(2) there exist d elements $x_j \in \mathbb{C}^m$ ($1 \le j \le d$), where $d = \dim \ker T$, such that the general solution of $Tx = y$ (if such exists) is of the form $x' + \sum_j a_j x_j$, where $a_j \in \mathbb{C}$ are arbitrary and x' is a particular solution;
(3) $d - d^* = m - n$.

A Fredholm operator (or "normally solvable" operator) is a mapping $T \in L(X, Y)$, where X, Y are Banach spaces, for which the statements (1)–(3) remain valid (see Appendix E.1 for the formal definition). This concept, originating from the investigation of linear integral equations beginning with the famous articles of Ivar Fredholm of 1900–1903, proves itself useful for the study of spectral properties of various types of operators, including the Toeplitz and Hankel operators. The reason of this efficacy can perhaps be found in the fact that "Fredholm theory" is a spectral theory "up to a compact operator" (see Theorem E.7.5 for details).

The standard information on compact operators can be found in Appendix D; others (less standard) are presented below. Recall (see Appendices C and D) that the spectrum of an operator $T \in L(X)$ is

$$\sigma(T) = \{\lambda \in \mathbb{C} : \lambda I - T \text{ is not invertible}\}.$$

The spectrum is always non-empty and compact.

A few historical remarks on Fredholm theory can be found in the biography of Erik Ivar Fredholm below, as well as in Appendix E.10.

Erik Ivar Fredholm (1866–1927), a Swedish mathematician, was the founder of *Fredholm theory* (1903), and author of the famous "Fredholm alternative" (1900) and of the theory of determinants of infinite matrices (1903). His entire academic career was spent between the University of Uppsala, where he obtained his doctorate (1893) and his *Habilitation* in mathematics (1898), and the University of Stockholm (known as Stockholms Högskola from 1878 to 1960) where he *de facto* worked under the direction of Gösta Mittag-Leffler. The principal theme of his research was potential theory and the associated PDEs. However, the results that made him famous were obtained in the year 1899, which he spent in Paris studying the Dirichlet problem alongside Poincaré, Picard, and Hadamard. These results, published in 1900 and 1903 (*Acta Mathematica*), immediately became "classics": they were presented (by Erik Holmgren) at the Hilbert seminar in Göttingen (capital of mathematics at the time) where Hilbert perceived them as a major breakthrough opening a new epoch in analysis. Effectively, it was these results that led Hilbert to the concept of "Hilbert spaces" and transformed integral equations into a major theme of research in the first quarter of the twentieth century: "Fredholm's great paper on integral operators," according to Reed and Simon ([1972], Chapter VI), came 70 years later. According to Lars Gårding, these results also served to "improve the self-esteem of Swedish mathematicians who up to then worked in the shadow of the continental cultural empires of Germany and France."

According to witnesses, Fredholm had a melancholic temperament, spoke with a soft voice, and worked slowly. He left relatively few articles (his *Complete Works* contain only 160 pages), and published nothing after 1910 (by coincidence, he married in 1911 at the age of 45 and had one child). Fredholm also had a talent and a passion for craftsmanship, first building with his own hands a violin from half a coconut (!) – he was a gifted violinist – and finishing with mechanical machines to generate solutions of certain differential equations.

Several mathematical objects bear Fredholm's name: the *analytic Fredholm theorem*, the *Fredholm alternative*, the *Fredholm determinant*, the *Fredholm equation*, the *Fredholm kernel*, the *Fredholm module*, the *Fredholm operator, Fredholm theory*, etc.

Fredholm was a member of the Finnish Society of Sciences, of the Accademia dei Lincei, laureate of the V. A. Wallmarks Prize (1903) and the Prix Poncelet of the Académie des Sciences (1908), and also *Doctor Honoris Causa* of the University of Leipzig (1909).

In what follows, we suppose that *X and Y are Banach spaces* satisfying the following *approximation property*: the finite rank operators of X to Y are dense in the space of compact operators.

E.1 Fredholm Operators

An operator $T \in L(X, Y)$ is said to be *Fredholm* if it is *normally solvable* (i.e. the range $\text{Ran}(T) = TX$ is a closed subspace of Y) *and* $\dim \ker T < \infty$, $\dim \ker T^* < \infty$. The difference

$$\text{ind} T = \dim \ker T - \dim \ker T^*$$

is called the *Noether index of* T (for the history of this concept see Appendix E.10, as well as the biography of Fritz Noether). We let $\text{Fred}(X, Y)$ denote the set of Fredholm operators from X to Y, and more succinctly, $\text{Fred}(X, X) = \text{Fred}(X)$.

Observe that $\ker T^* = (TX)^\perp$ according to Appendix D.2, and hence the above matricial "golden rule" remains valid for the Fredholm operators: if $T \in \text{Fred}(X, Y)$, then the equation

$$Tx = y$$

is resolvable if and only if $y \in Y$ satisfies d^* conditions of resolvability $\langle y, f_j \rangle = 0$, $1 \leq j \leq d^* = \dim \ker T^*$ (the f_j are a basis of $\ker T^*$), and in this case the general solution x is of the form $x = x' + \sum_{j=1}^{d} a_j x_j$ where the (x_j) denote a basis of $\ker T$ and x' is a particular solution; the difference between the number of free parameters and the number of conditions of resolvability is $d - d^* = \operatorname{ind} T$.

E.2 Invertibility and the Fredholm Property

An operator $T \in \operatorname{Fred}(X)$ is invertible if and only if $\ker T = \{0\}$, $\ker T^* = \{0\}$, and if and only if $\operatorname{ind} T = 0$ and $\ker T = \{0\}$ (or $\ker T^* = \{0\}$).

Indeed, $TX = \operatorname{clos}(TX) = (\ker T^*)_\perp = X$ and $\ker T = \{0\}$, which implies $T \in L(X)^{-1}$. ∎

E.3 The Fredholm Spectrum (Essential Spectrum)

Let $T \in L(X)$. The set

$$\sigma_F(T) = \{\lambda \in \mathbb{C}: \lambda I - T \notin \operatorname{Fred}(X)\}$$

is called the *essential (or Fredholm) spectrum of T*. Clearly $\sigma_F(T) \subset \sigma(T)$. The principal aim of Fredholm theory is to provide a description and the properties of the essential spectrum; in particular, we will see that $\sigma_F(T) = \sigma_{\text{ess}}(T)$, the essential spectrum as defined in the Calkin algebra: see Appendix D.6.

E.4 Operators Between Finite-Dimensional Spaces

Suppose $\dim X < \infty$, $\dim Y < \infty$. Then $\operatorname{Fred}(X, Y) = L(X, Y)$, and for every $T \in L(X, Y)$ we have

$$\operatorname{ind} T = \dim X - \dim Y.$$

Indeed, according to linear algebra, $\dim(TX) + \dim \ker T = \dim X$, and by duality (see Appendix D.2) $\dim \ker T^* = \dim(TX)^\perp = \dim Y^* - \dim(TX)$. Since $\dim Y^* = \dim Y$, we obtain the index formula. Moreover, TX is closed, since it is a finite-dimensional subspace (see Appendix A.3). ∎

E.5 Fredholm Operators and Isomorphisms

Let $T \in L(X, Y)$ and let $V_1 \in L(W, X)$ and $V_2 \in L(Y, Z)$ be isomorphisms. Then

$$T \in \operatorname{Fred}(X, Y) \Leftrightarrow V_2 T V_1 \in \operatorname{Fred}(W, Z).$$

Moreover, $\operatorname{ind} T = \operatorname{ind}(V_2 T V_1)$.

Indeed, the property is clear directly from the definition.

$$\begin{array}{ccc} X & \xrightarrow{T} & Y \\ V_1 \downarrow & & \downarrow V_2 \\ W & \xrightarrow{V_2 T V_1} & Z \end{array}$$

E.6 Direct Sums

Let $X = X_1 + X_2$ and $Y = Y_1 + Y_2$ be decompositions into direct sums of subspaces such that, for every $x_1 \in X_1$, $x_2 \in X_2$,

$$c(\|x_1\| + \|x_2\|) \le \|x_1 + x_2\| \le C(\|x_1\| + \|x_2\|),$$

and similarly for Y, and let $T_j \in L(X_j, Y_j)$, $j = 1, 2$. Then

(1) $T(x_1 + x_2) = T_1 x_1 + T_2 x_2$ is a bounded operator $X \to Y$,
(2) $T \in \text{Fred}(X, Y) \Leftrightarrow T_j \in \text{Fred}(X_j, Y_j)$ $(j = 1, 2)$,
(3) and if these properties hold (i.e. if $T \in \text{Fred}$), $\text{ind}T = \text{ind}T_1 + \text{ind}T_2$.

Indeed, for the closure of $\text{Ran}T = \text{Ran}T_1 + \text{Ran}T_2$ see Appendix A.3, and the rest follows from the fact that $\ker T = \ker T_1 + \ker T_2$, $\text{Ran}T = \text{Ran}T_1 + \text{Ran}T_2$, $(\text{Ran}T)^\perp = (\text{Ran}T_1)^\perp + (\text{Ran}T_2)^\perp$. ∎

E.7 Fundamentals of Fredholm Operators

Theorem E.7.1 (Fredholm operators of index 0) *Let X and Y be two Banach spaces, and let $T \in L(X, Y)$. The following assertions are equivalent.*

(1) *There exists an isomorphism $V \in L(X, Y)$ and a finite rank operator $A \in \mathfrak{F}(X, Y)$ such that $T = V + A$.*
(2) *There exists an isomorphism $V \in L(X, Y)$ and a compact operator $A \in \mathfrak{S}_\infty(X, Y)$ such that $T = V + A$.*
(3) $T \in \text{Fred}(X, Y)$ *and* $\text{ind}T = 0$.

In particular, X and Y are isomorphic if and only if there exists $T \in \text{Fred}(X, Y)$, $\text{ind}T = 0$.

Proof The implication (1) $\Rightarrow$ (2) is evident.

(2) $\Rightarrow$ (3) Let $T = V + A$ where $V: X \to Y$ is an isomorphism and $A \in \mathfrak{S}_\infty(X, Y)$, and let $B \in L(X, Y)$ be a finite rank operator such that $\|A - B\| < \|V^{-1}\|^{-1}$.

By Appendix C.1, $U = V + A - B$ is an isomorphism, and hence $T = UT'$ where

$$T' = I + F \colon X \to X \text{ and } \operatorname{rank}(F) = \operatorname{rank}(U^{-1}B) < \infty.$$

Remark that $\ker F$ is a closed subspace of X of finite codimension (since $\ker F = (\operatorname{Ran} F^*)_\perp$), and hence by Appendix A.3 there exists a subspace $L \subset X$, $\dim L < \infty$, such that

$$X = L + \ker F \quad \text{(a direct sum)}.$$

Thus $\operatorname{Ran} T' = T'L + T' \ker F = T'L + \ker F$, where $\dim T'L < \infty$. Then clearly there exists a subspace $L' \subset \ker F$, $\dim L' < \infty$ such that $T'L \subset L + L' =: X_1$, and hence $T'X_1 \subset X_1$. By again applying Appendix A.3 we obtain a subspace $X_2 \subset \ker F$ such that $\ker F = L' + X_2$ (a direct sum). Clearly $T'|X_2 = \operatorname{id}$.
We now have $X = X_1 + X_2$ (a direct sum), $T'X_j \subset X_j$, $j = 1, 2$, and $\dim X_1 < \infty$, $T'|X_2 = \operatorname{id}$. By Appendix E.6, T' is Fredholm and $\operatorname{ind} T' = \operatorname{ind}(T'|X_1) + \operatorname{ind}(T'|X_2) = \operatorname{ind}(T'|X_1) = 0$ by Appendix E.4. The result follows from the factorization $T = UT'$, where U is an isomorphism, and the property Appendix E.5.

(3) $\Rightarrow$ (1) By Appendix A.3, there exist closed subspaces $L \subset X$ and $M \subset Y$ such that $X = L + \ker T$, $Y = \operatorname{Ran} T + M$ (direct sums) and $\dim \ker T - \dim M = \operatorname{ind} T = 0$. Define $V \in L(X, Y)$ by $V|L = T|L$ (an isomorphism of L to $\operatorname{Ran} T$) and with $V \colon \ker T \to M$ an arbitrary linear bijection. Clearly V is an isomorphism $X \to Y$ and $\operatorname{rank}(T - V) < \infty$. ∎

See Appendix E.8 for a few concrete examples illustrating Theorem E.7.1.

Theorem E.7.2 (regularizers) *Let X, Y be two Banach spaces and let $T \in L(X, Y)$. The following assertions are equivalent.*

(1) $T \in \operatorname{Fred}(X, Y)$.
(2) *There exists $R \in L(Y, X)$ such that $RT - I \in \mathfrak{F}(X)$ and $TR - I \in \mathfrak{F}(Y)$.*
(3) *There exist R_l, $R_r \in L(Y, X)$ such that $R_l T - I \in \mathfrak{S}_\infty(X)$ and $TR_r - I \in \mathfrak{S}_\infty(Y)$.*

The operators R_l, R_r of (3) are called the regularizers of T (left and right, respectively).

Proof (1) $\Rightarrow$ (2) We use the same reasoning as (3) $\Rightarrow$ (1) of Theorem E.7.1 (with the same notation): since $T|L$ is an isomorphism of L to $\operatorname{Ran} T$, it suffices to define R by $(T|L)^{-1}$ on $\operatorname{Ran} T$ and $R = 0$ on M.

(2) $\Rightarrow$ (3) Clear.

(3) ⇒ (1) By Theorem E.7.1, the operators R_1T and TR_2 are Fredholm, and hence $\dim \ker T < \infty$, $\dim \ker T^* < \infty$. Moreover, $T(X) \supset TR_2(Y)$ and this last subspace is closed and of finite codimension. Hence, $T(X) = TR_2(Y) + L$, where $\dim L < \infty$ (see Appendix A.3), is also closed. ∎

Theorem E.7.3 (homomorphy of the index; Atkinson (1948)) *Let X, Y, Z be Banach spaces and let $T \in \mathrm{Fred}(X, Y)$, $R \in \mathrm{Fred}(Y, Z)$. Then, $RT \in \mathrm{Fred}(X, Z)$ and*

$$\mathrm{ind}(RT) = \mathrm{ind}(R) + \mathrm{ind}(T).$$

Hence, $\mathrm{Fred}(X)$ *is a multiplicative semigroup and* $\mathrm{ind}: \mathrm{Fred}(X) \to \mathbb{Z}$ *is a group homomorphism.*

Proof By Theorem E.7.2, there exist operators Q, S such that $QR - I \in \mathfrak{S}_\infty$, $RQ - I \in \mathfrak{S}_\infty$, $ST - I \in \mathfrak{S}_\infty$, $TS - I \in \mathfrak{S}_\infty$. Hence $SQ(RT) - I \in \mathfrak{S}_\infty$ and $RT(SQ) - I \in \mathfrak{S}_\infty$, so that $RT \in \mathrm{Fred}$.

For the formula of the index, first consider the case of finite dimensions: by Appendix E.4, we have

$$\mathrm{ind}(RT) = \dim X - \dim Z,$$
$$\mathrm{ind}(T) = \dim X - \dim Y,$$
$$\mathrm{ind}(R) = \dim Y - \dim Z,$$

and the formula is evident.

In the general case, we use the same method of "decomposition ker–Ran" as in Theorems E.7.1 and E.7.2. There exist closed subspaces $L \subset X$, $M \subset Y$ such that $X = L + \ker T$, $Y = M + \ker R$ and $T|L$ and $R|M$ are isomorphisms on their ranges $TL = \mathrm{Ran}T$, $RM = \mathrm{Ran}R$. As M and $\mathrm{Ran}T$ are closed subspaces of finite codimension, so is the intersection $M' = M \cap \mathrm{Ran}T$. Hence there exists a finite-dimensional subspace $M'' \subset M$ such that

$$M = M' + M'' \quad \text{(direct sum)},$$

consequently

$$Y = M' + M'' + \ker T = M' + Y_1 \quad \text{(direct sum)}$$

where $Y_1 = M'' + \ker T$, $\dim Y_1 < \infty$. This implies $\mathrm{Ran}R = RM' + RM''$ (direct sum, $\mathrm{Ran}R$ a subspace of finite codimension), and then

$$Z = RM' + Z_1 \quad \text{(direct sum)}$$

where $\dim Z_1 < \infty$, $Z_1 \supset RM''$. The same holds for the space X: by setting $L' = (T|L)^{-1}M'$ and $L'' = (T|L)^{-1}M''$, we have $L = L' + L''$ (direct sum) and then

$$X = L' + X_1 \quad \text{(direct sum)},$$

where $X_1 = L'' + \ker T$ (a finite-dimensional subspace).

Finally, $T|L'$ is an isomorphism of L' to M', and $R|M'$ an isomorphism of M' to RM', whereas $TX_1 = TL'' = M'' \subset Y_1$ and $RY_1 = RM'' \subset Z_1$. This means that the mapping

$$RT: L' + X_1 \to M' + Y_1 \to RM' + Z_1$$

splits into a direct sum of isomorphisms

$$(R|M')(T|L'): L' \to M' \to RM'$$

and finite-dimensional mappings

$$(R|Y_1)(T|X_1): X_1 \to Y_1 \to Z_1.$$

The formula thus follows from Appendix E.6 and the part already justified. ∎

Frederick V. Atkinson (1916–2002) was a British and Canadian mathematician who made important contributions to different domains of mathematics: the theory of the "zeta" function, the spectral theory of differential operators, functional analysis, orthogonal polynomials, etc. Atkinson was the author of two reference texts, *Discrete and Continuous Boundary Problems* (1964) and *Multiparameter Eigenvalue Problems* (1972). The first immediately became a classic (even before its publication, with the proofs in hand, Richard Bellman, creator of control theory, said "This book will become a classic": see [Mingarelli, 2005]). The theme of this text – the spectral theory of ordinary differential operators – was Atkinson's favorite subject; as he said himself while joking about the central paradigm of the theory (of Weyl) "I am a member of the Limit-Point, Limit-Circle Classification

Society." Among the results that became classics are his theorems on the Fredholm operators and the index, presented in Appendix E (Theorems E.7.1–E.7.5). As noted by his biographer and disciple Angelo Mingarelli [2005], this masterpiece, published in a single article in one of the principal Russian mathematical journals *Matematicheskii Sbornik* (submitted in 1948, but only appearing in 1951 (!) and written in Russian by Atkinson himself: he mastered a dozen languages perfectly, self-taught!), remains somewhat isolated among his creations, and its motivation remains a bit mysterious.

Atkinson's career was quite surprising: after attending the prestigious St. Paul's School in West Kensington, London (the same school attended by J. E. Littlewood, G. N. Watson, and the poet John Milton), and doing his doctoral studies at The Queen's College, Oxford (directed by Edward Titchmarsh), he submitted his thesis (1939) on the "average" behavior of the "zeta" function. During the Second World War, Atkinson agreed to work for the Intelligence Corps, first at the Government Code and Cypher School at Bletchley Park (1940–1943) for the decryption of German secret messages (probably alongside the famous Alan Turing), and then (1943–1946) in India for cryptanalysis of Japanese codes. At the end of his service with the Intelligence Corps, Atkinson had acquired the rank of Major in the British Army, but when he returned to his academic career, he only succeeded in obtaining minor teaching positions. In 1948, it was the British Government (Office of the Under Secretary of State) that offered him a highly advantageous Full Professorship at the University College of Ibadan in Nigeria (a British colony at the time). On his return to Britain in 1955, he once again had problems finding employment in mathematics, and for some time he supplemented his income by translating Russian mathematical articles. Fate once again dealt him a cruel hand: on receiving an offer from the famous Institute for Advanced Study (Princeton, USA) for the year 1955/1956 (at the invitation of Leopold Infeld, a close collaborator of Einstein), Atkinson was not able to obtain an entry visa for the United States.

From this time onwards his career began to run more smoothly: he obtained the position of professor in Australia (1955–1960), and then at the University of Toronto until his retirement in 1981. Atkinson's last exceptional period began just afterwards (when he was more than 65 years old!), a veritable explosion of scientific productivity: it was during the years 1981–1991 that almost half of his research publications appeared!

His career was crowned with numerous honors: he was Fellow of the Royal Society of Canada (1967), Honorary Fellow of the Royal Society of Edinburgh (1975), Royal Society of Edinburgh's Makdougall–Brisbane Prize recipient (1974–1976), 29th President of the Canadian Mathematical Society (1989–1991), and winner of an Alexander Von Humboldt Research Award (1992).

Atkinson married Dusja Haas (of Hungarian origin) and had three children. He was also gifted in music: a superb pianist, he also played the piano-accordion at a professional level. His 10 last years turned out to be very sad: on his return from the triumphal reception of the Humboldt Prize, he suffered a massive stroke, and remained paralysed and unable to utter a single word until the end of his life.

Corollaries E.7.4
(1) A description of regularizers. *If $T \in \text{Fred}(X,Y)$, then every left regularizer R is also a right regularizer (and vice versa) and $R \in \text{Fred}(Y,X)$; if R_1 is a regularizer, then so is R_2 if and only if $R_1 - R_2 \in \mathfrak{S}_\infty(Y,X)$.*

Indeed, every regularizer of T is Fredholm by Theorem E.7.2. Moreover, if $R = R_l$ and R_r are regularizers of T, then $R_l T R_r = (I + A)R_r = R_r + B$ and $R_l T R_r = R_l(I + C) = R_l + D$ where $A, B, C, D \in \mathfrak{S}_\infty$, giving

$$R_l - R_r \in \mathfrak{S}_\infty(Y,X). \qquad \blacksquare$$

(2) *Let $T \in \text{Fred}(X,Y)$ and $R \in \text{Fred}(Y,X)$ its regularizer, then $\text{ind} T = -\text{ind} R$.*

Indeed, by definition, $RT = I + A$ where $A \in \mathfrak{S}_\infty(X)$, and hence by Theorem E.7.3, $\text{ind} R + \text{ind} T = \text{ind}(RT) = \text{ind}(I + A) = 0$ (by Theorem E.7.1).
$\blacksquare$

(3) The stability of Fred and of the index. *Fred(X,Y) is an open set in $L(X,Y)$ and the function*

$$T \mapsto \text{ind}(T)$$

is locally constant on Fred(X,Y).

Indeed, let $T \in \text{Fred}(X,Y)$ and let R be a regularizer of T, i.e. $RT - I \in \mathfrak{S}_\infty(X)$, $TR - I \in \mathfrak{S}_\infty(Y)$. If $A \in L(X,Y)$ such that $\|R\| \cdot \|A\| < 1$, then $R(T + A) = I + C + RA$ where $C \in \mathfrak{S}_\infty(X)$, whereas $B =: I + RA \in L(X)^{-1}$, which implies $R(T + A) = B(I + B^{-1}C)$, and hence $B^{-1}R$ is a left regularizer of $T + A$.

E.7 Fundamentals of Fredholm Operators

In the same manner, there exists a right regularizer, so that $T + A \in \text{Fred}(X, Y)$. Moreover, by (2) and Theorems E.7.3 and E.7.1, we have

$$-\text{ind}T + \text{ind}(T + A) = \text{ind}R + \text{ind}(T + A) = \text{ind}B + \text{ind}(I + B^{-1}C) = 0. \quad \blacksquare$$

(4) Invariance of the index by homotopy in Fred. *If the operators T_0, $T_1 \in \text{Fred}(X, Y)$ are homotopic (i.e. if there exists a continuous mapping $f : [0, 1] \to \text{Fred}(X, Y)$ such that $f(0) = T_0$, $f(1) = T_1$), then $\text{ind}T_0 = \text{ind}T_1$.*

Indeed, the function $t \mapsto \text{ind}(f(t))$, locally constant on $[0, 1]$, is constant. $\blacksquare$

(5) The stability of Fred and of the index (part II). *If $T \in \text{Fred}(X, Y)$ and $A \in \mathfrak{S}_\infty(X, Y)$, then $T + A \in \text{Fred}(X, Y)$ and $\text{ind}(T + A) = \text{ind}T$.*

Indeed, every regularizer R of T is a regularizer of $T + A$, hence by Theorem E.7.2, $T + A \in \text{Fred}(X, Y)$. Then, by (2) and a double application of Theorem E.7.3, $\text{ind}T = -\text{ind}R = \text{ind}(T + A)$ ensues. $\blacksquare$

Remark. Reminders on the Calkin algebra

As defined in Appendix D.6, the Calkin algebra is the quotient algebra of $L(X)$ by the bilateral ideal $\mathfrak{S}_\infty(X)$,

$$\mathcal{K} = L(X)/\mathfrak{S}_\infty(X).$$

Retaining the notation of Appendix D, we let Π be the canonical projection of $L(X)$ onto $\mathcal{K}$,

$$\Pi(T) = T^\bullet =: T + \mathfrak{S}_\infty(X), \quad T \in L(X).$$

The *essential spectrum* of an operator $T \in L(X)$ is defined as that of $\Pi(T)$ in the algebra $\mathcal{K}$,

$$\sigma_{\text{ess}}(T) =: \sigma_\mathcal{K}(\Pi(T));$$

see Appendices C and D for more information.

Theorem E.7.5 (Fredholm spectrum and essential spectrum) *Let $T \in L(X)$.*

(1) $\sigma_F(T) = \sigma_{\text{ess}}(T)$.
(2) *In particular, $T \in \text{Fred}(X)$ if and only if $\Pi(T)$ is invertible in $\mathcal{K}$.*
(3) *If $T \in \text{Fred}(X)$, then $\Pi(T)^{-1} = \Pi(R)$ for every regularizer of T.*
(4) $\|\Pi(T)^{-1}\| = \inf \|R\|$ *where R runs over the set of regularizers of T.*
(5) $\text{ind}(\lambda I - T)$ *is constant on every connected component of $\mathbb{C} \setminus \sigma_{\text{ess}}(T)$. In particular, such a component ω is included in the spectrum $\sigma(T)$ if $\text{ind}(\lambda_0 I - T) \neq 0$ for some $\lambda_0 \in \omega$.*

Proof Since Π is a homomorphism (see Appendix D.6), for every $R, T \in L(X)$ we have $\Pi(RT) = \Pi(R)\Pi(T)$, and hence $\Pi(T)$ is invertible if and only if there exist R_1, R_2 such that $\Pi(R_1 T) = I^\bullet$, $\Pi(T R_2) = I^\bullet$, i.e. $R_1 T - I \in \mathfrak{S}_\infty(X)$, $T R_2 - I \in \mathfrak{S}_\infty(X)$. By Theorem E.7.2, this reasoning shows (2), (1), and (3).

For (4), clearly $\|\Pi(T)^{-1}\| = \|\Pi(R)\| = \inf\{\|R + A\| : A \in \mathfrak{S}_\infty(X)\}$ for every regularizer R. However, by Corollary E.7.4(1), $\{R + A : A \in \mathfrak{S}_\infty(X)\}$ is the set of regularizers of T.

Property (5) follows from Corollary E.7.4(3). ∎

Before passing to the final theorem of the Appendix, recall that according to a classical theorem of Banach (see Appendix D.3), a mapping $A \in L(X, Y)$ is invertible if and only if there exists $\epsilon > 0$ such that $\|Ax\| \geq \epsilon \|x\|$ ($\forall x \in X$) and $\ker A^* = \{0\}$ (the last condition is, of course, a consequence of $\|A^* f\| \geq \epsilon \|f\|$ ($\forall f \in Y^*$)). To replace "invertible" with "regularizable" (hence Fredholm) we require a lemma from linear algebra.

Lemma E.7.6 *Let X, Y be two vector spaces and let $T : X \to Y$ be a linear mapping. If $E \subset X$ is a subspace with finite codimension such that $\ker(T|E) = \{0\}$, then*

$$\dim \ker T \leq \operatorname{codim} E < \infty.$$

Proof Let $L \subset X$ be a complement of E, $X = E + L$ (direct sum), $\dim L = \operatorname{codim} E$, and $\mathcal{P}_{L\|E}$ the projection onto L parallel to E. By hypothesis, the restriction $\mathcal{P}_{L\|E} | \ker T$ is injective, thus

$$\dim \ker T = \dim(\mathcal{P}_{L\|E}(\ker T)) \leq \dim L.$$

∎

Theorem E.7.7 (Fred and subspaces of finite codimension) *Let $T \in L(X, Y)$. The following assertions are equivalent.*

(1) $T \in \operatorname{Fred}(X, Y)$.

(2) *There exist subspaces $E \subset X$ and $E_* \subset Y^*$ of finite codimension such that*

$$\|Tx\| \geq \epsilon \|x\| \quad (\forall x \in E) \quad \text{and} \quad \|T^* f\| \geq \epsilon \|f\| \quad (\forall f \in E_*).$$

(3) $\dim \ker T^* < \infty$ *and there exists a subspace $E \subset X$ of finite codimension such that*

$$\|Tx\| \geq \epsilon \|x\| \quad (\forall x \in E).$$

Proof (1) $\Rightarrow$ (2) If R is a regularizer of T such that $RT - I \in \mathfrak{F}(X)$ and $TR - I \in \mathfrak{F}(Y)$, let $E = \ker(RT - I)$ and $E_* = \ker(R^* T^* - I)$. Then $\operatorname{codim} E < \infty$, and for every $x \in E$ we have $\|x\| = \|RTx\| \leq \|R\| \cdot \|Tx\|$, which gives the first lower bound. Doing the same for T^* completes the proof.

(2) ⇒ (3) Since $\ker(T^*|E_*) = \{0\}$, we can apply Lemma E.7.6, giving $\dim \ker T^* < \infty$, and hence (3).

(3) ⇒ (1) By Lemma E.7.6, $\dim \ker T < \infty$. Moreover, by the hypothesis, $T|E$ is an isomorphism to its range $T(E)$, and thus $T(E)$ is closed. However, since $\mathrm{codim} E < \infty$, there exists a complement L, $L \subset X$, $\dim L < \infty$ such that $X = E + L$ (direct sum). We have $\dim T(L) < \infty$, hence $T(L)$ is closed, and by Appendix A.3, $T(X) = T(E) + T(L)$ is also closed, thus $T \in \mathrm{Fred}(X, Y)$. ∎

Remark In a Hilbert space, the conditions (2) and (3) can be realized on a *nested sequence of subspaces given in advance*, which is interesting from the point of view of the finite section method (see §5.4). More precisely, let H be a Hilbert space and let (H_n) be a sequence of closed subspaces of H such that

$$H \supset H_1 \supset H_2 \supset \cdots \supset H_n \supset \cdots, \quad \cap_n H_n = \{0\}, \quad \mathrm{codim} H_n < \infty \quad (\forall n).$$

Then,

(a) for any $C \in \mathfrak{S}_\infty(H)$, $\lim_n \|C|H_n\| = \lim_n \|CP_{H_n}\| = 0$ (where P_{H_n} is an orthogonal projection onto H_n),

(b) for every $T \in L(H)$, $\|T\|_{\mathrm{ess}} = \lim_n \|T|H_n\|$,

(indeed, on one hand, $T(I - P_{H_n}) \in \mathfrak{S}_\infty(H) \Rightarrow \|T\|_{\mathrm{ess}} \leq \lim_n \|T|H_n\|$, and, on the other hand, $\|T|H_n\| \leq \|T + C\| + \|C|H_n\|$ (for any $C \in \mathfrak{S}_\infty(H)$) giving $\lim_n \|T|H_n\| \leq \|T\|_{\mathrm{ess}}$ thanks to (a)),

(c) $T \in \mathrm{Fred}(H) \Leftrightarrow (\exists \epsilon > 0, \exists n \geq 1$ such that $\|Tx\| \geq \epsilon \|x\|$ and $\|T^*x\| \geq \epsilon \|x\|$, $\forall x \in H_n$).

Indeed, the implication ⇐ follows directly from the theorem above, and the converse ⇒ follows the proof of (1) ⇒ (2) of the same theorem: it suffices to remark that for $x \in H_n$, we have

$$\|R\| \cdot \|Tx\| \geq \|RTx\| \geq \|x\| - \|(RT - I)x\|$$

and $\lim_n \|(RT - I)|H_n\| = 0$ (and the same for T^*). ∎

E.8 Some Examples

(1) The Fredholm alternative [Riesz, 1916]. Let $T = \lambda I - A$ where $\lambda \in \mathbb{C} \setminus \{0\}$ and $A \in \mathfrak{S}_\infty(X)$ for a Banach space X. Then,

- either $\ker T = 0$, and then $T \in L(X)^{-1}$,

- or $\ker T \neq \{0\}$ and then $\dim \ker T = \dim \ker T^* < \infty$ and the equation $Tx = y$ is resolvable if and only if $y \perp \ker T^*$ (in the sense of the duality between X and X^*), and the general solution of the equation $Tx = y$ is $x = x' + \ker T$, where x' is a particular solution.

This is a direct consequence of Theorem E.7.1. ∎

Remark on the historical form of the Fredholm alternative [Fredholm, 1900, 1903]. In fact, Fredholm examined only the integral operators of the example Appendix D.4 on the spaces $L^p(a,b)$ or $C[a,b]$,

$$Af(x) = \int_a^b K(x,y) f(y)\, dy,$$

where $K \in C([a,b]^2)$ or $K \in L^2([a,b]^2)$ and $p = 2$ (see Appendix D.8).

(2) Approximation by compact operators [Yosida, 1940]. *Let $A \in L(X)$ and define the "Fredholm radius" of A as*

$$r_F(A) = \lim_n \mathrm{dist}_{L(X)}(A^n, \mathfrak{S}_\infty(X))^{1/n}$$

[Radon, 1919]. Then, for every $|\lambda| > r_F(A)$, $\lambda I - A \in \mathrm{Fred}(X)$.

Indeed, $\sigma_F(T) = \sigma_{\mathrm{ess}}(T)$ by Theorem E.7.5, and $\mathrm{dist}_{L(X)}(A^n, \mathfrak{S}_\infty(X)) = \|\Pi(A)^n\|_{\mathcal{K}}$, where $\mathcal{K}$ is the Calkin algebra. Hence $r_F(A) = \max_{\lambda \in \sigma_F(A)} |\lambda|$ (Gelfand's formula, Appendix C.3), and the result follows. ∎

(3) The adjoint operator and Fred. *Let $T \in L(X,Y)$. Then, $T \in \mathrm{Fred}(X,Y) \Leftrightarrow T^* \in \mathrm{Fred}(Y^*, X^*)$, and if $T \in \mathrm{Fred}(X,Y)$ then $\mathrm{ind} T = -\mathrm{ind}(T^*)$.*

By a theorem of Banach (see Appendix D.3), if one of the $\mathrm{Ran}\, T$ or $\mathrm{Ran}\, T^*$ is closed, then so is the other, and moreover $(\ker T)^\perp = \mathrm{Ran}\, T^* = (\ker T^{**})_\perp$, hence $\dim \ker T = \mathrm{codim}(\mathrm{Ran}\, T^*) = \dim \ker T^{**}$. Thus, if one of T or T^* is Fredholm, then both $\mathrm{Ran}\, T$, $\mathrm{Ran}\, T^*$ are closed, and $\dim \ker T^* < \infty$ and $\dim \ker T = \dim \ker T^{**} < \infty$. ∎

(4) Normal operators on a Hilbert space. *Let $N \colon H \to H$ be a normal operator on a Hilbert space H (see Appendix D.7). Then $N \in \mathrm{Fred}(H)$ if and only if either N is invertible or 0 is an isolated point of the spectrum $\sigma(N)$ and $\dim \ker N < \infty$.*

The sufficiency is evident. For the necessity, since $\|Nx\| = \|N^* x\|$ ($\forall x$) for every normal operator (see Appendix D.7), then $\ker N = \ker N^*$, and hence any normal operator $N \in \mathrm{Fred}(H)$ with $\ker N = \{0\}$ is necessarily invertible. Applying this conclusion to $N|(\ker N)^\perp$, we obtain the result. ∎

(5) An isometry $V: X \to X$ **on a Banach space** *is Fredholm if and only if* $\dim \ker V^* < \infty$.

Clearly, $\operatorname{Ran} V$ is always closed and $\ker V = \{0\}$. ∎

(6) Multiplication operators. *Let* (Ω, μ) *be a measure space and* $1 \le p \le \infty$, $\varphi \in L^\infty(\Omega, \mu)$. *The multiplication operator* $M_\varphi: L^p(\Omega, \mu) \to L^p(\Omega, \mu)$, $M_\varphi f = \varphi f$, *is Fredholm if and only if there exists a finite set* $\sigma \subset \Omega$ *such that* $1/\varphi \in L^\infty(\Omega \setminus \sigma, \mu)$, *and in this case*, $\operatorname{ind}(M_\varphi) = 0$.

Clearly, $M_\varphi^* = M_\varphi$ (but on the space $L^{p'}(\Omega, \mu)$; we use the bilinear duality between $L^p(\Omega, \mu)$ and $L^{p'}(\Omega, \mu)$, $1/p + 1/p' = 1$, $p < \infty$ – for $p = \infty$, we treat M_φ as the adjoint of M_φ on L^1: see (3) above).

Moreover,

$$\ker M_\varphi = \{f \in L^p(\Omega, \mu): f\varphi = 0\} = \{f \in L^p(\Omega, \mu): f = 0 \text{ } \mu\text{-a.e. on } \Omega \setminus \sigma\},$$

where $\sigma = \{x \in \Omega: \varphi(x) = 0\}$ (defined uniquely μ modulo 0). Clearly the formula provides $\dim \ker M_\varphi = \dim \ker M_\varphi^*$. Next, $0 < \dim \ker M_\varphi < \infty$ if and only if σ is finite and $\mu(\sigma) > 0$. Writing $L^p(\Omega, \mu) = L^p(\sigma, \mu) + L^p(\Omega \setminus \sigma, \mu)$ (direct sum; we consider $L^p(\sigma, \mu)$ as a subspace of $L^p(\Omega, \mu)$ by extending $f \in L^p(\sigma, \mu)$ by zero on $\Omega \setminus \sigma$), we apply the theorem of Appendix E.6 to obtain that M_φ is Fredholm if and only if the restrictions $M_\varphi | L^p(\sigma, \mu)$ and $M_\varphi | L^p(\Omega \setminus \sigma, \mu)$ are also Fredholm, hence if and only if σ is finite and $M_\varphi | L^p(\Omega \setminus \sigma, \mu)$ is invertible. The result follows. ∎

E.9 By Way of Conclusion

The above examples are few in number and particularly illustrative because – among other reasons – Theorem E.7.1 affirms that every Fredholm operator of index 0 is an isomorphism plus a compact operator $V + C = V(I + V^{-1}C)$, and the operators of the form $I + V^{-1}C$ are the subject of Appendix E.8(1). The situation is richer for the Toeplitz operators (see Chapter 2) and for non-bounded operators (differential and pseudo-differential): see the comments in Appendix E.10 and §3.5.

Staying within the context of bounded operators, we conclude with a short list of (tiny) technical traps that await the solitary voyager in the land of Fredholm (these are also exercises, as we skip over the details and provide only a few short hints).

(1) The definition. *Any two of the conditions in the definition Appendix E.1 (i.e.* $\dim \ker T < \infty$, $\dim \ker T^* < \infty$, *and* $\operatorname{Ran}(T)$ *closed) do not necessarily imply the third.*

Hint Consider the multiplication operators of Appendix E.8(6) or E.9(7) below.

(2) The Ker and Coker. *In general, neither* $\dim \ker(\lambda I - T)$ *nor* $\dim \ker(\lambda I - T)^*$ *remain constant on a connected component of* $\mathbb{C} \setminus \sigma_{\text{ess}}(T)$ *(compare with Theorem E.7.5(5)).*

Hint Consider a diagonal operator $T: \ell^2 \to \ell^2$, $T(x_k) = (x_k \lambda_k)$ where $\lambda_k \to 0$ for which $\sigma_{\text{ess}}(T) = \{0\}$.

(3) The left inverse. *For* $T \in L(X, Y)$, *the condition* $\|Tx\| \geq \epsilon \|x\|$ ($\forall x \in X$) *does not, in general, imply the existence of a left inverse (an* $R \in L(Y, X)$ *such that* $RT = I_X$).

Hint Consider the embedding mapping $j: c_0 \to \ell^\infty$, $j(x) = x$, knowing that $c_0 = \{x = (x_j)_{j \geq 0} : \lim_j x_j = 0\}$ is not complementable in ℓ^∞ (there is no bounded projection $P: \ell^\infty \to \ell^\infty$ satisfying $P(\ell^\infty) = c_0$).

See also Appendix D.3 for the properties of the Ran$T$, ker$T$, and the inverses.

(4) Non-trivial index. *The function* $i: T \mapsto \text{ind} T$ *is not necessarily "non-trivial" on* Fred(X): i *is not identically zero if and only if the space* X *is isomorphic to at least one closed proper subspace of finite codimension.*

Hint [Yood, 1951] Consider the direct decompositions $X = \ker T + L$ and $X = M + TX$ with dim Ker T – dim $M \neq 0$).

(5) Connected components of Fred(X). *The converse of Corollary E.7.4(4) is not necessarily correct. Let* X *be a Banach space,* $n \in \mathbb{Z}$ *and* Fred$_n(X) = \{T \in \text{Fred}(X): \text{ind} T = n\}$. *Then* Fred$_n(X)$ *is an open connected component (by arcs) if and only if the linear group* $GL(X) =: L(X)^{-1}$ *is connected (by arcs)* [Coburn and Lebow, 1966]. *In particular, this is the case if* $X = H$ *is a Hilbert space* [Cordes and Labrousse, 1963].

Remark $L(X)^{-1}$ is connected by arcs (and even *contractible*: there exists a continuous mapping $\varphi: L(X)^{-1} \times [0, 1] \to L(X)^{-1}$ such that, for every $T \in L(X)^{-1}$, we have $\varphi(T, 0) = T$, $\varphi(T, 1) = I_X$) for a Hilbert space $X = H$ (found by Kuiper in 1965), for $X = \ell^p$, L^p (Neubauer, 1967), and for many other spaces, see [Mityagin, 1970]. There exist Banach spaces with $L(X)^{-1}$ non-connected [Douady, 1965].

(6) Index Formula of Fedosov–Dynin–Hörmander. *Let* $T \in \text{Fred}(X, Y)$ *and let* $R \in L(Y, X)$ *be a regularizer such that* $RT - I \in \mathfrak{F}(X)$ *and* $TR - I \in \mathfrak{F}(Y)$. *Then*

$$\text{ind} T = \text{Trace}(I_X - RT) - \text{Trace}(I_Y - TR)$$

[Fedosov, 1970; Hörmander, 1971].

Hint Use the decompositions $X = \ker T + L$ and $Y = M + TX$ of Theorems E.7.1–E.7.2.

(7) Multiplication operators on $H^p(\mathbb{D})$. *A restriction of a Fredholm operator is not necessarily Fredholm.*

More precisely, let $\varphi \in H^\infty(\mathbb{D})$ and let $M_\varphi f = \varphi f$ be a multiplication operator $H^p(\mathbb{D}) \to H^p(\mathbb{D})$. Then M_φ is Fredholm if and only if $\varphi = BF$, where B is a finite Blaschke product and $F \in H^\infty(\mathbb{D})^{-1}$.

In particular,

(a) *if $\varphi = z - \lambda$ ($\lambda \in \mathbb{C}$), then $M_\varphi \in \text{Fred}(H^p) \Leftrightarrow |\lambda| \neq 1$, and $\text{ind} M_\varphi = -1$ if $|\lambda| < 1$ and $\text{ind} M_\varphi = 0$ if $|\lambda| > 1$,*
(b) *for an inner function φ that is not a finite Blaschke product, we have $M_\varphi \in \text{Fred}(L^p(\mathbb{T}))$ but $M_\varphi | H^p \notin \text{Fred}(H^p)$.*

Hint See Appendix F for information on Hardy spaces. If $\varphi = VF$ is the inner–outer factorization of φ, then $M_\varphi = M_V M_F$ where M_V is an isometry of $H^p(\mathbb{D})$ and $\dim \ker M_V^* = \deg(V)$ (see the comments in §2.4 on Exercise 2.3.6). Next, $M_F H^p$ is dense in H^p, and if it is closed, then $F \in H^\infty(\mathbb{D})^{-1}$.

(8) Fred and the product of operators. *If $A \in$ Fred and B is a bounded operator, then $AB \in$ Fred (or $BA \in$ Fred) $\Leftrightarrow B \in$ Fred.*

Hint For $\Leftarrow$ see Theorem E.7.3. For the converse, if R is a regularizer of A, then $R(AB) = (I + K_1)B = B + K_2$ where $K_i \in \mathfrak{S}_\infty$, hence B and $R(AB)$ are Fredholm – or not – simultaneously.

E.10 Notes and Remarks

The very first index theorem is none other than the "golden rule" of linear algebra cited in Appendix E.1. First-year students throughout the world are familiar with it by various names: the *Rouché–Fontené theorem* (France), the *Fontené–Rouché–Frobenius theorem* (Germany), the *Rouché–Capelli theorem* (English-speaking countries and Italy), the *Rouché–Frobenius theorem* (Spain and Latin America), or the *Kronecker–Capelli theorem* (Russia and Eastern Europe).

The first attribution (Rouché–Fontené) seems to be justified.

- Eugène Rouché (1832–1910), a French mathematician, was the very first to publish the theorem, in a note in the *Comptes Rendus de l'Académie des sciences* (Paris) in 1875.

- Georges Fontené (1848–1923), another French mathematician, riposted to Rouché's publication with an article in the *Nouvelles Annales de Mathématiques* (1875), providing evidence that he already knew the theorem beforehand.

- Alfredo Capelli (1855–1910), an Italian mathematician, in his 1886 book (with Giovanni Garbieri), presented the same result, and identified rank(A) (previously defined by Georg Frobenius) with dim(Ran(A)), with the aid of minors of the matrix A.

- Leopold Kronecker (1823–1891), a German mathematician, gave several courses on the theory of determinants at the University of Berlin in the years 1862–1874 (he was not a professor at this university but he was already a member of several Academies of Science), where he developed applications to systems of linear equations. Published posthumously in 1903 under the title *Vorlesungen über die Theorie der Determinanten*, this course had great influence on the final form of the theory of linear equations with a finite number of variables.

- Georg Frobenius (1849–1917), a German mathematician, clarified the question of attributions, on one hand recognizing Fontené's role, and on the other hand citing in this connection certain results of Kronecker on systems of linear equations dating back to 1874.

One way or another, but around 1903, the theory of equations of a finite number of independent variables was supplanted by a "superior" theory, that of Fredholm (see also the biography at the beginning of Appendix E). Fredholm [1900, 1903] (in *Acta Mathematica*) gave a burst of acceleration to the theory of functional equations (whose influence is felt even today) by proving the theorem of Appendix E.8 for the integral operators A,

$$Af(x) = \int_a^b K(x,y)f(y)\,dy, \quad a \le x \le b,$$

in the spaces $L^p(a,b)$. (The general case of $T = \lambda I - A$ where $A \in \mathfrak{S}_\infty(H)$ was resolved later by Frigyes Riesz [1916] and Schauder [1930].) These articles contain what is known today as the "Fredholm alternative," and also the important new technique of *Fredholm determinants*. The latter is a rich generalization of the techniques of matrix determinants aimed at distinguishing the cases where $\ker(\lambda I - A) = \{0\}$ (and hence $\lambda I - A$ is an isomorphism) from the cases where $\ker(\lambda I - A) \ne \{0\}$. More precisely,

Fredholm considered the above integral operator A on the space $C[a, b]$ and showed [Fredholm, 1903] that (under the hypothesis of a kernel $K \in C([a, b]^2)$), for every $\lambda \in \mathbb{C} \setminus \{0\}$,

$$\ker(\lambda I - A) = \{0\} \text{ if and only if } \det\left(I - \frac{1}{\lambda}A\right) \neq 0,$$

where by definition

$$\det\left(I - \frac{1}{\lambda}A\right) = \sum_{n \geq 0} \frac{(-1)^n a_n}{n! \lambda^n},$$

$$a_n = \int_{[a,b]^n} \det\left(K(x_i, x_j)_{1 \leq i, j \leq n}\right) dx_1 \ldots dx_n;$$

the series converges for every $\lambda \in \mathbb{C}$ by the *Hadamard inequality* for the determinant of a matrix M of dimension $n \times n$:

$$|\det M| \leq \prod_{k=1}^{n} \|Me_k\|,$$

where (e_k) is an orthonormal basis (see Exercise 5.7.0(a(3))). The coefficients a_n can also be found with the following recurrence (with the aid of a sequence $b_n = b_n(x, y)$ defining, conjointly with $\det(I - (1/\lambda)A)$, the resolvent of A)

$$a_0 = 1, \quad b_0(x, y) = K(x, y), \quad a_n = \int_a^b b_{n-1}(x, x) \, dx,$$

$$b_n(x, y) = K(x, y) a_n - n \int_a^b K(x, t) b_{n-1}(t, y) \, dt \quad (n \geq 1).$$

Even to this day, the theory of determinants is an indispensable tool in the theory of operators and in mathematical physics. In particular, it extends far beyond the trace class operators A, $A \in \mathfrak{S}_1(X)$ (or "nuclear," the natural framework, where it was developed by Grothendieck [1956]), or the Schatten–von Neumann classes $\mathfrak{S}_p(X)$ (regularized determinants initiated by [Hilbert, 1904] and [Poincaré, 1910b]). The literature on Fredholm determinants is immense, and we limit ourselves to the high points of the subject. For a "genetic" presentation following the lines of Fredholm [1903], we cite [Riesz and Szőkefalvi-Nagy, 1955, §74], and [Gohberg, Goldberg, and Kaashoek, 1990], and for more modern treatments, [Simon, 2005a], [Gohberg, Goldberg, and Krupnik, 1996], [Ruston, 1986], and [Reed and Simon, 1978] (especially §13.17 as well as the "Historical Remarks" at the end of Chapter 13; for a historical aspect

see also [Pietsch, 2007]). An overview appears in the Wikipedia entry for "Determinant" (https://en.wikipedia.org/wiki/Determinant).
We return to the history of Fredholm theory.

- Frigyes (or Friedrich) Riesz [1916] proved the Fredholm alternative (Appendix E.8(1)) for $T = \lambda I - A$, $A \in \mathfrak{S}_\infty(X)$, where X is the space $C[a,b]$ or a Hilbert space.

- Fritz Noether [1921] was the first to study the class of operators $M_a + bS$, where S is the Cauchy singular integral operator and a, b are continuous functions; *it was here that the notion of index appeared* explicitly (*de facto*), with arbitrary integer values. See also the biography of Noether as well as the history of the "index" below.

Fritz Noether (1884–1941) was a German mathematician. He was the younger brother of the famous Emmy Noether (an algebraist and theoretical physicist, 1882–1935), and the son and father of mathematicians. Most of Noether's publications (the complete list consists of only 36) concern applied mathematics and mechanics, including his doctoral thesis (1909) written under the inspiration of the physicist Arnold Sommerfeld. The results of his thesis constitute the principal portion of the last of the four volumes of the classic *Theory of the Gyroscope* (in German) by Klein and Sommerfeld (1910). Remaining faithful to applied mathematics throughout his life, Noether enters into the history of mathematics thanks to what is practically his only theoretical work (*Math. Annalen* **82** (1921), 42–63), where he proved the very first index theorem, in particular by introducing the index of the symbol of an operator (without using the word). The history of the "index" is described below (Appendix E.10); here we briefly retrace the tragic odyssey of the life of Fritz Noether.

In 1911, Noether submitted his *Habilitation* thesis, and then obtained his first position as a Privatdozent and was married (to Regina Würth; they had two sons (1910, 1915)). He fought in the First World War on the French–German front, was wounded in April 1917, and decorated with the Iron Cross for his courage. His years as a professor between 1922 and 1933 at the Technische Hochschule of Breslau (Wrocław) were rudely interrupted by his eviction due to the anti-Jewish Nazi law of April 7, 1933. In principle, Noether should not have been affected as he had fought in the war, but he was ousted, at the demand of the "Students' Society"; the students of the German universities at the time were even more anti-Semitic than the professors. Even in Göttingen, a boycott of Edmund Landau's course was organized in 1933 by Oswald Teichmüller, a young and otherwise brilliant mathematician. The same fate befell Fritz's sister Emmy Noether (1882–1935; see the photo of them together), famous as the founder of modern algebra and author of a fundamental theorem of theoretical physics on the relationships between symmetries and conservation laws. Being resolutely on the political left, they both decided to leave Nazi Germany for Stalin's Soviet Union. Losing patience (happily!) with the bureaucratic delays of the Soviets, Emmy accepted a modest position as guest lecturer at Bryn Mawr College, Pennsylvania (where, unfortunately, she fell ill and died suddenly of complications from an operation on an ovarian cyst). By contrast, and in spite of his wife's opposition, Fritz accepted a position as professor at the University of Tomsk (in Siberia, some 3600 kilometers from Moscow, historically the place where political and ethnic exiles were sent – the "kulaks," Jews, Poles, Lithuanians, etc.).

On his arrival in Tomsk in 1934, he breathed life into a seminar in mathematics (with Stefan Bergman who was there for much the same reason), founded a journal (*Bulletin of the Tomsk State University* (*Zapiski Tomskogo Universiteta*)), and began to study Russian. In 1935 he visited Germany, where he received two pieces of bad news – first the death of his sister and then the suicide of his wife, who left him with two young children. In 1936, the International Congress of Mathematicians (ICM) was held in Oslo (where the very first Fields Medal was awarded), but the participation of Soviet mathematicians was forbidden by the Bolsheviks. Nonetheless, Noether profited from his German passport to go, and was immediately suspected of "treason" (according to later witnesses, an unbelievable circumstance played an important role in this affair: by an unhappy coincidence, Leon Trotsky, Stalin's number one enemy, was in Oslo at the time). Noether was put under watch by the NKVD (predecessor of the KGB) for a year, and then arrested on September 22, 1937 and accused of espionage and sabotage for the Germans (who had kicked him out!) and even for the preparation of terrorist acts such as introducing German submarines to the delta of the Ob River in Siberia. His book on Bessel functions (ready for publication) was pulled, and his children expelled – first to Sweden, and then to the United States.

Noether was condemned to 25 years of imprisonment and sent to the Central Prison in Orel (*Orel* = "Eagle"; see the photo), a city in the European part of Russia. On April 28, 1938, Einstein wrote a letter in his defense to Maxim Litvinov, Stalin's Minister of Foreign Affairs, and then on October 3, 1939 Hermann Weyl wrote to the Georgian mathematician Nikoloz Muskhelishvili, asking him to act through the intermediary of "his friend Lavrentiy Beria" (according to Weyl), who had become head of the NKVD in 1938 – but all in vain. After the invasion of the Soviet Union by the Third Reich, when the front drew near to Orel, on September 10, 1941 the NKVD summarily executed hundreds of prisoners of the Central Prison, including Fritz Noether (there is testimony stating that the list of prisoners executed was approved by Stalin himself). According to recollections of former employees of the prison, the "technique" of execution was particularly sadistic: the victims were led to a room under the pretext of an "interrogation," then tied up, their mouths stuffed with a dishcloth, a short sentence was pronounced, and they were finished off with a pistol shot to the back of the neck. The author of the very first index theorem was among them.

The fact of the execution and the dates were confirmed by the Russian Supreme Court in the 1990s, using the archives of the KGB, as was the fact that all of the accusations against Noether were falsified. Nonetheless, according to Boris Schein (there is a photo of his report in *Zentralblatt für Mathematik und ihre Grenzgebiete*, Zbl 0449.01015, in the book *Emmy Noether, 1882–1935* by Auguste Dick), Fritz Noether was seen alive in Moscow at the end of 1941 (presumably to play a role in an NKVD project, never realized, to create an "International Jewish anti-Hitlerian Committee"). For more information on Fritz Noether see the special issue of *Integral Equations and Operator Theory* (**8:5** (1985)) dedicated to his memory, as well as Siegmund-Schultze ([2009], page 97), and the website http://berkovich-zametki.com/2010/Starina/Nomer2/Berkovich1.php.

- Theophil H. Hildebrandt [1928], Juliusz Schauder [1930], and Stefan Banach [1932] justified the Fredholm alternative for $\lambda I - A$, $A \in \mathfrak{S}_\infty(X)$ on Banach spaces.

- Felix Hausdorff [1932] identified and formally defined the class of "normally solvable" equations (operators). It is interesting to note that for the range $T(X)$ of an operator $T \in L(X, Y)$ (X, Y are Banach spaces), the (purely algebraic) condition $\dim(Y/T(X)) < \infty$ already implies that $T(X)$ is closed (which is not the case for an arbitrary vector subspace: for example, in every infinite-dimensional Banach space Y there exist hyperplanes $Y_1 \subset Y$ with $\dim(Y/Y_1) = 1$ that are not closed).

Felix Hausdorff (1868–1942), a German mathematician, was one of the founders of the topology and set theory of the twentieth century, as well as measure theory and functional analysis. His classical text *Grundzüge der Mengenlehre* (1914), translated into several languages and published in a number of editions, played a major role in establishing the modern axiomatic approach to topology, metric spaces, and the general theory of equations. Hausdorff is known for the notion of *Hausdorff space*, for the *Hausdorff–Young inequality*, for the solution of the *Hausdorff moment problem*, for his "paradoxical" decomposition of the sphere S^2 (solution in the negative to the "general problem of measure" – the non-existence of a "natural" non-trivial measure on the total σ-algebra of S^2), for the invention of *Hausdorff measures* associated with an arbitrary monotone function, and for the *Hausdorff dimension* (1919) – a distant precursor of fractals (in particular, he showed that the dimension of the Cantor "missing thirds" set is $\ln 2/\ln 3$).

As a youth, Hausdorff attempted to become a composer, but under family pressure he continued his studies in astronomy and astrophysics at the University of Leipzig. His doctoral thesis (1891) and *Habilitation* thesis (1895) were devoted to these subjects. But his principal interests, at the time, were in literature and philosophy: between 1897 and 1904 Hausdorff published (under the pseudonym Paul Mongré) several stories, poems, philosophical essays, and plays (*Der Arzt seiner Ehre* was staged in 1912, with great success). He started working in topology in 1901 (on partially ordered sets, a concept he introduced), but perhaps had not definitively chosen a career in mathematics, as in 1902 he refused an offer as "extraordinary professor" at Göttingen in order to stay in Leipzig. Later, he was professor at the University of Bonn (1910–1913, 1921–1935). Under the rise of the Nazi regime in Germany, Hausdorff continued to work for some time (the 1933 law of eviction of Jews initially did not apply to civil servants already in place before the 1914 war), but his efforts to remain loyal to Hitler's *Neue Ordnung* (in particular, in 1934 he took an oath of fidelity to the regime) did not make much difference in the end: he was fired in 1935. Hausdorff made several attempts to obtain a position in the United States, but it was too late. In 1941 he and his wife, Charlotte Sara Goldschmidt, whom he had married in 1899, were obliged to wear the "Jewish yellow star"; under the threat of imminent deportation to the extermination camps, they committed suicide using barbiturates at the end of January 1942.

The university library of Bonn possesses approximately 26 000 pages of Hausdorff's manuscripts. A team of editors has been put in place, comprising 16 mathematicians and logicians, five historians of mathematics and astronomy, and three philosophers and Germanists. At the end of 2011, six of the planned nine volumes (the literary and philosophical works) were published by Springer. In Hausdorff's honor, his name has been given to a street in Bonn, to the Hausdorff Research Institute for Mathematics in Bonn, and to the asteroid 24947.

- Sergei M. Nikol'skyï [1943] proved Theorem E.7.1 in full generality.

- Jean Dieudonné [1943], in an article published under the Occupation during the Second World War (and for this reason overlooked until 1951), showed the stability in Corollary E.7.4(3) (a portion of this result corresponding to the index 0 is also found in [Nikol'skyï, 1943]).

- Frederick Atkinson [1951] (in an article submitted in 1948 and published in Russian) gave the final form to the theory of Fredholm operators, as presented in Appendices E.1–E.9, with the exception of the terminology; in particular, he referred to the elements of Fred as "generalized Fredholm operators" (keeping the name "Fredholm operators" for T with $T \in$ Fred, ind$T = 0$). He did not use the word "index" at all (see the history of the index below), but proved its log-homomorphic nature, its continuity, and the stability with respect to compact perturbations.

- Israel Gohberg [1951a], independently of Atkinson and Dieudonné, obtained a portion of the results of [Atkinson, 1951], including the stability theorems of Corollaries E.7.4(3) and E.7.4(5) (the latter also appeared in [Yood, 1951]).

- *The "post-mature" history of Fredholm operators.* After the "canonization" of the theory in a survey article by Gohberg and Krein [1957], its history revolves mainly around "index theorems" for elliptic differential operators, Toeplitz operators, and other pseudo-differential operators. The principle, discovered for a special case by Noether (1921), states that such an operator T can be associated with a function (on a manifold) sym$T$, or with an element of a fairly simple algebra (generally commutative), that plays the role of an "indicator" of the operator, in order to then establish that T is Fredholm if and only if sym$T$ is invertible, and that the index ind$T$ (also called the "analytical index") coincides with the Cauchy "topological index" of sym$T$ (this latter usually corresponds to the "number of obstacles" to overcome for an invertible function to become an exponential).

The "milestone" publications of this circle of ideas are those of Mikhlin [1948] on differential operators (it was Mikhlin who first used the expression the "symbol of an operator" in the 1930s), Krein and Krasnoselskii [1952] (an extension of the theory to unbounded operators), Gohberg and Krein [1957] on the Toeplitz operators where the technique of Banach algebras was systematically used, Gelfand [1960] with a program of studies of the index of elliptic operators, and Atiyah and Singer [1963]. The latter is famous for the announcement of the result universally known as the

Atiyah–Singer index theorem (the solution to a problem posed by Gelfand [1960]): the index of an elliptic operator on a compact differential manifold is equal to its topological index. The proof announced and sketched out in the article was never published, but was reconstructed from its description by the participants of the Henri Cartan seminar in 1963–1964 and then published in the Palais seminar in 1965. Later, Atiyah and Singer found a proof using other promising techniques (those of K-theory), whereas the first short proof belongs to Fedosov [1970] (for a particular case), then Getzler (in 1983), but this is already another history (for which the reader could begin with the (competent) Wikipedia article http://en.wikipedia.org/wiki/Atiyah-Singer_index_theorem).

We conclude by adding that the index theorem sparked a veritable torrent of generalizations, applications, and adaptations. In particular, there exist versions for elliptic pseudo-differential operators, for the "Fredholm complexes" (where the index of principal operator coincides with the Euler characteristic of the complex: see [Hörmander, 1985]), for von Neumann algebras, etc., even including versions for Toeplitz operators on manifolds, in several variables, etc.: see [Hörmander, 1985] and [Boutet de Monvel and Guillemin, 1981], to mention only a very few.

- *An abridged history of the terminology.* Terminology is an important part of the language of mathematics. That is why it is interesting to track down the emergence of terms that have become true landmarks for domains of research, such as the term "index."

For the Cauchy "topological index" of a curve, see Definitions 3.1.2 above, as well as the comments about Definitions 3.1.2 in §3.5). The history of its operator counterpart is more intriguing. It is said that "What does not have a name does not exist" (a historical sophism), but nevertheless the *analytical index* of an operator $\operatorname{ind} T = \dim \ker T - \dim \operatorname{coker} T$ circulated in the literature without being named for 30 years, from its introduction (without a name) by Noether [1921] up to its late baptism by Gohberg [1951b], Yood [1951], and Krein and Krasnoselskii [1952] (independently). Then, these "godparents" behaved very differently with respect to the "newborn": Yood, in his subsequent publications (1954–1955), carefully avoided using the term, as if it had escaped from his pen (in his preceding article), whereas Krein and his school spread it with all of their power, before its definitive "canonization" by Gohberg and Krein [1957]. It is curious to mention that in the years 1921–1951 (and even today), for the term "analytical index," specialists refer to Noether (1921) (sometimes even calling it the "Noetherian operator index"). Indeed the article contains the word

E.10 Notes and Remarks

"index," but applied to the Cauchy index of a curve or a function (i.e. to the symbol of the singular integral operator $T = aI + bS$). It is thus the topological index $\mathrm{ind}_t(aI + bS)$ that is the "Noetherian index" [Noether, 1921] (the "number of turns" of a symbol around 0, or "winding number"). As for the analytical index $\mathrm{ind} T = \dim \ker T - \dim \mathrm{coker} T$, it only "legally" obtained its name from [Gohberg, 1951b], [Yood, 1951], [Krein and Krasnoselskii, 1952], [Gohberg and Krein, 1957], and especially from [Gelfand, 1960], but the shift of the word "index" from "topological index" to "analytical index" began much earlier. See also Gohberg's recollections in his letter to a Dutch publication [Gohberg, 2004].

The adjective "topological" in the expression "topological index" was only introduced in the work of Atiyah and Singer [1963]; previously one spoke of the "Cauchy index" of a symbol (especially on manifolds of dimension 1, i.e. on curves), "number of turns," "winding number," etc.

The term "normally solvable" is 20 years older than the "index" and dates back to Hausdorff [1932].

The "regularizers" of a linear operator, as well as the (almost) synonymous "parametrix" (especially in the theory of PDEs and pseudo-differential operators) and "pseudo-inverses," have been used for a long time – the traces date back to Hilbert (a usage not precisely localized, but probably arising at the time of the treatment of the Riemann–Hilbert problem).

The "essential spectrum" (linked to the invertibility up to a compact operator) was initially defined by Weyl [1910a] for certain differential operators (PDOs) on a Hilbert space as "the spectrum that does not depend on the boundary conditions"; thus the "Weyl spectrum" is in particular

$$\sigma_W(T) = \cap_{A \in \mathfrak{S}_\infty} \sigma(T + A),$$

i.e. $\lambda \in \mathbb{C} \setminus \sigma_W(T)$ if and only if $\lambda I - T$ is Fredholm with index 0. For the *self-adjoint* (or *normal*) operators on a Hilbert space, with which Weyl worked, we have $\sigma_W(T) = \sigma_{\mathrm{ess}}(T)$ (= the spectrum of T minus the isolated eigenvalues of finite multiplicity). Moreover, for self-adjoint (or normal) operator T on a Hilbert space X, there is another characterization (by Weyl) of the essential spectrum: $\lambda \in \sigma_{\mathrm{ess}}(T)$ if and only if there exists a non-precompact sequence (x_k) such that $\lim_k \|Tx_k - \lambda x_k\| = 0$ (in fact, this is an immediate consequence of Theorem E.7.7); according to another terminology, it is a point in the *continuous spectrum* $\sigma_c(T)$; see also Exercise 3.4.1 above. The question of terminology, as before, is not so easy, especially in the domain of the classification of subsets of the spectrum where there is much confusion and incoherence. We refer the reader to [Riesz and

Szőkefalvi-Nagy, 1955], [Reed and Simon, 1978], and the Wikipedia article http://en.wikipedia.org/wiki/Essential_spectrum. With precautions, we can affirm that the expression "essential spectrum" was used for the first time by Hartman and Wintner [1950b] (but concerning $\sigma_W(T)$). For the Weyl spectrum of certain non-normal operators, see [Coburn, 1966] and [Nikolski, 2002b], Appendices C.2.5.3 and C.2.6.

The expression "Fredholm alternative" for Appendix E.8(1) was proposed by Riesz [1916].

The approximation property stated just before Appendix E.1 is a consequence of the *Grothendieck approximation property* (AP), but is not equivalent to it. Recall that, by definition, $X \in$ (AP) if, for every compact $K \subset X$ and for any $\epsilon > 0$, there exists $T \in \mathfrak{F}(X)$ such that $\|Tx - x\| < \epsilon$, $\forall x \in K$. The following properties are known.

(1) $Y \in$ (AP) $\Leftrightarrow$ ($\mathfrak{F}(X, Y)$ is dense in $\mathfrak{S}_\infty(X, Y)$, $\forall X$).
(2) $X^* \in$ (AP) $\Leftrightarrow$ ($\mathfrak{F}(X, Y)$ is dense in $\mathfrak{S}_\infty(X, Y)$, $\forall Y$).
(3) $X^* \in$ (AP) $\Rightarrow X \in$ (AP) (the converse is false).
(4) If X has a Schauder basis, then $X \in$ (AP).
(5) There exist Banach spaces X not having the property (AP), in particular $L(H) \notin$ (AP) where H is a Hilbert space, $\dim H = \infty$ (found by Szankowski in 1978). See [Lindenstrauss and Tzafriri, 1977, 1979] for details and references.

Appendix F
A Brief Overview of Hardy Spaces

All the information below can be found in [Nikolski, 2019]; see also Appendix F.7.

F.1 Hardy Spaces on the Disk and on the Circle

We make the identification $\mathbb{T} = \mathbb{R}/2\pi\mathbb{Z}$ ($e^{it} \equiv t \pmod{2\pi}$). Let m denote the normalized Lebesgue measure on $\mathbb{T}$, defined by

$$m(\{e^{it}: \alpha \leq t \leq \beta\}) = (\beta - \alpha)/2\pi, \quad \text{for } 0 \leq \beta - \alpha \leq 2\pi,$$

and $L^p(\mathbb{T}) = L^p(\mathbb{T}, m)$. For $f \in L^p(\mathbb{T})$, the nth Fourier coefficient is defined as

$$\widehat{f}(n) = \int_{\mathbb{T}} f\bar{z}^n \, dm, \quad n \in \mathbb{Z}.$$

The *Hardy spaces* are defined as

$$H^p(\mathbb{T}) = \{f \in L^p(\mathbb{T}): \widehat{f}(n) = 0 \text{ for every } n < 0\}, \quad 1 \leq p \leq \infty,$$

$$H^p(\mathbb{D}) = \left\{f \in \text{Hol}(\mathbb{D}): \sup_{0<r<1} \|f_{(r)}\|_{L^p(\mathbb{T})} < \infty\right\}, \quad 0 < p \leq \infty,$$

where $f_{(r)}(z) = f(rz)$ for $|z| < 1/r$.

- *Identification.*
  (i) Let $1 \leq p \leq \infty$: $\forall f \in H^p(\mathbb{D})$ there exists $bf \in H^p(\mathbb{T})$ such that

  $$\lim_{r \to 1} \|f_{(r)}|\mathbb{T} - bf\|_p = 0$$

  (if $1 \leq p < \infty$, and weak $*$ limit if $p = \infty$); the mapping $f \mapsto bf$ is an isometric bijection between $H^p(\mathbb{D})$ and $H^p(\mathbb{T})$. Moreover, $\forall f \in H^p(\mathbb{D})$ we have $f_{(r)} = (bf)_r = bf * P_r$ (the *Poisson formula* representing f in terms of bf).

Here, the convolution on $\mathbb{T}$ is defined by

$$f * g(s) = \int_{\mathbb{T}} f(s\bar{t})g(t)\,dm(t), \quad s \in \mathbb{T},$$

and the *Poisson kernel* by

$$P_r(x) = P(r\,e^{ix}) = \frac{1 - r^2}{|1 - r\,e^{ix}|^2} = \sum_{j \in \mathbb{Z}} r^{|j|} e^{ijx}.$$

(ii) Fatou's theorem (1906). For all $f \in H^p(\mathbb{D})$, there exists $\lim_{r \to 1} f(r\zeta) = (bf)(\zeta)$ m-a.e. on $\mathbb{T}$. Moreover, for every complex measure $\mu \in \mathcal{M}(\mathbb{T})$ and every Lebesgue point $\zeta \in \mathbb{T}$ of μ (where there exists a density $(d\mu/dm)(\zeta)$: see §4.1.2(3) above or [Nikolski, 2019] for more details), the Poisson integral of μ,

$$P * \mu(z) = \int_{\mathbb{T}} \frac{1 - |z|^2}{|t - z|^2}\,d\mu(t), \quad z \in \mathbb{D},$$

admits a limit $\lim_{r \to 1} P * \mu(r\zeta) = (d\mu/dm)(\zeta)$ (and even a "non-tangential limit"). In particular, this identity is correct m-a.e.

Given theorems (i) and (ii), we can identify $f \equiv bf$, $f \in H^p(\mathbb{D})$.

- *The case of $C(\mathbb{T})$, the disk algebra.* The same correspondence $f \equiv bf$ identifies

$$C_a(\mathbb{D}) =: H^\infty(\mathbb{D}) \cap C(\overline{\mathbb{D}}), \quad \text{and}$$

$$C_a(\mathbb{T}) = \{f \in C(\mathbb{T}): \widehat{f}(n) = 0 \text{ for every } n < 0\}.$$

F.2 Basic Properties

- $f, \overline{f} \in H^1(\mathbb{T}) \Rightarrow f = \text{const}$.

- *Uniqueness theorem [Riesz and Riesz, 1916].* $f \in H^1(\mathbb{T})$, $\log|f| \notin L^1(\mathbb{T}) \Rightarrow f = 0$.

- *The reproducing kernel of $H^2(\mathbb{D})$.* The space $H^2(\mathbb{D})$ is RKHS (see Appendix B.6 above): $\forall f \in H^2(\mathbb{D})$, $\lambda \in \mathbb{D}$ we have $f(\lambda) = (f, k_\lambda)_{H^2}$ where $k_\lambda(z) = (1/(1 - \overline{\lambda}z))$ (Szegő kernel).

F.2 Basic Properties

- *Inner functions.* $\varphi \in H^\infty$ is said to be *inner* if $|\varphi| = 1$ m-a.e. on $\mathbb{T}$. An inner function is a product of three factors $\varphi = \lambda BV$ where $\lambda \in \mathbb{T}$,

$$B = B_{\{\lambda_k\}} = \prod_{k \geq 1} b_{\lambda_k} \quad \text{(Blaschke product)},$$

$$b_\lambda =: \frac{\lambda - z}{1 - \bar{\lambda}z} \cdot \frac{\bar{\lambda}}{|\lambda|} \quad (\lambda \in \mathbb{D} \setminus \{0\}),$$

$$b_0 = z \quad \text{and} \quad \lambda_k \in \mathbb{D}, \quad \sum_k (1 - |\lambda_k|) < \infty,$$

and

$$V(z) = V_\mu(z) =: \exp\left(-\int_\mathbb{T} \frac{\zeta + z}{\zeta - z} d\mu(\zeta)\right), \quad z \in \mathbb{D},$$

where $\mu \geq 0$ is a measure on $\mathbb{T}$ singular with respect to m ($\mu \perp m$).

- *Spectrum of an inner function.* Let φ be an inner function, $\varphi = \lambda BV$ its factorization as above. The *spectrum of* φ is defined by

$$\sigma(\varphi) = \text{supp}(\mu) \cup \text{clos}\{\lambda_k\},$$

and it can be shown that $\zeta \in \overline{\mathbb{D}} \setminus \sigma(\varphi)$ if and only if $\underline{\lim}_{z \to \zeta} |\varphi(z)| > 0$, and if and only if $1/\varphi$ can be holomorphically extended in a neighborhood of ζ.

- *Outer functions.* Given f with $\log|f| \in L^p(\mathbb{T})$, an *outer function (of Szegő)* with absolute value $|f|$ is defined by

$$[f](z) = \exp\left(\int_\mathbb{T} \frac{\zeta + z}{\zeta - z} \log|f(\zeta)| dm(\zeta)\right), \quad z \in \mathbb{D}.$$

If $f \in L^p(\mathbb{T})$, then $[f] \in H^p$, and for every $f \in H^p(\mathbb{D})$ and every $z \in \mathbb{D}$, $|f(z)| \leq |[f](z)|$.

- *Canonical factorization (Smirnov, 1928, Riesz, and Nevanlinna).* Every $f \in H^1(\mathbb{D})$ can be represented in the form $f = f_{\text{in}} f_{\text{out}}$, where f_{in} is inner and $f_{\text{out}} = [f]$.

- *Corollary.* If $f \in H^1(\mathbb{D})$ and $1/f \in H^p(\mathbb{D})$ ($p > 0$), then $f = \lambda[f]$ ($\lambda \in \mathbb{C}$) (f is outer).

- *The scale of the spaces* $H^p(\mathbb{T})$. If $1/p = 1/q + 1/r$, then $H^p = H^q \cdot H^r$, and moreover $\forall f \in H^p$ there exists $f_1 \in H^q$, $f_2 \in H^r$ such that $f = f_1 f_2$ and $\|f\|_p^p = \|f_1\|_q^q = \|f_2\|_r^r$. If, in addition, $f \in C_a(\mathbb{D})$, then we can choose $f_j \in C_a(\mathbb{D})$, $j = 1, 2$.

- *Invariant subspaces (Beurling, 1949).* Let $E \subset H^p(\mathbb{T})$ be a closed subspace invariant under multiplication, $M_z E \subset E$, $E \neq 0$, then there exists a unique (up to a unimodular constant factor) inner function θ such that

$$E = \theta H^p = \{\theta f : f \in H^p\}.$$

In particular, $E_f = \text{span}_{H^p}(z^n f : n \geq 0) = f_{in} H^p$, and $f \in H^p$ is outer if and only if $E_f = H^p$ (Smirnov, 1928).

F.3 Harmonic Conjugation (Hilbert Transform)

For a real function $u \in L^p(\mathbb{T})$, $1 < p < \infty$, the mapping of *harmonic conjugation* $\mathbb{X}$ (also called *the Hilbert operator*") is defined by

$$u + i\mathbb{X}u \in H^p(\mathbb{T}), \quad \widehat{\mathbb{X}u}(0) = 0.$$

Recall that we have already defined and examined the operator $\mathbb{X}$ in a slightly different form, namely as

$$\mathbb{H}u = i\mathbb{X}u, \quad \text{and also } \mathbb{X}u = \tilde{u};$$

see § 2.3.4 on page 64 and § 4.1.3 on page 186.

Here, we summarize a few basic properties.

- $\mathbb{X}$ is linear (over the field $\mathbb{R}$), and if it is "complexified" by defining $\mathbb{X}(u + iv) = \mathbb{X}u + i\mathbb{X}v$ for every function $u + iv \in L^p(\mathbb{T})$, where u, v are real, then $\mathbb{X}$ becomes linear over the field $\mathbb{C}$ and bounded $\mathbb{X}: L^p(\mathbb{T}) \to L^p(\mathbb{T})$ (Riesz, 1927).

- We have

$$\mathbb{X}f = \frac{1}{i}(P_+ f - P_- f) - \frac{1}{i}\widehat{f}(0),$$

where

$$P_+\left(\sum_{k \in \mathbb{Z}} \widehat{f}(k) z^k\right) = \sum_{k \geq 0} \widehat{f}(k) z^k$$

is the Riesz projection and $P_- = \text{id} - P_+$. Hence, in the space $L^2(\mathbb{T})$, $\|\mathbb{X}f\|_2 \leq \|f\|_2$ ($\forall f \in L^2(\mathbb{T})$) and $\|\mathbb{X}f\|_2 = \|f\|_2$ ($\forall f \in L^2(\mathbb{T}), \widehat{f}(0) = 0$).

- *Herglotz operator and the Poisson conjugate kernel.* For a real function f, we have $\mathbb{X}f = \text{Im}(\Gamma f)$, where Γ is the *Herglotz operator*,

$$\Gamma f(z) = \int_\mathbb{T} \frac{\zeta + z}{\zeta - z} f(\zeta) \, dm(\zeta), \quad z \in \mathbb{D},$$

and thus, for $z = re^{i\theta}$, $0 < r < 1$,

F.3 Harmonic Conjugation (Hilbert Transform)

$$\mathbb{X}f(re^{i\theta}) = \int_0^{2\pi} Q_r(t-\theta)f(e^{it})\,dt/2\pi,$$

where

$$Q_r(t) = \operatorname{Im}\left(\frac{e^{it}+z}{e^{it}-z}\right) = \frac{2r\sin t}{1-2r\cos t + r^2} \quad \text{(Poisson conjugate kernel).}$$

There exist boundary limits a.e. in $\mathbb{T}$, $\lim_{r\to 1} \mathbb{X}f(re^{i\theta})$, as well as limits of the integral on the right-hand side of the equality, which is equal to the "Cauchy principal value" of the following integral:

$$\mathbb{X}f(re^{i\theta}) = \text{p.v.} \int_0^{2\pi} f(e^{it}) \operatorname{ctg}\left(\frac{t-\theta}{2}\right) dt/2\pi,$$

the Hilbert form of the harmonic conjugate. Recall that the *Cauchy principal value* of an integral (at a point θ) is defined by

$$\text{p.v.} \int F(t)\,dt =: \lim_{\epsilon \to 0} \int_{|t-\theta|>\epsilon} F(t)\,dt.$$

- *Kolmogorov's theorem (1925).* If $u \in L^\infty(\mathbb{T})$ is a real function, then $e^{t\mathbb{X}u} \in L^1(\mathbb{T})$ if $t\|u\|_\infty < \pi/2$, where $t > 0$; the bound $\pi/2$ is sharp (it suffices to consider $u = \arg(1-z)$).

- *Lindelöf's theorem (1915).* If $f \in H^\infty(\mathbb{T})$ and if there exist $l_\pm = \lim_{t\to \pm 0} f(e^{it})$, then $l_+ = l_-$.

- *Muckenhoupt condition (1973).* For a positive function $w \in L^1(\mathbb{T})$, the following conditions are equivalent:
  (a)

$$\sup_I \left(\frac{1}{|I|}\int_I w\,dm\right)\cdot\left(\frac{1}{|I|}\int_I \frac{1}{w}\,dm\right) < \infty$$

  ($I \subset \mathbb{T}$ ranges over the arcs of $\mathbb{T}$).
  (b) $w = e^{\tilde{u}+v}$ where $u, v \in L^\infty(\mathbb{T})$ are real and $\|u\|_\infty < \pi/2$ (Helson–Szegő condition).
  (c) $\mathbb{H}$ and $\mathbb{X}: L^2(\mathbb{T}, w) \to L^2(\mathbb{T}, w)$ are bounded.
  (d) $(z^k)_{k\in\mathbb{Z}}$ is a Schauder basis of $L^2(\mathbb{T}, w)$.

F.4 Polynomial Approximation

Theorem (Szegő 1920, 1921a; Verblunsky 1936). Let μ be a Borel measure on $\mathbb{T}$, with $\mu = wm + \mu_s$, $w = d\mu/dm$, its Radon–Nikodym decomposition. Then,

$$d(\mu)^2 =: \mathrm{dist}_{L^2(\mu)}(1, z\mathcal{P}_a)^2 = \inf_{p \in \mathcal{P}_a} \int_\mathbb{T} |1 - zp|^2 \, d\mu = \exp\left(\int_\mathbb{T} \log\left|\frac{d\mu}{dm}\right| dm\right).$$

For the proof (dependent on a study of M_z-invariant subspaces of $L^2(\mu)$) see [Nikolski, 2019] or another text on Hardy spaces (see Appendix F.7 below); the case of μ absolutely continuous $\mu = wm$ was solved by Szegő [1920, 1921a] and the general case by Verblunsky [1936]. At the time, this latter publication went almost unnoticed by the experts, and the result for arbitrary μ is often attributed to Kolmogorov [1941] (this work contains only the case where $d(\mu) = 0$, hence the case of the completeness of polynomials) or to Szegő (the formula exists in [Grenander and Szegő, 1958]). The complete history is retraced in [Simon, 2005b], "Notes" in §2.3.

F.5 Transfer of $H^2(\mathbb{T})$ onto the Real Axis $\mathbb{R}$

- *A unitary mapping of $L^2(\mathbb{T})$ onto $L^2(\mathbb{R})$.* Let $\mathbb{C}_+ = \{z \in \mathbb{C} \colon \mathrm{Im}\, z > 0\}$ and let $\omega \colon \mathbb{D} \to \mathbb{C}_+$ be the conformal mapping $\omega(z) = i((1+z)/(1-z))$. The restriction of the inverse $\omega^{-1}(w) = ((w-i)/(w+i))$ on the boundary $\partial \mathbb{C}_+ = \mathbb{R}$ defines the mapping U,

$$Uf(x) = \frac{1}{\sqrt{\pi}(x+i)} \cdot f\left(\frac{x-i}{x+i}\right), \quad x \in \mathbb{R},$$

which is a unitary isomorphism between the spaces $L^2(\mathbb{T})$ and $L^2(\mathbb{R})$, $U \colon L^2(\mathbb{T}) \to L^2(\mathbb{R})$.

- *Transfer of $H^2(\mathbb{T})$.* We have $UH^2(\mathbb{T}) = H^2(\mathbb{R})$, where

$$H^2(\mathbb{R}) =: \mathrm{span}_{L^2(\mathbb{R})}\left(\frac{1}{x - \bar{\mu}} \colon \mathrm{Im}\, \mu > 0\right).$$

We identify $H^2(\mathbb{R})$ with the boundary values of the space (by an ω-transfer of Fatou's theorem)

$$H^2(\mathbb{C}_+) = \left\{f \in \mathrm{Hol}(\mathbb{C}_+) \colon \sup_{y>0} \int_\mathbb{R} |f(x+iy)|^2 \, dx < \infty\right\},$$

and we thus write $H^2(\mathbb{R}) = H^2(\mathbb{C}_+)$.

- *Theorem of Paley and Wiener [1934].*

$$H^2(\mathbb{C}_+) = \mathcal{F}^{-1} L^2(\mathbb{R}_+) = \mathcal{F} L^2(\mathbb{R}_-),$$

where $\mathbb{R}_+ = (0, \infty)$, $\mathbb{R}_- = (-\infty, 0)$, and a space $L^2(E)$, $E \subset \mathbb{R}$, is regarded as the subspace of $L^2(\mathbb{R})$ containing the functions of $L^2(\mathbb{R})$ which vanish on the complement $\mathbb{R} \setminus E$.

Here, $\mathcal{F}$ and $\mathcal{F}^{-1}$ are respectively the Fourier and the inverse Fourier transforms (in particular, their extensions to $L^2(\mathbb{R})$ by Plancherel),

$$\mathcal{F}f(z) = \frac{1}{\sqrt{2\pi}} \int_\mathbb{R} f(x) e^{-ixz} dx, \quad \mathcal{F}^{-1}f(z) = \frac{1}{\sqrt{2\pi}} \int_\mathbb{R} f(x) e^{ixz} dx.$$

F.6 Vector-Valued H^p Spaces

All of the basic definitions of Appendix F.1 can easily be modified for the case of functions with values in a Banach space E:

$$H^p(\mathbb{T}, E) = \{f \in L^p(\mathbb{T}, E) \colon \widehat{f}(n) = 0, \forall n < 0\}.$$

Moreover, for the case $E = \mathbb{C}^n$ and $p = 2$, the definition simply reduces to the sum of copies of the scalar space $H^2(\mathbb{T})$; for example, for

$$f = (f_1, \ldots, f_n) \in L^2(\mathbb{T}, \mathbb{C}^n), \quad \|f\|^2_{L^2(\mathbb{T}, \mathbb{C}^n)} = \sum_{j=1}^n \|f_j\|^2_{L^2(\mathbb{T})}.$$

The same holds for the Fourier coefficients $\widehat{f}(k)$, and for many other properties. We do not delve into the details, but note that, for example, with the same proof as in the scalar case, we can show that a bounded operator $T \colon L^2(\mathbb{T}, \mathbb{C}^n) \to L^2(\mathbb{T}, \mathbb{C}^n)$ commutes with translation (shift) $M_z f = zf$ if and only if $T = M_\varphi$ – the multiplication operator $M_\varphi f = \varphi f$, $f \in L^2(\mathbb{T}, \mathbb{C}^n)$ by a matrix-valued function $\varphi \in L^\infty(\mathbb{T}, L(\mathbb{C}^n))$, and $\|M_\varphi\| = \|\varphi\|_\infty = \operatorname{ess\,sup}_{\zeta \in \mathbb{T}} \|\varphi(\zeta)\|_{L(\mathbb{C}^n)}$. Moreover,

$$M_\varphi H^2(\mathbb{T}, \mathbb{C}^n) \subset H^2(\mathbb{T}, \mathbb{C}^n) \Leftrightarrow \varphi \in H^\infty(\mathbb{T}, L(\mathbb{C}^n)),$$

etc. We refer to [Böttcher and Silbermann, 1990], [Böttcher, Karlovich, and Spitkovsky, 2002], and [Rosenblum and Rovnyak, 1985], as well as [Burbea and Masani, 1984] for the generalities.

F.7 Notes and Remarks

There exists a large number of excellent texts/monographs on the subject, for example [Hoffman, 1962], [Koosis, 1980], [Garnett, 1981], [Rosenblum and Rovnyak, 1985], [Sarason, 1973a], [Stein, 1993], and [Zhu, 2007].

See also [Nikolski, 2019, 2002a] for presentations close to those used in the present text. For a proof of Lindelöf's theorem of Appendix F.3 that uses the Hankel operators (as well as for references to the original proof related to the geometric E.9(7) theory of conformal mappings), see [Power, 1982] and [Nikolski, 2002a].

References

A

G. Acosta and R. Durán (2004), An optimal Poincaré inequality in L^1 for convex domains. *Proc. Amer. Math. Soc.* **132:1**, 195–202.

D. R. Adams and L. I. Hedberg (1996), *Function Spaces and Potential Theory*. Springer.

V. M. Adamyan, D. Z. Arov, and M. G. Krein (1968), On infinite Hankel matrices and the generalized problems of Carathéodory–Fejér and F. Riesz (in Russian). *Funkcional. Analiz i Prilozhen.* **2**, 1–19. English translation: *Funct. Anal. Appl.* **2**, 1–18.

V. M. Adamyan, D. Z. Arov, and M. G. Krein (1971), Infinite Hankel block matrices and some related continuation problems (in Russian). *Izvestia Akad. Nauk Armyan. SSR Ser. Mat.* **6**, 87–112. English translation: *Amer. Math. Soc. Transl.* (II) **111** (1978), 133–156.

N. I. Akhiezer (1965), *Lectures on Approximation Theory* (in Russian), second edition. Nauka, Moscow. English translation: *Approximation Theory*, Dover, New York (1992).

N. I. Akhiezer and M. G. Krein (1938), *Some Questions in the Theory of Moments* (in Russian). GONTI, Kharkov. English translation: vol. 2 of Translations of Mathematical Monographs, American Mathematical Society, Providence, RI (1962).

A. Aleman, S. Pott, and M. C. Reguera (2013), Sarason conjecture on the Bergman space. arXiv:1304.1750v1

A. Aleman and D. Vukotic (2009), Zero products of Toeplitz operators. *Duke Math. J.* **148:3**, 373–403.

G. R. Allan (1968), Ideals of vector-valued functions. *Proc. London Math. Soc.* (3) **18**, 193–216.

H. Alzer (2002), Inequalities for the constants of Landau and Lebesgue. *J. Comput. Appl. Math.* **139**, 215–230.

D. Anosov and A. Bolibruch (1994), *The Riemann–Hilbert problem*, vol. E22 of Aspects of Mathematics. Vieweg, Braunschweig.

N. Arcozzi, R. Rochberg, E. T. Sawyer, and B. D. Wick (2010), Bilinear forms on the Dirichlet space. *Anal. PDE* **3:1**, 21–47.

N. Arcozzi, R. Rochberg, E. T. Sawyer, and B. D. Wick (2011), The Dirichlet space: a survey. *New York J. Math.* **17**, 45–86.
N. Aronszajn (1950), Theory of reproducing kernels. *Trans. Amer. Math. Soc.* **68**, 337–404.
G. Arsene and A. Gheondea (1982), Completing matrix contractions. *J. Oper. Theory* **7**, 179–189.
W. Arveson (1975), Interpolation problems in nest algebras. *J. Funct. Anal.* **20:3**, 208–233.
M. Atiyah and I. Singer (1963), The index of elliptic operators on compact manifolds. *Bull. Amer. Math. Soc.* **69:3**, 322–433.
F. V. Atkinson (1951), Normal solvability of linear equations on normed spaces (in Russian). *Mat. Sbornik* **28:1**, 3–14.
S. Axler, A. S.-Y. Chang, and D. Sarason (1978), Products of Toeplitz operators. *Integral Equations Oper. Theory* **1**, 285–309.

B

M. Bakonyi and D. Timotin (2001), On an extension problem for polynomials. *Bull. London Math. Soc.* **33**, 599–605.
M. Bakonyi and H. Woerdeman (2011), *Matrix Completion, Moments, and Sums of Hermitian Squares*. Princeton University Press.
S. Banach (1932), *Théorie des opérations linéaires*. Monografje Matematyczne, Warsaw.
A. Baranov, R. Bessonov, and V. Kapustin (2011), Symbols of truncated Toeplitz operators. *J. Funct. Anal.* **261:12**, 3437–3456.
J. Barria and P. Halmos (1982), Asymptotic Toeplitz operators. *Trans. Amer. Math. Soc.* **273:2**, 621–630.
H. Bart, T. Hempfling, and M. A. Kaashoek, eds (2008), *Israel Gohberg and Friends: On the Occasion of his 80th Birthday*. Birkhäuser, Basel and Boston.
G. Baxter (1963), A norm inequality for a "finite section" Wiener–Hopf equation. *Illinois J. Math.* **7**, 97–103.
G. Baxter and I. I. Hirshman (1964), An explicit inversion formula for finite-section Wiener–Hopf operators. *Bull. Amer. Math. Soc.* **70**, 820–823.
E. F. Beckenbach and R. Bellman (1961), *Inequalities*. Springer.
H. Bercovici, C. Foias, and A. Tannenbaum (1988), On skew Toeplitz operators, I. In vol. 29 of *Operator Theory: Advances and Applications*, pp. 21–45. Birkhäuser, Basel.
H. Bercovici, C. Foias, and A. Tannenbaum (1998), On skew Toeplitz operators, II. In vol. 104 of *Operator Theory: Advances and Applications*, pp. 23–35. Birkhäuser, Basel.
F. A. Berezin (1972), Covariant and contravariant symbols of operators (in Russian). *Izvestia Akad. Nauk. SSSR Ser. Mat.* **36**, 1134–1167.
F. A. Berezin (1989), *The Method of Second Quantization*, revised (augmented) second edition. Kluwer. First English edition: Academic Press (1966).
C. Berg, J. Christensen, and P. Ressel (1984), *Harmonic Analysis on Semigroups: Theory of Positive Definite and Related Functions*. Springer, New York.

C. A. Berger and B. I. Shaw (1973), Intertwining, analytic structure, and the trace norm estimate. In *Proc. Conf. on Operator Theory (Halifax, 1973)*, vol. 345 of Lecture Notes in Mathematics, pp. 1–6. Springer, Berlin and Heidelberg.

G. Bertrand (1921), Equations de Fredholm à intégrales principales au sens de Cauchy. *CR Acad. Sci. Paris* **1972**, 1458–1461.

R. V. Bessonov (2014), Duality theorems for coinvariant subspaces of H^1. arXiv:1401.0452v1

D. Bini (1995), Toeplitz matrices, algorithms and applications. *ECRIM News online edition* **22**, July 1995. www.ercim.eu/publication/Ercim_News/enw22/toeplitz.html

G. D. Birkhoff (1909), Singular points of ordinary linear differential equations. *Trans. Amer. Math. Soc.* **10**, 436–470.

G. D. Birkhoff (1913), The generalized Riemann problem for linear differential equations and the allied problems for linear difference and q-difference equations. *Proc. Amer. Acad. Art Sci.* **49**, 521–568.

O. Blasco (2004), Remarks on operator BMO spaces. *Revista Union Mat. Argentina* **45:1**, 63–78.

P. M. Bleher (1979), Inversion of Toeplitz matrices (in Russian). *Trudy Moscow Math. Soc.* **40**, 207–240. English translation: *Trans. Moscow Math. Soc.* (1981), 201–224.

F. F. Bonsall (1984), Boundedness of Hankel matrices. *J. London Math. Soc.* (2) **29**, 289–300.

A. Borodin and A. Okounkov (2000), A Fredholm determinant formula for Toeplitz determinants. *Integral Equations Oper. Theory* **37:4**, 386–396.

A. Böttcher (1994), Pseudospectra and singular values of large convolution operators. *J. Integral Equations Appl.* (6) **3**, 267–301.

A. Böttcher (1995), The Onsager formula, the Fischer–Hartwig conjecture, and their influence on research into Toeplitz operators. *J. Statist. Phys. (Lars Onsager Festschrift)* **78**, 575–584.

A. Böttcher (2001), Featured review on Borodin and Okounkov [2000]. *Math. Reviews* MR1780118 (2001g:47042a).

A. Böttcher and S. M. Grudsky (1996), Toeplitz operators with discontinuous symbols: phenomena beyond piecewise continuity. In vol. 90 of Operator Theory: Advances and Applications, pp. 55–118. Birkhäuser, Basel.

A. Böttcher and S. M. Grudsky (2005), *Spectral Properties of Banded Toeplitz Matrices*. SIAM, Philadelphia.

A. Böttcher, J. M. Bogoya, S. M. Grudsky, and E. A. Maximenko (2017), Asymptotic formulas for the eigenvalues and eigenvectors of Toeplitz matrices (in Russian). *Mat. Sbornik* **208**, 4–28. English translation: *Sbornik Math.* **208** (2017), 1578–1601.

A. Böttcher and Y. Karlovich (1997), *Carleson Curves, Muckenhoupt Weights, and Toeplitz Operators*, vol. 154 of Progress in Mathematics. Birkhäuser, Basel.

A. Böttcher, Y. Karlovich, and I. Spitkovsky (2002), *Convolution Operators and Factorization of Almost Periodic Matrix Functions*, vol. 131 of Operator Theory: Advances and Applications. Birkhäuser, Basel.

A. Böttcher and B. Silbermann (1990), *Analysis of Toeplitz Operators*. Akademie, Berlin, and Springer.

A. Böttcher and B. Silbermann (1999), *Introduction to Large Truncated Toeplitz Matrices*. Springer, New York.

A. Böttcher and I. M. Spitkovsky (2013), The Factorization Problem: Some Known Results and Open Questions. In *Advances in Harmonic Analysis and Operator Theory* (ed. A. Almeida, L. Castro, and F.-O. Speck), vol. 229 of Operator Theory: Advances and Applications, pp. 101–122. Birkhäuser, Basel.

A. Böttcher and H. Widom (2006), Szegő via Jacobi. *Linear Alg. Appl.* **419**, 656–667.

A. Böttcher and H. Widom (2007), From Toeplitz eigenvalues through Green's kernels to higher-order Wirtinger–Sobolev inequalities. In vol. 171 of Operator Theory: Advances and Applications, pp. 73–87. Birkhäuser, Basel.

N. Bourbaki (1967), *Eléments de mathématique*, book XXXII: *Théories spectrales*. Hermann, Paris.

L. Boutet de Monvel and V. Guillemin (1981), *The Spectral Theory of Toeplitz Operators*. Princeton University Press.

L. G. Brown, R. G. Douglas, and P. A. Fillmore (1973), Unitary equivalent modulo the compact operators and extensions of C^*-algebras. In *Proc. Conf. on Operator Theory (Halifax, 1973)*, vol. 345 of Lecture Notes in Mathematics, pp. 58–128. Springer, Berlin and Heidelberg.

A. Brown and P. Halmos (1963), Algebraic properties of Toeplitz operators. *J. Reine Angew. Math.* **213**, 89–102.

A. Brown, P. R. Halmos and A. L. Shields (1965), Cesàro operators. *Acta Sci. Math. Szeged* **26**, 125–137.

J. Burbea and P. R. Masani (1984), *Banach and Hilbert Spaces of Vector-Valued Functions: Their General Theory and Applications to Holomorphy*. Pitman.

C

C. Carathéodory (1907), Über den Variabilitätsbereich der Koeffizienten von Potenzreihen, die gegebene Werte nicht annehmen. *Math. Ann.* **64**, 95–115.

C. Carathéodory (1911), Über den Variabilitätsbereich der Fourierschen Konstanten von positiven harmonischen Funktionen. *Rend. Circ. Mat. Palermo* **32**, 193–217.

C. Carathéodory and L. Fejér (1911), Über den Zuzammenhang der Extremen von harmonischen Funktionen mit ihren Koeffizienten und über den Picard–Landauschen Satz. *Rend. Circ. Mat. Palermo* **32**, 218–239.

A. L. Cauchy (1841), *Exercices d'analyse et de physique mathématique*, vol. 2, second edition. Bachelier, Paris.

R. H.-F. Chan and X.-Q. Jin (2007), *An Introduction to Iterative Toeplitz Solvers*, vol. 5: *Fundamentals of Algorithms*. SIAM.

K. Chandrasekharan, ed. (1986), *Weyl Centenary Symposium, 1885–1985*. Springer.

S.-Y. A. Chang (1976), A characterization of Douglas subalgebras. *Acta Math.* **137**, 81–89.

C. Chevalley and A. Weil (1957), Hermann Weyl. *Enseignement Math.* **3:3**, 157–187.

M.-D. Choi (1983), Tricks or treats with the Hilbert matrix. *Amer. Math. Monthly* **90:5**, 301–312.

K. Clancey and I. Gohberg (1981), *Factorization of Matrix Functions and Singular Integral Operators*, vol. 3 of Operator Theory: Advances and Applications. Birkhäuser, Basel.

L. A. Coburn (1966), Weyl's theorem for nonnormal operators. *Mich. Math. J.* **13**, 285–288.
L. A. Coburn (1967), The C^*-algebra generated by an isometry I. *Bull. Amer. Math. Soc.* **73**, 722–726.
L. A. Coburn (1969), The C^*-algebra generated by an isometry II. *Trans. Amer. Math. Soc.* **137**, 211–217.
L. Coburn and R. G. Douglas (1969), Translation operators on the half-line. *Proc. Nat. Acad. Sci. USA* **62**, 1010–1013.
L. Coburn and A. Lebow (1966), Algebaic theory of Fredholm operators. *Indiana Univ. Math. J.* **15**, 577–584.
P. Cohen (1961), A note on constructive methods in Banach algebras. *Proc. Amer. Math. Soc.* **12:1**, 159–163.
R. R. Coifman, R. Rochberg, and G. Weiss (1976), Factorization theorems for Hardy spaces in several variables. *Ann. Math.* **103**, 611–635.
R. R. Coifman and G. Weiss (1977), *Transference Methods in Analysis*, vol. 31 of CBMS Regional Conference Series in Mathematics. American Mathematical Society, Providence, RI.
Y. Colin de Verdière (1995), Une introduction aux opérateurs de Toeplitz. *Séminaire de Théorie spectrale et géométrie* **13**, 135–141.
H. O. Cordes and J. P. Labrousse (1963), The invariance of the index in the metric space of closed operators. *J. Math. Mech.* **12**, 693–719.
M. Cotlar and C. Sadosky (1979), On the Helson–Szegő theorem and related class of modified Toeplitz kernels. In *Harmonic Analysis in Euclidean Spaces (Williamstown, MA, 1978)*, part 1 (ed. G. Weiss and S. Wainger), vol. 35 of Proceedings of Symposia in Pure Mathematics, pp. 387–407. American Mathematical Society, Providence, RI.
R. Courant, K. Friedrichs, and H. Lewy (1928), Über die partiellen Differenzengleichungen der mathematischen Physik. *Math. Ann.* **100**, 32–74.
R. B. Crofoot (1994), Multipliers between invariant subspaces of the backward shift. *Pacific J. Math.* **166:2**, 225–246.

D

K. Davidson (1988), *Nest Algebras*, vol. 191 of Pitman Research Notes Series. Longman Scientific & Technical.
P. J. Davis (1994), *Circulant Matrices*. Chelsea Publishing, New York.
P. Deift (2000), *Orthogonal Polynomials and Random Matrices: A Riemann–Hilbert Approach*, Courant Lecture Notes. American Mathematical Society.
P. Deift, A. Its, and I. Krasovsky (2013), Toeplitz matrices and Toeplitz determinants under the impetus of the Ising model: some history and some recent results. *Comm. Pure Appl. Math.* **66:9**, 1360–1438.
A. Devinatz (1964), Toeplitz operators on H^2 spaces. *Trans. Amer. Math. Soc.* **112**, 304–317.
A. Devinatz (1967), On Wiener–Hopf operators. In *Functional Analysis (Irvine, CA, 1966)* (ed. B. R. Gelbaum), pp. 81–118. Thompson, Washington, DC.
A. Devinatz and M. Shinbrot (1969), General Wiener–Hopf operators. *Trans. Amer. Math. Soc.* **145**, 467–494.

R. A. DeVore and G. G. Lorentz (1993), *Constructive Approximation*. Springer, Berlin and Heidelberg.
J. Dieudonné (1943), Sur les homomorphismes d'espaces normés. *Bull. Sci. Math.* (2) **67**, 72–84.
J. Dieudonné (1975), Jules-Henri Poincaré. In *Dictionary of Scientific Biography*, vol. XI (ed. C. C. Gillespie). Scribner, New York.
J. Dieudonné (1981), *History of Functional Analysis*. vol. 49 of North-Holland Math. Studies. North-Holland, Amsterdam.
J. Dixmier (1996), *Les C^*-algèbres et leurs représentations*. Gauthier-Villars, Paris.
A. Douady (1965), Un espace de Banach dont le groupe linéaire n'est pas connexe. *Indag. Math.* **68**, 787–789.
J. Douglas (1931), Solution of the problem of Plateau. *Trans. Amer. Math. Soc.* **33**:1, 263–321.
R. G. Douglas (1968a), On the spectrum of a class of Toeplitz operators. *J. Math. Mech. Indiana Univ.* **18**, 433–436.
R. G. Douglas (1968b), Toeplitz and Wiener–Hopf operators in $H^\infty + C$. *Bull. Amer. Math. Soc.* **75**, 895–899.
R. G. Douglas (1969a), On the operator equations $S^*XT = X$ and related topics. *Acta Sci. Math. Szeged* **30(1–2)**, 19–32.
R. G. Douglas (1969b), On the spectrum of Toeplitz and Wiener–Hopf operators. In *Abstract Spaces and Approximation Theory* (ed. P. L. Butzer and B. Szőkefalvi-Nagy), pp. 53–66. Birkhäuser, Basel and Stuttgart.
R. G. Douglas (1972), *Banach Algebra Techniques in Operator Theory*. Academic Press, New York and London.
R. G. Douglas (1973), *Banach Algebra Techniques in the Theory of Toeplitz Operators*, vol. 15 of CBMS Regional Conference Series in Mathematics. American Mathematical Society, Providence, RI.
R. G. Douglas (1980), *C^*-Algebra Extensions and K-Homology*. Princeton University Press.
R. G. Douglas and W. Rudin (1969), Approximation by inner functions. *Pacif. J. Math.* **31**, 313–320.
R. G. Douglas and D. Sarason (1970), Fredholm Toeplitz operators. *Proc. Amer. Math. Soc.* **26**, 117–120.
R. G. Douglas and J. L. Taylor (1972), Wiener–Hopf operators with measure kernels. In *Hilbert Space Operators and Operator Algebras (Tihany, 1970)*, vol. 5 of Colloquia Mathematica Societatis János Bolyai, pp. 135–141. North-Holland, Amsterdam.
R. G. Douglas and H. Widom (1970), Toeplitz operators with locally sectorial symbols. *Indiana Univ. Math. J.* **20**, 385–388.
R. Duduchava (1979), *Integral Equations with Fixed Singularities*. Teubner, Leipzig.
N. Dunford and J. Schwartz (1958), *Linear Operators*, part 1: *General Theory*. Wiley (Interscience), New York.
J. Duoandikoetxea (2001), *Fourier Analysis*. American Mathematical Society, Providence, RI.
J. Durbin (1960), The fitting of time series models. *Revue de l'Institut International de Statistique* **28**, 233–44.

H. Dym and H. P. McKean (1976), *Gaussian Processes, Function Theory and the Inverse Spectral Problem*. Academic Press, New York.

E

R. E. Edwards and G. I. Gaudry (1977), *Littlewood–Paley and Multiplier Theory*. Springer, Berlin.

T. Ehrhardt (2003), A generalization of Pincus' formula and Toeplitz operator determinants. *Arch. Math. (Basel)* **80**, 302–309.

F

P. Fatou (1906), Séries trigonométriques et séries de Taylor. *Acta Math.* **30**, 335–400.

B. V. Fedosov (1970), A direct proof of the index formula for an elliptic system on Euclidean space. *Funct. Anal. Appl.* **4:4**, 83–84.

C. Fefferman (1971), Characterizations of bounded mean oscillation. *Bull. Amer. Math. Soc.* **77:4**, 587–588.

C. Fefferman and E. Stein (1972), H^p spaces of several variables. *Acta Math.* **129:3–4**, 137–193.

S. Ferguson and M. Lacey (2002), A characterization of product BMO by commutators. *Acta Math.* **189**, 143–160.

E. Fischer (1911), Über das Carathéodorische Problem, Potenzreihen mit positiven reellen Teil betreffend. *Rendiconti Circ. Mat. Palermo* **32**, 240–256.

M. Fisher and R. Hartwig (1968), Toeplitz determinants: some applications, theorems, and conjectures. *Adv. Chem. Phys.* **15**, 333–353.

C. Foias and A. E. Frazho (1990), *The Commutant Lifting Approach to Interpolation Problems*, vol. 44 of Operator Theory: Advances and Applications. Birkhäuser, Basel.

C. Foias, A. E. Frazho, I. Gohberg, and M. A. Kaashoek (1998), *Metric Constrained Interpolation, Commutant Lifting and Systems*. Birkhäuser, Basel.

I. Fredholm (1900), Sur une nouvelle méthode pour la résolution du probléme de Dirichlet. *Kong. Vetenskaps-Akad. Fbrh. Stockholm*, 39–46.

I. Fredholm (1903), Sur une classe d'équations fonctionnelles. *Acta Math.* **27**, 365–390.

G

F. D. Gakhov (1963), *Boundary Value Problems*. Dover, New York (1990). Russian original: Fizmatgiz, Moscow (1963).

F. D. Gakhov and Y. I. Chersky (1978), *Equations of Convolution Type*. Fizmatgiz, Moscow.

F. R. Gantmacher (1966), *The Theory of Matrices* (in Russian), second edition. Nauka, Moscow. English translation: Chelsea Publishing, New York (1960).

J. B. Garnett (1981), *Bounded Analytic Functions*. Academic Press, New York.

S. R. Garcia and W. T. Ross (2012), Recent progress on truncated Toeplitz operators. arXiv:1108.1858v4

A. Garsia (1971), A presentation of Fefferman theorem. Unpublished.

I. M. Gelfand (1960), On elliptic equations (in Russian). *Uspekhi Mat. Nauk* **15:3**, 121–132. English translation: *Russ. Math. Surv.* **15:3** (1960), 113–123.

I. M. Gelfand, D. A. Raikov, and G. E. Shilov (1960), *Commutative Normed Rings* (in Russian). Fizmatgiz, Moscow. English translation: Chelsea Publishing, New York (1964).

I. M. Gelfand and N. Y. Vilenkin (1961), *Generalized Functions*, vol. 4: *Some Applications of Harmonic Analysis* (in Russian). Fizmatgiz, Moscow. English translation: Academic Press, New York (1964).

M. Georgiadou (2004), *Constantin Carathéodory: Mathematics and Politics in Turbulent Times*. Springer, Berlin and Heidelberg.

J. S. Geronimo and K. M. Case (1979), Scattering theory for polynomials orthogonal on the unit circle. *J. Math. Phys.* **20**, 299–310.

I. M. Glazman and Y. I. Lyubich (1969), *Finite-dimensional Linear Analysis* (in Russian). Nauka, Moscow. English translation: *Finite-dimensional Linear Analysis: A Systematic Presentation in Problem Form*, MIT Press, Cambridge, MA (1974).

I. C. Gohberg (1951a), On linear equations in normed spaces. *Dokl. Akad. Nauk USSR* **76**, 477–480.

I. C. Gohberg (1951b), On linear operators analytically depending on a parameter. *Dokl. Akad. Nauk USSR* **78:4**, 620–632.

I. Gohberg (1964), Factorization problem in normed rings, functions of isometric and symmetric operators, and singular integral equations. *Uspekhi Mat. Nauk* **19:1**, 71–124.

I. Gohberg (2004), The Atiyah–Singer index formula. *Nieuwe Archief voor Wiskunde* 5/5, no. 4, December 2004, p. 319 (Letters to the editor).

I. C. Gohberg and I. A. Feldman (1967), *Convolution Equations and Projection Methods for their Solutions*. Moldavian Academy, Kishinev (1967), and Nauka, Moscow (1971). English translation: vol. 41 of Translations of Mathematical Monographs, American Mathematical Society, Providence, RI (1974).

I. C. Gohberg and I. A. Feldman (1968), On Wiener–Hopf integro-difference equations (in Russian). *Doklady Akad. Nauk SSSR* **183:1**, 25–28. English translation: *Soviet Math. Doklady* **9** (1968), 1312–1316.

I. Gohberg, S. Goldberg, and M. A. Kaashoek (1990), *Classes of Linear Operators*, vol. I. Birkhäuser, Basel.

I. Gohberg, S. Goldberg, and N. Krupnik (1996), Traces and determinants of linear operators. *Integral Equations Oper. Theory* **26:2**, 136–187.

I. Gohberg, M. A. Kaashoek, and I. M. Spitkovsky (2003), An overview of matrix factorization theory and operator applications. In *Factorization and Integrable Systems* (ed. I. Gohberg, N. Manojlovic, and A. Ferreira dos Santos), vol. 141 of Operator Theory: Advances and Applications, pp. 1–102. Birkhäuser, Basel.

I. C. Gohberg and M. G. Krein (1957), The basic propositions on defect numbers, root numbers and indices of linear operators (in Russian). *Uspekhi Mat. Nauk* **12:2**, 43–118. English translation: *Amer. Math. Soc. Transl.* (2) **13** (1960), 185–264.

I. C. Gohberg and M. G. Krein (1958), Systems of integral equations on a half-line with kernel depending upon the difference of the arguments (in Russian). *Uspekhi Mat. Nauk* **13:2(80)**, 3–72. English translation: *Amer. Math. Soc. Transl.* **14** (1960), 217–287.

I. Gohberg and N. Y. Krupnik (1969), On the algebra generated by Toeplitz matrices (in Russian). *Funktsional. Analiz i Prilozhen* **3:2**, 46–56. English translation: *Funct. Anal. Appl.* **3:2** (1969), 119–127.

I. Gohberg and N. Krupnik (1973), *One-dimensional Linear Singular Integral Equations* (in Russian). Shtiintsa, Kishinev. English translation, vols I and II: Birkhäuser, Basel (1992).

I. Gohberg and A. Sementsul (1972), On the inversion of finite Toeplitz matrices and their continuous analogues (in Russian). *Matem. Issledovania (Kishinev)* **7:2**, 272–283.

B. Golinskii and I. Ibragimov (1971), On Szegő's limit theorem. *Math. URSS Izvestia* **5:2**, 421–446.

G. H. Golub and C. F. Van Loan (1996), *Matrix Computations*, third edition. Johns Hopkins University Press, Baltimore.

L. Grafakos (2008), *Classical Fourier Analysis*, second edition. Springer.

R. M. Gray (2005), Toeplitz and circulant matrices: a review. *Found. Trends Commun. Inform. Theory* **2:3**, 155–239.

U. Grenander and G. Szegő (1958), *Toeplitz Forms and Their Applications*. University of California Press, Berkeley.

A. Grothendieck (1956), La théorie de Fredholm. *Bull. Soc. Math. France* **84**, 319–384.

H

J. Hadamard (1893), Sur le module maximum que puisse atteindre un déterminant. *CR Acad. Sci. Paris* **116**, 1500–1501. See also pp. 239–245 of *Œuvres de Jacques Hadamard*, vol. I, CNRS, Paris (1968).

G. H. Hardy (1908), The theory of Cauchy's principal values. *Proc. London Math. Soc.* **7:2**, 181–208.

G. H. Hardy, J. E. Littlewood, and G. Pólya (1934), *Inequalities*. Cambridge University Press.

E. M. Harrell II (2004), A short history of operator theory. www.mathphysics.com/opthy/OpHistory.html

P. Hartman (1958), On completely continuous Hankel matrices. *Proc. Amer. Math. Soc.* **9**, 862–866.

P. Hartman and A. Wintner (1950a), On the spectra of Toeplitz matrices. *Amer. J. Math.* **72**, 359–366.

P. Hartman and A. Wintner (1950b), On the essential spectra of singular eigenvalue problems. *Amer. J. Math.* **72**, 545–552.

P. Hartman and A. Wintner (1954), The spectra of Toeplitz's matrices. *Amer. J. Math.* **76**, 867–882.

F. Hausdorff (1932), Zur Theorie der linearen metrischen Räume. *J. Reine Angew. Math.* **167**, 294–311.

V. P. Havin and N. K. Nikolski (2000), Stanislav Aleksandrovich Vinogradov, his life and mathematics. In *Complex Analysis, Operators, and Related Topics: The S. A. Vinogradov Memorial Volume* (ed. V. Havin and N. Nikolski), vol. 113 of Operator Theory: Advances and Applications, pp. 1–18. Birkhäuser, Basel.

H. Hedenmalm, B. Korenblum, and K. Zhu (2000), *Theory of Bergman Spaces*. Springer, New York and Heidelberg.

G. Heinig and K. Rost (1984), *Algebraic Methods for Toeplitz-like Matrices and Operators*. Birkhäuser, Basel.

H. Helson (1964), *Lectures on Invariant Subspaces*. Academic Press, New York.

H. Helson (2010), Hankel forms. *Studia Math.* **198**, 79–84.

H. Helson and G. Szegő (1960), A problem of prediction theory. *Ann. Mat. Pura Appl.* **51**, 107–138.

J. W. Helton and R. Howe (1973), Integral operators: commutators, traces, index, and homology. In *Proc. Conf. on Operator Theory (Halifax, 1973)*, vol. 345 of Lecture Notes in Mathematics, pp. 141–209. Springer, Berlin and Heidelberg.

G. Herglotz (1911), Über Potenzreihen mit positiven reellen Teil im Einheitskreise. *Berichte Verh. Kgl.-sächs. Gesellsch. Wiss. Leipzig, Math.-Phys. Kl.* **63**, 501–511.

C. Hermite, H. Poincaré, and E. Rouché, eds (1898), *Œuvres de Laguerre*, vol. 1: *Algèbre, Calcul intégral* (reprint of the 1898 edition). Bronx, New York (1972).

D. Hilbert (1894), Ein Beitrag zur Theorie des Legendre'schen Polynom. *Acta Math.* **18**, 155–159.

D. Hilbert (1904), Grundzüge einer allgemeinen Theorie der linearen Integralgleichungen, Erste Mitteilung. *Nachr. Acad. Wiss. Göttingen Math. Phys. Kl.* **II**, 49–91.

D. Hilbert (1912), *Gründzüge einer allgemeinen Theorie der linearen Integralgleichungen*. Teubner, Leipzig.

D. Hilbert (1998), *Œuvres choisies*, vol. 2: *Analyse, Physique, Problèmes, Personalia* (in Russian translation) (ed. A. Parshin), Moscow.

T. Hildebrandt (1928), Über vollstetige lineare Transformationen. *Acta Math.* **51**, 311–318.

I. I. Hirschman and D. V. Widder (1955), *The Convolution Transform*. Princeton University Press (Dover edition, 2005).

K. Hoffman (1962), *Banach Spaces of Analytic Functions*. Prentice Hall, Englewood Cliffs, NJ.

K. Hoffman and I. M. Singer (1960), Maximal algebras of continuous functions. *Acta Math.* **103**, 217–241.

L. Hörmander (1960), Estimates for translation invariant operators in L^p spaces. *Acta Math.* **104**, 93–140.

L. Hörmander (1971), On the index of pseudo-differential operators. In *Elliptische Differentialgleichungen*, vol. II, pp. 127–146. Akademie, Berlin.

L. Hörmander (1985), *The Analysis of Linear Partial Differential Operators*, vol. III: *Pseudo-differential Operators*. Springer, Berlin.

A. Hurwitz (1901), Sur le problème des isopérimètres. *CR Acad. Paris* **132**, 401–403.

T. P. Hytönen (2009), A framework for non-homogeneous analysis on metric spaces, and the RBMO space of Tolsa. arXiv:0909.3231v1

I

I. S. Iokhvidov (1974), *Hankel and Toeplitz Matrices and Forms: Algebraic Theory.* Nauka, Moscow.
R. S. Ismagilov (1963), On the spectrum of Toeplitz matrices. *Dokl. Akad. Nauk SSSR* **149:4**, 769–772.
A. R. Its (2003), The Riemann–Hilbert problem and integrable systems. *Notices Amer. Math. Soc.* **50:11**, 1389–1400.

J

K. Johansson (1988), On Szegő's asymptotic formula for Toeplitz determinants and generalizations. *Bull. Soc. Math.* (2) **112**, 257–304.
F. John (1961), Rotation and strain. *Comm. Pure Appl. Math.* **14:3**, 391–413.
F. John and L. Nirenberg (1961), On functions of bounded mean oscillation. *Comm. Pure Appl. Math.* **14:3**, 415–426.
G. Julia (1944), Sur la représentation analytique des opérateurs bornés ou fermés de l'espace hilbertien. *CR Acad. Sci. Paris* **219**, 225–227.

K

M. Kac (1954), Toeplitz matrices, translation kernels and a related problem in probability theory. *Duke Math. J.* **21**, 501–509.
T. Kailath and A. H. Sayed, eds (1999), *Fast Reliable Algorithms for Matrices with Structure.* SIAM, Philadelphia.
Y. Katznelson (1976), *An Introduction to Harmonic Analysis.* Dover, New York.
H. Kesten (1961), Random walk with absorbing barriers and Toeplitz forms. *Illinois J. Math.* **5**, 267–290.
S. V. Khrushchev (1977), Entropy meaning of summability of the logarithm (in Russian). *Zapiski Nauchn. Seminarov LOMI (Steklov Math. Inst.)* **73**, 152–187. English translation: *J. Soviet Math.* **34:6** (1986), 212-233.
I. Klemes (2001), Finite Toeplitz matrices and sharp Littlewood conjectures. In memoriam: Thomas H. Wolff. *Algebra i Analiz (St. Petersburg Math. J.)* **13:1**, 39–59.
A. N. Kolmogorov (1941), Stationary sequences in Hilbert space (in Russian). *Bull. Moscow Univ. Math.* **2:6**, 1–40.
P. Koosis (1980), *Introduction to H^p Spaces.* Cambridge University Press.
A. V. Kozak (1973), A local principle in the theory of projection methods (in Russian). *Dokl. Acad. Sci. URSS* **212:6**, 1287–1289. English translation: *Soviet Math. Dokl.* **14** (1974), 1580–1583.
I. Kra and S. Simanca (2012), On circulant matrices. *Notices Amer. Math. Soc.* **59:3**, 368–377.
M. Krein (1947a), The theory of self-adjoint extensions of semi-bounded Hermitian operators and applications, I. *Mat. Sbornik* **20**, 431–495.
M. Krein (1947b), The theory of self-adjoint extensions of semi-bounded Hermitian operators and applications, II. *Mat. Sbornik* **21**, 365–404.

M. Krein (1958), Integral equations on a half-line with kernel depending upon the difference of the arguments (in Russian). *Uspekhi Mat. Nauk* **13:5**, 3–120. English translation: *Amer. Math. Soc. Transl.* (2) **22** (1962), 163–288.

M. Krein (1966), On some new Banach algebras and Wiener–Lévy type theorems for Fourier series and integrals (in Russian). *Matem. Issledovania (Kishinev)* **1** (1966), 82–109. English translation: *Amer. Math. Soc. Transl.* **93** (1970), 177–199.

M. Krein and M. Krasnoselskii (1952), Stability of the index of an unbounded operator (in Russian). *Mat. Sbornik* **30:1**, 219–224.

L. Kronecker (1881), Zur Theorie der Elimination einer Variablen aus zwei algebraischen Gleichungen. *Monatsber. Königl. Preussischen Akad. Wiss. (Berlin)* (1881), 535–600.

N. Y. Krupnik (1984), *Banach Algebras with Symbol and Singular Integral Operators* (in Russian). Shtiintsa, Kishinev. English translation: Birkhäuser, Basel (1987).

N. Y. Krupnik and I. A. Feldman, (1985), Relations between factorization and invertibility of finite Toeplitz matrices (in Russian). *Izvestia AN Moldavskoi SSR Ser. Phys.-Tech. i Math.* **3**, 20–26.

L. Kuipers and H. Niederreiter (1974), *Uniform Distribution of Sequences*. Wiley (Dover edition, 2006).

L

M. Lacey (2013a), The two weight inequality for the Hilbert transform: a primer. arXiv:1304.5004

M. Lacey (2013b), Two weight inequality for the Hilbert transform: a real variable characterization, II. arXiv:1301.4663

M. Lacey and E. Terwilleger (2009), Hankel operators in several complex variables and product BMO. *Houston J. Math.* **35**, 159–183.

E. Landau (1913), Abschätzung der Koeffizientensumme einer Potenzreihe. *Arch. Math. Phys.* **21**, 42–50, 250–255.

N. S. Landkof (1972), *Foundations of Modern Potential Theory*. Springer.

G. Landsberg (1910), Theorie der Elementarteiler linearer Integralgleichungen. *Math. Ann.* **69**, 227–265.

J. B. Lawrie and I. D. Abrahams (2007), A brief historical perspective of the Wiener–Hopf technique. *J. Engrg. Math.* **59:4**, 351–358.

H. Lebesgue (1906), *Leçons sur les séries trigonométriques*. Gauthier-Villars, Paris.

Y. W. Lee (1960), *Statistical Theory of Communications*. Wiley, New York.

N. Levinson (1947), The Wiener RMS error criterion in filter design and prediction. *J. Math. Phys.* **25**, 261–78.

L. M. Libkind (1972), On asymptotics of the eigenvalues of Toeplitz forms (in Russian). *Mat. Zametki* **11:2**, 151–158.

E. Lindelöf (1905), *Le calcul des résidues et ses applications à la théorie des fonctions*. Gauthier-Villars, Paris (Chelsea Publishing edition, 1947).

J. Lindenstrauss and L. Tzafriri (1977), *Classical Banach Spaces*, vol. I. Springer, Berlin.

J. Lindenstrauss and L. Tzafriri (1979), *Classical Banach Spaces*, vol. II. Springer, Berlin.

C. M. Linton and P. McIver (2001), *Handbook of Mathematical Techniques for Wave/Structure Interactions*. CRC Press.
G. S. Litvinchuk and I. Spitkovsky (1987), *Factorization of Measurable Matrix Functions*. Akademie, Berlin, and Birkhäuser, Basel.
L. H. Loomis (1953), *An Introduction to Abstract Harmonic Analysis*. Van Nostrand, Toronto.
L. Lorch (1954), The principal term in the asymptotic expansion of the Lebesgue constants. *Amer. Math. Monthly* **61**, 245–249.
N. N. Luzin (1913), Sur la convergence des séries trigonométriques de Fourier. *CR Acad. Sci. Paris* **156**, 1655–1658.
N. N. Luzin (1915), *L'intégrale et les séries trigonométriques* (in Russian). University of Moscow (reproduced in *Mat. Sbornik* **30:1** (1916), 1–242). Second edition: Izd. Akad. Nauk SSSR (1951).

M

D. E. Marshall (1976), Subalgebras of L^∞ containing H^∞. *Acta Math.* **137**, 91–98.
V. Maz'ya and I. Verbitsky (2002), The Schrödinger operator on the energy space: boundedness and compactness criteria. *Acta Math.* **188**, 263–302.
T. Mei (2007), *Operator Valued Hardy Spaces*, vol. 881 of Memoirs of the American Mathematical Society. American Mathematical Society, Providence, RI.
S. G. Mikhlin (1936), Composition of singular integrals. *Doklady Akad. Nauk SSSR* **2(II):1(87)**, 3–6.
S. G. Mikhlin (1948), Singular integral equations. *Uspekhi Mat. Nauk* **3:3**, 29–112.
S. G. Mikhlin (1962), *Multivariate Singular Integrals and Integral Equations* (in Russian). GIFML, Moscow. English translation: Pergamon Press, Oxford (1965).
A. B. Mingarelli (2005), A glimpse into the life and times of F. V. Atkinson. *Math. Nachrichten* **278:12–13**, 1364–1387.
B. S. Mityagin (1970), Homotopy structure of the linear group of a Banach space. *Uspekhi Mat. Nauk* **25:5**, 63–106.
G. J. Murphy (1990), C^*-*algebras and Operator Theory*. Academic Press, Boston.
N. I. Muskhelishvili (1947), *Singular Integral Equations: Boundary Problems of Function Theory and their Application to Mathematical Physics* (in Russian), first edition Moscow. English translation: Dover, second edition (1992).

N

F. Nazarov (1997), A counterexample to Sarason's conjecture.
www.math.msu.edu/~fedja/prepr.html
F. Nazarov, G. Pisier, S. Treil, and A. Volberg (2002), Sharp estimates in vector Carleson imbedding theorem and for vector paraproducts. *J. Reine Angew. Math.* **542**, 147–171.
F. Nazarov, S. Treil, and A. Volberg (2004), Two weight estimate for the Hilbert transform and Corona decomposition for non-doubling measures. arXiv:1003.1596
Z. Nehari (1957), On bounded bilinear forms. *Ann. of Math.* **65**, 153–162.

C. Neumann (1877), *Untersuchungen über das logarithmische und Newton'sche Potential*. Teubner, Leipzig.
L. N. Nikol'skaya and Y. B. Farforovskaya (2003), Toeplitz and Hankel matrices as Hadamard–Schur multipliers. *Algebra i Analiz (St. Petersburg Math. J.)* **15:6**, 141–160.
N. Nikolski (1985), Ha-plitz operators: a survey of some recent results. In *Operators and Function Theory*, vol. 153 of NATO ASI Series, Math. Phys. (ed. S. Power), pp. 87–138. Reidel, Dordrecht.
N. Nikolski (1986), *Treatise on the Shift Operator*. Springer, Berlin.
N. Nikolski (2002a), *Operators, Functions, and Systems*, vol. 1. American Mathematical Society, Providence, RI.
N. Nikolski (2002b), *Operators, Functions, and Systems*, vol. 2. American Mathematical Society, Providence, RI.
N. Nikolski (2002c), *On the Norm of a Finite Toeplitz Matrix*. Lecture notes, Michigan State University (unpublished).
N. Nikolski (2012), *Espaces de Hardy*. Belin.
N. Nikolski (2019), *Hardy Spaces*. Cambridge University Press.
S. M. Nikol'skyï (1943), Linear equations in linear normed spaces (in Russian). *Izvetia Akad. Nauk SSSR Ser. Mat.* **7:3**, 147–166.
B. Noble (1998), *Methods Based on the Wiener–Hopf Technique for the Solution of Partial Differential Equations*. Chelsea Publishing (editions 1958, 1988, 1998).
F. Noether (1921), Über eine Klasse singulärer Integralgleichungen. *Math. Ann.* **82**, 42–63.

O

J. Ortega-Cerdà and K. Seip (2012), A lower bound in Nehari's theorem on the polydisc. *J. Anal. Math.* **118:1**, 339–342.

P

L. B. Page (1970), Bounded and compact vectorial Hankel operators. *Trans. Amer. Math. Soc.* **150**, 529–539.
S. Parrott (1978), On a quotient norm and the Sz.-Nagy–Foias lifting theorem. *J. Funct. Anal.* **30**, 311–328.
J. Peetre (1989), *The Berezin Transform and Ha-plitz Operators*. Math. Institute Reports, Lund University.
V. V. Peller (1980), Hankel operators of class $\mathfrak{S}_p$ and their applications (rational approximation, Gaussian processes, the problem of majorization of operators) (in Russian). *Mat. Sbornik* **113(155):4**, 538–581. English translation: *Math. USSR Sbornik* **41** (1982), 443–479.
V. V. Peller (2003), *Hankel Operators and their Applications*. Springer, New York.
S. Petermichl (2000), Dyadic shifts and a logarithmic estimate for Hankel operators with matrix symbol. *CR Acad. Sci. Paris Ser. I Math.* **330**, 455–460.

K. E. Petersen (1977), *Brownian Motion, Hardy Spaces and Bounded Mean Oscillation.* Cambridge University Press.

É. Picard (1927), *Leçons sur quelques types simples d'équations aux dérivées partielles.* Paris.

A. Pietsch (2007), *History of Banach Spaces and Linear Operators.* Birkhäuser, Boston.

J. Plemelj (1908a), Ein Ergänzungssatz zur Cauchyschen Integraldarstellung analytischer Funktionen, Randwerte betreffend. *Monatsheft für Math. Phys.* **XIX**, 205–210.

J. Plemelj (1908b), Riemannsche Funktionenscharen mit gegebener Monodromiegruppe. *Monatsheft für Math. Phys.* **XIX**, 211–245.

H. Poincaré (1895), La méthode de Neumann et le problème de Dirichlet. *Acta Math.* **20** (1896–1897), 59–142. Reprinted in Œuvres de Henri Poincaré, vol. 9, pp. 202–272, Gauthier-Villars, Paris.

H. Poincaré (1908), *Science et méthode.* Flammarion, Paris.

H. Poincaré (1910a), *Leçons de Mécanique Céleste*, vol. 3. Gauthier-Villars, Paris.

H. Poincaré (1910b), Remarques diverses sur l'équation de Fredholm. *Acta Math.* **33**, 57–86.

G. Pólya (1914), Question 4340. *L'Intermédiaire des mathématiciens* **21**, 27.

G. Pólya and G. Szegő (1925), *Aufgaben und Lehrsatze aus der Analysis*, vols 1, 2. Springer, Berlin. Russian translation: GITTL, Moscow (1948). English translation: Springer, Berlin and New York (1972).

S. C. Power (1980), Hankel operators on Hilbert space. *Bull. London Math. Soc.* **12**, 422–442.

S. C. Power (1982), *Hankel Operators on Hilbert Space*, vol. 64 of Pitman Research Notes in Mathematics. Pitman, Boston.

S. C. Power (1984), Quasinilpotent Hankel operators. In *Linear and Complex Analysis Problem Book* (ed. V. Havin and N. Nikolski), vol. 1043 of Lecture Notes in Mathematics, pp. 259–261. Springer.

I. I. Privalov (1919), *L'intégrale de Cauchy* (in Russian). Thesis, University of Saratov.

I. I. Privalov (1941), *Boundary Properties of Analytic Functions* (in Russian). Second edition: GITTL, Moscow (1950). German translation: Deutscher Verlag, Berlin (1956).

S. Prössdorf and B. Silbermann (1991), *Numerical Analysis for Integral and Related Operator Equations.* Birkhäuser, Basel.

R

J. Radon (1919), Über lineare Funktionaltransformationen und Funktionalgleichungen. *Sitzungsber. Akad. Wiss. Wien, math.-naturw. Kl., Abt. IIa* **128**, 1083–1121.

P. Rambour and J.-M. Rinkel (2007), Un théorème de Spitzer–Stone fort pour une matrice de Toeplitz à symbole singulier défini par une classe de fonctions analytiques. *Ann. Fac. Sci. Toulouse Math.* **6:16**, 331–367.

P. Rambour and A. Seghier (2012), Inversion des matrices de Toeplitz dont le symbole admet un zéro d'ordre rationnel positif, valeur propre minimale. *Ann. Fac. Sci. Toulouse Math.* (6) **21:1**, 173–211.

I. M. Rappoport (1948a), On a class of singular integral equations (in Russian). *Doklady Akad. Nauk SSSR* **59:8**, 1403–1406.

I. M. Rappoport (1948b), On a class of infinite systems of algebraic linear equations (in Ukrainian). *Doklady Ukrainian Akad. Nauk Fiz.-Mat. and Chemical Sci.* **3**, 6–10.

M. Reed and B. Simon (1972), *Methods of Modern Mathematical Physics*, vol. I: *Functional Analysis*. Academic Press, New York.

M. Reed and B. Simon (1978), *Methods of Modern Mathematical Physics*, vol. IV: *Analysis of Operators*. Academic Press, New York.

C. Reid (1970), *Hilbert, With an Appreciation of Hilbert's Mathematical Work by Hermann Weyl*. Springer, Berlin and Heidelberg.

B. Riemann (1876), *Gesammelte mathematische Werke und wissenschaftlicher Nachlass* (ed. R. Dedekind and H. Weber), Teubner, Leipzig. http://archive.org/details/bernardrgesamm00riemrich

F. Riesz (1916), Über lineare Funktionalgleichungen. *Acta Math.* **41**, 71–98.

F. Riesz and M. Riesz (1916), Über die Randwerte einer analytische Funktion. In *Quatrième Congrès des Math. Scand.*, Stockholm, pp. 27–44.

F. Riesz and B. Szőkefalvi-Nagy (1955), *Leçons d'analyse fonctionnelle*. Akadémiai Kiado, Szeged.

M. Riesz (1949), L'intégrale de Riemann–Liouville et le problème de Cauchy. *Acta Math.* **81**, 1–223.

R. Rochberg (1987), Toeplitz and Hankel operators on the Paley–Wiener space. *Integral Equations Oper. Theory* **10:2**, 187–235.

R. Rochberg and Z. Wu (1993), A new characterization of Dirichlet type spaces and applications. *Illinois J. Math.* **37:1**, 101–122.

L. Rodman, I. Spitkovsky, and H. Woerdeman (2002), *Abstract Band Method via Factorization, Positive and Band Extensions of Multivariable Almost Periodic Matrix Functions, and Spectral Estimation*, vol. 762 of Memoirs of the American Mathematical Society. American Mathematical Society, Providence, RI.

M. Rosenblum (1960), The absolute continuity of Toeplitz's matrices. *Pacif. J. Math.* **10**, 987–996.

M. Rosenblum (1965), A concrete spectral theory of self-adjoint Toeplitz operators. *Amer. J. Math.* **87**, 709–718.

M. Rosenblum and J. Rovnyak (1985), *Hardy Classes and Operator Theory*. Oxford University Press.

Y. A. Rozanov (1963), *Stationary Stochastic Processes* (in Russian). Fizmatgiz, Moscow. English translation: Holden-Day, San Francisco (1967).

W. Rudin (1959), Weak almost periodic functions and Fourier–Stieltjes transforms. *Duke Math. J.* **26**, 215–220.

W. Rudin (1962), *Fourier Analysis on Groups*. Wiley, New York.

W. Rudin (1986), *Real and Complex Analysis*. McGraw-Hill, New York.

W. Rudin (1991), *Functional Analysis*. McGraw-Hill, New York.

C. Runge (1885), Zur Theorie der eindeutigen analytischen Funktionen. *Acta Math.* **6**, 229–244.

A. F. Ruston (1986), *Fredholm Theory in Banach Spaces*. Cambridge University Press.

S

E. B. Saff and V. Totik (1997), *Logarithmic Potentials with External Fields*. Springer, Berlin and Heidelberg.

D. Sarason (1967), Generalized interpolation in H^∞. *Trans. Amer. Math. Soc.* **127:2**, 179–203.

D. Sarason (1973a), Algebras of functions on the unit circle. *Bull. Amer. Math. Soc.* **79**, 286–299.

D. Sarason (1973b), On products of Toeplitz operators. *Acta Sci. Math. Szeged* **35**, 7–12.

D. Sarason (1975), Functions of vanishing mean oscillation. *Trans. Amer. Math. Soc.* **207**, 391–405.

D. Sarason (1977a), Toeplitz operators with piecewise quasicontinuous symbols. *Indiana Univ. Math. J.* **26**, 817–838.

D. Sarason (1977b), Toeplitz operators with semi-almost periodic symbols. *Duke Math. J.* **44**, 357–364.

D. Sarason (1978), *Function Theory on the Unit Circle*. Virginia Polytechnic Institute and State University.

D. Sarason (1994), Products of Toeplitz operators. In *Linear and Complex Analysis Problem Book 3*, part 1 (ed. V. Havin and N. Nikolski), vol. 1573 of Lecture Notes in Mathematics, pp. 318–319. Springer, Berlin and Heidelberg.

D. Sarason (2007), Algebraic properties of truncated Toeplitz operators. *Operators and Matrices (Springer)* **1:4**, 491–526.

J. Schauder (1930), Über lineare, vollstetige Funktionaloperationen. *Studia Math.* **2:1**, 183–196.

I. Schur (1917), Über die Potenzreihen, die in Innern des Einheitskreises beschränkt sind, I. *J. Reine Angew. Math.* **147**, 205–232.

I. Schur (1918), Über die Potenzreihen, die in Innern des Einheitskreises beschränkt sind, II. *J. Reine Angew. Math.* **148**, 122–145.

M. A. Shubin (1987), *Pseudo-differential Operators and Spectral Theory*. Springer. First Russian edition: Nauka, Moscow (1978).

R. Siegmund-Schultze (2009), *Mathematicians Fleeing from Nazi Germany: Individual Fates and Global Impact*. Princeton University Press.

B. Simon (2005a), *Trace Ideals and their Applications*, vol. 120 of Mathematical Surveys and Monographs. American Mathematical Society, Providence, RI.

B. Simon (2005b), *Orthogonal Polynomials on the Unit Circle*, vol. 1: *Classical Theory*, vol. 54 of Colloquium Publications. American Mathematical Society, Providence, RI.

I. B. Simonenko (1960), Riemann boundary problem with measurable coefficients (in Russian). *Doklady Akad. Nauk SSSR* **135:3**, 538–541.

I. B. Simonenko (1961), Riemann boundary problem for n pairs of functions with continuous coefficients. *Izvestia Vyssh. Uchebn. Zaved. Mat.* **1**, 140–145.

I. B. Simonenko (1964), A new general method for studying linear operator equations of the type of singular integral equations (in Russian). *Doklady Akad. Nauk SSSR* **158**, 790–793.

I. B. Simonenko (1965a), A new general method for studying linear operator equations of the type of singular integral equations, I (in Russian). *Izvestia Akad. Nauk SSSR Ser. Mat.* **29:3**, 567–586.

I. B. Simonenko (1965b), A new general method for studying linear operator equations of the type of singular integral equations, II (in Russian). *Izvestia Akad. Nauk SSSR Ser. Mat.* **29:4**, 757–782.

I. B. Simonenko (1968), Some general questions of the Riemann boundary problem (in Russian). *Izvestia Akad. Nauk SSSR Ser. Mat.* **32:5**, 1138–1146. English translation: *Math USSR Izvestia* **2**, 1091–1099.

Y. V. Sokhotsky (1873), Ob opredelennykh integralakh i funktsiakh upotreblyaemykh pri razlozheniakh v ryady (On definite integrals and functions used for serial expansions). Habilitation thesis, University of St. Petersburg.

I. M. Spitkovsky (1976), The problem of the factorization of measurable matrix-valued functions (in Russian). *Dokl. Akad. Nauk SSSR* **227:3**, 576–579. English translation: *Soviet. Math. Dokl.* **17:2** (1976), 481–485.

I. M. Spitkovsky (1980), Multipliers that do not influence factorizability (in Russian). *Mat. Zametki* **27:2**, 291–299. English translation: *Math. Notes* **27** (1980), 145–149.

F. L. Spitzer and C. J. Stone (1960), A class of Toeplitz forms and their applications to probability theory. *Illinois J. Math.* **4**, 253–277.

M. Spivak (1971), *Calculus on Manifolds: A Modern Approach to Classical Theorems of Advanced Calculus*. Westview Press.

H. Stahl and V. Totik (1992), *General Orthogonal Polynomials*. Cambridge University Press.

E. Stein (1993), *Harmonic Analysis: Real-variable Methods, Orthogonality, and Oscillatory Integrals*. Princeton University Press.

E. Stein and G. Weiss (1971), *Introduction to Fourier analysis on Euclidean spaces*. Princeton University Press.

M. H. Stone (1932), *Linear Transformations in Hilbert Space and their Applications to Analysis*, vol. 15 of Colloquium Publications. American Mathematical Society, Providence, RI.

G. Szegő (1915), Ein Grentzwertsatz über die Toeplitzschen Determinanten einer reellen positiven Funktion. *Math. Ann.* **76**, 490–503.

G. Szegő (1920), Beiträge zur Theorie der Toeplitzsche Formen, I. *Math. Zeit.* **6**, 167–202.

G. Szegő (1921a), Beiträge zur Theorie der Toeplitzsche Formen, II. *Math. Zeit.* **9**, 167–190.

G. Szegő (1921b), Über die Randwerte einer analytischen Funktion. *Math. Ann.* **84:3/4**, 232–244.

G. Szegő (1952), On certain Hermitian forms associated with the Fourier series of a positive function. In *Festskrift Marcel Riesz (Lund, 1952)*, Comm. Sém. Math. Univ. Lund, suppl. vol., pp. 222–238.

G. Szegő (1954), On a theorem of C. Carathéodory. In *Studies in Mathematics and Mechanics: Studies Presented to Richard von Mises*, pp. 62–66. Academic Press, New York.

G. Szegő (1959), *Orthogonal Polynomials*, second edition, vol. XXIII of Colloquium Publications. American Mathematical Society, New York.
B. Szőkefalvi-Nagy and C. Foias (1967), *Analyse harmonique des opérateurs de l'espace de Hilbert*. Akadémiai Kiado, Budapest.

T

J. L. Taylor (1973), *Measure Algebras*. American Mathematical Society, Providence, RI.
O. Toeplitz (1911a), Zur Theorie der quadratischen und bilinearen Formen von unendlichvielen Veränderlichen, I: Theorie des L-Formen. *Math. Ann.* **70**, 351–376.
O. Toeplitz (1911b), Über die Fouriersche Entwickelung positiver Funktionen. *Rend. Circ. Mat. Palermo* **32**, 191–192.
V. A. Tolokonnikov (1981), Estimates in Carleson's corona theorem, ideals of the algebra H^∞, the problem of Szőkefalvi-Nagy (in Russian). *Zapiski Nauchn. Semin. LOMI* **113**, 178–198. English translation: *J. Soviet Math.* **22:6** (1983), 1814–1828.
L. N. Trefethen and M. Embree (2005), *Spectra and Pseudospectra: The Behavior of Nonnormal Matrices and Operators*. Princeton University Press.
S. Treil (1987), Invertibility of Toeplitz operators does not imply applicability of the finite section method (in Russian). *Dokl. Akad. Nauk SSSR* **292**, 563–567.
S. Treil (2002), Estimates in the corona theorem and ideals of H^∞: a problem of T. Wolff; dedicated to the memory of Thomas Wolff. *J. Anal. Math.* **87**, 481–495.
S. Treil (2012), A remark on the reproducing kernel thesis for Hankel operators. arXiv:1201.0063v2
S. Treil and A. Volberg (1994), A fixed point approach to Nehari's problem and its applications. In vol. 71 of Operator Theory: Advances and Applications, pp. 165–186. Birkhäuser, Basel.
W. F. Trench (1964), An algorithm for the inversion of finite Toeplitz matrices. *J. Soc. Indust. Appl. Math.* (12) **3**, 515–522.

V

S. Verblunsky (1936), On positive harmonic functions (second paper). *Proc. London Math. Soc.* (2) **40**, 290–320.
F. Verhulst (2012), *Henri Poincaré: Impatient Genius*. Springer.
A. Volberg (1982), Two remarks concerning the theorem of S. Axler, S.-Y. A. Chang and D. Sarason. *J. Operator Theory* **7**, 209–218.
A. Volberg (2003), *Calderón–Zygmund Capacities and Operators on Nonhomogeneous Spaces*, vol. 100 of CBMS Regional Conference Series in Mathematics. American Mathematical Society, Providence, RI.

V. Volterra (1882), Sopra alcune condizioni caratteristiche delle funzioni di una variabile complessa. *Annali di Mat. Pura Applicata* (2) **11**, 1–55.
V. Volterra (1896), Sulla inversione degli inegrali definiti. *Rend. Accad. Lincei* **5**, 177–185, 289–300.
V. Volterra (1931), *Leçons sur la théorie mathématique de la lutte pour la vie*. Gauthier-Villars, Paris, 1931 (second edition, 1990).

W

A. Weil (1940), *L'intégration dans les groupes topologiques et ses applications*, vol. 869 of Actualités Scientifiques et Industrielles. Hermann, Paris.
J. Wermer (1953), On algebras of continuous functions. *Proc. Amer. Math. Soc.* **4**, 866–869.
H. Weyl (1910a), Über gewöhnliche Differentialgleichungen mit Singularitäten und die zugehörigen Entwicklungen willkürlicher Funktionen. *Math. Ann.* **68**, 220–269.
H. Weyl (1910b), Über die Gibbs'sche Erscheinung und verwandte Konvergenzphänomene. *Rendiconti Circ. Mat. Palermo* **30**, 377–407.
H. Weyl (1944), David Hilbert and his mathematical work. *Bull. Amer. Math. Soc.* **50**, 612–654.
H. Weyl (1984), *Selected Works: Mathematics and Theoretical Physics*, (ed. V. Arnold and A. Parshin). Nauka, Moscow.
H. Widom (1960a), Inversion of Toeplitz matrices, II. *Illinois J. Math.* **4**, 88–99.
H. Widom (1960b), Inversion of Toeplitz matrices, III. *Notices Amer. Math. Soc.* **7**, 63.
H. Widom (1960c), Singular integral equations in L^p. *Trans. Amer. Math. Soc.* **97**, 131–160.
H. Widom (1964), On the spectrum of Toeplitz operators. *Pacif. J. Math.* **14**, 365–375.
H. Widom (1973), Toeplitz determinants with singular generating functions. *Amer. J. Math.* **95**, 333–383.
H. Widom (1976), Asymptotic behavior of block Toeplitz matrices and determinants, II. *Adv. Math.* **21**, 1–29.
N. Wiener (1933), *The Fourier Integral and Certain of its Applications*. Cambridge University Press, New York.
N. Wiener (1966), *Norbert Wiener, 1894–1964*. Special issue of *Bull. Amer. Math. Soc.* (ed. F. Browder, E. Spanier, and M. Gerstenhaber) **72:1**, part II.
N. Wiener and E. Hopf (1931), Über eine Klasse singulären Integralgleichungen. *S.-B. Preuss Akad. Wiss. Berlin, Phys.-Math. Kl.* **30/32**, 696–706.
W. Wirtinger (1897), Beiträge zur Riemann's Integrationsmethode für hyperbolische Differentialgleichungen, und deren Anwendungen auf Schwingungsprobleme. *Math. Ann.* **48**, 364–389.
T. H. Wolff (1983), Counterexamples to two variants of the Helson–Szegő theorem. Report 11, CalTech, Pasadena. Published in *J. Anal. Math.* **88:1** (2002), 41–62.
Z. Wu (1998), Function theory and operator theory on the Dirichlet space. In *Holomorphic Spaces* (ed. S. Axler, J. E. McCarthy, and D. Sarason), vol. 33 of MSRI Publications, pp. 179–199. Cambridge University Press.

X

D. Xia (1983), *Spectral Theory of Hyponormal Operators*. Birkhäuser, Basel.

Y

B. Yood (1951), Properties of linear transformations preserved under addition of a completely continuous transformation. *Duke Math. J.* **18**, 599–612.

K. Yosida (1940), Quasi-completely-continuous linear functional operations. In *Collected Papers, Faculty of Science, Osaka Imperial University, Ser. A. Math.* **7** (1939 (1940)), 297–301.

Z

K. Zhu (2007), *Operator Theory in Function Spaces*, second edition. American Mathematical Society, Providence, RI.

A. Zygmund (1959), *Trigonometric Series*, vols I and II. Cambridge University Press.

Notation

Sets and Measures

$A =: B$ - definition by an equality, §2.1.1
A^{-1} - Appendix C.1
$\mathbb{C}$ - the complex plane, $\mathbb{T} = \{z \in \mathbb{C}: |z| = 1\}$
$\mathbb{C}_+ = \{z \in \mathbb{C}: \text{Im}(z) > 0\}$, $\mathbb{C}^+ = \{z \in \mathbb{C}: \text{Re}(z) > 0\}$
$D(z, r) = \{\zeta \in \mathbb{C}: |z - \zeta| < r\}$, $\mathbb{D} = D(0, 1)$
$\text{Ran}_{\text{ess}}(\varphi)$ - Definitions §3.1.2, $\text{Ran}(T)$ - Appendix D.2
m - normalized Lebesgue measure, §2.1.1 and Appendix F.1
$\mathcal{P}_a, \mathcal{P}$ - §2.1.1; $\mathcal{P}_n$ - §5.1.3
$\Sigma(\varphi) = \text{conv}(\text{Ran}_{\text{ess}}(\varphi))$ - Theorem 5.6.6
$\sigma(T), \sigma(x)$ - Appendix C.1; $\sigma_F(T) = \sigma_{\text{ess}}(T)$ - Appendices D.6, E.1, E.3
$\sigma(\theta)$ - spectrum of a function, Appendix F.2
$\mu_a = wm, \mu_s$ - Appendix F.4

Spaces and Set Operations

$\text{alg}(\mathcal{T}_X)$ - Definitions 3.1.2
$AP(\mathbb{R})$ - Appendix C.5
BMO - Exercise 2.3.4
$\text{Circ}(\mathbb{C}^N)$ - §5.5.2
$\text{coker}(T)$ - Appendix E.9(2)
$C_a(\mathbb{D})$ - Appendix F.1 and Example 2.1.7(3)
$\text{clos}_X(A) = \text{clos}(A)$ - Appendix A.1
$\mathcal{D}(\mathbb{T})$ (Dirichlet space) and $\mathcal{D}_a(\mathbb{T})$ - §5.6.5.1
E_f - Appendix F.2
$E(\varphi)$ - Theorem 5.6.8(2)
$\text{Fred}(X, Y), \text{Fred}(X)$ - Appendix E.1

$\mathcal{F}(X, Y), \mathcal{F}(X)$ - Appendix D.5
$H^2 = H^2(\mathbb{T})$ - §2.1.1 and Appendix F.1
$H^p = H^p(\mathbb{T})$ - Appendix F.1
$H^\infty = H^\infty(\mathbb{T})$ - S2.1.1 and Appendix F.1
$H^p(\mathbb{T}, E)$ - Appendix F.6
$H^2_- = L^2 \ominus H^2$ - §2.2.2; $K_\theta = H^2 \ominus \theta H^2$ - Exercise 3.4.15
$H^2(\mathbb{C}_+), H^\infty(\mathbb{C}_+)$ - §4.2.1, Appendix F.5
$\mathcal{K} = L(X)/\mathfrak{S}_\infty(X)$ - Appendix D.6
$K(\mathbb{T})$ - §5.6.5.1
$L(X, Y), L(X)$ - Appendix D.1
$\mathcal{M}(\Omega)$ - Appendix C.5, Appendix C.4(c)
$\mathfrak{M}(A)$ - Appendix C.4
Mult(X) - Exercise 2.3.3
PC($\mathbb{T}$) - Examples 3.2.2; QC -Exercise 2.3.5, Exercise 3.4.3(e)
$\mathcal{P}M(\Omega)$ - §4.2.2
$\mathcal{P}_n = \text{Vect}(z^k : 0 \le k \le n)$ - §5.1.3(1)
$\text{span}_X(A) = \text{span}(A)$ - Appendix A.1
$\mathfrak{S}_\infty(H)$ - Appendix D.5; $\mathfrak{S}_p(H)$ - Appendix D.8
$[T_\varphi, T_\psi), [T_\varphi, T_\psi]$ - Exercise 2.3.9
Vect(A) linear (vector) hull of A - §2.1.1, Appendix A.1
VMO - Exercise 2.3.5
$W = W(\mathbb{T})$ - Theorem 5.3.1, §5.5.1(4); $W_a(\mathbb{D})$ - §5.6.2
T^* adjoint operator - Appendix D.2
$X^*, \sigma(X, X^*), \sigma(X^*, X)$ - Appendix A.1

Functions, Constants, Transforms

$A \gg 0$ - Theorem 5.1.1
$A(L, M) = A_H(L, M)$ - Lemma 4.3.1 (angle between L and M)
$b_T(\cdot)$ - Exercise 2.3.1 (Berezin transform of T)
codim(E) - Appendix A.3
C_n - §5.5.1 (cyclic "shift")
det($I + A$) - Appendix D.8
$D_n = D_n(\varphi) = \det(T_{\varphi,n}), D_n(\mu)$ - §5.1.4(1)
dist$_\lambda$ - Theorem 3.2.1; wind(φ) - Definitions 3.1.2
$f_{\text{in}}, f_{\text{out}}$ - Appendix F.2 (inner/outer parts of a function f)
$\mathcal{F}$ ($\mathcal{F}_P$) - §4.2.1, Appendix F.5 (Fourier, Fourier–Plancherel)
$f^\vee = 2\pi \mathcal{F}^{-1}$ - §4.2.4
$\mathcal{F}_d$ - §5.5.1(3) (Fourier on $(\epsilon^k)_0^n$)

$f * g$ - Appendix F.1
$G(\varphi), G(\mu)$ - §5.6
$\mathbb{H} = i\mathbb{X}, \mathbb{X}u = \tilde{u}$ - Appendix F.3, §2.3.4, §4.1.2 (Hilbert)
H^d - Example 2.1.7 (Hilbert discrete)
$H_\varphi, \Gamma_\varphi$ - Example 2.2.2 (Hankel)
ind(T) - Appendix E.1
$l_k(x), \Lambda_k(x), \mathcal{L}_k(x)$ - Lemma 4.3.1; $L_k(x)$ - Exercise 4.5.4
M_φ - Lemma 2.1.4, Appendix E.8(6)
P_+ - §2.1.1, Appendix F.3
P_n - §5.1.3(1); $P_{l,m}$ - §5.2.2
$P_r(e^{ix})$ - Appendix F.1
P_E - Appendix B.2
$P_{L\|M}$ - §4.2.1
rank(A) - Appendix D.5
$R_\lambda(x) = (\lambda e - x)^{-1}$ - Appendix C.1
$\mathbb{S}, \mathbb{X}, \mathbb{Y}$ - Theorem 4.1.2, §4.1.3
S_n - §5.1.3(3)
Sym(T) - Theorem 3.1.3
$T^\bullet = \Pi(T)$ - Appendix D.6, §3.1
Trace(A) - Appendix D.8
T_φ - Example 2.1.3; T_φ^θ - Exercise 3.4.15
$T_{a,n} = T_{\varphi,n} = P_n T_\varphi P_n$ - §5.1.3(1)
$T_n(p), T_n(c)$ - §5.5.2
W_k - Example 2.1.7, §4.2.2
Γf - Appendix F.3 (Herglotz)
τ_s - §4.2.2, §4.4.1
χ_A - indicator function of A ($\chi_A(x) = 1$ if $x \in A$, otherwise $\chi_A(x) = 0$)
$\|\cdot\|_{\text{ess}}$ - Appendix D.6

Some Acronyms

SLiC - strong Littlewood conjecture, §5.8
RKHS - reproducing kernel Hilbert space (RKHS), Appendix B.6
IFSM - inversion by the finite section method, §5.4.1
SIO - singular integral operator, §4.1
WHO - Wiener–Hopf operator, §1.3.2, §4.2.2
TMP - truncated moment problem, §5.1
RHP - Riemann–Hilbert problem, §1.1.2, §4.1
RKT - reproducing kernel thesis, Exercise 2.3.1

Index

AAK step-by-step extension algorithm, 87
Abel, Niels, 105
Abrahams, I. David, 172, 407
Acosta, Gabriel, 318, 395
Adams, David R., 86, 395
Adamyan, Vadym M., 27, 47, 48, 87, 93, 176, 313, 395
Akhiezer, Naum I., 20, 232, 237, 238, 309, 310, 325, 338, 395
Aleksandrov, Pavel, 15, 184, 186
Aleman, Alexandru, 77, 98, 100, 395
Algebra
    AP($\mathbb{R}$) of almost periodic functions, 149
    C^*, 340
    Calkin, 101
    Dirichlet, 162
    disk, 388
    Douglas, 93, 164, 167
    Eberlein, 226
    Hardy, 33
    Krein (Krein–Dirichlet), 230, 324
    Sarason QC, 69
    Toeplitz, 101, 102, 145
    Wiener, 155, 266
Algorithm
    step-by-step extension AAK, 87
Allan, Graham R., 160, 395
Almagià, Virginia, 6
Alzer, Horst, 312, 395
Anosov, Dmitri, 12, 395
Appell, Paul, 15
Approximation property (AP), 351
Arcozzi, Nicola, 86, 395, 396
Arnold, Vladimir, 11, 274, 345, 414
Aronszajn, Nachman, 310, 396

Arov, Damir Z., 27, 47, 48, 87, 93, 176, 313, 395
Arsene, Grigore, 313, 396
Arveson, William, 250, 253, 313, 396
Askey, Richard, 246
Asymptotic density of spectra, 277
Asymptotic distribution of spectra, 230, 276, 280, 306, 317, 319–321, 327
Asymptotics of Toeplitz determinants, 318
Atiyah, Michael, 342, 344, 345, 383–385, 396
Atkinson, Frederick V., xxi, 359, 365, 366, 383, 396
Audin, Michèle, 91, 92
Axler, Sheldon, 77, 100, 145, 166, 168, 396, 415

Bakonyi, Mihály, 310–312, 396
Balanchine, George, 344
Banach, Stefan, 18, 192, 349, 370, 381, 396
Band extension, 311, 313
Baranov, Anton, xix, 175, 227, 396
Bari, Nina, 184, 186
Barria, José, 81, 396
Bart, Harm, 109, 396
Basis
    unconditional, 332
    Riesz, 332
    Schauder, 331, 351, 386
Basis (ONB)
    Laguerre, 26
    of Laguerre fractions, 190, 201, 202
    of Laguerre functions, 40, 178, 200, 202
Basor, Estelle, 307, 323, 324
Baxter, Glen, 230, 256, 263, 313–315, 396
Beckenbach, Edwin, 396

Bell, Eric, 15
Belliver, André, 14
Bellman, Richard, 232, 366, 396
Bercovici, Hari, 174, 396
Berezansky, Yurii, 48
Berezin, Felix A., xxi, 31, 60, 61, 80, 94, 396
Berg, Christian, 309, 396
Berger, C. A., 291, 397
Bergman, Stefan, 45, 379
Beria, Lavrentiy, 380
Bernkopf, Michael, 201
Bernstein, Felix, 7
Bernstein, Sergei, 8, 186, 299
Bertrand, G., 158, 222, 397
Besicovich, A., 222
Bessonov, Roman V., 175, 227, 312, 396, 397
Betti, Enrico, 4, 6
Beurling, Arne, 53, 390, 397
Bieberbach, Ludwig, 92
Bini, Dario, 327, 397
Birkhoff, George D., xv, xxi, 1, 8, 11, 12, 22, 99, 101, 170–172, 193, 194, 220, 224, 397
Blaschke
 condition, 298
 product, 176, 389
 product (finite), 54
Blasco, Oscar, 96, 397
Bleher, Pavel M., 316, 397
Bôcher, Maxime, 171
Bochner, Salomon, 309
Bogoya, Johan M., 321, 397
Bohl, Piers, 318
Bohr, Harald, 8, 58, 93, 149
Bolibruch, Andrei, 11, 12, 212, 395
Bolzano, Bernard, 28, 105
Bonsall, Frank F., 96, 397
Borel, Armand, 15
Borel, Émile, 15, 184, 185
Born, Max, 7, 8, 17
Borodin, Alexei, 307, 323, 397
Böttcher, Albrecht, xviii, 95, 99, 114, 140, 142, 160–162, 168, 169, 173, 174, 222, 224, 225, 227, 228, 265, 278, 307, 308, 311, 315, 316, 318, 321–325, 327, 328, 393, 397, 398
Bourbaki, Nicolas, 223, 398
Boutet de Monvel, Louis, 80, 81, 384, 398
Boutroux, Pierre, 14
Brelot, Marcel, 15
Brouwer, Luitzen, 72, 273
Browder, Felix, 15, 415

Brown, Arlen, 76, 82, 84, 99, 158, 161, 358, 398
Brown, Lawrence G., 325, 352, 398
Bugaev, Nikolai, 184
Burbea, Jacob, 393, 398
Bush, Vannevar, 193

Calculus
 symbolic, 16, 23, 93, 101, 158, 220, 222
Calderón, Alberto, 94, 99, 227
Cantor, Georg, 71, 72
Capelli, Alfredo, 376
Carathéodory, Constantin, xxi, 230–232, 241, 247, 299, 308, 309, 312, 325, 398
Carathéodory, Euphrosyne, 234
Carleman, Torsten, 79, 156, 220
Carleson condition (C), 176
Carleson, Lennart, 185
Carnot, Lazare, 104
Carroll, Lewis (Charles Dodgson), 3
Cartan, Élie, 273
Cartan, Henri, 384
Case, K. M., 307, 323, 402
Cauchy complex singular integral, 187
Cauchy, Augustin L., xxi, 101, 104, 105, 159, 326, 398
Cayley, Arthur, 71, 200
Cézanne, Paul, 344
Chan, Raymond H.-F., 327, 398
Chandrasekharan, Komaravolu, 274, 398
Chang, Sun-Yung A., 77, 93, 100, 166, 167, 396, 398
Charpentier, Éric, xviii
Chausson, Ernest, 91
Chausson, Marianne, 91
Chebyshev, Pafnuty, 20, 27, 84
Chersky, Y. I., 220, 402
Chevalley, Claude, 274, 399
Chevassus-au-Louis, Nicolas, 92
Choi, Man-Duen, 326, 399
Christensen, Jens, 309, 396
Church, Alonzo, 8
Circulant (matrix), 80, 250, 258, 265, 267–269, 280, 317, 321
Clancey, Kevin, 99, 162, 172, 399
Classes
 Schatten–von Neumann, 354, 377
Coburn, Lewis A., 24, 106, 107, 158, 159, 169, 374, 386, 399
Cohen, Paul, 141, 167, 399
Coifman, Ronald R., 85, 95, 399
Colin de Verdière, Yves, 30, 399

Index

Commutators, 99
Commutators $[P_+, M_\varphi]$, 73
Completion of matrices, 313
Condition
  Blaschke, 298
  Helson–Szegő, 124
  Muckenhoupt, 97, 124, 127, 391
Condition numbers, 327
Constant(s)
  Douglas–Wolff–Tolokonnikov, 166
  Glaisher–Kinkelin, 326
  Landau, 248, 312
  Lebesgue, 248, 312
  Poincaré, 318
Convergence weak-*, 330
Convexity (polynomial), 122
Cordes, H. O., 374, 399
Cotlar, Mischa, 98, 399
Courant, Richard, 8, 17, 123, 316, 399
Crofoot, Robert B., 175, 399
Curie point, 23
Cyclic shift (translation), 265

Dantzig, T., 15
Darboux, Gaston, 14, 15, 92
David, G., 97
Davidson, Kenneth, 313, 357, 399
Davis, Philip J., 317, 399
Dedekind, Richard, 3, 4, 71, 72, 410
Deift, Percy, 25, 29, 172, 321, 323, 324, 399
Demidov, Sergei, 186
Denjoy, Arnaud, 185
Derivative
  Radon–Nikodym, 270
Determinant(s)
  Cauchy, 304, 326
  Fredholm, 323, 326, 357
  Hilbert, 305, 326
  of an operator, 294, 357, 377
  Toeplitz, 238, 270
Devinatz, Allen, 25, 26, 81, 124, 142, 157, 162, 167, 221, 223, 226, 228, 264, 302, 315, 325, 400
DeVore, Ronald, 319, 325, 400
Dick, Auguste, 381
Dieudonné, Jean, 14, 15, 156, 338, 383, 400
Dikyi, L., 342
Dirichlet, Per Gustav, 4, 71
Distance
  Arveson's formula, 250
  local, 113
Dixmier, Jacques, 111, 400

Douady, Adrien, 374, 400
Douglas, Jesse, 324, 400
Douglas, Ronald G., 24, 69, 81, 93, 96, 99, 109, 111, 113, 114, 139, 140, 151, 152, 158–161, 165, 167–169, 173, 174, 220, 221, 225, 228, 325, 352, 398–401
Dreyfus, Alfred, 15
Du Bois-Reymond, Emil, 123
Du Bois-Reymond, Paul, 123
Duality (Fefferman), 66, 95
Duduchava, Roland, 222, 401
Dugac, Pierre, 15, 186
Duhamel, Jean-Marie, 105
Dunford, Nelson, 63, 332, 347, 358, 401
Duoandikoetxea, Javier, 95, 209, 211, 227, 228, 401
Durán, Ricardo, 318, 395
Durbin, James, 314, 401
Duren, Peter, 345
Dym, Harry, 162, 401
Dynkin, Eugene, 61

Edwards, Robert E., 95, 401
Egorov, Dmitri, 184
Ehrhardt, T., 326, 401
Einstein, Albert, 2, 13, 170, 171, 274, 367, 380
Elizabeth II, Queen, 345
Embree, Mark, 278, 327, 413
Enflo, Per, 331, 351
Equation(s)
  Bézout, 314
  of Hankel operators, 42
  of Toeplitz matrices, 237
  of Toeplitz operators, 36
  Wiener–Hopf (WHE), 22, 157, 194, 222, 224
Equidistribution
  according to Weyl, 230, 271, 318
Eremenok, P., 49
Esenin-Volpin, Alexander, 343
Essential range, 103
Euler, Leonhard, 104

Factorization
  Birkhoff (or Wiener–Hopf), 12, 97, 101, 124, 126, 161, 170–172
  Birkhoff–Wiener–Hopf matricial, 169, 172, 256
Faddeev, Ludvig, 60
Farforovskaya, Yuliya B., 310, 408
Fatou, Pierre, 89, 91, 214, 228, 401

Fedosov, Boris V., 374, 384, 401
Fefferman, Charles, 66, 86, 95, 401
Fejér, Lipót, 233, 247, 248, 312, 398
Fekete, Michael, 45
Feldman, I. A., 158, 159, 169, 173, 174, 205,
    227, 228, 256, 263, 264, 313–316, 325,
    402, 406
Ferguson, Sarah, 85, 401
Fillmore, Peter A., 352, 398
Finsler, Paul, 233
Fischer, Ernst, 19, 401
Fisher, Michael, 316, 323, 324, 401
Florensky, Pavel, 186
Foias, Ciprian, 90, 174, 175, 312, 313, 325,
    396, 401, 413
Fontené, Georges, 376
Ford, Charles, 186
Form(s)
  Hankel (bilinear), 81, 86
  Laurent (quadratic), 6, 16
  Toeplitz (quadratic), 16, 19, 82, 224
Formula(s)
  Arveson's distance, 230, 250, 313
  Berger–Shaw (trace), 230, 291
  Borodin–Okounkov, 307, 323, 327
  Fedosov–Dynin–Hörmander, 374
  Gelfand, 341, 372
  Helton–Howe, 230, 291, 306, 326
  Jesse Douglas, 285
  Libkind–Widom, 230, 306, 326
  Poincaré–Bertrand, 79, 93, 222
  Sokhotsky–Plemelj, 23, 178, 211, 212, 228
  Szegő asymptotics, 271
  Szegő–Verblunsky, 270, 392
  Widom
    for T, Γ, 76
    for the semi-commutators, 301
    strong Szegő theorem, 286
Frazho, Arthur E., 312, 313, 325, 401
Fredholm
  alternative, 360, 371, 378, 381, 386
  theory, 101, 139, 142, 149, 222, 378
Fredholm, Erik Ivar, xxi, 79, 156, 158, 174,
    225, 359–361, 372, 376, 377, 401
Freud, Sigmund, 185
Freudenthal, Hans, 3
Friedrichs, Kurt, 316, 399
Frobenius, Georg, 376
Fuchs, D., 342
Fuchs, Lazarus, 10, 156
Function(s)
  dilated Laguerre, 217

Euler Gamma, 133, 211
Green, 317
inner, 389
Laguerre, 32, 35, 40, 188, 200, 202, 218,
    226
of the Schur class, 230
outer, 389
regulated, 116
weakly almost periodic, 226
Functional calculus
  local, 341
  of C_n, 266

Gakhov, Fyodor D., 114, 161, 220, 223, 402
Galerkin, Boris, 78
Galkin, P., 312
Gantmacher, Felix R., 301, 307, 402
Garcia, Stephan R., 175, 227, 402
Gårding, Lars, 360
Garnett, John, 46, 67, 69, 86, 92, 93, 95, 96,
    164, 167, 176, 223, 312, 394, 402
Garsia, Adriano, 31, 64–66, 95, 402
Gaudry, Garth I., 95, 401
Gauss, Carl Friedrich, 1–4
Gelbaum, B., 400
Gelfand, Israel M., xxi, 61, 225, 309,
    339–345, 347, 383–385, 402
Gelfond, Alexander, 186
Geoffroy Saint-Hilaire, Étienne, 15
Georgiadou, Maria, 234, 402
Geronimo, Jeff S., 307, 323, 402
Gerstenhaber, M., 415
Getzler, E., 384
Gheondea, Aurelian, 313, 396
Gillespie, Charles C., 400
Glazman, Israel M., xvi, 48, 402
Godefroy, Gilles, xix, 30
Gödel, Kurt, 9, 14, 274
Gohberg, Israel C., xxi, 24, 26, 48, 99, 101,
    107–109, 142, 151, 158, 159, 162,
    167–169, 172–174, 220–222, 224, 225,
    228, 230, 255, 256, 263, 264, 307,
    312–316, 325, 357, 377, 383–385, 399,
    401–403
Goldberg, Seymour, 99, 169, 173, 307, 325,
    357, 377, 402
Goldschmidt, Charlotte, 382
Golinskii, B., 284, 293, 322, 403
Golub, Gene H., 314, 403
Gordan, Paul, 8
Graev, Mark, 342

Index

Grafakos, Loukas, 95, 211, 403
Grassmann, Hermann, 71
Gray, J., 15
Gray, Robert M., 317, 319, 327, 403
Grenander, Ulf, 244, 310, 317, 319, 321, 327, 392, 403
Griffiths, Phillip, 15
Gronwall, Thomas, 312
Grothendieck, Alexander, 351, 377, 403
Grudsky, Sergei M., 265, 311, 315, 321, 327, 397
Grynszpan, Herschel, 18
Guillemin, Victor, 80, 81, 384, 398

Haar, Alfréd, 8, 245
Haas, D., 368
Hadamard, Jacques, 5, 15, 184, 193, 325, 360, 403
Halmos, Paul, xxi, 31, 76, 81–84, 99, 158, 161, 358, 396, 398
Hamburger, Hans, 233
Hankel(-Dippe), Marie, xxi, 1, 28
Hankel, Hermann, xxi, 1, 16, 26–28, 84
Hardy, Godfrey H., 84, 92, 93, 193, 222, 312, 317, 318, 403
Harmonic (Poisson) extension, 65
Harmonic conjugation, 64, 68, 180, 390
Harrell II, Evans, 156, 403
Hartman, Philip, 26, 47, 82, 93, 128, 163, 164, 168, 386, 403, 404
Hartwig, Robert, 316, 323, 324, 401
Hausdorff, Felix, xxi, 273, 359, 381, 382, 385, 404
Havin, Victor, 94, 404, 409, 411
Heaviside, Oliver, 22
Hecke, Erich, 8
Hedberg, Lars I., 86, 395
Hedenmalm, Hakan, 81, 94, 404
Heine, Eduard, 72
Heinig, Georg, 313, 327, 404
Heisenberg, Werner, 7
Hellinger, Ernst, 8, 17, 273
Helson, Henry, 53, 85, 161, 162, 172, 404
Helton, J. William, 291, 293, 306, 325, 326, 404
Hempfling, Thomas, 109, 396
Herglotz, Gustav, 232, 299, 308, 325, 404
Hermite, Charles, 15, 91, 201, 226, 404
Hilbert
 SIO (singular integral operator) , 199
 transform $\mathbb{H}$, 84, 180, 390

Hilbert, David, xv, xxi, 1, 2, 4–12, 17, 19, 22, 26, 29, 30, 71, 72, 84, 92, 123, 156, 157, 170, 174, 179, 193, 212, 220, 221, 224, 225, 228, 245, 272, 273, 326, 360, 377, 385, 404
Hildebrandt, Theophil H., 381, 404
Hirschman, Isidore I., 220, 230, 256, 313, 314, 396, 404
Hoffman, Kenneth, 93, 141, 167, 223, 394, 404
Holmgren, Erik, 360
Hopf, Eberhard, xxi, 1, 21, 22, 84, 124, 157, 161, 169, 172, 178, 194, 195, 224, 415
Hopf, Heinz, 194
Hörmander, Lars, 79–81, 94, 95, 374, 384, 404, 405
Howe, Roger, 291, 293, 306, 325, 326, 404
Humbert, Pierre, 91
Hurwitz, Adolf, 8, 317, 405
Hytönen, Tuomas, 95, 405

Ibragimov, Ildar, 284, 293, 322, 403
Ilyashenko, Y., 11
Index
 Cauchy topological, 101, 103, 158, 383
 Noether index of an operator, 359, 361, 378, 384
Inequality
 Hadamard, 295, 325, 377
 Hilbert, 45, 92
 Poincaré, 318
 von Neumann, 55–57, 93
 Weyl, 357
 Wirtinger, 269
Wirtinger integral, 270, 317
Infeld, Leopold, 367
Inversion by the finite section method (IFSM), 230, 258, 315
Iokhvidov, Iosif, 313, 405
Ising, Ernst, 23, 24, 30
Ismagilov, Rustem, 163, 405
Its, Alexander, 12, 25, 29, 172, 223, 321, 323, 324, 399, 405

Jacobi, C., 307
Janson, S., 77, 100
Jarosch, K., 8
Jin, Xiao-Qing, 327, 398
Johansson, K., 25, 293, 405
John, Fritz, 68, 95, 405
Joseph, Helen, 274
Jourdain, xiii, 22

Journée, J.-L., 97
Julia set, 89
Julia, Gaston, xxi, 31, 89, 91, 92, 253, 313, 405

Kaashoek, Marinus A., 99, 109, 169, 173, 225, 307, 312, 325, 357, 377, 396, 401–403
Kac, Mark, 322, 326, 405
Kailath, Thomas, 327, 405
Kakutani, Shizuo, 25
Kapustin, Vladimir, 175, 227, 396
Karlovich, Yuri, 99, 162, 169, 173, 174, 222, 224, 225, 227, 228, 325, 328, 393, 397, 398
Katsnelson, Victor, 58
Katznelson, Yitzhak, 346, 347, 405
Keldysh, Lyudmila, 184
Kellogg, Oliver, 8, 11
Kernel
 Cauchy, 78, 187, 189
 conjugate Poisson, 181, 390, 391
 Dirichlet, 249, 319
 Hilbert, 187, 221
 of a Hankel operator , 176
 Poisson, 65, 181, 388
 Szegő reproducing, 164, 388
Kesten, Harry, 316, 405
Khinchin, Aleksandr, 184, 186, 194
Khrushchev, Nikita, 61
Khrushchev, Sergei V., 223, 405
Kirillov, Alexandre, 60
Klein, Felix, 13, 200, 212, 233, 274, 378
Klemes, Ivo, 319, 320, 405
Knopp, Konrad, 245
Koch, Elise, 4
Kohn Treibich, A., 11
Kolman, Ernst, 185
Kolmogorov, Andrey, 184, 186, 193, 194, 343, 345, 392, 406
Konyagin, Sergei, 319
Koosis, Paul, 66, 67, 69, 86, 93, 95, 223, 394, 406
Korenblum, Boris, 81, 94, 404
Kotelnikov, Vladimir, 192
Köthe, Gottfried, 18
Kozak, A. V., 114, 316, 406
Kra, Irwin, 317, 406
Krasnoselskii, M., 383–385, 406
Krasovsky, Igor, 25, 321, 324, 399
Krein, Mark G., xxi, 20, 24, 26, 27, 31, 47–49, 87, 89, 93, 157, 158, 161, 162, 172, 173, 176, 220, 221, 228, 237, 238, 285, 309, 310, 313, 324, 383–385, 395, 403, 406

Kronecker, Leopold, xxi, 28, 31, 70–72, 98, 123, 376, 406
Kronrod, Alexander, 184
Krupnik, Naum Y., 80, 99, 142, 152, 158, 162, 167–169, 173, 174, 205, 215, 221, 222, 224, 227, 377, 402, 403, 406
Kuiper, N., 374
Kuipers, Lauwerens, 319, 406
Kummer, Ernst, 71
Kutateladze, Semën, 186

Labrousse, J. P., 374, 399
Lacey, Michael, 85, 97, 98, 124, 401, 406
Laguerre, Edmond, xxi, 178, 200, 201, 219, 226, 404
Landau, Edmund, 4, 184, 193, 245, 248, 312, 379, 406
Landau, Lev, 344
Landkof, Naum N., 86, 406
Landsberg, Georg, 318, 407
Langlands, Robert, 14
Lasker, Emanuel, 8
Lattès, S., 89
Laugwitz, Detlef, 3, 4
Lavrentiev, Mikhail, 184
Lawrie, Jane B., 172, 407
Lax, Peter, 49, 245
Lebesgue point, 182
Lebesgue, Henri, 91, 182, 184, 185, 193, 312, 407
Lebow, Arnold, 374, 399
Lee, Yuk Wing, 223, 407
Legendre, Adrien-Marie, 105
Leibniz, Gottfried, 193
Leites, Dmitry, 60
Lemma
 Coburn, 107
 Douglas and Sarason, 114
 Four-block, 87, 230, 252, 313
 Julia, 89, 230, 253, 313
 Sarason, 46, 92
Lenz, Wilhelm, 23, 24, 30
Lenz–Ising model, 23–25, 30, 245, 284, 316, 321, 323, 324
Levinson, Norman, 193, 314, 407
Levitan, Boris, 342
Levshin, Boris, 186
Lewy, Hans, 316, 399
Libkind, L. M., 306, 322, 407
Lichtenstein, Leon, 195
Lidskii, Victor B., 357
Lindelöf, Ernst, 159, 407

Index

Lindenstrauss, Joram, 332, 358, 386, 407
Linton, Christopher M., 172, 407
Liouville, Joseph, 105, 156
Littlewood, John E., 84, 92, 185, 193, 317, 318, 367, 403
Litvinchuk, Georgii S., 162, 172–174, 407
Litvinov, Maxim, 380
Livshits, Mikhail, 48
Localization
  on the center of a C^*-algebra, 160
Logarithmic capacity, 86
Loomis, Lynn H., 223, 347, 407
Lorch, Lee, 312, 407
Lorentz, George G., 186, 319, 325, 400
Lorentz, Hendrik, 13
Lowdenslager, David, 162, 172
Lusternik, Lazar, 184
Luzin, Nikolaï N., xxi, 130, 178, 182, 184–186, 221, 222, 407
Lyapunov, Aleksandr, 184
Lyubich, Yuri I., xvi, 402

Mac Lane, Saunders, 171
Mandelbrot, Benoit, 89
Mandelstam, Osip, xvi
Mann, Thomas, 344
Mapping Sym, 23, 105
Marcinkiewicz, Józef, 79, 94, 222
Marshall, Donald E., 93, 167, 407
Masani, Pesi R., 172, 393, 398
Maslov, Viktor, 60
Matisse, Henri, 344
Matrix
  band extension, 249
  Cauchy-like, 326
  circulant, 19, 80, 230, 250, 258, 265, 267–269, 280, 317, 321
  generalized Toeplitz, 327
  Gram, 238, 271, 294, 336, 338
  Ha-plitz, xv
  Hankel, xiv, 41, 45, 81, 84
  Julia, 89
  of an operator, 34
  positive definite Toeplitz, 230
  Toeplitz, xiv, 35, 230, 231
  transpose, 254
Matrix factorization
  Birkhoff–Wiener–Hopf, 152
Matsaev, Vladimir, 48, 58
Maximal function
  Hardy–Littlewood, 222

Maximenko, Egor A., 321, 397
Maz'ya, Vladimir, 79, 86, 407
McCarthy, John E., 415
McGehee, Oscar, 319
McIver, Philip, 172, 407
McKean, Henry P., 162, 401
Mean motion (perpetual), 346
Measure
  Carleson, 86
  Haar, 223
Mei, Tao, 96, 407
Mekhlis, Lev, 185
Menshov, Dmitrii, 184, 186
Method of rotation, 207
Mikhlin, Solomon G., xxi, 16, 23, 31, 78, 79, 84, 94, 99, 158, 173, 220, 222, 223, 227, 383, 407
Milman, David, 48
Milne, Edward, 21
Milnor, John, 91
Milton, John, 367
Mingarelli, Angelo B., 366, 407
Minkowski, Hermann, 233
Minlos, Robert, 61
Mittag-Leffler, Gösta, 14, 360
Mityagin, Boris S., 374, 407
Möbius, August, 28
Model space (or subspace), 153, 227, 316
Moiseyev, Nikita, 114
Molière, xiv
Monge, Gaspard, 104
Mongré, Paul, 382
Montel, Paul, 185
Moore, E. H., 171
Mozart, Wolfgang Amadeus, 344
Multiplier
  convolutor of ℓ^p, 63
  of IFSM, 261, 262, 315
  preserving Fred(H^2)., 145
Müntz, Herman, 298, 325
Murphy, Gerard J., 340, 341, 408
Murray, Francis, 90, 92
Muskhelishvili, Nikoloz I., 220, 222, 223, 380, 408

Nadis, Steve, 171
Naimark, Mark, 48, 340–342
Nazarov, Fedor, 96, 97, 124, 408
Nehari, Zeev, xxi, 27, 31, 44, 85, 159, 408
Neményi, Anne, 245
Neubauer, G., 374
Neumann, Carl, 156, 408

Nevanlinna, Rolf, 389
Niederreiter, Harald, 319, 406
Nikol'skaya, Ludmila N., 310, 408
Nikolski, Nikolaï, xvi, 57, 58, 60, 62–66, 68,
  69, 73, 77, 78, 81, 84, 85, 87, 89, 92–94,
  96–100, 124, 140–142, 148, 158,
  160–162, 164, 165, 167–169, 175–177,
  180, 181, 189, 190, 209, 214, 223, 224,
  228, 234, 249, 270, 310, 313–315, 338,
  346, 386–388, 392, 394, 404, 408,
  409, 411
Nikol'skyĭ Sergei M., 383, 408
Nirenberg, Louis, 68, 95, 405
Noble, Ben, 172, 408
Noether, Emmy, 8, 378, 379, 381
Noether, Fritz, xxi, 10, 79, 156, 158, 159, 220,
  359, 361, 378, 380, 381, 383–385, 408
Norm
  Frobenius, 354
  essential, 352
  Garsia, 64–66, 95
  Hilbert–Schmidt, 354
  Hilbert–Schmidt (or Frobenius), 282
  of a Toeplitz matrix, 310
  Schatten–von Neumann, 319, 354
Novikov, Pyotr, 184

Okounkov, Andrei, 307, 323, 397
Onsager, Lars, 23–25, 245,
  284, 322
Operator
  Ha-plitz, 29
  aggregate (composite), 102, 143, 158
  backward shift, 34
  Calderón–Zygmund, 222
  compact, 351, 354
  compact W_k^a, 219
  convolution, 130
  directional on $\mathbb{R}^n$, 208
  finite difference, 198
  finite trace, 354
  Fredholm, 359, 361, 383
  Hankel, xv, 41
    compact, 47, 68
    finite rank, 70
    Hilbert–Schmidt, 72, 286
    on $\mathbb{R}_+$, 228
  Herglotz, 390
  Hilbert, 131, 208, 217, 390
  Hilbert–Schmidt, 72, 354, 355
  Laurent, 16–18
  local type, 113

locally Fredholm, 116
matricial Wiener–Hopf, 228
multiplication M_a, 23
multiplication M_φ, 27, 37, 39, 73, 75, 99,
  134, 373, 375, 393
normally solvable, 359, 361, 385
RHP, 179
Schrödinger, 86
self-adjoint, 50
singular integral (SIO), 1, 129, 178, 180,
  220
Toeplitz, xv, 16, 24, 29, 31, 35–37, 39, 81
  normal, 77
  on ℓ^p, 62
  TMO, 1
  truncated, 153, 169, 174, 226, 236
  trace class, 356
  translation (shift), 33, 34, 52, 239
  truncated Wiener–Hopf, 203
  Volterra, 207
  WH on a finite interval, 153, 174, 203
  WH with causal measure, 198
  Wiener–Hopf (WHO), 1, 21, 26, 99, 220
  with simple spectrum, 234, 354
Ortega-Cerdà, Joaquim, 85, 408
Ostrowski, Alexander, 78

Page, Lavon B., 92, 93, 409
Palamodov, Viktor, 60
Paley, Raymond, 393, 409
Parrott, Stephen, 87, 89, 252, 313, 409
Parshin, Alexey, 274, 404, 414
Peano, G., 6
Peetre, Jaak, 94, 409
Peller, Vladimir V., 78, 85, 87, 92, 96,
  98–100, 176, 228, 409
Perelman, Grigori, 13
Petermichl, Stefanie, 95, 409
Petersen, Karl E., 95, 409
Petrovsky, Ivan, 61
Phenomena
  Wiener–Pitt–Schreider, 225, 347
Phillips, Ralph, 49
Piatetskii-Shapiro, Ilya, 60
Picard, Émile, 10, 91, 360, 409
Pichorides, S. K., 210
Picone, Mauro, 78
Pietsch, Albrecht, 332, 338, 351, 378, 409
Pigno, Louis, 319
Pincherle, Salvatore, 89
Pincus, J., 327
Pisier, Gilles, 96, 408

Planck, M., 123
Plemelj, Josip, xxi, 11, 23, 161, 170, 178, 212, 220, 224, 228, 409
Plessner, Abraham, 222
Poincaré, Henri, xxi, 1, 12–16, 72, 79, 156, 158, 171, 201, 220, 222, 226, 272, 360, 377, 404, 409
Poincaré, Raymond, 15
Polar representation (factorization), 353
Pólya, George, 84, 92, 244, 245, 312, 317, 318, 326, 403, 409
Polynomials
 Laguerre, 26, 178, 218, 228
 orthogonal, 230, 237, 240, 310, 336
Poncelet, Jean-Victor, 105
Pontryagin, Lev, 186, 343
Positive definite sequence, 231
Potapov, Vladimir, 48
Pott, Sandra, 98, 395
Poulain d'Andecy, Louise, 15
Power, Stephen C., 73, 77, 99, 100, 160, 168, 228, 394, 408, 409
Privalov, Ivan I., 187, 213, 221, 222, 325, 410
Problem
 normally solvable RHP, 11
 of monodromy groups, 10, 11
 Riemann (RP), 2, 220
 Riemann–Hilbert (RHP), 1, 7, 11, 178, 220
 trigonometric moment (TMP), 19, 230, 231
Product
 of two Toeplitz operators, 76
 semi-inner, 333
Prössdorf, Siegfried, 313, 316, 325, 410
Pseudo-measure, 191, 192
Pseudo-spectrum (ϵ-pseudo-spectrum), 277, 278, 327

Queffélec, H., 92

Röhrl, Helmut, 11
Rademacher, Hans, 18, 233
Radon, Johann, 372, 410
Raikov, D. A., 225, 340–342, 347, 402
Rambour, Philippe, 316, 324, 410
Rappoport, I. M., 24, 26, 157, 220, 410
Rayleigh, Lord (John William Strutt), 22
Reed, Michael, 358, 360, 377, 386, 410
Reguera, Maria Carmen, 98, 395
Regularizer, 364
 Toeplitz, 112
Reid, Constance, 8, 9, 410

Reproducing kernel
 Hilbert space (RKHS), 94, 98, 337
 thesis (RKT), 59, 60, 94
Ressel, Paul, 309, 396
Riemann problem (RP), 4–6
Riemann, Bernhard, xv, xxi, 1–4, 28, 157, 159, 410
Riesz projection, 33, 126, 390
Riesz, Frigyes, 156, 225, 337, 338, 351, 371, 376–378, 386, 388, 389, 410
Riesz, Marcel, 211, 227, 388, 410
Rinkel, Jean-Marc, 316, 410
Ritz, Walter, 78
Rochberg, Richard, 85, 86, 175, 204, 227, 310, 395, 396, 399, 410
Rodman, Leiba, 313, 410
Rosenblum, Marvin, 25, 26, 84, 157, 163, 221, 223, 226, 228, 325, 393, 394, 411
Ross, William T., 175, 227, 402
Rost, Karla, 313, 327, 404
Rouché, Eugène, 201, 226, 375, 376, 404
Rovnyak, James, 84, 157, 163, 223, 228, 325, 393, 394, 411
Rozanov, Yuri A., 162, 172, 411
Rudin, Walter, 44, 46, 56, 62, 63, 73, 92, 93, 103, 111, 122, 128, 144, 161, 167, 182, 191, 225, 226, 309, 329, 339, 347, 348, 400, 411
Runge, Carl, xxi, 101, 122, 123, 161, 411
Russell, Bertrand, 193
Ruston, Anthony F., 377, 411
Rutherford, Ernest, 13

Sadosky, Cora, 98, 399
Saff, Edward B., 21, 411
Sarason, Donald, 46, 57, 68, 69, 73, 77, 92, 93, 96–100, 111, 113–115, 140, 144, 154, 158, 159, 165–167, 169, 174, 175, 220, 221, 226, 394, 396, 400, 411, 415
Sawyer, Eric T., 86, 395, 396
Sayed, Ali, 327, 405
Schatten–von Neumann classes, 99
Schauder, Juliusz, 349, 351, 376, 381, 411
Schein, Boris M., 381
Schlesinger, Ludwig, 11
Schmidt, Erhard, 8, 156, 194, 273
Schrödinger, Erwin, 7, 273
Schur test, 131, 132, 350
Schur, Issai, 55, 194, 308, 312, 325, 412
Schwartz, Jacob, 63, 332, 347, 358, 401
Schwartz, Laurent, 191, 342
Schwarz, Hermann, 233, 245

Schwarzschild, Karl, 233
Second term of the asymptotics, 284, 306, 326
Seghier, Abdellatif, 324, 410
Seidel, Wladimir, 233
Seip, Kristian, 85, 408
Sementsul, Arkadii, 230, 255, 313, 314, 403
Semi-commutator $[T_f, T_g]$, 96, 101, 103, 164
Shannon, Claude, 192, 193
Sharp Littlewood conjecture (SLiC), 319
Shaw, B. I., 291, 397
Shields, Allen L., 358, 398
Shifman, M., 61
Shilov, Georgiy E., 225, 340–342, 347, 402
Shinbrot, Marvin, 81, 264, 302, 315, 325, 400
Shnirelman, N., 184, 186
Shubin, Mikhail A., 60, 80, 81, 412
Siegmund-Schultze, Reinhard, 381, 412
Sierpinski, Wacław, 185, 318
Silbermann, Bernd, 95, 114, 140, 142, 160,
    161, 168, 173, 174, 222, 224, 278, 313,
    315, 316, 324, 325, 327, 328, 393, 398,
    410
Simanca, Santiago, 317, 406
Simon, Barry, 21, 293, 306, 310, 316, 325,
    326, 357, 358, 360, 377, 386, 392, 410,
    412
Simonenko, Igor B., xxi, 99, 101, 112, 113,
    142–144, 151, 152, 158, 159, 162, 167,
    168, 172–174, 221, 316, 412
Sinai, Yakov, 60
Singer, Isadore, 93, 141, 167, 342, 344,
    383–385, 396, 404
Smirnov, Vladimir I., 79, 389, 390, 412
Smith, Brent, 319
Sobolev, Sergei, 186
Sokhotsky, Yulian V., 23, 212, 220, 228, 412
Sommerfeld, Arnold, 378
Sonin, Nikolai Y., 226, 229
Space(s)
    Bergman, 80, 98, 136
    BMO, 64, 68, 95, 96, 312
    Dirichlet, 81, 85, 86, 285
    H^p vector-valued, 393
    Hardy, 29, 31, 32, 387
    Lebesgue, 32
    Paley–Wiener, 174, 226
    VMO and QC, 68–70, 95, 96,
        129, 140
Spanier, E., 415
Spectral
    inclusion, 118, 137
    theory of the WHO, 196, 224

Spectrum
    asymptotic distribution, xvii
    continuous, 137
    essential (Fredholm), 158, 352, 362, 369
    of a Toeplitz operator, 117
    of an element, 339
    point, 137
    Weyl, 385
Spitkovsky, Ilya M., 48, 99, 162, 168, 169,
    172–174, 222, 224, 225, 227, 228, 313,
    325, 328, 393, 398, 403, 407, 410, 412
Spitzer, Frank L., 316, 317, 412
Spivak, Michael, 293, 412
Stahl, Herbert, 310, 413
Stalin, Joseph, 186, 379, 380
Stechkin, Sergey, 94, 318
Stein, Elias, 86, 95, 99, 211, 222, 394, 401,
    413
Steinhaus, Hugo, 8
Stieltjes, Thomas, 20, 27, 84
Stone, C. J., 316, 317, 412
Stone, Marshall H., 232, 235, 309, 413
Stravinsky, Igor, 344
Structure
    almost block-diagonal, 288
Subspace
    finite codimension, 331
Symbol
    almost periodic, 149, 169
    Fisher–Hartwig, 323
    holomorphic, 138
    homotopic, 110
    local principle, 112, 113
    locally sectorial, 263
    matrix-valued, 96, 151, 169
    of a Hankel operator, 41
    of a Toeplitz operator, 37
    of a Wiener–Hopf operator, 195
    of an operator, 23, 31, 78, 383
    sectorial, 117, 262
    simplified local principle, 143
    taking on two values, 128, 143
Szász, Otto, 299, 325
Szőkefalvi-Nagy, Béla, 90, 175, 313, 338,
    377, 386, 400, 410, 413
Szankowski, A., 386
Szarek, Stanisław, 351
Szegő, Gábor, xxi, 1, 19–21, 24, 25, 29, 82,
    84, 161, 162, 172, 230, 232, 241, 242,
    244, 245, 270, 276, 277, 280, 284, 302,
    308, 310, 312, 317–319, 321, 322,
    325–327, 392, 403, 404, 409, 413

Index

Tamarkin, Jacob D., 245
Tannenbaum, Allen, 174, 396
Taylor, Joseph L., 225, 347, 400, 413
Tazzioli, Rossana, 4
Teichmüller, Oswald, 379
Terwilleger, Erin, 85, 406
Theorem
  Atkinson, 365
  Baxter–Hirschman, 256
  Bohr, 149
  Carathéodory, 241
  Carathéodory–Fejér, 247
  Carathéodory–Toeplitz, 231
  Devinatz–Shinbrot, 302
  Devinatz–Widom, 124
  Douglas–Sarason $T_\varphi \in$ Fred, 111
  Fatou, 388
  Gohberg–Coburn, 107
  Gohberg–Douglas, 109
  Gohberg–Sementsul, 255
  Gram–Schmidt orthogonalization, 336
  Hartman–Wintner, 128
  Kolmogorov, 391
  Krupnik–Feldman, 205
  Lindelöf, 391
  Luzin, 130, 182
  Müntz–Szász, 297
  Nehari, 43
  Paley–Wiener, 190, 393
  Privalov, 187
  Rochberg, 204
  Rouché–Fontené, 375
  Schur, 55, 56, 300
  Simonenko, 113
  spectral mapping, 340
  strong Szegő, 286
  von Neumann spectral, 234
Theory of the struggle for life, 5, 22
Timotin, Dan, 310–312, 396
Titchmarsh, Edward, 367
TMP truncated, 235, 241
Toeplitz
  conjecture, 18
  forms (quadratic), 308, 309
  matrix norm, 246
Toeplitz operator
  IFSM, 261
Toeplitz, Otto, xv, xxi, 1, 6, 8, 16–21, 80, 82, 84, 224, 231, 273, 277, 280, 299, 308, 309, 317–321, 325, 413
Toeplitz/Hankel complementarity, 49
Tolmé, Fanny, 123

Tolokonnikov, Vadim A., 92, 165, 413
Totik, Vilmos, 21, 310, 411, 413
Toulouse, Édouard, 14
Transform
  Berezin, 59, 94
  Crofoot, 175
  Fourier, 191, 192, 202
  Fourier discrete, 33
  Fourier–Plancherel, 190
  Gelfand, 341
  Hilbert $\mathbb{H}$, 78
  Hilbert discrete, 40
  Laplace, 132
  normalized Fourier, 223
  Riesz, 211, 222, 227
Trefethen, L. Nicholas, 278, 327, 413
Treil, Sergei, 92, 96, 97, 124, 166, 265, 315, 316, 408, 413, 414
Trench, William F., 314, 414
Trotsky, Leon, 380
Turing, Alan, 367
Tzafriri, Lior, 332, 358, 386, 407

Uchiyama, Akihito, 314
Ulam, Stanisław, 194
Unitary dilation, 83, 202, 313

Valiron, Georges, 15
Van Loan, Charles F., 314, 403
Veblen, Oswald, 274
Verbitsky, Igor, 86, 407
Verblunsky, Samuel, 392, 414
Verhulst, Ferdinand, 15, 414
Vilenkin, Naum Y., 309, 402
Vinogradov, Ivan, 343
Vitushkin, Anatoli, 345
Volberg, Alexander, 92, 96–98, 124, 166, 227, 310, 408, 414
Volterra, Vito, xxi, 1, 5, 22, 156, 414
von Neumann, John, 8, 55, 57, 78, 81, 83, 90, 93, 94, 234, 245, 274, 319, 354, 377, 384
von Neumann, Max, 245
Vorovich, Iosif, 114
Vukotic, Dragan, 77, 100, 395

Wainger, S., 399
Watson, George N., 367
Weber, Heinrich, 71, 410
Weber, Wilhelm Eduard, 3
Weierstrass, Karl, 4, 28, 71, 72, 123
Weil, André, 223, 274, 309, 347, 399, 414
Weiss, Guido, 85, 95, 222, 399, 413

Wermer, John, 167, 414
Weyl criterion, 273, 385
Weyl, Hermann, xxi, 7, 8, 72, 164, 230, 271, 272, 274, 318, 366, 380, 385, 414
Whitney, Hassler, 48
Whittaker, Edmund, 13
Wick, Brett D., 86, 395, 396
Widder, David, 220, 404
Widom, Harold, 25, 76, 97, 99, 124, 161, 162, 167, 173, 174, 220, 221, 227, 283, 284, 286, 288, 301, 306–308, 316, 318, 321–325, 327, 398, 401, 414, 415
Wiener filter, 195
Wiener, Norbert, xiv, xix, 1, 21, 22, 26, 39, 84, 124, 157, 159, 161, 169, 171, 172, 178, 192–195, 224, 393, 409, 415
Wiener, Peggy, 194
Wilhelm II, Kaiser, 9, 90
Wintner, Aurel, 26, 82, 128, 163, 164, 168, 386, 404

Wirtinger, Wilhelm, 156, 269, 415
Woerdeman, Hugo, 312, 313, 396, 410
Wolff, Thomas H., 165, 314, 415
Wu, T. T., 324
Wu, Zhijian, 86, 410, 415
Würth, Regina, 379

Xia, Daoxing, 327, 415

Yau, Shing-Tung, 171
Yood, Bertram, 374, 383–385, 415
Yosida, Kosaku, 372, 415
Youschkevitch, Adolf, 186
Yudovich, Victor, 114

Zalcman, Lawrence, 92
Zariski, Oscar, 171
Zdravkovska, S., 345
Zermelo, Ernst, 8
Zhu, Kehe, 81, 94, 164, 394, 404, 415
Zygmund, Antoni, 94, 221, 227, 312, 415

Printed in the United States
by Baker & Taylor Publisher Services